Technological Developments in Industry 4.0 for Business Applications

Luis Ferreira
Polytechnic Institute of Cávado and Ave, Portugal

Nuno Lopes
Polytechnic Institute of Cávado and Ave, Portugal

Joaquim Silva
Polytechnic Institute of Cávado and Ave, Portugal

Goran D Putnik
University of Minho, Portugal

Maria Manuela Cruz-Cunha
Polytechnic Institute of Cávado and Ave, Portugal

Paulo Silva Ávila
Polytechnic of Porto, Portugal

A volume in the Advances in Logistics, Operations, and Management Science (ALOMS) Book Series

IGI Global
DISSEMINATOR OF KNOWLEDGE

Published in the United States of America by
IGI Global
Business Science Reference (an imprint of IGI Global)
701 E. Chocolate Avenue
Hershey PA, USA 17033
Tel: 717-533-8845
Fax: 717-533-8661
E-mail: cust@igi-global.com
Web site: http://www.igi-global.com

Library of Congress Cataloging-in-Publication Data

Names: Ferreira, Luis, 1967 December 28- editor. | Lopes, Nuno, 1977- editor. | Silva, Joaquim, editor.
Title: Handbook of research on emerging developments in industry 4.0 / Luis Ferreira, Nuno Lopes, Joaquim Silva, Goran D. Putnik, Maria Manuela, Cruz-Cunha, and Paulo Silva Avila, editors.
Description: Hershey : Business Science Reference, [2018]
Identifiers: LCCN 2017032933| ISBN 9781522549369 (hardcover) | ISBN 9781522549376 (ebook)
Subjects: LCSH: Technological innovations--Management. | Information technology--Management.
Classification: LCC HD45 .H2937 2018 | DDC 658.5/14--dc23 LC record available at https://lccn.loc.gov/2017032933

This book is published in the IGI Global book series Advances in Logistics, Operations, and Management Science (ALOMS) (ISSN: 2327-350X; eISSN: 2327-3518)

British Cataloguing in Publication Data
A Cataloguing in Publication record for this book is available from the British Library.

All work contributed to this book is new, previously-unpublished material. The views expressed in this book are those of the authors, but not necessarily of the publisher.

For electronic access to this publication, please contact: eresources@igi-global.com.

Advances in Logistics, Operations, and Management Science (ALOMS) Book Series

John Wang
Montclair State University, USA

ISSN:2327-350X
EISSN:2327-3518

Mission

Operations research and management science continue to influence business processes, administration, and management information systems, particularly in covering the application methods for decision-making processes. New case studies and applications on management science, operations management, social sciences, and other behavioral sciences have been incorporated into business and organizations real-world objectives.

The **Advances in Logistics, Operations, and Management Science** (ALOMS) Book Series provides a collection of reference publications on the current trends, applications, theories, and practices in the management science field. Providing relevant and current research, this series and its individual publications would be useful for academics, researchers, scholars, and practitioners interested in improving decision making models and business functions.

Coverage

- Operations management
- Risk Management
- Information management
- Decision analysis and decision support
- Finance
- Marketing engineering
- Computing and information technologies
- Networks
- Political science
- Organizational behavior

IGI Global is currently accepting manuscripts for publication within this series. To submit a proposal for a volume in this series, please contact our Acquisition Editors at Acquisitions@igi-global.com or visit: http://www.igi-global.com/publish/.

The Advances in Logistics, Operations, and Management Science (ALOMS) Book Series (ISSN 2327-350X) is published by IGI Global, 701 E. Chocolate Avenue, Hershey, PA 17033-1240, USA, www.igi-global.com. This series is composed of titles available for purchase individually; each title is edited to be contextually exclusive from any other title within the series. For pricing and ordering information please visit http://www.igi-global.com/book-series/advances-logistics-operations-management-science/37170. Postmaster: Send all address changes to above address.

Titles in this Series

For a list of additional titles in this series, please visit: www.igi-global.com/book-series

Information Technology as a Facilitator of Social Processes in Project Management and Collaborative Work
Timothy C. Bagwell (Walden University, USA) Robert A. Cropf (Saint Louis University, USA) and Sheryl L. Foster-Gadkari (Strategic Impact, USA)
Business Science Reference • copyright 2018 • 217pp • H/C (ISBN: 9781522534716) • US $185.00 (our price)

Handbook of Research on Cross-Cultural Business Education
Chandan Maheshkar (University of Indore, India) and Vinod Sharma (Christ Institute of Management, India)
Business Science Reference • copyright 2018 • 463pp • H/C (ISBN: 9781522537762) • US $235.00 (our price)

Handbook of Research on Supply Chain Management for Sustainable Development
Ulas Akkucuk (Bogazici University, Turkey)
Business Science Reference • copyright 2018 • 439pp • H/C (ISBN: 9781522557579) • US $265.00 (our price)

Globalization and Trade Integration in Developing Countries
Ahu Coşkun Özer (Marmara University, Turkey)
Business Science Reference • copyright 2018 • 320pp • H/C (ISBN: 9781522540328) • US $205.00 (our price)

Handbook of Research on Intrapreneurship and Organizational Sustainability in SMEs
Rafael Perez-Uribe (EAN University, Colombia) Carlos Salcedo-Perez (EAN University, Colombia) and David Ocampo-Guzman (EAN University, Colombia)
Business Science Reference • copyright 2018 • 475pp • H/C (ISBN: 9781522535430) • US $295.00 (our price)

Entrepreneurship, Collaboration, and Innovation in the Modern Business Era
Mehdi Khosrow-Pour, D.B.A. (Information Resources Management Association, USA)
Business Science Reference • copyright 2018 • 367pp • H/C (ISBN: 9781522550143) • US $215.00 (our price)

Developing Organizational Maturity for Effective Project Management
Gilbert Silvius (LOI University of Applied Sciences, The Netherlands & University of Johannesburg, South Africa) and Gamze Karayaz (Isik University, Turkey)
Business Science Reference • copyright 2018 • 349pp • H/C (ISBN: 9781522531975) • US $225.00 (our price)

Organizational Leadership for the Fourth Industrial Revolution Emerging Research and Opportunities
Peter A.C. Smith (The Leadership Alliance Inc., Canada) and John Pourdehnad (University of Pennsylvania, USA)
Business Science Reference • copyright 2018 • 125pp • H/C (ISBN: 9781522553908) • US $165.00 (our price)

IGI Global
DISSEMINATOR OF KNOWLEDGE

701 East Chocolate Avenue, Hershey, PA 17033, USA
Tel: 717-533-8845 x100 • Fax: 717-533-8661
E-Mail: cust@igi-global.com • www.igi-global.com

Table of Contents

Detailed Table of Contents

Cloud computing (CC) is generating new computing and business models because of its service-based nature, which enables collaboration and data exchange at higher level, more flexibility with better efficiency and parallel decreasing costs. Manufacturing environments can also benefit from cloud technology and follow fast changes in market demands. In these new scenarios, interoperability has vital importance in the operation and in the interaction among industrial realizations of cyber-physical systems. The chapter introduces the different cloud models and the interoperability issues concerning connected enterprise information systems. Various standardization frameworks have been developed for homogeneous integration of IT models in industrial environments: the IIRA and the RAMI 4.0 are the best-known ones. The chapter introduces both architectures; their methodological approach to industrial integration efforts and how integration can be realized through the OPC Unified Architecture. Finally, the authors propose a basic conceptual model for cloud manufacturing.

An overview of the work under development within the EU-funded collaborative project MAESTRI is presented in this chapter. The project provides a framework of new Industrial methodology, integrating several tools and methods, to help industries facing the fourth industrial revolution. This concept, called the MAESTRI Total Efficiency Framework (MTEF), aims to advance the sustainability of manufacturing and process industries by providing a management system in the form of a flexible and scalable platform and methodology. The MTEF is based on four pillars: a) an effective management system targeted at continuous process improvement; b) Efficiency assessment tools to support improvements, optimization

Preface

INTRODUCTION

About the Subject

Nowadays, almost everything turns around internet. New technologies and novel electronic devices surprise us each moment. Their processing capacity and interoperability goes further increasingly, and information systems are forced to align with that. The required foundations for an innovative internet environment, where all participating "things" can be easily connected and interoperable, and new services can emerge on top of that, are in plain sight. Industrial engineering and manufacturing areas are evidences of this. The planning, management and decision capacity, the control and monitorization of current production states, the capacity: i) to quickly reconfigure in an effort to correct or prevent defects and; ii) to realign to ever more global, distinct and dynamic requirements and stakeholders, are goals that interest to be continuously and jointly rethought and redefined, towards more competitive results. This scenario represents the framework of the Industry 4.0.

The Industry 4.0 has been announced as the next relevant milestone in continuous industrial transformation: 1) water and steam power production mechanization; 2) mass electric power production; 3) automation and computers; and 4) global industrial digitalization. Furthermore, if the previous took several years to make a relevant geo-localized impact, this one has come so fast and with so large a scope of impact, that industries, processes and humans, are not sufficiently prepared to deal with and get the most of it.

The short and dynamic product life cycle, the continuous technology progress, the aging work force, the shortage and heterogeneity of resources, the individualized product requirements, and high level of quality competition, are handicaps that hinders the required conversion of existent industries towards smart industries, the ones that fit Industry 4.0 paradigm.

The capacity to have production processes with: 1) interacting physical elements, processes and its controlling computational systems; 2) immediate data processing and analysis and, if required; 3) settings adjustment, to prevent or correct problems, substantiating a cyber-physical system, is the essence to the successful deployment of Industry 4.0.

Information and communications technology (ICT) is the key to the coming Industry 4.0 challenges. Artificial and augmented intelligence, advanced analytics, and predictive learning are relevant buzzwords that represent advanced processes for data mitigation and knowledge generation. Innovative products, processes and services should be on top of those processes, as well as on new and emerging technologies.

The expected uncertainty can be analyzed to help create processes based on effective collaboration and cooperation services. This human-centered behavior allows people work that only people can do, with automation responsible for the remaining processes. Thus, innovation can be turned to a competitive advantage.

The Mission

The mission of this book is to present the main issues, challenges, opportunities and trends related to the explosive range of new developments and applications of Industry 4.0, which has caused a permanent evolution, and impacts every organization and society as a whole.

The overall objectives are: to discuss the importance of this concept and their supporting technologies; to present recent developments; to introduce the state-of-the-art technologies; to discuss organizational preparedness; and to analyze the impact on society and on people.

This book is intended to support a professional audience of top managers and IT professionals and an academic audience (teachers, researchers and students, mainly of post-graduate studies).

ORGANIZATION OF THE BOOK

This book is organized in 15 chapters, co-authored by 55 contributors belonging to 27 different organizations ranging from the academia to different private sectors including industry and services.

The first chapter, "Teamwork Behavior: A Review to Interconnect Industry 4.0 Entities," focuses on the integration of teamwork into existing models by providing a review identifying key factors that affect teamwork behavior to both human and agent-based models.

In Chapter 2, "Digital transformation Towards a New Context of Labor: Enterprise 4.0," the authors present a comprehensive view on a new context of labour through the adoption of new technology enablers to improve organizational processes and Information Systems.

In Chapter 3, "The Role of Universities in Industry 4.0 Era: Entrepreneurship and Innovation Perspectives," the authors take a focus on the relation between universities and industry, through the review of the entrepreneurial and research models applied in several countries.

In Chapter 4, "Technologies for Industry 4.0 Data Solutions," the authors overview the requirements necessary to build an integrated data solution using existing isolated technologies like Industrial Control System and Information and Communication Technology, including work force without specific ICT background.

In Chapter 5, "Smart DC Microgrid: A Cyber-Physical System Perspective," the authors present a cyber-physical system example of microgrid technologies, where producers can be at the same time consumers, discussing the relation between the physical domain, governed by the laws of physics, and the cyber domain, where computers run algorithms and make control decisions by communicating with actuators.

In Chapter 6, "Cyber-Physical Systems Framework for Smart Built-Environments," the authors tackle the gap between digital building model (BIM) and automated data capturing (ADC) to promote the collaboration among the multiple dimensions of construction entities, creating a new cyber-physical system that while integrating the construction and logistic processes, manages the progress of construction site activities in close to real time.

In Chapter 7, "Smart Gateways for IOT-Factory Integration: Trends and Use Case," the authors explore the integration of cyber-physical systems with industrial internet of things, presenting a case study developing smart gateways for different shop floor equipment. Furthermore, the authors also propose the tight collaboration between industry, university and startups to foster SME development.

In Chapter 8, "Cloud-Based Manufacturing (CBM) Interoperability in Industry 4.0," the authors present the interoperability challenges between cyber-physical systems and cloud computing based services, presenting frameworks for IT model integration in industrial environment like IIRA and RAMI 4.0.

In Chapter 9, "Towards Industry 4.0: Efficient and Sustainable Manufacturing Leveraging MTEF – MTEF-MAESTRI Total Efficiency Framework," the authors present an overview of the collaborative project MAESTRI that provides a total efficiency framework to advance the sustainability of the manufacturing industries based on four pillars: effective management, efficiency assessment tools, industrial symbiosis paradigm from energy and waste exchange, and an IOT infrastructure.

In Chapter 10, "A Practical Approach to Manufacturing Execution Systems at Bosch AvP: Scope, Structure, and Implementation," the authors discuss the use of Manufacturing Execution Systems to maintain competiveness and present a transition between the theoretical concept and a practical applied system controlling the shop floor operations.

In Chapter 11 "The Logistic Model Based on Positions (LoMoBaP [MoLoBaC]) and Industry 4.0," the authors study the impact of a logistics model based on positions, which is globally applicable to any organization, when adapted to the Industry 4.0 business approach.

In Chapter 12, "Ceramic Industry 4.0: Paths of Revolution in Traditional Products," the authors present the case study of a traditional industrial sector, the Portuguese ceramic industry, which is typically not related with high-tech. This study shows a pathway for traditional industries to assess, plan and lead the Industry 4.0 adoption.

In Chapter 13, "Digital Transformation in the Utilities Industry: Industry 4.0 and the Smart Network Water," the authors propose a proof of concept platform for the water resources. This platform will achieve a more efficient management of related resources by improving existing processes through the application of digital transformation.

In Chapter 14, "Industry 4.0 from the Supply Chain Perspective: Case Study in the Food Sector," the authors present a case study in the food sector from the point of view of the supply chain perspective, using a classification scheme of technologies, features and extensions.

In Chapter 15, "Enterprise Integration With the Structural Services Architectural Style," the author presents a comparison of two architectures for enterprise application integration, SOA and REST, observing that the architectures are either state or behavior oriented. A new architecture style and service-oriented programming language are proposed to combine the best features of both architectures.

We hope you find it useful. Enjoy your reading.

Luís Ferreira
Polytechnic Institute of Cávado and Ave, Portugal

Nuno Lopes
Polytechnic Institute of Cávado and Ave, Portugal

Joaquim Silva
Polytechnic Institute of Cávado and Ave, Portugal

Goran D. Putnik
University of Minho, Portugal

Maria Manuela Cruz-Cunha
Polytechnic Institute of Cávado and Ave, Portugal

Paulo Silva Ávila
Polytechnic of Porto, Portugal

Acknowledgment

We express our gratitude to IGI Global for the opportunity to edit this book, and for the excellent support of their team of professionals. We would like to thank all the members of the Scientific Committee, for their commitment and for sharing their knowledge and experience in the support of the decision-making processes. Finally, we would like to show our appreciation and gratitude to all the authors for their excellent contributions: this book encloses the result of their relevant work and deepens knowledge on the Industry 4.0 domain.

Chapter 1
Teamwork Behavior:
A Review to Interconnect Industry 4.0 Entities

Paraskevi Tsoutsa
University of Applied Sciences of Thessaly, Greece

Panos Fitsilis
University of Applied Sciences of Thessaly, Greece

Omiros Ragos
University of Patras, Greece, Greece

ABSTRACT

Teamwork has become an important research field and its contribution to organizational performance has attracted attention of researchers from several disciplines. The development and application of newly emerged technologies such as Industry 4.0, Internet of Things, and cyber physical systems create additional concerns for teamwork which claim to be integrated into existing models. The objective of this chapter is to advance research on teamwork, by facilitating researchers with a review which identifies the key factors that affect teamwork behavior both in human and in agent-based teamwork models, while indicating if and how they are inter-related. A review of related studies was conducted, and, as a result, a range of factors that affect teamwork behavior to both human and agent-based models was identified and analyzed. From the analysis, stand out factors that gain attention while newly-appeared factors are determined from recent studies about models shift towards dynamic and realistic environments. These discoveries point to new aspects of teamwork behavior.

1. INTRODUCTION

Team working has been thoroughly studied by psychologists and human resource experts over the last decades. During these years, many researchers studied why humans succeed or fail in joint activities and a variety of models have been developed that follow social - psychological approaches for human team formation while focusing on concepts that involve different strategies and approaches. In early studies,

DOI: 10.4018/978-1-5225-4936-9.ch001

authors proposed models focused on specific characteristics that team members should have such as personality, functional expertise, competencies, goal orientations, teamwork orientations, etc. (Delarue et al., 2008; Mathieu et al., 2008). As the team research matured, research moved firmly from dealing with single characteristics that teams and members have to a variety of behaviors that team members expose (Rousseau et al., 2006; Salas et al., 2008). Ultimately, in recent studies member's tasks and behaviors are grouped into distinct roles within the system that are aligned with the expertise of team members (Kozlowski et al., 2015). Increasingly, researchers propose that teammates, jointly with the operational tasks that they perform in a team, they also have to play some other role such that of coordinator, contributor, idea generator, etc. Indeed, as it is witnessed by empirical studies, these approaches are effective for certain contexts, tasks and domains (Berlin et al., 2012).

Apart from human team working, another area where similar problems have been studied and team work theories are applied is the area of software agents. As with humans, a group of software agents must accomplish given tasks by organizing themselves according to their individual characteristics and their role within the overall system. Furthermore, complex teams that comprise of humans and agents were studied, since in the real-world scenarios, agents play a variety of roles in their interactions with humans; they support humans in tasks that they jointly perform or independently work to carry out tasks for which they are responsible (Van Wissen et al., 2012). Examples of applications to human–agent teams include autonomous agent pilots that work together with people to complete a joint mission, mission execution assistants for astronauts (Tambe, 1997), personal assistants interacting with users, e.g. Siri for iPhone and Cortana for Windows Phone (Kato et al., 2014) and hybrid teams consisting of humans, robots and virtual characters situated in a flexible industrial production environment (Schwartz, 2016). Considering that humans and agents have different capacity and capabilities in order to combine their strengths, it is crucial to have good coordination (Bradshaw et al., 2012) and other behaviors that humans do. Indeed, the studies on agent teamwork have been motivated by concepts and theories that were developed on human teamwork. In doing so, common characteristics with human team working are used, e.g. planning, coordination, collaboration.

While past studies on team formation focused on interaction only between human or agents, in recent years studies research heterogeneous teams of humans, agents and agents in robotic settings, where agents or robots acting autonomously alongside humans as peers. In such heterogeneous teams, most studies are based on the assumption that the non-human team members need to act in the same manner that human ones do. Examples of such teams are teams of robots in rescue operations (Beck et al., 2016) and teams of humans, robots, and virtual agents cooperating in production settings (Schwartz et al., 2016).

Today's industry environment is so complex that work is done in teams composed of members that are either humans, agents or robots where each one is specialized in specific tasks. For example, humans together with various e-commerce cooperative agents created by different companies having different standards need to collaborate and the successful result for the endeavor depends on the team rather than the individual effort of the member. When referring to team with such content, in this work the term agent-based teams is used. In the context of Industry 4.0 (Data, 2011), embedded systems will be linked with organizational processes to transform industries and this will enable the real time connection between humans, machines and smart objects. In other words, Industry 4.0 (i4.0) envisions a future where, in industries, humans and Cyber Physical Systems (CPS) are effortlessly connected while exhibiting teaming behavioral attitudes. In order for these entities to effectively be connected and working as a team, models referring to team working in literature are investigated.

Indeed, over the last decades, teamwork has become a thriving area of research and academic efforts, with a substantial body of research work produced. Several surveys that focus on teamwork have been published but many of these surveys were conducted at the earlier stage of teamworking (before 2007) and recent advancements in the technology were not covered. Additionally, these surveys overview and analyze teamwork behavior only from specific perspectives and lack a holistic view of teamwork.

The aim of this chapter is to survey the existing major models and frameworks proposed on human and agent teamwork behavior and contribute by (i) providing an integrated table of major human roles that gives a holistic view of the current models on humans, (ii) presenting the major models-frameworks of team behavior proposed in agent based team working and (iii) identifying and analyzing the different teamwork factors among variant types of teams in agent based models while indicating if and how they are inter-related. Furthermore, the authors attempt to form a classification from the teamwork factors revealed in to four clusters emerged, which indicate the dominant teamwork behavior of members. The rationale is to lay the groundwork to teamwork about entities that are get together to form teams in dynamic environments like i4.0, by employing major roles that are used in human teams and habilitate their behavior with the key factors identified at agent-based team working studies.

Section 2 of this chapter presents a background work for human based teamwork models and a mapping of major human team roles reviewed to a base taxonomy. Section 3 presents leading agent-based collaboration models. In Section 4, the agent-based team working factors revealed from the reviewed studies are discussed and a running example is presented to make more understandable the need for teamwork roles. Finally, the conclusions obtained from the review and future trends are discussed in the last section.

2. HUMAN BASED TEAMWORK MODELS

Several authors aimed at identifying and measuring the factors that are likely to increase the effectiveness of teamwork (Tannenbaum et al, 1992; Mathieu et al., 2008; Berlin et al., 2012; Parker, 2008). The literature review indicates that the research on teamwork has been fragmented through the years in two main trends. These are (i) modeling teamwork according to team and team process characteristics and (ii) modeling teamwork according to role – behavioral attitude of team members. In the first category the studies witness that many different team characteristics, such as team structure, team size, team composition, physical distribution and team process characteristics, such as communication, coordination, shared knowledge, adaptability and planning, are critical to the success of a team. In the second category, which is central to this review, team models are based on role behaviors of participants. A role behavior is defined as the patterns of the behavioral characteristics of the ways in which one team member interacts with others in order to facilitate the progress of the team as a whole (Belbin, 1981). The existence of team members that play predefined roles such as a coordinator, organizer, promoter etc. in a team is a necessity for its performance. In addition to role existence, studies defend the idea that high performing teams need to have a balanced representation of all proposed team roles.

As already indicated, the research involved in this chapter draws upon studies where human teamwork models solely focus on roles and role behaviors. This is done so, because the authors are interested in developing a review table with the series of key roles suggested in the various proposed models that can potentially be applicable to cases where heterogeneous non-human teams need to be formed in industry environments. Since the focus is on the second category, where teamwork is modeled according to role and behavioral attitude, the authors proceed by giving definitions about teamwork behaviors and teamwork roles.

Teamwork behavior is a multifaceted concept that has been difficult to conceptualize (Rousseau et al., 2006). A team member's behavior involves (i) all the operations-related activities to be performed while working in the team and (ii) all activities that are supportive to the existence of the work team. According to Aritzeta et al. (2007), the role concept can be defined from a psychosocial perspective as the behavior expected from an individual occupying a specific position. The cognition and expected behavior associated with the position are fundamentally important to success of the role. Similarly, Belbin defines "the tendency to behave, contribute and interrelate with others in a particular way" as a team role (Belbin, 1981). Consequently, a role that member of team play can be described by the behaviors they express during the occupation of the specific position and this is the way that the terms role and behavior are used in this review. In general, teamwork roles follow the categorization of behaviors and they are divided into two main categories, (i) roles that are task-oriented and are concerned to solve operational tasks in the team and (ii) roles that are team-oriented and they are concerned with promoting group-centered behavior (Aritzeta et al., 2007). Since teamwork behavior is required for effective team performance of team members, teams rely on different members to play a role and fulfill different needs through their behaviors (Aritzeta et al., 2007).

There were many attempts to classify roles and behaviors in the context of human team working. In fact, as it is witnessed in Driskell et al. (2017), there is a seemingly unlimited number of roles that have been described as relevant to human team performance. For this reason, they are only selected and mapped five pivotal studies of the field. The complication, which in general is faced during the classification, is that different studies have often used different names to refer to the same role or similar names to refer to different roles and behaviors. The most common ways of organization used to previous reviews include classification according to dimensions of teamwork behaviors through a subset of teamwork characteristics that can potentially be grouped in to a specific teamwork behavior. In the reviewed studies, not all researchers state explicitly that use roles or behaviors but it is subsumed so. More specifically, Rousseau et al., (2006) present the diversity of ways that teamwork behaviors are conceptualized.

The human teamwork models examined are the following:

1. The model of McCann et al., (1989) and Margerison et al., (1990) is consisting of eight different types of team roles that are: (a) explorer–promoter, (b) assessor–developer, (c) thruster–organizer, (d) concluder–producer, (e) checker–inspector, (f) upholder–maintainer, (g) reporter–advisor, and (h) creator–innovator. The authors recognize that individual work-style preferences that humans bring to teamwork are key factors influencing individual and team performance. When human teams are constructed by humans that are put together on an ad hoc basis with no choice of members, then the teams have potential weaknesses since some roles are not present. They recognize that team behavior is related to the tasks and the roles that members play. Additionally, high team performance is connected with teams that are well balanced with respect to the team roles of the various team members along with good linking skills of members.
2. The work of Belbin (1981) on team roles contained eight clusters of behavior that individuals adopt when participating in a team for constructing successful management teams. The model consists of eight roles, where each role is defined as the member's tendency to behave, contribute and

interrelate with others in a particular way. In subsequent research (Belbin, 1993) renames some team roles, and a ninth role, the Specialist, is added. These nine roles are (a) Plant, (b) Resource investigator, (c) Coordinator, (d) Shaper, (e) Monitor evaluator, (f) Implementer, (g) Team worker, (h) Completer–Finisher and (i) Specialist. Belbin also defends the idea that high performing teams need to have a balanced representation of all team roles. The team role balance hypothesis assumes that if all team roles are present and follow a specific distribution in a team, then it will perform better than other teams do not have this balance. The team role that a member plays is distinguished from the functional role that refers to the technical skills and operational knowledge relevant to the member's job. Consequently, several people may have the same functional role but vary greatly in their team roles (Aritzeta et al., 2007).

3. Parker (2008) developed a set of team player styles, which fall into four types: (a) Contributor, (b) Collaborator, (c) Communicator, and (d) Challenger. The styles are categories in which human with similar characteristics are found. A Contributor is a task - oriented team member who enjoys providing the team with good technical information and data. After doing his work, he pushes the team to set high performance standards and to use their resources wisely. A Collaborator is a member of the team who is flexible and open to new ideas, is willing to work outside the defined role and is able to share with other team members. A Communicator is a process - oriented member that is an effective listener and a positive person in general. A Challenger is a member who questions the goals, methods, and even the ethics of the team, is willing to disagree with the team leader and other members and encourages the team to take well – conceived risks. While Margerison's and Belbin's team roles are based on Jungian tradition, Parker argued that his style-based team roles provide more latitude for members in taking different team roles and feedback for improvement (Mathieu et al., 2015).
4. According to Salas et al., (2005), team effectiveness takes a holistic perspective in considering not only whether the team performed and completed the team task but also how the team interacted to team processes to achieve the team outcome. They argue that, regardless of the team task that is examined, a set of teamwork components will be required to complete the task and present these components as the "Big Five" of teamwork. These five components are team leadership, mutual performance monitoring, back up behavior, adaptability and team orientation. Moreover, they suggest that these factors improve performance as long as three supporting mechanisms are also present within the team, which are shared mental models, closed loop communication and mutual trust.
5. Mathieu et al., (2015) reviewed and synthesized previous team research and argued that, a team can be composed by members occupying six different team roles, which are referred to as Team Role Experience and Orientation (TREO) dimensions. The roles of each member are defined by taking into account the combination of its previous experiences and orientations. The authors notice that team composition serves as the foundation upon which other team factors are built, and represents a key enabling feature of teams. They synthesize the previous work of Belbin, Margerison's and other researcher's work that focus on team roles and propose a six-dimensional framework consisting of the following roles: (i) Organizer, (ii) Doer, (iii) Challenger, (iv) Innovator, (v) Team Builder and (vi) Connector. Moreover, they validated the TREO model by measuring the outcomes and its predictive validity.

Although the existing literature has suggested numerous team roles, this review adopts Margerison et al. taxonomy (Margerison et al., 1990), since their study was among the first ones that proposed role behaviors in the extent that they are still used; the first version of the model was released during 1984 (Margerison et al., 1984). In the following, there is a description of the eight necessary roles for a successful team composition, as they are recognized by Margerison's study. The roles are defined according to the key work elements need to be addressed in order for a team to be successful. The tasks and responsibilities of team members are related to work elements addressed and they are all covered by the critical roles construct the team management wheel.

1. **Creator–Innovator:** People that play this role are future oriented and enjoy thinking up new ideas and new ways of doing things. They come up with new strategies and different approaches to tasks, creating new ideas and experimenting with them in order to handle various situations and challenges. They are looking to tomorrow rather than worrying about today. An organization needs creators-innovators to invent new products or services or ways to improve existing ones. They aim at creating fundamental ideas that will move forward the organization. Research and development departments need to be stuffed with employees that play this kind of role.
2. **Explorer–Promoter:** People that play this role enjoy being with other people and they are excellent at taking ideas and promoting them. The promoter helps bridge, connect members of the team with members outside of the team working in a twofold aspect by searching for new opportunities and exploring possibilities inside, and outside of the domain. They are good at bringing back contracts and resources that can help the organization move forward. They are excellent at seeing the "big picture" and they can capture opportunities. While innovators think up ideas, promoter's task is to sell the ideas for the organization. Many good ideas may be lost simply because they are purely promoted from the organization since no member of the team is playing this role.
3. **Assessor–Developer:** People that play this role have characteristics of both explorers and organizers. They are excellent at making someone else's idea work in practice. They can take an idea and match it to an opportunity without caring about organizations' constraints. They respond well to new challenges and they can push forward an idea into a workable scheme. After an idea has been sold, the work activity follows often is based to the developers. Doing the project's work often requires further development, which may involve prototype testing or checking alternative versions of the idea before implementation.
4. **Thruster–Organizer:** People that play this role enjoy making things happen. They are analytical decisions makers, always doing what is best for the task. Their aim is to implement ways and means of making things work on time and on budget. They monitor the global schedule and keep track of accomplishments and how teams are progressing relative to goals and timelines. They usually prefer to work according to an established plan and in a structured manner so that objectives are clearly set out and everyone in the team knows what has to be achieved and by when. They organize others so that deadlines are met. They are task oriented, they set objectives, establish plans and then press for action. In an organization, once a project is ready to be implemented, plans have to be made, budgets approved, and schedules established so that the project is implemented according to deadlines. These activities are part of the organizers tasks.
5. **Concluder–Producer:** People that play this role are practical and can be counted on to carry out things through to the end. Their most important goal is to work on a systematic way to produce work outputs. Their strength is in setting up plans so that output can be achieved on a regular basis.

For this role, it is a challenge the moto "job well done". They are usually more patient than other members and have the ability to work in a reliable and stable manner to deliver results. Since the production of the organization has to be done on a regular basis according to standards and effectiveness, producer roles are appropriate for the production work function.

6. **Checker–Inspector:** People that play this role enjoy the detailed side of work and like working with facts and figures. They are usually careful and spend long periods of time on a particular task. They check and audit that systems and contracts are working focusing on the detailed and controlling aspects of work. They are comfortable working within rules and regulations that others have established to ensure that quality is maintained in an organization inspecting activities that are important. It often happens in organization for some team projects to fail due to weakness in inspecting. Members that play the inspector role are ideal to end such weaknesses.
7. **Upholder–Maintainer:** People that play this role have high concerns for other people and they strongly support others that share the same ideas with them. They prefer to work in a supportive environment where they make sure things are done according to their standards. Their goal is to ensure that standards and processes are upheld and to maintain team excellence. They prefer to take an advisor role rather than an executive role. All teams need to have a solid infrastructure so that all other activities can proceed in the most efficient manner. Personnel in administrative positions need to have maintaining activities through their supporting functions.
8. **Reporter–Advisor:** People that play this role are excellent in gathering information and putting it together in an understandable way. They are patient people and prefer to make sure they have all the information before they take action. They foresee conflicts might happen and they usually move to defuse the conflict themselves. People work for information offices need to play this role. They spent most of their work time in advising activities, collecting data and passing them on for action to others in the team.

This research is conducted within human and agent based teamwork models and is targeted at coupling human and agent behavior to address the challenges associated while interconnected different entities that are met in industry domains. To have a holistic view of major human team roles, the authors draw upon the five substantive studies described in the second section, and by using the classification of roles that is proposed in Margerison et al. as a base taxonomy, they map the roles and behaviors from the above-mentioned studies to the base taxonomy in Table 1.

The variance of the number of roles proposed by each model is noticeable in the above table. The smallest number of roles is proposed in Parker's study and they are four, while the study that proposes the most roles is that of Belbin where nine roles exist. Mathiew et al. (2008) fuse the nine roles of Belbin's model to six, while Salas et al. (2005) propose six behavioral attitudes. This number is near to the average of roles proposed in general. In such a context, Aritzeta et al. (2007) raise the question about the existence of well-differentiated team roles in proposed models and, whether these team roles are potentially better differentiated using some other grouping.

The next observation regards to the overlapping behaviors that exist among roles in different models. The challenger role, for example, of Mathiew et al. (2008) model presents a behavior that matches both to explorer's and to reporter's behavior of the Margerison (1990) model. Apparently, in models that the number of roles is small, the roles are more concentrated while in models that a large number of roles is proposed, there is a more detailed overall description for behaviors appearing in the team and the necessary expertise of members to play these roles is distinguished.

Table 1. Table with mapped roles of selected models to Margerison's model

Role name according to Margerison et al., (2015)	Belbin (1981, 1993)	Parker (2008)	Salas et al. (2005)	Mathieu et al. (2008)
Explorer–Promoter	Resource Investigator	Challenger	Team orientation	Challenger, Connector
Assessor–Developer	Implementer, Team Worker	Collaborator		Team Builder
Thruster–Organizer	Co-ordinator, Shaper	Collaborator, Communicator	Team leadership, Backup behavior - (Delegation), Adaptability, Team orientation	Organizer
Concluder–Producer	Plant, Completer-Finisher, Implementer	Contributor		Doer
Checker–Inspector	Completer-Finisher, Monitor Evaluator	Communicator	Mutual performance Monitoring, Backup behavior - (Coaching)	Organizer
Upholder–Maintainer			Backup behavior - (Delegation), adaptability	Organizer
Reporter–Advisor	Specialist	Contributor	Backup behavior - (Assist)	Challenger
Creator–Innovator	Idea generator	Contributor	Adaptability, team orientation	Innovator

3. AGENT BASED COLLABORATION MODELS

The term agent is used in different disciplines to refer to different entities or artifacts. In artificial intelligence, agents are defined as autonomous entities that observe and act upon an environment and direct their activities toward achieving goals (Russell et al., 2005). Agents are either developed to work isolated or in collaboration with other entities to achieve a goal. This covers a variety of possible uses for agents, from agents that act auxiliary with limited autonomy to agents that respond to particular tasks in a manner that humans do (Chen et al., 2013). A lot of research effort was devoted to determining the basic dimensions that define agent team working in order to reach an optimal teamwork behavior (Jennings et al., 1998). In dynamic environments, a team of agents is more than a group of intelligent entities working towards achieving a goal. An agent, in order to reach a certain level of intelligence, autonomy and team work behavior, has certain properties such as reactivity, proactivity, social ability, decision making ability, etc. Like human team members, they need to be coordinated using share knowledge, collaborate by working together on common tasks and exhibit more teamwork characteristics.

The requirement for agents that interact with humans in a practical manner demands that agent capabilities need to be augmented with techniques that facilitate effective team abilities. In particular, agent-based teamwork, where human and software agents interact as peers, is concerned with joint commitments and joint responsibility, and requires establishing situation awareness collaboratively, developing shared mental models as situation evolves, adapting to mixed-initiative activities, etc. Furthermore, successful agent technologies take advantage of the diversity within teams and the differences between human and agent strengths, i.e. human reasoning has different characteristics than algorithmic reasoning (Bradshaw et al., 2012). Although many theoretical and applied techniques have been applied and researchers have attempted to build on and integrate the work of others, due to the problems inherent complexity, we notice that most studies only focus to some of the identified challenges and do not handle the problem holistically. Additionally, models in agent-based team working should be adapted and take into account concerns that arise from the development and application of new emerged technologies such as IoT, i4.0, CPS, etc.

In the following, some major teamwork concepts and theories concerning the agent-based teamwork research areas are presented that have contributed to the research of the agent team working problem.

One of the first theories, namely team cognition theory, was used for explaining team functioning (Cannon-Bowers et al., 1990). In this theory, the concept of Shared Mental Model (SMM) is fundamental in both the context of human teams and heterogeneous teams of humans and agents. A mental model is a mechanism where humans are able to generate descriptions of system's purpose, form explanations of system functioning, observe system states and predict future system states. The idea behind SMM is that team performance is improved if team members have a shared understanding of the task to be performed, knowledge and beliefs, goals and intentions and plans of the involved teamwork. It actually refers to an overlapping understanding among members of the team, either humans or agents, regarding their objectives (Yen et al., 2006). This is achieved by developing knowledge structures held by members of the team that enables the adaptation of their behaviors to the demands of the team. Jonker et al. (2011) present a more detailed definition of the concept of SMM by formalizing it with a mental model ontology. Several researchers follow the theory of SMM to build teamwork models (Sycara et al., 2006; Jonker et al., 2011).

Another approach is Joint Intention theory (JIT) that focuses on a team's mental state joint by all members, which is called a 'joint intention'. Using the notion of JIT, Cohen et al. (1997) propose a formal approach to build artificial collaborative agents. Their view of JIT is the persistent commitment of the team to a shared goal and implies this commitment of all members to a mutual belief about the state of the goal.

One influential theory of teamwork is the one of Jennings et al. (1998). They define join commitment towards the group goal and point to the vital role of intentions in teamwork like the commitment that agents demonstrate and how focus on handling unexpected failures. In order to enter a joint commitment, a team member that is in a specified mental state, communicates to establish appropriate mutual beliefs and coordinate with actions towards the shared plan. Members inform the team when they reach the conclusion about whether the goal is achievable or impossible.

Similarly, Shared Plans theory originated from Pollack's mental state of plans model (Pollack, 1990). It was extended to provide formalization in order to accommodate group activity that handles complex actions while it allows plans to be partial and handles evolution of shared plans over time (Grosz et al., 1996; Grosz et al., 1999). According to Pollack's observation, there are different types of plans. More specifically, in agent planning, there is difference between those plans that agents know that are applicable and potentially can be followed and those plans that agents adopt because they know the recipe to follow them and subsequently guide their action. Having a plan is not only knowing how to execute an action, but also having the intention to execute the actions required. The term "recipe" is used to describe how to execute an action while the term "plan" implies knowing and having a plan to execute the action. One of the focuses of shared plans is on specifying the capabilities and mental attitudes (beliefs and intentions) required for the participants of a group in an activity to have a collaborative plan. A shared plan for a group action implies that member's individual and mutual beliefs about the action and its constituent sub actions are to be executed.

However, agents may differ in their beliefs about the ways to perform an action and in their assessments of the ability and intention to make an action. To better infer the plan it adopts, it is necessary for an agent to analyze the nature of the mental state of the plan it has and to be aware about intents of other agents with whom it discourses. These activities require plan-based reasoning that arises from the participants' attitudes of intentions toward the actions of others and of the group as a whole. To model these accurately, the adoption of a mental state view of plans is essential. Using mental state is possible to accommodate differences among the beliefs of individual participants and distinguish between knowledge about action performance and intention to act. The theory provides a detailed account of an agent's motivations for initiation of communication that will lead to do an action. The extended model of shared plans deals with agents that initially have only partial knowledge of how to perform an action and distinguishes between partial and complete plans. Furthermore, it provides the rationale for the process of revising beliefs and intentions since both individual and team plans evolve over time.

The formalization of shared plans has strong logical foundations and is defined in terms of modal operators, meta-predicates, i.e. abbreviations for complex formulas involving other predicates, modal operators and actions. It also can be used to define the propositional attitude of intending to do an action and introduces the attitude of intending that a proposition hold.

Ad Hoc team working is a relatively new research trend (Bowling et al, 2005). In this case, a member's fundamental objective is to collaborate without pre-coordination. Ad hoc teams were studied by Agmon et al. (2011), Stone et al. (2010), Barret et al. (2011). Ad hoc teams are heterogeneous teams of humans, agents or robots that are able to cooperate with multiple potential teammates without prior coordination in order to reach an optimal team behavior. Stone et al. (2010) defined an ad hoc team setting as a problem in which team coordination strategies cannot be developed a priori. In their study, they present an algorithm for determining the optimal strategy for an ad hoc agent to lead, with minimal cost, the best response agent to perform the action yielding optimal joint utility.

Finally, another approach of modeling agent-based teamwork, central to this paper, is the employment of the role theory. This theory deals with collaboration and co-ordination and is exploited in modeling systems in different domains and applications. The concept of roles is ubiquitous in social sciences and role-based modeling has been proposed in business process modeling (Ertugrul et al., 2015), object-oriented software engineering (Andersen, 1997), agent engineering (Kazik, 2010) and service engineering (Tsoutsa et al., 2017). Roles are considered as a technique of packaging behaviors and capabilities that team members have and they are related with the tasks of team members in achieving the team goal. In agent engineering, roles abstract over agents describing capabilities for specific tasks (Genter et al., 2011). Each team member implicitly follows a role that accomplishes a specific task or exhibits a particular behavior. The role assignment to agents may be done according to different scenarios. One approach for example is to assume for each agent a role at which it is most capable of playing, while other approaches take into consideration previous agent behavior and consult learning algorithms to find the appropriate role to play.

4. DISCUSSION ON TEAMWORKING FACTORS

4.1 Taxonomy of Factors Affecting Team Working in Agent Based Domains

In the following, the team factors that merit particular attention in the reviewed studies on agent-based teamwork literature are presented; the specific factors regard the behavioral attitude with reference to models concerning agent-based systems. Moreover, in an attempt to classify them, a wheel is presented describing the four major clusters gear toward four primary dimensions that determine agent-based team role behavior which are: (a) Synergy related, (b) Strategy related, (c) Coordination and (d) Organizing factors. It should be noted that this is not an exhaustive list of factors, but a list derived from the literature giving an indication about the expected team working behavior of systems where agent members are collaborating in a way that remind human collaboration. The fourteen team factors revealed are discussed in relative detail and their implication during their activity is examined in the next section by using an example in order to gain further insight into the team role structure and performance.

Synergy Related Factors

In this category they are grouped the factors that are needed by members of a group in order to work as a team to achieve a shared goal. These factors are considered as the fundamental characteristics that a member should have in order to be able to participate within a team.

1. **Ad Hoc (ADH):** The ad hoc factor determines the behavior that a team member exposes by which it is capable of cooperating with multiple potential unknown members without pre-coordination in order to achieve a common goal and reach an optimal team behavior. Potential members could be heterogeneous, with different capabilities and have no prior agreement regarding how to work together. In realistic environments, many teams are ad hoc, such as teams met in rescue, military and industrial domains. Applications include rescue robots brought to an earthquake site from different parts of the world, or e-commerce cooperative agents created by different companies with different standards (Stone et al., 2010).
2. **Autonomy (AUT):** Generally, autonomous systems are systems that have decision-making ability and does not depend directly on another coordinating system. There exist different levels of autonomous systems where, at the highest autonomy level, the system decides everything in order to act autonomously, and at the lowest autonomy level, the system offers no assistance while human members must take all decisions and actions. Fully autonomous systems are capable of more functionality. In such systems, the tasks usually are not assigned centrally by the system, but they are chosen by the team members according to their individual skills and capabilities to take them over. In this review, were took into consideration only studies that an agent participating in a team have a minimum level of autonomy, being able to take simple or more complex decisions based on its current state. Some of the studies investigate team members autonomy are Harbers (2012), Wu (2011), Schwartz (2016).

3. **Delegation (DLG):** Delegation is a substitution mechanism of all or a subset of member's activities to other members and its aim is to increase flexibility. For the better performance of the entire team, some members often need to have the ability to choose to cancel out some of their tasks. The delegation behavior also is essential for the members that assign responsibility of tasks and are able to choose what to delegate from a member to other members and what to leave. While delegation can allow substitution of a member and solves the unavailability of a member problem successfully, it might also potentially lead to undesirable situations if it is applied incorrectly (Saidani et al., 2006).
4. **Joint-Intention (INT):** The common intention that systems have toward the group goal; the commitment and focus they demonstrate in order to handle unexpected failures. Joint intention is a central concept during the whole process of teamwork and changes dynamically during planning. Managing the maintenance of joint intention to all individual involved through the whole process in changing environments is necessary according to Dunin et al. (2011).
5. **Knowledge of Teammate's Capabilities (KCT):** The teamwork literature has identified the need for agent members, except from knowing their own capabilities, to also be aware about capabilities of their teammates. There are systems that enable agents to learn about capabilities of their teammates, model them and then use that knowledge to improve collaboration. Kaminka et al. (2005) present a model that allow agents to predict the future behavior of other agents in order to coordinate the activities by observing those agents and building a model of their behavior. Kennedy et al. (2008) exploit cognitive science approaches to produce an autonomous robotic system that "mentally" simulates the decision-making of its teammate.
6. **Knowledge Sharing (KNL):** For teamwork to succeed, participants need to establish and share a common view for specific issues related to the environment. Performance is improved when members have a shared understanding of the task, goals, intentions and plans of the involved to be performed. According to Dunin et al. (2011) the awareness of the group, is of vital importance. This could be established through knowledge sharing and help the group to predict future system states.
7. **Trust (TRS):** For agents to be successful as team members, it is imperative to act appropriately to human's behavior, meeting people's expectations of how they would behave. Not only human but also agent's behavior in group settings has been shown to depend on sociological and psychological aspects, such as trust and fairness (Van Wissen et al., 2012). In autonomous systems, the issue of trust is taken into account and the factors that help to build trust relationships encompass both the social and ethical aspects. As intelligent systems become increasing sophisticated and are capable of accessing information from other networks, it is necessary to examine the implications of these capabilities on trust of the systems (Chen et al., 2013). Researchers suggest that agent designers should include the issue of trust generation in agents' decision-making models in order to promote the development of trust relationships between participants when they act as team members (Van Wissen et al., 2012).

Strategy Related Factors

This category concentrates on characteristics that are fundamental for designing the strategy behavior of the team. Not all team members need to have such abilities, but it is imperative to exist some members that could lead the team following the developed strategy.

1. **Planning (PLN):** A key issue in agent system teamwork is the planning task, with this to include recognizing and resolving conflicts among agents and within plans. In classical planning, the planning problem involves modeling explicitly the states of the world and search for a sequence of actions that change the world state from an initial state to a goal state. After a plan with a sequence of actions is found, the actions are assigned to the members of the team. They execute the actions in order to reach the final state with the requested goal.
2. **Decision making (DM):** Decision making process and problem solving is essential for effective team performance. Team members usually collaborate according to decisions that make for a plan to follow. In recent years, researchers have designed agents that make decisions to support team behaviors. Korsah et al. (2012), presented a multi-robot control system, called xBots, that could support optimal task-allocation planning and flexible execution of the plans. The system optimizes the plans based on temporal constraints in the tasking environments (Chen et al., 2013). Fan (2006) developed the R-CAST agent system where agents adapt their decisions by detecting and responding to situations that are anomalies in comparison to the original expectancy (i.e., what is observed/ informed conflicts with what was expected); such anomalous situations can frequently happen in dynamic environments.
3. **Teamwork Policies (POL):** Having the vision of heterogeneous teams for humans and agents loosely or more tightly bound, e.g. hybrid teams, ad hoc teams, teamwork policies and mechanisms were developed. As an example, in Bradshaw et al. (2004) policies are developed assuring that unsupervised autonomous agent behavior is kept within safe and secure limits of prescribed bounds, even if buggy, poorly designed, unsophisticated, or malicious agent code is developed. In addition, researchers attempt to use policies to assure effective and natural human-agent team interaction, without individual agents having to be specifically programmed with the knowledge to do so.

Coordination Factors

The factors in this category foster coordination.

1. **Coordination (COO):** From early studies on agent teamwork, the problem of coordination is tackled as an essential requirement. Teamwork implies autonomy, but at the same time, interdependence among the members is often the case; the problem arises when members jointly perform a task and one team member is dependent on other team members for achieving a subtask. When the task involves interdependence among the team members, coordination is required to achieve good team performance (Harbers et al., 2012). According to Barange et al. (2014) the coordination between human-agent team members is the problem where agents need to adopt the flexibility of human behavior, as humans may not strictly follow the rules of coordination. In order to achieve coordination in team context, members should share team knowledge and establish common grounding and mutual awareness. Team leaders that have the authority to guide their teams affect the dynamics of coordination. One of the most well-known heuristics in team coordination was originally formulated by Cohen et al. (1997) where any team member who discovers privately that a goal is impossible, it is already has been achieved or it is irrelevant should not define this fact known to the team as a whole. For example, if a robot was asked to get a wrench from the garage so it can help fix a sink, and the robot finds the garage door locked, it would be expected to inform the other partner about this (Bradshaw et al., 2008).

2. **Evaluation (EVL):** Evaluation methodologies and metrics are used to determine the performance and effectiveness of teams and in what degree they achieve their intended effects. The team effectiveness is the result of the evaluation of the outcomes relative to some set of criteria that determines in what degree the team is workable under that context. This evaluation usually measures performance from the perspective of team formation alternatives, interaction efficiency (Scholtz et al., 2004), cost efficiency, trust issues, ad hoc capabilities (Barrett et al., 2011; Stone et al., 2010), time efficiency and plans evaluation mechanisms. For example, when using an evaluation strategy to select potential members during team formation, a team setting is distinguished from another not only according to member's capability actions but also according to the history of member's collaboration or other relevant factors that are used in order to approximate the optimal team behavior.

Organizing Factors

The factors in this category foster organization of the group.

1. **Collaboration (COL):** The establishment of mutual awareness and common grounding among team members that work together performing a task or carrying out the team activities needed to satisfy a shared goal. Multi agent systems require that agents have the ability to plan and act collaboratively and avoid conflicting actions. In contrast with human collaboration, collaboration in agent teamwork poses many important challenges since in many cases there exist no global resource that agent team members can rely on to share their knowledge. In recent years, a recognized challenge regarding collaboration about multi agent systems is the collaboration among agents without prior coordination (Stone et al., 2010).
2. **Communication (COM):** Communication plays important role in multi agent action. It provides means for agents and humans working together to achieve shared objectives. Establishing communication between collaborating members is an activity that usually requires shared knowledge, situation awareness and planning. Within the team context, through communication, agents send messages that represent their goals and planning mechanisms for fulfilling these goals. Therefore, an agent is aware of the objectives it can plan for and the tasks it can perform. In multi agent systems, agents use different means of communication across different channels (e.g. verbal, text, nonverbal). As an example, dialogues and speech acts have been frequently used to model communication (Poslad et al., 2002).

The distribution of factors affecting team working in agent-based domains into four clusters is presented in the team work agent wheel in Figure 1. The wheel has four sectors labeled as: (a) Synergy related, (b) Strategy related, (c) Coordination and (d) Organizing factors, which are the aforementioned clusters. The outer sectors of the wheel describe the factors that indicate the behavior exhibited by a member of the team. Our future goal is to add a third external layer that will contain the primitive behaviors that have to be exhibited from team members.

After investigating the factors that merit particular attention in agent based teamwork literature, by adopting a multidimensional perspective, the authors map each reviewed study to the aforementioned teamwork factors and present them in Table 2. The reviewed studies are marked with an asterisk in the reference list.

Figure 1. Agent team wheel containing teamwork factors located on four clusters

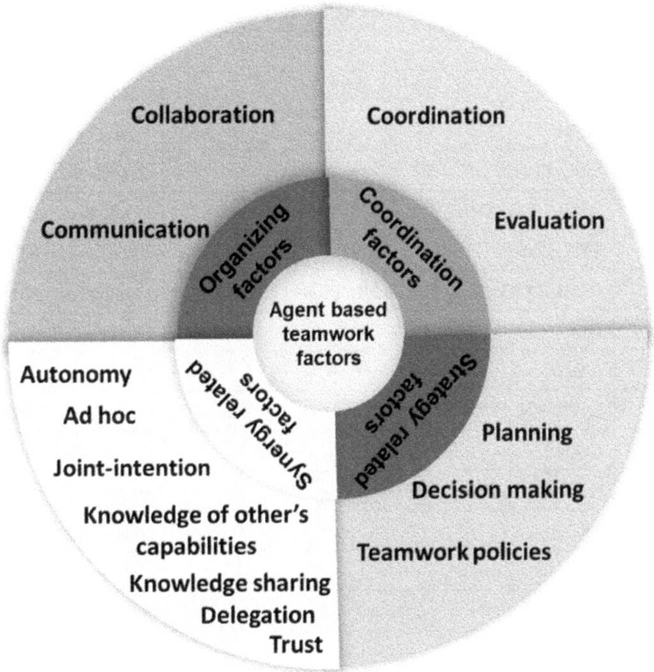

In Table 3, it is presented for each teamwork factor recognized in the review, a brief description based to the corresponding studies that factors appear.

4.2 A Running Example

In this Section an example application is considered that briefly describes a real business scenario, which demonstrates the need for a high-level teamwork among the participants. The idea behind this is to depict how one could apply team roles habilitated with teamwork behavior, enhanced by the agent-based teamwork factors, in order to take further insight into team role structure and discuss their implication and the performance gained. The example is based on SAP "Order-to-Cash" business process (Monk et al., 2012).

According to this scenario, a customer could request information from the company, in respect to their products or services, without obligation to purchase. A quotation is created to address questions of the potential customer and a sales order originates from it. In general, sales order can originate from a variety of documents and activities, e.g. customer phone, internet, email. When the order is created, the system must check availability and determine the delivery date for the order. On the date of delivery, the system checks the order and the relative materials to determine if the delivery is possible, and a credit management is made that alerts the sales representative about the customer's credit situation that arises and if necessary, it blocks orders and deliveries to this customer. After customer's credit situation is checked, the picking and packing process of the goods is through. Then, the goods are loaded and delivered to the customer accompanied with the billing documents that are created. Finally, cash or check is received to close the customer's account.

Table 2. Encoded Teamwork Factors appear in each study reviewed

AA	ENCODED TEAMWORK FACTORS	STUDY
1	ADH, KCT	Agmon et al. (2011)
2	KNL, COM, COL, INT, PLN, COO	Barange et al. (2014)
3	ADH	Barret et al. (2011)
4	POL, PLN, COL, AUT	Bradshaw et al. (2004)
5	AUT, COO, TRS	Bradshaw et al. (2008)
6	TRS, DM	Chen et al. (2013)
7	COM, COO, COL, INT, KNL, PLN	Dunin et al. (2011)
8	DM	Fan et al. (2006)
9	ADH	Genter et al. (2011)
10	INT, PLN, COM	Grant et al. (2005)
11	COM	Hanna et al. (2013)
12	KNL, AUT, COO, TRS	Harbers et al. (2012)
13	INT	Jonker et al. (2011)
14	KNL	Kamar et al. (2013)
15	AUT, COO	Kaminka et al. (2003)
16	KCT	Kennedy et al. (2008)
17	PLN, DM	Korsah et al. (2012)
18	INT PLN, COO, COL	Leng et al. (2006)
19	KCT, DLG	Nalbandyan (2008)
20	COM, PLN, AUT, DLG	Schwartz et al. (2016)
21	KCT, PLN, INT	Skubch et al. (2010)
22	ADH, EVL	Stone et al. (2010)
23	INT	Tambe (1997)
24	INT, TRS	Van Wissen et al. (2012)
25	COL, AUT	Wu et al. (2011)
26	INT, EVL	Yen et al. (2006)

Figure 2 depicts the distributed services that could participate in the example. Indeed, in the context of i4.0, software agents or robots have the ability to automatically execute all these services. Among other issues, the industry domain in this scenario needs (i) to avoid the manual search in different web registries, where there exist myriads of web services records in order to find appropriate services for implementing specific tasks, (ii) to have the ability to use local and remote services as if they are local, i.e. to act like they are fully connected in the same domain by dynamically integrating them to the system, (iii) to monitor the composition process, (iv) to explore new plans for the composition, etc.

The ultimate goal of all sale activities appearing in Figure 3 in the simplified "Order-to-Cash" business process is to equip all actors involved with all the necessary information to complete the potential sale and receive the payment for it. Services that are involved, could be executed from customer representatives, retailers, planners, transportation specialists, carriers, manufacturing plants, logistic centers and

Teamwork Behavior

Table 3. Teamwork factors emerged from the corresponding reviewed studies

AA	Teamwork factor	Description	Study
1	Ad hoc team setting (ADH)	A team member's behavior that is able to cooperate with multiple potential unknown and heterogeneous members without pre-coordination in order to obtain a common goal and reach an optimal team behavior.	Agmon et al. (2011); Barret et al. (2011); Genter et al. (2011); Stone et al. (2010)
2	Autonomy (AUT)	The ability of the member to assist the team by having a degree of reasoning and taking decisions.	Bradshaw et al. (2004); Bradshaw et al. (2008); Harbers et al. (2012); Wu et al. (2011); Schwartz et al. (2016); Kaminka et al. (2003)
3	Collaboration (COL)	The establishment of the ability to act collaboratively with mutual awareness among team members that work together performing a task or carrying out the team activities needed to satisfy a shared goal.	Barange et al. (2014); Bradshaw et al. (2004); Dunin et al. (2011); Leng et al. (2006)
4	Communication (COM)	Through communication, members provide to other members the essential information about working together to achieve shared objectives.	Barange et al. (2014); Dunin et al. (2011); Grant et al. (2005); Hanna et al. (2013); Schwartz et al. (2016)
5	Coordination (COO)	Through coordination the group face the interdependence problem among team members when they jointly perform a task.	Barange et al. (2014); Bradshaw et al. (2008); Harbers et al. (2012); Kaminka et al. (2003); Leng et al. (2006)
6	Delegation (DLG)	The ability of a member to choose the tasks to delegate to other team members in order to increase team performance when unavailability to perform a task occurs.	Schwartz et al. (2016)
7	Evaluation (EVL)	Team members use evaluation methodologies and metrics to determine the performance and the team effectiveness and in what degree they achieve the intended effects.	Stone et al. (2010); Yen et al. (2006)
8	Joint-intention (INT)	The common intentions, the commitment and focus members demonstrate in order to handle unexpected failures and reach the group goal.	Barange et al. (2014); Grant et al. (2005); Leng et al. (2006); Skubch et al. (2010)
9	Knowledge of teammates capabilities (KCT)	In order to improve collaboration, members learn about capabilities of their teammates, model them and use that knowledge to improve team performance.	Kennedy et al. (2008)
10	Knowledge sharing (KNL)	The shared understanding of the task to be performed, common knowledge about shared goals, intentions and plans, that the involved team members have which help them to reason and predict future system states.	Barange et al. (2014)
11	Planning (PLN)	The ability of some members to plan, recognize and resolve conflicts to scheduled plans.	Bradshaw et al. (2004); Grant et al. (2005); Schwartz et al. (2016); Korsah et al. (2012); Fan et al. (2006)
12	Decision making (DM)	Decision making process and problem solving is essential for effective team performance, thus reasoning agents exist in teams and they are responsible for organizing the system's decision-making process.	Korsah et al. (2012); Fan et al. (2006); Chen et al., 2013
13	Teamwork policies (POL)	Policies assuring the appropriate functionality of members according to domain and other safety rules.	Bradshaw et al. (2004)
14	Trust (TRS)	All team members take into account the issue of trust and the factors that help to build trusted relationships that encompass both the social and ethical aspects in order to adjust their behavior.	Bradshaw et al. (2008); Harbers et al. (2012); Van Wissen et al. (2012); Chen et al., 2013

Figure 2. Distributed services appearing in industry domains, cloud and www

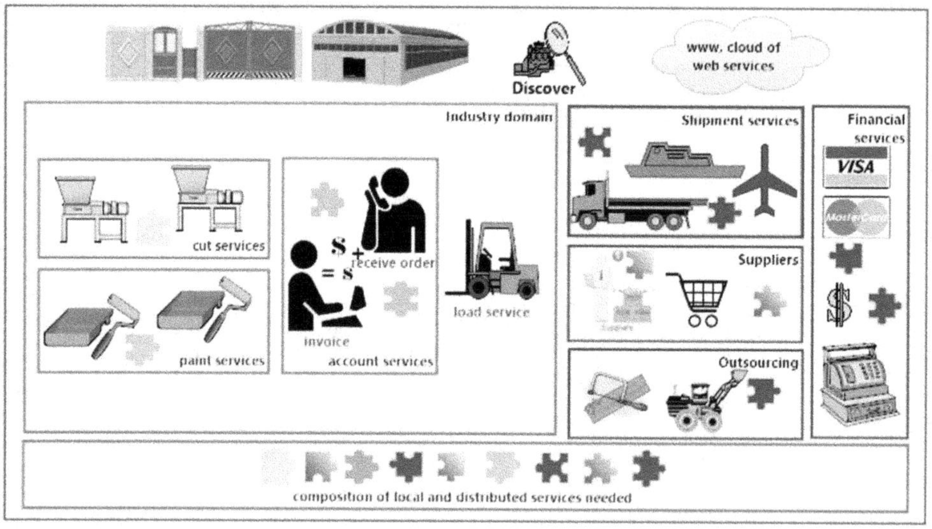

others. These entities share information with one another to receive sale orders, procure or manufacture, deliver goods to customers and receive payment.

The service cooperation in industry domains can be modeled as a role model (RM) (Tsoutsa et al., 2017). The RM diagram can be identified as the intuitive model that describes the way that services interact with other services. A role model consists of roles and their collaboration. A role model encloses all possible aspects of a given role collaboration with other roles within the domain. A role describes the observable behavioral aspect of a participant of the domain. Potential local and distributed services that are interested in a specific domain could become participants of the domain and could act according to

Figure 3. Simplified Order-to-Cash business process for the composite service (Monk et al., 2012)

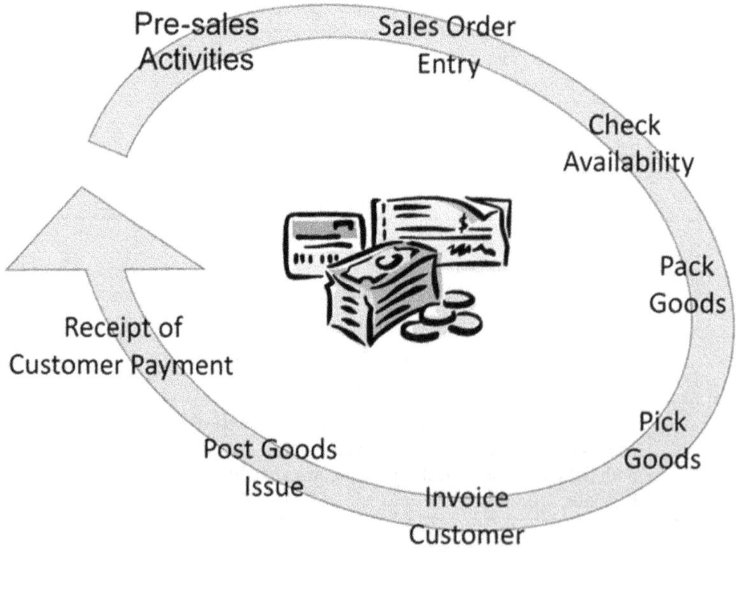

different roles. At any time, the roles of the domain are assigned to services; a service is a participant that plays the roles that are assigned to it. Consequently, a role model can depict the description of services collaborations using roles.

In Figure 4, we depict part of the role model of the domain. The actors of a domain, in general, have some impact on the overall process with their behavior. According to the scenario, a composite service needs to be executed for the Order-to-Cash business process. Composite services are composed of multiple component services that play an important role in the entire execution process. If one of the processes controlled by a service malfunctions, this can significantly affect the execution, e.g. order's fulfillment time, cash flow, etc. To coordinate and execute all the services that are necessary for a customer to receive an order is sometimes difficult and goes out of the schedule, as warehouses can run out of material during the process, manufacturing plants could have failures in their system, transportation delays could occur and so forth.

Assuming that an emergent happens during the execution of the composite service Order to Cash and the composition process head on problems. Imagine that there is a missing component service at run time, during the execution of the order and delivery composite service. What will happen? Is it somebody there to react? According to the traditional composition method, where the processes extracted from the plan would be executed sequentially by the services they are assigned to do so, the composition would failure. A new plan would be searched and when provided, the composition process would start from the beginning in case it is possible due to time constraints, security reasons, etc.

Rather, solutions could be provided by conducting teamwork behavior executed by some participant services that play a teamwork role during the composition process. When the composition process is monitored by teamwork roles these roles are ready to capture emergent behavior and would intervene to alter the result. The detection of a missing service for example, as it is stated above, would cause the reaction of the inspector role. The inspector role and its behavior could be defined according to the agent based teamwork factors emerged and the specified domain that it operates. One intervening behavior

Figure 4. Part of the role model of the domain

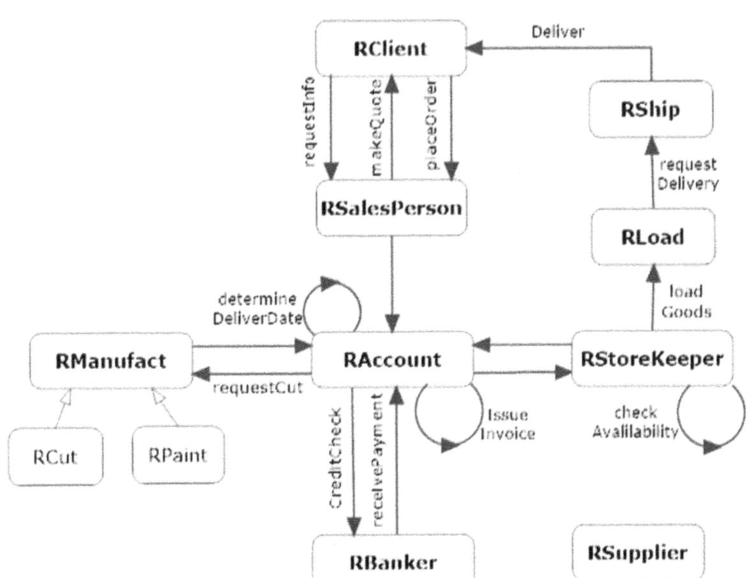

for this role is to find a replacing atomic or composite service that would substitute the behaviors of the removed service were detected during runtime. It would first look to potentially identify more missing behaviors that are related to the missing service. After finding all the gaps in the path execution, it constitutes a list with all behaviors went missing and look for new participants that are able to execute such behaviors so that the replacement services would substitute the missing ones and recover the stopped composition process.

Another intervening behavior of the inspector role is when a service participating in the composition fails and there is no possibility to replace it with another atomic service. In this case, the composition needs to be modified in such a way that the lost functionality can be recovered. This functionality (behavior) could be split, in order to be covered by various services, or may be regrouped with other part of the composition.

Consequently, teamwork roles could have intervening behavior which is expressed when they catch errors or when they are faced with emergent situations happen in a domain. Hiring existing or new participants to play teamwork roles where emergent might occur could greatly increase the efficiency of the domain.

In this example, we outlined the principles of an agent-based teamwork role, and this is a first step to make a bridge, that mediates between teamwork behavior in human and teamwork factors embedded in agents. We presented some first steps aiming at the integration of them, which is a fundamental building block for the envisioned industry 4.0 where human, agents and robots will work together.

CONCLUSION

The development and application of new emerged technologies such as i4.0, Internet of Things (IoT) and Cyber Physical Systems create additional concerns in teamwork that claim to be integrated into existing models. Robots are increasingly used on factory floors and researchers are looking for ways to help humans work better with their new teammates (Kyrarini et al., 2016). In such environments, teams composed of services, agents and other entities requires teamwork abilities to solve complex problems. Team research that is not directly applied to industry domains is potentially useful for developing intelligent teams whose members can behave as human teammates do. This research is conducted within human and agent-based teamwork and is targeted at coupling human-based behavior and agent based behavior to address the challenges associated with interconnection of entities appear in industry domains.

In sum, the main goals of this chapter were to survey existing major models and frameworks to bring together theories and models within the research literature, on both human based teamwork and agent-based teamwork. From the survey, an integrated table of major human roles that gives a holistic view of the current models on humans is derived and a core set of agent team work factors revealed that indicate primary behaviors which can be used by industry entities in order to obtain the teamwork abilities needed. Furthermore, the authors attempting to classify the teamwork factors revealed in to four clusters emerged, which indicate the dominant teamwork behavior of members.

According to authors opinion this is an attempt toward using a common language in discussing team behaviors for heterogeneous teams in agent-based domains. Although, at this point, the most identified factors are solely met to distinct studies, the long-term goal of this research, is to define the analytical primitive behaviors that should be extracted from each team role containing they key factors needed according to the four dimensions emerged for effective teamwork. The rationale is to lay the groundwork to teamwork for entities are called to form groups in dynamic environments like i4.0, by employing some major roles that are used in human teams, adapt them and propose novel roles as central pillars to agent-based team working habilitating their behaviors with the key factors identified at agent based teamworking studies.

The results of this review are a step towards our vision to define a model that will enable heterogeneous intelligent entities to interact with each other autonomously, to crawl within their domain to discover other trusted services and devices, to delegate their assignments to other similar trustworthy services, and state explicitly their capabilities within their domain. We are currently investigating the implementation algorithms that will configure each role with specific factors and implement the primary behaviors in order to optimize team effectiveness and subsequently team composition.

REFERENCES

Agmon, N., & Stone, P. (2011). Leading multiple ad hoc teammates in joint action settings. In *Interactive decision theory and game theory* (pp. 2–8). ACM.

Aritzeta, A., Swailes, S., & Senior, B. (2007). Belbin's team role model: Development, validity and applications for team building. *Journal of Management Studies*, *44*(1), 96–118. doi:10.1111/j.1467-6486.2007.00666.x

Barange, M., Kabil, A., De Keukelaere, C., & Chevaillier, P. (2014, June). Task-oriented conversational behavior of agents for collaboration in human-agent teamwork. In *International Conference on Practical Applications of Agents and Multi-Agent Systems* (pp. 25–37). Springer;. doi:10.1007/978-3-319-07551-8_3

Barrett, S., Stone, P., & Kraus, S. (2011). Empirical evaluation of ad hoc teamwork in the pursuit domain. *Autonomous Agents and Multi-Agent Systems*, 2(May), 567–574. Retrieved from http://dl.acm.org/citation.cfm?id=2031678.2031698

Beck, Z., Teacy, L., Rogers, A., & Jennings, N. R. (2016, May). Online planning for collaborative search and rescue by heterogeneous robot teams. In *Proceedings of the 2016 International Conference on Autonomous Agents & Multiagent Systems* (pp. 1024-1033). International Foundation for Autonomous Agents and Multiagent Systems.

Belbin, M. (1981). *Management Teams, Why They Succeed or Fail.* London: Heinemann.

Belbin, R. M. (1993). Team roles at work: A strategy for human resource management. *Zitiert in: Teamarbeit und Teamentwicklung, 321.*

Berlin, J. M., Carlstrom, E. D., & Sandberg, H. S. (2012). Models of teamwork: Ideal or not? A critical study of theoretical team models. *Team Performance Management: An International Journal*, *18*(5–6), 328–340. doi:10.1108/13527591211251096

Bradshaw, J. M. Acquisti, A., Allen, J., Breedy, M., Bunch, L., Chambers, N., … Field, M. (2004). Teamwork-Centered Autonomy for Extended Human-Agent Interaction in Space Applications. Retrieved from http://www.aaai.org/Papers/Symposia/Spring/2004/SS-04-03/SS04-03-026.pdf

Bradshaw, J. M., Feltovich, P. J., Johnson, M., Bunch, L., Breedy, M., Jung, H., & Uszok, A. (2008). Coordination in Human-Agent-Robot Teamwork. In *International Symposium on Collaborative Technologies and Systems CTS 2008* (pp. 467-476). IEEE.

Bradshaw, J. M., Feltovich, P. J., & Johnson, M. (2012). Human–Agent Interaction. In *The Handbook of Human-Machine Interaction: A Human-Centered Design Approach.* Ashrate Publishing, Ltd.

Bowling, M., & McCracken, P. (2005, July). *Coordination and adaptation in impromptu teams* (Vol. 5, pp. 53–58). AAAI.

Cannon-Bowers, J. A., Salas, E., & Converse, S. A. (1990). Cognitive psychology and team training: Training shared mental models and complex systems. *Bulletin - Human Factors Society*, *33*, 1–4.

Cohen, S. G., & Bailey, D. E. (1997). What makes teams work? Group effectiveness research from the shop floor to the executive suite. *Journal of Management*, *23*(3), 239–290. doi:10.1177/014920639702300303

Chen, J. Y., & Barnes, M. J. (2013). *Human-agent teaming for multi-robot control: A literature review (No. ARL-TR-6328).* Army research lab, Aberdeen Proving Ground, MD. doi:10.21236/ADA583900

Data, B. (2011). Let's start with the definition of Industry 4. 0. What does it mean? Retrieved from http://www.opendatasrl.it/index.php/en/

Delarue, A., Van Hootegem, G., Procter, S., & Burridge, M. (2008). Teamworking and organizational performance: A review of survey-based research. *International Journal of Management Reviews*, *10*(2), 127–148. doi:10.1111/j.1468-2370.2007.00227.x

Driskell, T., Driskell, J. E., Burke, C. S., & Salas, E. (2017). Team Roles: A Review and Integration. *Small Group Research.*

Dunin-Keplicz, B., & Verbrugge, R. (2011). *Teamwork in multi-agent systems: A formal approach* (Vol. 21). John Wiley & Sons.

Andersen, E.P. (1997). Conceptual Modeling of Objects [PhD Thesis]. University of Oslo, Oslo, Norway.

Ertugrul, A. M., & Demirors, O. (2015). An Exploratory Study on Role-based Collaborative Business Process Modeling Approaches. In *Proceedings of the 7th International Conference on Subject-Oriented Business Process Management* (pp. 14:1-14:5). 10.1145/2723839.2723857

Fan, X., & Yen, J. (2004). Modeling and simulating human teamwork behaviors using intelligent agents. *Physics of Life Reviews*, *1*(3), 173–201. doi:10.1016/j.plrev.2004.10.001

Genter, K. L., Agmon, N., & Stone, P. (2011, August). Role-Based Ad Hoc Teamwork. In Plan, Activity, and Intent Recognition (pp. 1782–1783). Association for the Advancement of Artificial Intelligence.

Grant, J., Kraus, S., & Perlis, D. (2005). A logic-based model of intention formation and action for multi-agent subcontracting. *Artificial Intelligence*, *163*(2), 163–201. doi:10.1016/j.artint.2004.11.003

Grosz, B. J., & Kraus, S. (1996). Collaborative plans for complex group action. *Artificial Intelligence*, *86*(2), 269–357. doi:10.1016/0004-3702(95)00103-4

Grosz, B. J., & Kraus, S. (1999). The evolution of SharedPlans. In *Foundations of rational agency* (pp. 227–262). Springer Netherlands. doi:10.1007/978-94-015-9204-8_10

Jennings, N. R., Sycara, K., & Wooldridge, M. (1998). A roadmap of agent research and development. *Autonomous Agents and Multi-Agent Systems, 38*(1), 7–38. doi:10.1023/A:1010090405266

Jonker, C., van Riemsdijk, M. B., & Vermeulen, B. (2011). Shared Mental Models. In *Proceedings of 9th Workshop on COIN* (pp. 132-151). Springer. 10.1007/978-3-642-21268-0_8

Kamar, E., Gal, Y., & Grosz, B. J. (2013). Modeling information exchange opportunities for effective human-computer teamwork. *Artificial Intelligence, 195*, 528–550. doi:10.1016/j.artint.2012.11.007

Kaminka, G. A., & Frenkel, I. (2005). Flexible teamwork in behavior-based robots. In *Proceedings Of The National Conference On Artificial Intelligence* (Vol. 20, No. 1, p. 108). London, UK: AAAI Press.

Kato, J., Sakamoto, D., Igarashi, T., & Goto, M. (2014). Sharedo: to-do list interface for human-agent task sharing. In *Proceedings of the second international conference on Human-agent interaction* (pp. 345-351). ACM. 10.1145/2658861.2658894

Kazik, O. (2010). Role-based Approaches to Development of Multi-Agent Systems: A Survey. In WDS' 10 Proceedings of Contributed Papers, Part I (pp. 19–24).

Kennedy, W. G., Bugajska, M. D., Adams, W., Schultz, A. C., & Trafton, J. G. (2008). Incorporating Mental Simulation for a More Effective Robotic Teammate. In *Proceedings of the AAAI Conference on Artificial Intelligence* (pp. 1300-1305).

Kozlowski, S. W. J., Grand, J. A., Baard, S. K., Pearce, M., & Durso, F. (2015). Teams, teamwork, and team effectiveness: Implications for human systems integration. In Handbook of Human Systems Integration (pp. 1–30).

Korsah, G. A., Browning, B., Stentz, A., & Dias, M. B. (2012). xBots: An Approach to Generating and Executing Optimal Multi-Robot Plans with Cross-Schedule Dependencies. In *2012 IEEE International Conference on Robotics and Automation (ICRA)* (pp. 115-122). IEEE.

Kyrarini, M., Leu, A., Ristić-Durrant, D., Gräser, A., Jackowski, A., Gebhard, M., & Schlick, C. M. (2016). Human-robot synergy for cooperative robots. *Automatic Control and Robotics, 15*(3), 187–204.

Hanna, N., & Richards, D. (2013). A Collaborative Activity for Evaluating HAT-COM: Human-agent Teamwork Communication Model. In *Proceedings of the 2013 International Conference on Autonomous Agents and Multi-Agent Systems*, St. Paul (pp. 1369-1370). Retrieved from http://dl.acm.org/citation.cfm?id=2484920.2485229

Harbers, M., Bradshaw, J. M., Johnson, M., Feltovich, P., Van Den Bosch, K., & Meyer, J. J. (2012). Explanation in human-agent teamwork. In Coordination, Organizations, Institutions, and Norms in Agent System VII (pp. 21-37). Springer. doi:10.1007/978-3-642-35545-5_2

Leng, J., Fyfe, C., & Jain, L. (2006, October). Teamwork and simulation in hybrid cognitive architecture. In *International Conference on Knowledge-Based and Intelligent Information and Engineering Systems* (pp. 472-478). Springer.

Margerison, C., & McCann, D. (1984). Team mapping: A new approach to managerial leadership. *Journal of European Industrial Training*, *8*(1), 12–16. doi:10.1108/eb002166

Margerison, C. J., & McCann, D. (1990). *Team management: Practical new approaches*. Mercury Books.

Mathieu, J. E., Maynard, T., Rapp, T., & Gilson, L. L. (2008). Team effectiveness 1997-2007: A review of recent advancements and a glimpse into the future. *Journal of Management*, *34*(3), 410–476. doi:10.1177/0149206308316061

Mathieu, J. E., Tannenbaum, S. I., Kukenberger, M. R., Donsbach, J. S., & Alliger, G. M. (2015). Team Role Experience and Orientation: A Measure and Tests of Construct Validity. *Group & Organization Management*, *40*(1), 6–34. doi:10.1177/1059601114562000

Monk, E., & Wagner, B. (2012). *Concepts in enterprise resource planning* (4th ed.). Cengage Learning, US.

Harris, P. R., & Harris, D. L. (1989). High Performance Team Management. *Leadership and Organization Development Journal*, *10*(4), 28–32. doi:10.1108/EUM0000000001141

Nalbandyan, N. (2008). A mutual assistance protocol for agent teamwork [MSc thesis]. Yerevan State University.

Parker, G. M. (2008). *Team players and teamwork: New strategies for the competitive enterprise*. San Francisco, CA: Jossey-Bass.

Pollack, M. E. (1990). Plans as complex mental attitudes. In *Intentions in communication* (pp. 77–103). London, UK: The MIT Press.

Poslad, S., Buckle, P., & Hadingham, R. (2000). The FIPA-OS agent platform: Open source for open standards. In *Proceedings of the 5th International Conference and Exhibition on the Practical Application of Intelligent Agents and Multi-Agents* (Vol. 355, p. 368).

Russell, S., & Norvig, P. (2005). Artificial Intelligence a modern approach. *Learning*, *2*(3), 4.

Rousseau, V., Aube, C., & Savoie, A. (2006). Teamwork Behaviors: A Review and an Integration of Frameworks. *Small Group Research*, *37*(5), 540–570. doi:10.1177/1046496406293125

Saidani, O., & Nurcan, S. (2006). A role-based approach for modeling flexible business processes. *BPMDS*, *6*, 111–120.

Salas, E., Sims, D. E., & Burke, C. S. (2005). Is there a "big five" in teamwork? *Small Group Research*, *36*(5), 555–599. doi:10.1177/1046496405277134

Salas, E., Cooke, N. J., & Rosen, M. A. (2008). On Teams, Teamwork, and Team Performance: Discoveries and Developments. *Human Factors*, *50*(3), 540–547. doi:10.1518/001872008X288457 PMID:18689065

Stone, P., Kaminka, G. A., Kraus, S., & Rosenschein, J. S. (2010). Ad Hoc Autonomous Agent Teams: Collaboration without Pre-Coordination. In *Twenty-Fourth AAAI Conference on Artificial Intelligence*, July (pp. 1504-1509).

Schwartz, T., Feld, M., Burckert, C., Dimitrov, S., Folz, J., Hutter, D., & Straube, S. (2016). Hybrid teams of humans, robots, and virtual agents in a production setting. In *Proceedings - 12th International Conference on Intelligent Environments, IE 2016* (pp. 234-237). 10.1109/IE.2016.53

Scholtz, J., Young, J., Drury, J. L., & Yanco, H. A. (2004, April). Evaluation of human-robot interaction awareness in search and rescue. In *2004 IEEE International Conference on Robotics and Automation ICRA'04* (Vol. 3, pp. 2327-2332). IEEE. 10.1109/ROBOT.2004.1307409

Skubch, H., Wagner, M., Reichle, R., Triller, S., & Geihs, K. (2010). Towards a Comprehensive Teamwork Model for Highly Dynamic Domains. *ICAART*, *10*(2), 121–127.

Sycara, K., & Sukthankar, G. (2006). Literature review of teamwork models (Technical Report CMU-RI-TR-06-50). Carnegie Mellon University.

Tambe, M. (1997). Towards flexible teamwork. *Journal of Artificial Intelligence Research*, *7*, 83–124.

Tannenbaum, S. I., Beard, R. L., & Salas, E. (1992). Team building and its influence on team effectiveness: An examination of conceptual and empirical developments. In K. Kelley (Ed.), *Issues, theory, and research in industrial/organizational psychology* (pp. 117–153). Amsterdam: Elsevier. doi:10.1016/S0166-4115(08)62601-1

Tsoutsa, P., Fitsilis, P., & Ragos, O. (2017). Role Modeling of IoT Services in Industry Domains. In *Proceedings of the 2017 International Conference on Management Engineering, Software Engineering and Service Sciences* (pp. 290–295). 10.1145/3034950.3035002

Van Wissen, A., Gal, Y., Kamphorst, B. A., & Dignum, M. V. (2012). Human-agent teamwork in dynamic environments. *Computers in Human Behavior*, *28*(1), 23–33. doi:10.1016/j.chb.2011.08.006

Wu, F., Zilberstein, S., & Chen, X. (2011). Online planning for ad hoc autonomous agent teams. In *IJCAI International Joint Conference on Artificial Intelligence* (pp. 439–445). doi:10.5591/978-1-57735-516-8/IJCAI11-081

Yen, J., Fan, X., Sun, S., Hanratty, T., & Dumer, J. (2006). Agents with shared mental models for enhancing team decision makings. *Decision Support Systems*, *41*(3), 634–653. doi:10.1016/j.dss.2004.06.008

Chapter 2
Digital Transformation Towards a New Context of Labour:
Enterprise 4.0

Maria João Ferreira
Portucalense University, Portugal & University of Minho, Portugal

Fernando Moreira
Portucalense University, Portugal & University of Aveiro, Portugal

Isabel Seruca
Portucalense University, Portugal & University of Minho, Portugal

ABSTRACT

Information Systems are the core of every business and cut across almost all aspects of organizational life. The adoption of technology enablers, by itself, does not guarantee such an organizational transformation. The new technology enablers allow the production, sharing and management of information and knowledge within the organization between peers and other stakeholders, and they also allow the improvement of organizational processes, requiring the updating of the supporting IS. Taking advantage of these technologies for organizations within the context of digital transformation requires a comprehension exercise in how to demonstrate their usefulness with regard to the creation, access and sharing of contents and IS improvements in a safe way. To this end, this chapter provides a comprehensive view of a new context of labour faced by the DT of organizations, which we term Enterprise 4.0 and which we propose to be implemented through the m_CSDIT framework, so as to improve the organizational well-being considering the collective intelligence and agility dimensions.

DOI: 10.4018/978-1-5225-4936-9.ch002

INTRODUCTION

It is widely acknowledged that organizations have suffered a large evolution at the social, economic and technological levels where the traditional barriers of transferring information and knowledge have been progressively eliminated. This evolution allowed the elimination of silos, the breaking down of hierarchies, the connection of internal and external stakeholders and the empowering of employees (Berkman, 2014). Furthermore, the use of technology enablers such as Big Data and associated analytics, Cloud Computing, Mobile Connectivity, and Social, the four pillars of digital transformation (DT), within business practice can enable significant organizational competitive advantage (Uhl, & Gollenia, 2016).

According to Earley Information Science (2016) digital transformation (DT) is today a top priority for executives, being that (1) 125000 enterprises expect revenue from their digital initiatives to increase by 80% by 2020; (2) DT initiatives will more than double by 2020, from 22% to almost 50% and, (3) only 27% of businesses have a coherent digital strategy for creating customer value in place.

From the organizations' point of view, DT can be seen as a deep and accelerating transformation with regard to processes, activities, competencies and models, in order to take advantage of the changes and opportunities offered by the inclusion of digital technologies into an organization. However, this advantage is only possible if the information systems of the organizations are also aligned with these new technologies. The main purpose of digital transformation is to redesign the organizational business through the introduction of digital technologies, achieving benefits such as productivity improvements, cost reductions and innovation (Hess et al., 2016). Nevertheless, as stated in Miller (2016), for these results to be achieved, a total organizational commitment is required. To this extent, Hinchcliffe (2016) points out that "…because digital itself is so intangible…. It's often even harder to understand the diverse needs, perspectives, and skill gaps of the people that have to change along with the technology". Hence, it may be assumed that continuous education/training is an imperative in this transformation organizational context.

In parallel with digital transformation and often referred as a major opportunity for promoting digital transformation in manufacturing, the paradigm of industry 4.0 has also arisen, whose goal is to achieve a higher level of operational efficiency and productivity, as well as a higher level of automatization (Thames & Schaefe, 2016). Industry 4.0 facilitates inter-connection and computerization into the traditional industry (Lu, 2017). Following this line of reasoning, in our research contribution we propose a framework to drive organizational digital transformation, where the former concept applied within the context of manufacturing is applied to any type of organization – Enterprise 4.0.

On the other hand, Information Systems (IS) are the core of every business and cut across almost all aspects of organizational life. They are used to support and improve all aspects of organizational functions and activities. In particular, under this context, the perception of increased product and services customization as a competitive advantage is universally shared (Kadiri et al., 2016).

Thus, it can be claimed that in order for DT to be successful in an organization, IS needs to be adapted/updated as well. IS must accommodate DT and must be aligned with the business in order to create value for the organization/business.

In this chapter, we provide a comprehensive view of the prevalent issues of a new context of labour faced by organizations on a digital transformation process: Enterprise 4.0 supported by mobile IST – m_CSDIT. Furthermore, we argue that this context will improve the well-being of these organizations through the collective intelligence and agility dimensions. The m_CSDT framework was formerly proposed in Ferreira, et al. (2014; 2015) at that stage, the framework was used as a basis to introduce and/

or systematize social business in organizations. We now propose an update of the m_CSDT framework to accommodate digital transformation and leading to Enterprise 4.0.

NEW TRENDS FOR ORGANIZATIONS

As already referred IS are nowadays central on any organization (Romero & Vernadat, 2016) regardless of its type, size, purpose or means, any organisation has an IS to support its internal operations and its interactions with the external environment. IS can be defined as "software systems for business management, encompassing modules supporting organizational functional areas such as planning, manufacturing, sales, marketing, distribution, accounting, financial, human resources management, project management, inventory management, service and maintenance, transportation and e-business" (Rashid et al., 2002) supported by computers, software, people, processes and data.

Organizations face daily pressures to demonstrate their ability to adapt quickly to unpredictable changes in their dynamics in terms of technology, social, legislative, competitiveness and globalization. Thus, to ensure their place in this difficult context, organizations must always be agile and must ensure their sustainability through continuous improvement of their IS (Imache et al., 2015). Therefore, organizational agility should be considered today as one of the main objectives of any organization (Imache et al., 2015).

According to Xu (2011) ISs have emerged in the last decades as a hopeful tool used for integrating and extending business processes across boundaries of business functions of an organization, at both intra and inter-organizational levels, in a worldwide economy with increasing global and competitive business operations. Moreover, the development of ICT including Industrial Informatics and the technological advances in ISs have provided a solution to the growing needs of information integration in businesses, supporting the operations of global supply networks. As stated in Imache et al. (2015), ISs have progressed in a constant interaction between a permanent and random change. In this context of organizational and ICT evolution, the emerging trends for organizations are presented during the next sub-sections.

Digital Transformation

The definition of digital transformation (DT) is not consensual among the stakeholders involved in this issue, particularly in organizations, leading to several perspectives of what it really means. These perspectives range from a focus on technology, to digital customer engagement, to new digital business models and so on. The lack of clarity often results in piecemeal initiatives, missed opportunities and false starts in the organization digitalization. According to Solis (2017) DT may be defined as "the realignment of, or new investment in, technology, business models, and processes to drive new value for customers and employees and more effectively compete in an ever-changing digital economy"

Following this line of reasoning, from the organizations' point of view, DT can be seen as a deep and accelerating transformation with regard to processes, activities, competences and models, in order to take advantage of the changes and opportunities offered by the inclusion of digital technologies into an organization.

On the other hand, Uhl and Gollenia (2016) enrich the DT concept, arguing that the adoption of technology-based change is focused on four technology enablers: (1) cloud, (2) mobile, (3) social, and (4) big data – analytics. Hence, DT draws on these four pillars to place a business context over the technologies, while taking advantage of them to support innovation. These technology enablers will be briefly described in the following sub-sections.

Big Data and Associated Analytics

As discussed in Chen et al. (2014), on the one hand Big data is considered an abstract concept and, on the other hand, the academic and industrial communities have different views on the definition (Team, 2011; Laney, 2001). However, some definitions have appeared over time, namely by Apache Hadoop, IBM, McKinsey & Company and IDC. The first definition was though introduced by Doug Laney in 2001, where he presents the 3Vs model (Volume – great volume, Velocity – rapid generation, and Variety – various modalities) (Gantz & Reinsel, 2011). This definition has undergone changes, and in 2011, IDC defined big data as "big data technologies describe a new generation of technologies and architectures, designed to economically extract value from very large volumes of a wide variety of data, by enabling the high-velocity capture, discovery, and/or analysis" (Labrinidis & Jagadish, 2012). This definition implied a change from the 3Vs model to a 4Vs model with the inclusion of Value (huge value but very low density).

Despite the added value that Big Data presents in responding to new requests, some of the existing challenges need to be addressed. In Agrawal, et al. (2012) eight challenges are presented. Within our contribution, we place special emphasis on the challenge identified by these authors as "Analytical mechanism".

Cloud Computing

The basic definition of cloud computing can be found in Armbrust, et al. (2010) and is given as "Cloud computing refers to both the applications delivered as services over the Internet and the hardware and systems software in the data centers that provide those services. The services themselves have long been referred to as Software as a Service (SaaS). Some vendors use terms such as IaaS (Infrastructure as a Service) and PaaS (Platform as a Service) to describe their products." While realizing that the definition may vary, Ochian et al. (2014) describe cloud computing "as a mechanism for generating ubiquitous access to a pool of convenient, on demand computing resources (compute, storage, platform, application and services) through a web interface with low administration overhead and the least intervention from a cloud service provider."

This new paradigm is supported by a set of vendors such as Amazon, Google, and Microsoft. Microsoft offers SaaS, PaaS and IaaS cloud packages to end users (Amato et al., 2013; Furht & Escalante, 2010). Some of the significant benefits (Furht & Escalante, 2010) of cloud computing include device and location independence, 24x7 support, lower total cost of ownership (TCO), reliability, scalability, sustainability, agile deployment, lower capital expenditure and a single infrastructure to fulfil all computing, networking and storage needs for various applications.

Regarding the advantages of the implementation of the cloud computing paradigm, Uhl and Gollenia (2016) put forward the following enhancers for organizations: "(i) Increase user mobility by a ubiquitous approval of information on any device; (ii) Flexibility and cost savings by eliminating the need for heavy local installations of end-user applications and high-powered client computers; (iii) Boosts of your business by the shift from fixed to variable costs."

Mobile Connectivity

Mobile devices and connections grew in 2013 to 7 billion, and will grow to 50 billion by 2020 (Reddi & Zhu, 2017). Not only the devices are targeted by this growth, but also the volume of mobile Web traffic, which often exceeds the "traditional" Web traffic.

This massive use of mobility devices through the vast amount of applications and services is based on Mobile computing. Mobile computing is defined in Bucki (2016) as "a generic term used to refer to a variety of devices that allow people to access data and information from where ever they are." Sometimes referred to as "human-computer interaction," mobile computing transports data, voice and video over a network via a mobile device."

The growth of the mobile computing market allows access to a wide range of technologies, and the integration of a variety of environments, such as cloud computing (Hsieh et al., 2013), social networks (Bellavista et al., 2013), big data computing and big data analytics (Hsieh et al., 2013).

Mobile connectivity summarizes the trend around mobile devices (smartphones, tablets, etc.) and communication technologies (NFC, Wi-Fi, etc.). The combination of devices and communication technologies will allow new forms of work and business, following the principle, any place, any time, and anywhere.

Regarding this technology enabler, Uhl and Gollenia (2016) identify the following enhancers for organizations: "(i) Direct connection and communication to stakeholders; (ii) Additional information about customers behavior; (iii) Customers-oriented offers; (iv) Mobile payment and ordering systems,"

Social Business and Social Media

According to Bharadwaj et al. (2013), digital technologies are transforming the relations of the digital structure, regarding the spaces of action of customers and companies, due to social media and social networks. The social business concept encompasses two directions, the resolution of social problems and, in a broader perspective, allowing a transformation of organizations so as to turn them more agile and more resilient.

Hence, Social Business may be regarded as a popular tendency that is revolutionizing organizational work and generating value for all of its elements, i.e. employees, customers, partners and suppliers. It means that all departments in an organization integrate their social capabilities into traditional business processes (Dorn et al., 2007) to change the way of working in order to create value. This view is reinforced by (Bharadwaj et al., 2013) as they state that "digital business strategy is simply that of organizational strategy formulated and executed by leveraging digital resources to create differential value." These authors also defined three indicators to assess this added value: "(1) going beyond the traditional view; (2) going beyond systems and technologies (Conner & Prahalad, 1996); and (3) explicitly linking digital business strategy to creating differential business value".

According to Kaplan and Haenlein (2010) Social Media "is a group of Internet-based application that build on the ideological and technological foundations of Web 2.0, and that allow the creation and exchange of user generated content." In this context, products, services and businesses are increasingly visible in the most used social networks (Facebook, YouTube, Twitter, Instagram, etc.), both by producers and consumers. While producers produce "information" in such a way that its dissemination is as positive as possible, aiming to promote a product or service, customers, business partners, and other stakeholders may also produce information about a product or service in the form of evaluation (positive or negative). This possibility allows, in principle, new levels of transparency in the supplier-customers relationship, as can be seen from the examples discussed in Ferreira, Moreira, and Seruca, (2015). Thus, social media forces organizations to rethink their traditional business models in order to engage the stakeholders in a more focused way.

In summary, Uhl and Gollenia (2016) present the following enhancers for organizations: "(i) Faster information sharing by direct and proactive communication with various stakeholders; (ii) Early brand and consumers feedback; (iii) Cost saving with the reduction of product lifecycles; and (iv) Transparency of organization processes."

Industry 4.0

Assuming that Industry 4.0 can be considered the 4th Industrial Revolution, it is worthwhile to note, briefly, its predecessors and the evolution of manufacturing and the industrial sector in general. The 1st Industrial Revolution occurred in England in the late eighteenth and early nineteenth centuries. This revolution was characterized by two important inventions that proposed a reorganization of the productive and transport sectors, with the appearance of the steam engine. The 2nd Industrial Revolution is characterized by the use of electric power, the use of the motor of explosion, and the invention of the telegraph, among others, provided the emergence of new markets and the acceleration of the industrial rhythm. The 3rd Industrial Revolution, also known as the Technical-Scientific Information Revolution, is characterized by advances in telecommunications and transportation systems, by the emergence and rapid expansion of information technology and automation, as well as robotics. Finally, as a reflection of the widespread adoption of information technology in all sectors of society, industry is currently witnessing the 4th Industrial or Technological Revolution, also known as "Industry 4.0".

The term "Industry 4.0" (Industry 4.0 Working Group, 2013) was adopted in 2011. The reference to ".0" stems from an analogy to the way software versions are usually called, meaning that the evolution of industry itself is also subjected to the rules of technological evolution of the information technology world.

According to the Industry 4.0 Working Group (2013), the current convergence between the physical (kinetic) world and the digital or virtual world (also called "cyberspace") makes it possible to create industrial ecosystems in which industrial processes arise from the networked interaction of objects, information and people. Thus, by using integrated digital systems for design, project, prototyping, component manufacturing, assembly and packaging, products can be planned and executed with minimal human intervention. This allows offering to the market, without cost increase, a wide range of products adapted to each individual customer. The flow of real-time and networked data between machines, robots and logistics systems will allow to anticipate failures, adapt production to new scenarios and integrate variables into the production process (with information from clients) that would otherwise be impossible.

Therefore, under Industry 4.0, there is the creation of cyber-physical systems that interrupt machines, warehouses and production lines, in a context in which all these manufacturing entities are endowed with (artificial) intelligence. The permanent connection to Internet of Things and the services of these cyber-physical systems allows production, logistics and marketing to be permanently at the service of the specific supply needs, regardless of the customer or the existing partner in the global market. For example, according to Uhl and Gollenia (2016), the automotive industry benefits greatly from the Industry 4.0 paradigm. Customers can change their orders (for example, car color) more easily, even when the production process has already been initialized.

Uhl and Gollenia (2016) show that Industry 4.0 can empower organizations in three positive aspects: "(i) Ad hoc reaction to customer changes (painting, delivery address, etc.) and changes in the supply chain (absent delivery can be replaced promptly); (ii) Optimization of the tasks of the staff members (reprogramming of machines); (iii) Find irregularities in the production process and correct them automatically."

DIGITAL TRANSFORMATION IN ORGANIZATIONS

There is a growing number of reports of DT in organizations. There are also different levels of maturity in the effort of DT by organizations. In this section, we provide a selection of examples that are focused on the evolution of companies' digital transformation efforts.

Social Media Tools and Social Business: Early DT Efforts

Starbucks is a multinational company with more than 20,000 stores worldwide, headquartered in Seattle, USA. The company created its name inspired by the Starbuck character of Moby Dick, and its logo is a sixteenth-century Scandinavian cut of a two-tailed mermaid.

In 2009, Starbucks, after a poor performance and share price value, which was cut in half, focused its digital strategy to re-establish the connection with its customers (Fitzgerald et al., 2013). One of the first measures introduced was to offer free Wi-Fi in all Starbucks stores, in addition to free access to The Economist's magazine contents.

Furthermore, an important step in the undergone digital transformation was to restructure teams to collaborate from the start of projects. By that time, the company was able to reduce 10 seconds in the processing of each card transaction, reducing the total service time by 900,000 hours, considering 3 million mobile payments per week.

The use of social media tools, in addition to other technologies, changed customer relationships, operations, and the business model, leading to increased customer relationships; as a result, the Starbucks company shares value increased from $8 in 2009, to nearly $73 in July 2013.

Internal communication also benefited from this strategy of use of social media tools; to that extent, the company's CIO remarked that "social business tools and processes had transformed the company's innovation culture, helping it develop products and understand risks as rapidly as new markets emerged." (Fitzgerald et al., 2013).

Hummel is a company founded in 1923 in Hamburg, Germany, by a young shoemaker, and has one of the oldest sportswear brands in the world. The company has suffered management changes and, in the late 1990s, was bought by a Danish lawyer. Since then, Hummel has maintained a growth of approximately 20% per year.

Hansen and Sia (2015) presented a case study on the digital transformation of the Hummel Company. This transformation began with the consolidation of the 22 local sites into only one global platform, which took local needs into account. In addition, the presence on the social network Facebook, where the company had about 25 different pages, with inconsistent brand expressions, went to a global Facebook page. The strategy was not only to use Facebook, but to be present on all relevant social media platforms, in order to maintain the most appropriate relationship with customers according to the profile of use in each of the various platforms (Instagram, etc.).

Besides its relationship with the outside world through social media platforms, the company developed a Digital Handbook so as to further align the online brand image, describing the set of norms and guidelines needed to be followed within online channels brand usage. This guide also provides guidance on how partners should sell products on channels such as their own sites, as well as policies for AdWords, etc. To cope with the evolution of communication channels and the emergence of new means of relationship with customers, the company regularly updates the manual every six months.

LEGO was founded in 1932 by Ole Kirk Kristiansen, a carpenter who made wooden toys; its headquarters are in Billund, Denmark, but the company has offices spread all over the world (U.S., U.K., China, etc.). It has more than 17,000 employees, with revenues of over $5 billion in 2015.

The company decided to make a digital transformation in its form of leadership and in all sectors of the organization's activities. Sawy et al., (2016) describe that the efforts made by the company were in the direction of using various digital channels, such as social media tools. For example, the company began to tell interactive stories within the "trailers" of online games to engage children with the new characters. Complementarily, the company has created a catalog of augmented reality products to meet the new technologies, and to not lose the market.

The increase of connectivity with customers through various communication channels (company website, social media tools, etc.) has put, however, new demands on the reorganization of the company's IT department, so as to adapt to this new reality, digital interaction with the consumer. Hence, the company learned the importance of social media tools in business and their value in discovering the concerns and needs of customers, and therefore allowing to create value.

Innovation Accelerators: New DT Directions

The results of introducing, among others, social media tools in business were very positive and resulted in the revitalization of many businesses and reference companies in their areas of intervention. These outcomes were presented and discussed in the previous section regarding Starbucks, Hummel and LEGO organizations. However, the rapid evolution of technology has led to the emergence of new solutions (cloud computing, increased use of mobile devices, etc.) that led to the emergence of the so called 3rd technology platform (early 2010), which is based, as previously described in four technology enablers or pillars (mobile devices, cloud computing, social networking and Big Data and associated analytics). This platform is the result of the evolution of the 1st platform (late 1950s), based on mainframes, and the 2nd platform (mid-1980s) based on the client-server model.

The more significant use of the pillars of digital transformation has been driven by innovation accelerators, which include, among other solutions, IoT, Robotics, 3D Printing, Artificial Intelligence, Augmented and Virtual Reality, Cognitive Systems and Next Generation (NextGen) Security as shown in Figure 1. To this extent, IDC expects that in, 2019, 3rd platform technologies and services will account for about 75% of IT spending (IDC, 2016).

The consideration of the 3rd platform and the innovation accelerators within Digital Transformation efforts will allow a large number of combinations that will result in diverse business models and implementations, as discussed in the next section. This evolution will leave many organizations with new challenges to meet the challenges they already had. Furthermore, digital transformation will entail a significant shift with which CIOs will have to confront and adapt. For example, systems will have to be addressed as services, IT agility will shift to business agility, and information will have to be transformed into innovation.

Figure 1. Digital transformation (Industry Week, 2016)

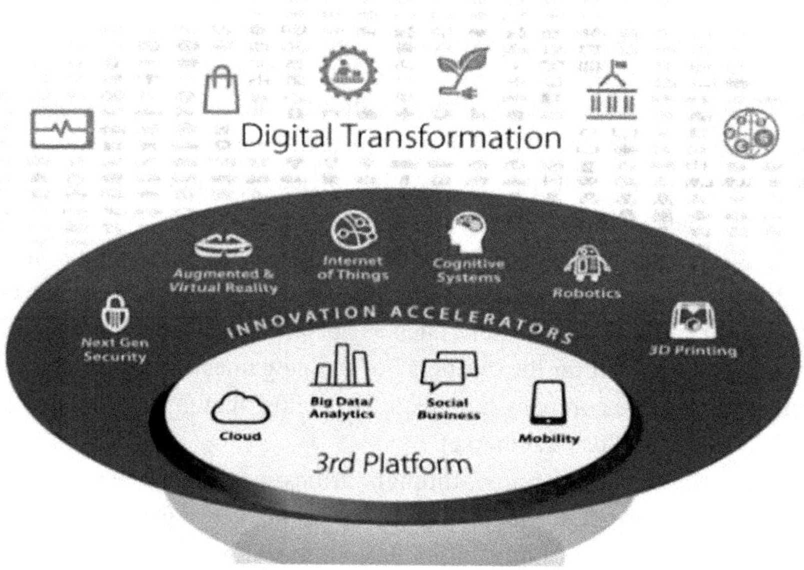

Thus, for companies to stay competitive in the market, ensuring more agility, efficiency and productivity in business, it is no longer enough to just invest in IT. It is necessary to recognize the strategic value of the innovation accelerators and to plan the IT infrastructure to support digital transformation.

One of the main advantages of digital transformation is the possibility of expanding physical to digital business. For example, one of the innovation accelerators, virtual reality technology, can be applied to the real estate market to conduct visits to fully digital properties, whereas in the medical field, the same resource can be applied to offer virtual training of surgeries and study of organisms.

The Rise of the DT Economy and the New Tech Industry

In the last years, we have witnessed the establishment and evolution of the four pillars of Digital Transformation as well as the maturation and mash-up of the so-called innovation accelerators (IDC, 2017). These two trends allowed the support of an expanding Digital Transformation economy and the first wave of players, services and solutions in all the industries.

Innovation plays a major role in the process and is considered one key domain of Digital Transformation (Rogers, 2016). Today's start-ups have shown us that digital technologies can enable a very different approach to innovation, one based on continuous learning through rapid experimentation.

As digital technologies make it easier and faster than ever to test ideas, market feedback can be gained from the very beginning of the innovation process, all the way through to launch and even afterwards. This new approach to innovation is focused on careful experiments and on minimum viable prototypes that maximize learning while minimizing costs. Assumptions are repeatedly tested, and design decisions are made based on validation by real customers. Products and services are developed iteratively through a process that saves time, reduces the cost of failures and improves organization learning.

The two case studies provided in this section describe real stories of business innovation through digital transformation, applied in start-up companies. Both cases target the use of intelligent systems in the conduction of the companies' operations and the benefits achieved, thus providing a new context of labour, which we term Enterprise 4.0.

The AddVolt Case

Founded in 2014 and incubated at the Science and Technology Park of the University of Porto (UPTEC), the Portuguese Start-Up company AddVolt (AddVolt, 2017) designs and develops electric controllers which manage intelligently and autonomously distinct sources of electric energy.

The company has created WeTruck, a product and technology designed to produce energy in trucks, targeted to transport and logistic companies operating in the temperature sensitive product supply chain, which have fleet equipped with refrigeration units that are powered by diesel engines with high level of fuel consumption, maintenance, noise and CO2 emissions.

The WeTruck system produces energy through photovoltaic panels installed on top of the truck and recovers energy during the vehicle braking and decelerations. It uses the energy generated to supply the refrigeration unit in electric mode.

The WeTruck System

WeTruck is a fresh and non-invasive solution for transport and logistic companies operating in the temperature sensitive product supply chain: it uses the solar and the kinetic energy of the vehicle to produce enough electricity to power the cooling system of refrigerated trucks. Through a portable electric equipment (capable of producing, recovering and storing electrical energy on board of the vehicle) and a controller, it has even become possible to recover energy during braking, without the need of structural modifications on the vehicles (Figure 2). The outcome is quite impressive: a reduction in fuel consumption, maintenance costs and noise, and less 87% of CO_2 emissions per year.

"There are 4.8 million refrigerated lorries and trucks circulating across the globe, a number that is expected to double by 2025. In Europe, there are approximately 1.1 million of these vehicles, where WeTruck can make a difference. AddVolt aims to improve fleet integration with the electrical network and with cities, thus contributing to more sustainable and effective transport, on a path towards 0% fuel consumption in the cooling of goods and foodstuffs", says Bruno Azevedo, Chief Executive Officer of

Figure 2. The WeTruck system (Addvolt, 2017)

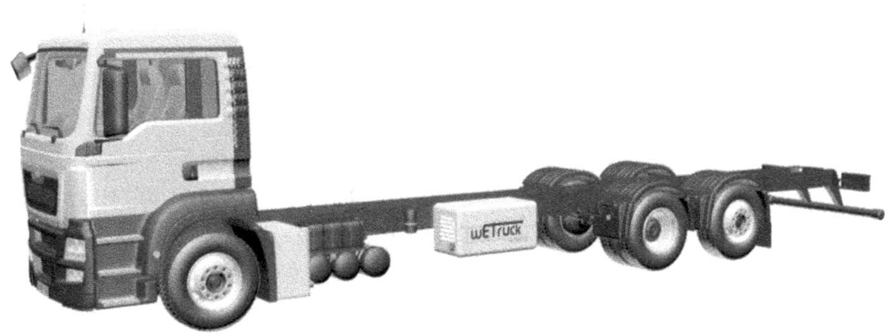

the company, which was one of the 66 start-ups selected to represent Portugal at Web Summit 2016 (AddVolt, 2017).

Intelligent Use of the Regenerative Braking Functionality Through Data Mining Techniques

In parallel with the WeTruck system, AddVolt is also developing a data driven optimization procedure to save energy and reduce costs while travelling. The data mining project uses both supervised and unsupervised machine learning techniques and aims to optimize the energy recovery process, using historical traffic data.

As already described, heavy vehicles that transport temperature-sensitive goods, generally use a fuel-needy dedicated diesel engine. The WeTruck system is an Energy Management System (EMS) capable of producing energy on-board of the vehicle. This recovery is possible due to the Regenerative Braking (RB) functionality, which consists in converting kinetic energy to electrical energy during a slowdown. The recovered energy is then stored in a set of batteries that supplies the refrigeration system when needed, allowing it to run in electrical mode.

Using data retrieved from the vehicle's operation and this management system, an opportunity towards the intelligent use of the regenerative braking functionality emerges. By introducing an intelligence layer on the energy management system, a decision on applying the RB functionality can be made based on the trip's energetic potential. This decision will optimize the battery usage and reduce the load and wear on the EMS components.

In order to calculate the energetic potential of a certain route, an estimation of the road is needed. In the modeling approach developed and implemented by AddVolt, a route is divided in several spatial segments and each segment is categorized among three pre-defined classes. A classification model is used to predict traffic using historical data as input. By using this modeling approach based on travel times, information on traffic flow and intersection queues are incorporated and by calculating the most likely sequence of states, a estimation of the road ahead is made.

Using the information of the modeled path, when the RB systems detects a situation where the functionality can be applied, a decision will be made by weighting the energetic potential of the path ahead and the energy need. When the algorithm sees fit, a higher torque may be applied to the generator, which will result in a larger quantity of energy recovered. Since this causes stress to the system, this functionality needs a robust intelligence layer.

Results and Internationalization Roadmap

Luis Simões has become the first company to implement the WeTruck technology, responding to the commitment of the company for R&D and environmental care (Portugal Startups, 2016). With almost 50 years of experience and more than 1800 employees, Luis Simões is one of the biggest logistic and transportation companies in Portugal.

This partnership has brought immediate effects in reducing gas refrigeration units generated in Luis Simões' trucks and trailers, allowing a reduction of up to 87% of carbon dioxide emissions during the transport of products subjected to temperature control.

After two years of research and more than 70,000 km of testing and trials in real operations, Luis Simões and AddVolt have shown that the technology contributes significantly in emissions, achieving a reduction of 844 kg per vehicle per month. The improvement also has an effect on a reduction of more than 30% of noise (about 30 dB) and increases the cooling efficiency of the products during transport.

This collaboration shows the potential of having big companies working with universities and startups to bring innovation to their activity.

In which concerns internationalization, AddVolt managed to attract funds from four investors who are now shareholders of the company, and aims to enter the German and Spanish markets (Portugal Ventures, 2017). The new investors will help AddVolt get its technology circulating on European streets and in cities where CO2 emissions and noise restrictions have driven the electrification of transport.

In Germany, a market where the biggest components and systems manufacturers for the transport sector are located, AddVolt already has strategic partnerships in place as well as the support of distributors to launch WeTruck. Entry into the Spanish market is also in the pipeline.

The Neomatix Case

Neomatix, an Israeli Start-Up company based in Tel Aviv, has invented a simple, seamless and cost-effective solution for automatically inspecting cars and trucks based on Computer Vision, Deep Learning and Big Data Analytics. In October 2016, Neomatix was selected as the winner in Shell's Bright Energy Ideas Challenge, out of 219 Smart Mobility start-ups from 36 countries (Neomatix, 2017).

The company offers an automated tire monitoring and asset management system which allows commercial vehicle fleets to save on operational expenses while increasing safety and reducing carbon emissions. This is achieved without any installation on the vehicles. The electro-optic tire sensor system monitors tire pressure and wear & tear remotely, based on advanced image processing, computer vision and machine learning algorithms. The solution greatly contributes to reduced CO_2 emissions, improved fuel efficiency and road safety by timely warning drivers of a low-pressure or torn tire.

Neomatix started with tire monitoring, devising a smart sensor for monitoring the status of tires and estimating tire pressure and tire wear using visual data from multiple cameras, and is now inspecting cars and trucks at fleet parking lots and logistic centers. The system enables seamless and cost effective predictive maintenance, bringing down the total cost of ownership, reducing carbon emissions, and making the roads safer for all.

The Neomatix TRAX Visual Intelligence Solution

Neomatix develops electro-optic tire monitors for automated fleet-level tire management (Figure 3). The Neomatic TRAX is an electro-optic tire monitoring and management solution, designed to work as a Tire Resource Automation Expert (TRAX).

The company's proprietary solutions are based on advanced computer vision and image processing algorithms, as well as on actionable business intelligence derived from big data analysis and machine learning.

The key benefits of automated external inspection provided by Neomatic TRAX are to allow vehicle fleet operators to get their tires effectively under control, increase fleet safety and up-time, reduce fuel waste and carbon emissions, and most importantly to significantly reduce their cost per kilometer. Thus, the solution helps vehicle fleet operators to save on operational expenses while increasing safety and reducing carbon emissions by automated tire monitoring and asset management.

Figure 3. The electro-optic tire monitoring system (neomatix, n.d.)

Intelligent Use of the Solution Through Data Mining Techniques

Visual Intelligence empowers organizations with instant business intelligence into areas that were previously out of reach. The Neoamtix TRAX system estimates tire pressure and tire wear using visual data from multiple cameras and uses a set of data mining techniques for classification, regression, localization and segmentation problems. The solution also uses 3D reconstruction from a stereo vision system and includes a real time wheel detection system.

Results and Internationalization Roadmap

Regardless of mobility solutions trending towards more shared, electric or autonomous driving, vehicle tires remain a constant factor in mobility. Innovations related to tires can substantially contribute to cleaner, safer and cheaper transportation.

Enabling V2V external inspections in the age of autonomous vehicles, Neomatix's patented technology, which uses computer vision and deep learning algorithms, is disrupting the way tires are managed in trucking fleets. According to Geert van de Wouw, managing director of Shell Technology Ventures (Business Wire, 2017), the winning startup was selected thanks to its potential effect on sustainability and road safety, both now and in the foreseeable future.

Kfir Wittmann, Founder and CEO of Neomatix, explains: "Fleets waste billions of dollars every year due to tire negligence, which affects roads safety as well as environmental issues. Despite the highly competitive landscape, no solution showed a high acceptance rate yet" (Business Wire, 2017).

Urubatan Helou Jr., fleet director at Braspress Transportes Urgentes (BTU), one of Brazil's leading parcel delivery companies that operates thousands of trucks throughout the country, and the first customer of Neomatix says "The beauty about Neomatix is that it is easy to adopt and implement, regardless of fleet size, and also very simple to maintain and use in the daily process of a fleet like ours". He also adds "We have tried every tire monitoring solution before adopting the Neomatix system. The information provided by the system is straightforward, actionable and very focused. We already see the effect on our bottom line. This is exactly the management solution we were looking for."

The company has already raised $3M from the Tel Aviv Angel Group and the Brazilian fleet management company SmartDrive and is an alumni of the Coca-Cola company's commercialization program "The Bridge", addressing the supply chain challenges of the popular beverage company.

ENTERPRISE 4.0 SUPORTED BY DIGITAL TRANSFORMATION

Digital transformation in organizations is already a reality that needs to be implemented and even, in some cases improved, as already discussed. Acknowledging this context, the m_CSDT framework was formerly proposed in Ferreira et al. (2015); at that stage, the framework was used to introduce and/or systematizes social business in organizations. We now propose an extension of the framework to accommodate digital transformation leading to the Enterprise 4.0. The main aim is to contribute to the improvement of business processes, relationships with stakeholders (business partners, customers) and continuous training of employees within organizations.

The following subsections present the updated approach, renamed as mobile Create, Share, Document, Improve and Training (m_CSDIT) and the rationale for its use, under a context of DT by an organization.

Mobile_Create, Share, Document, Improve and Training: m_CSDIT

The relevance of conducting Digital Transformation supported by the four enablers - Big Data and associated analytics, Cloud Computing, Mobile Connectivity, and Social tools is widely acknowledged and recognized by the scientific community and organizations, as discussed throughout the previous sections. However, there is a lack of approaches that allow the systematization and that guide its implementation within an organization, while improving IS and organizational processes.

To address this end, we propose the m_CSDIT approach as a three-layer framework with the following objectives:

1. Creation, sharing documentation of information and knowledge in and out of an organization, improvement of organizational processes and relationships with stakeholders, based on information and knowledge;
2. Training of organizational workers – ToOW model (Ferreira et al., 2017); and
3. Promotion of ad-hoc discussion.

As shown in Figure 4, a context for an Enterprise 4.0 implementation can be settled through the use of the four pillars of digital transformation, so as to achieve the well-being of the organization considering the agility and collective intelligence dimensions. Thus, emphasis is given in the production of value for the organizational ecosystem.

Creating value often occurs in different social contexts. This approach includes individual and organizational contextual factors. It identifies the crucial role that an organization plays in promoting a culture of lifelong learning, collaboration and innovation in order to achieve the organization's strategic objectives. The model identifies and presents the relationships and interactions namely between employees, technology and tools, the business strategy and business processes.

The approach highlights the collaboration issue and its relation with the accomplishment of the organizational goals. The employees are at the centre of the collaboration and are mostly members of social

Figure 4. The m_CSDIT framework

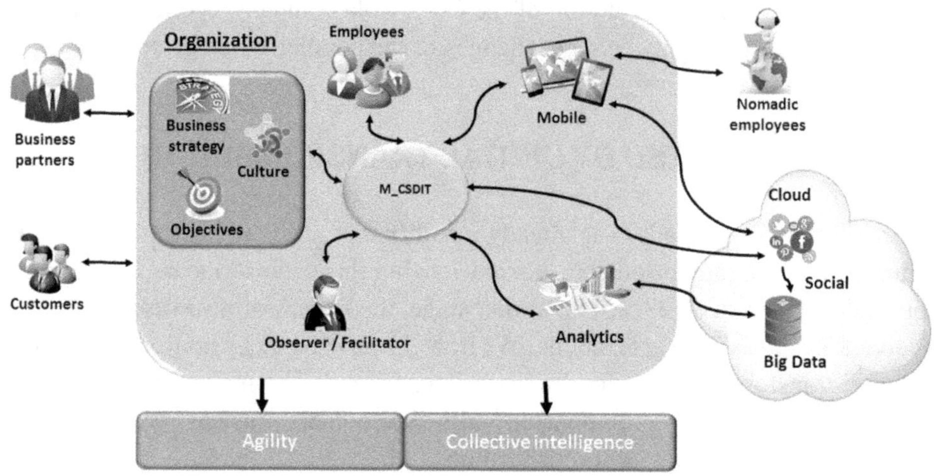

networks. Figure 4 illustrates the nature of the proposed collaboration in the workplace, where mobile devices are used, IS and underlying processes are largely organized and conducted by the organization, and which is based on the social context.

The approach may be briefly described as the collaboration in the workplace based on the four pillars of digital transformation. This means that the organization learns through the participation and involvement of its stakeholders, namely employees, through a network, connecting, interacting and collaborating to obtain or share information and / or knowledge in order to improve its organizational processes/IS. We propose that the collaboration in the workplace is achieved through the integration of suitable social media tools to the needs of organizational development and learning. To that end, we suggest a mixed form of peer tutoring with an instructor who acts as observer / facilitator. To achieve the potential benefits of collaboration, we recommend that the organization supports rather than restricts the adequate use of tools in the workplace. Thus, organizations have to define the best long-term strategies and implement action plans to take advantage of collaboration based on the DT pillars.

The Rationale for the Framework Proposal

As discussed throughout this chapter, the need for digital transformation of organizations, and consequently the need to introduce changes in the conduction of business and process improvement, are now recognized within the scientific community and by organizations. However, new approaches to address this paradigm are needed, so that guidance may be provided for its implementation within an organization.

Hence, a new contribution is hereby presented, originating from the work reported in Ferreira et al. (2015). The former approach was focused only on the changes introduced by social business, supported by social media tools. However, digital transformation, as discussed earlier, is based on four pillars: social, mobile, cloud and big data and associated analytics.

Thus, to address this objective – taking into account these four pillars and not just one – the former approach was extended; it consists of a three-tier framework (m_*CSDIT*), covering: (i) the creation, sharing, documentation of information and knowledge in and out of an organization in order to improving the business processes (that must be aligned to the organization strategy); (ii) training of organizational workers, and (iii) promotion of ad-hoc discussion.

As shown in Figure 4, it is possible to consider a generic approach for the implementation of digital transformation and IS improvements, based on the four DT pillars as well as in the Agility and Collective Intelligence dimensions.

The first layer of the m_CSDIT framework is presented in Figure 5 as a cyclic sequence of stages, aiming to use the four DT pillars in the definition of the creation, sharing, documentation of information and knowledge in and out of an organization and improving the business processes, in alignment with the organizational strategy.

As depicted in Figure 5, the m_*CSDIT* framework is designed to improve organizational processes based on information and knowledge provided in and out of organization (3). The processes updating strategies of the organization (2) are aligned with the organizational strategy (1) and analytical tools are used to evaluate the updating of processes, on the basis of their performance, according to the defined organizational strategy (4).

In order to enable a more flexible updating scheme, the model also considers updating actions proposed by employees; however, updating will always be compulsory according to the defined updating processes strategies (3). The definition/adjustment of the updating processes strategies (2) should be made in a periodical basis, so as to pace with the evolution / needs of the organization.

Regarding the use of DT enablers, the model considers a complete set of tools to be used within the updating processes activities, which can be made inside and outside the organization, that is, in differ-

Figure 5. First layer of the m_CSDIT framework

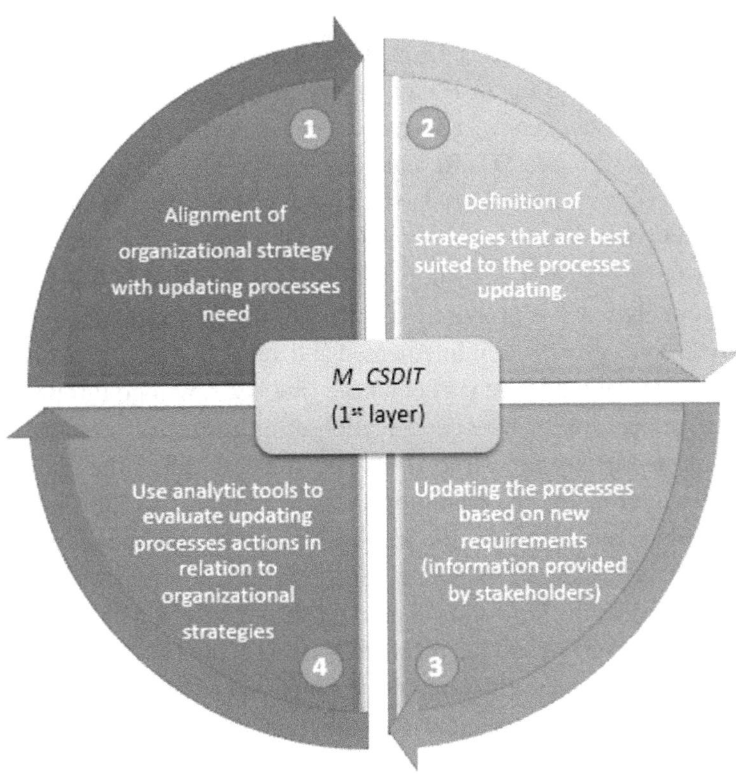

ent transformation (update) contexts. The model is designed in such a way as to enable the updating, at distance or in workplace context, in a formal or informal way. In the case of an informal context there will always be, as shown in Figure 2, an Observer / Facilitator, who will have a role of moderator on the ongoing activities.

As cloud is one of the pillars of digital transformation and social is one of the others, the use of social media tools located in the cloud, will allow to consider the *CSDIT* approach under collaborative work, as it enables internal transformation (Figure 2 – Employees) as well as external transformation (Figure 2 – Nomadic employees), enabling the interaction and collaboration among the participants, stakeholders, and, thus, the sharing of information and/or knowledge.

All the updates carried out will have to be evaluated so that it can be understood if the investment made meets the needs of the organization and the impact it may have on organizational processes' improvement. Hence, analytics tools – one of the pillars of digital transformation – will be used to monitor and compare updating results with the defined metrics goals to improve organizational processes.

The absence of a collaborative culture embedded in the organization's ecosystem may lead to the failure of strategies set for the organization in general. In order to avoid this failure scenario, the model hereby proposed aims to promote the development of a culture of lifelong training, adding value to organizational development and being central to achieve the objectives defined in the alignment of the organizational strategy with the organization needs.

FUTURE RESEARCH DIRECTIONS

Despite the successes and progress made to date, many important topics remain open for investigation with respect to providing appropriate approaches to undertake DT in organizations supported by the use of the Big Data and associated analytics, Cloud Computing, Mobile Connectivity, and Social tools.

Many of the results achieved so far in DT of organizations have come from joint work between researchers and organizations. Awareness, education and systematization remain three of the biggest issues to develop for those working in DT. Researchers need to have practical experience as well as a sound theoretical foundation. Practitioners need to be equipped with a variety of approaches to use where appropriate depending on what is best suited to a given situation. Customers need to understand the importance of the process, believe in it, and support the efforts involved in doing it right. It is therefore under this context that our approach is proposed. The m_CSDIT approach here proposed – a three layer framework – needs to be conceptualized, implemented and validated in an organizational context.

The conceptualization of the model, in particular the first layer, will go through a preliminary study supported by a maturity model for the use/identification of indicators of measurement. The indicators will be used to determine an activity pattern so as to select the best tools either producing organization's information "out" and in their input, i.e. what type and how the information should be made available to one or more stakeholders, and also how the feedback given by this/these stakeholder(s) should be analyzed in order to make the organization more competitive and improve its business processes. As far as out information is concerned, the indicators should guide the selection of tools to use against the objectives, the "drawing" of the content and the form of dissemination. Meanwhile, indicators for analysis of information indicate how to handle this same information in order to draw conclusions on the basis of the proposed objectives e.g. sentiment analysis within one or more social networks. In defining the indicators, qualitative and quantitative methods will be used.

For the 1st layer, as shown in Figure 6, the final/main goal is to instantiate the conceptual model in each organization according to its business domain and its objectives introducing whenever necessary the innovation accelerators (see section Innovation accelerators: New DT directions).

Regarding the conceptualization of the second layer of the framework, it is currently under development, i.e. studies (Ferreira et al., 2017) have been performed and a model has been outlined.

Finally, the third layer will not be conceptualized, because as it can be inferred by the name itself, it is intended that the promotion of ad- hoc discussion is unrestricted and not planned. However, the analysis of the results obtained will be performed through the in indicators set for the first layer.

CONCLUSION

DT is currently one of the major challenges faced by organizations and involves a set of highly complex activities. Within these activities, business improvement plays a fundamental role and requires particular attention. These activities are supported by information systems, which need to be aligned with organizations strategies.

Taking advantage of the DT technology enablers (Big Data and associated analytics, Cloud Computing, Mobile Connectivity, and Social) and realizing the need for updated approaches to address namely business processes improvement and organizational workers' training, we extended the m_CSDT framework to accommodate digital transformation leading to a new enterprise environment, that we termed Enterprise 4.0.

This chapter aimed to give an overview of the state of-the art issues underlying DT efforts undertaken by organizations. It then proposes the extended m_CSDIT approach consisting in a three-layer framework that covers the systematization of (i) the creation, sharing and documentation of information and knowledge in and out of an organization and improvement of organizational processes based on information and knowledge, (ii) the education and training of organizational workers and (iii) ad-hoc discussion.

Figure 6. Instantiations of the m_CSDIT framework

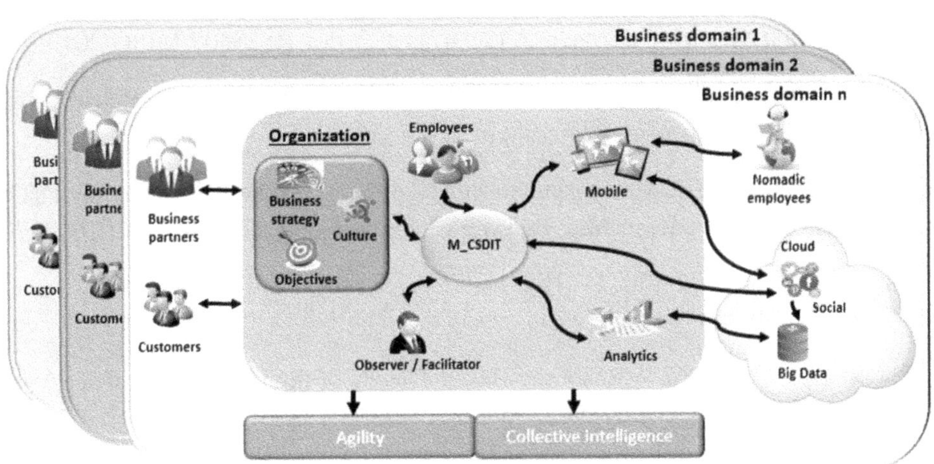

Even though the m_CSDIT approach seems to be a valuable contribution to offer a new context to the traditional way of performing business – the DT context, in order to achieve organizational well-being through the agility and collective intelligence dimensions, the approach needs a greater degree of formalization. A distinguishing feature of a traditional organization with and without DT is the involvement of customers, business partners and employees in the life of an organization. Both visions present benefits and drawbacks. We envisage that the m_CSDIT approach will allow the implementation of a new context of labour, which we term Enterprise 4.0. The approach is focused on the production of value for all the organizational stakeholders. Therefore, the involvement of the organization ecosystem is of paramount importance to achieve this goal. On the contrary, traditional approaches allow the conduction of business within an organization in a closed way.

REFERENCES

AddVolt. (2017). AddVolt company institutional web site. Retrieved from http://www.addvolt.com/en/

Bertino, E., Bernstein, P., Agrawal, D., Davidson, S., Dayal, U., Franklin, M., ... & Jadadish, H. V. (2011). Challenges and Opportunities with Big Data.

Amato, F., Mazzeo, A., Moscato, V., & Picariello, A. (2013). A Framework for Semantic Interoperability over the Cloud. In *27th IEEE International Conference on Advanced Information Networking and Applications Workshop* (pp. 1259–1264). 10.1109/WAINA.2013.218

Armbrust, M., Fox, A., Griffith, R., Joseph, A., Katz, R., Konwinski, A., ... Zaharia, M. (2010). A View of Cloud Computing. *Communications of the ACM*, *4*(53), 50–58. doi:10.1145/1721654.1721672

Bellavista, P., Montanari, R., & Das, S. K. (2013). Mobile social networking middleware: A survey. *Pervasive and Mobile Computing*, *4*(9), 437–453. doi:10.1016/j.pmcj.2013.03.001

Berkman, R. (2014). The social enterprise: Where does the information professional fit? *Online Searcher*, *6*, 45–48.

Bharadwaj, A., & Omar, A. (2013). Digital Business Strategy: Toward a next generation of insights. *Management Information Systems Quarterly*, *37*(2), 471–482. doi:10.25300/MISQ/2013/37:2.3

Bucki, J. (2016). Definition and Explanation of Mobile Computing. *The Balance*. Retrieved from https://www.thebalance.com/definition-of-mobile-computing-2533640

Business Wire. (2017). Stealth Tire Monitoring Startup Neomatix to Take 1st Place in Shell's Competition for Sustainable Mobility. Retrieved from http://www.businesswire.com/news/home/20161027006866/en/

Chen, M., Mao, S., & Liu, Y. (2014). Big Data: A Survey. *Mobile Network*, *19*(2), 171–209. doi:10.100711036-013-0489-0

Conner, K., & Prahalad, C. (1996). A Resource-Based Theory of the Firm: Know ledge Versus Opportunism. *Organization Science*, *5*(7), 477–501. doi:10.1287/orsc.7.5.477

Crăciun, D., & Bunoiu, M. (2015). Training Teachers for the Knowledge Society: Social Media in Science Education. *Broad Research in Artificial Intelligence and Neuroscience.*, *6*(3-4), 82–88.

DeNisco, A. (2017). Report: 56% of CEOs say digital transformation has increased profits. *Tech Republic*. Retrieved from http://www.techrepublic.com/article/report-56-of-ceos-say-digital-transformation-has-increased-profits/

Dorn, J., Grün, C., Werthner, H., & Zapletal, M. (2007). A survey of B2B methodologies and technologies: From business models towards deployment artifacts. In *40th Hawaii International Conference on Systems Science*, Waikoloa, Hawaii, January3-6. 10.1109/HICSS.2007.37

Earley Information Science. (2016) Building a Successful Digital Transformation Roadmap. Retrieved from https://cdn2.hubspot.net/hubfs/109038/EIS_Assets/EIS-Whitepaper-Building-a-Successful-Digital-Transformation-Roadmap.pdf

Ferreira, M. J., Moreira, F., & Seruca, I. (2014). *A traditional organization towards a new dimension of labour – social business. In Information systems and Technology for Organizational Agility, Intelligence, and Resilience* (Ch. 9, pp. 180–204). Hershey, PA: IGI Global. doi:10.4018/978-1-4666-5970-4.ch009

Ferreira, M. J., Moreira, F., & Seruca, I. (2015). Social Business: A way to promote organizational transformation. *International Journal of Information System Modeling and Design*, *6*(4), 57–81. doi:10.4018/IJISMD.2015100104

Ferreira, M. J., Moreira, F., & Seruca, I. (2017). Organizational training within digital transformation: The ToOW model. In *19th International Conference on Enterprise Information Systems (ICEIS)*, Porto, Portugal, April 26-29. 10.5220/0006366105260532

Fitzgerald, M., Kruschwitz, N., Bonnet, D., & Welch, M. (2013). Embracing Digital Technology: A New Strategic Imperative. *MIT Sloan Management Review*, 55(2), 1–12.

Furht, B., & Escalante, A. (2010). A Handbook of Cloud Computing. New York: Springer.

Gantz, J., Reinsel, D. 2011. Extracting value from chaos. *IDC iView, 1142*(2011), 1-12.

Hansen, R., & Sia, S. K. (2015). Hummel's Digital Transformation Toward Omnichannel Retailing: Key Lessons Learned. *MIS Quarterly Executive*, *14*(2), 51–66.

Hinchcliffe, D. (2016). Using online community for digital transformation. Retrieved from https://dionhinchcliffe.com/

Hess, T., Matt, C., Benlian, A., & Wiesböck, F. (2016). Options for formulating a digital transformation strategy. *MIS Quarterly Executive*, *15*(2), 123–139.

Hsieh, J. C., Li, A. H., & Yang, C. C. (2013). Mobile, cloud, and big data computing: Contributions, challenges, and new directions in telecardiology. *International Journal of Environmental Research and Public Health*, *11*(10), 6131–6153. doi:10.3390/ijerph10116131 PMID:24232290

IDC. (2016). Third platform. Retrieved from http://www.idc.com/promo/thirdplatform

IDC. (2017) Digital Transformation. Retrieved from https://www.idc.com/promo/thirdplatform/digital-transformation

Imache, R., Izza, S., & Ahmed-Nacer, M. (2015). Clustering-based urbanisation to improve enterprise information systems agility. *Enterprise Information Systems*, *8*(9), 861–877. doi:10.1080/17517575.2013.879210

Industry Week. (2016). IDC transformation [image]. Retrieved from http://www.industryweek.com/sites/industryweek.com/files/styles/article_featured_standard/public/uploads/2016/01/idc-transformation.jpg?itok=D1dkzVxX

Kadiri, E. S., Grabot, B., Thoben, K., Hribernik, K., Emmanouilidis, C., von Cieminski, G., & Kiritsis, D. (2016). Current trends on ICT technologies for enterprise information systems. *Computers in Industry*, *79*, 14–33. doi:10.1016/j.compind.2015.06.008

Lu, Y. (2017). Industry 4.0: A survey on technologies, applications and open research issues. *Journal Of Industrial Information Integration*, *6*, 1–10. doi:10.1016/j.jii.2017.04.005

Labrinidis, A., & Jagadish, H. V. (2012). Challenges and opportunities with big data. *Proc VLDB Endowment*, 5(12), 2032–2033. 10.14778/2367502.2367572

Laney, D. (2001). 3-d data management: controlling data volume, velocity and variety. *META Group Research Note*.

Mamaqi, X. (2015). The efficiency of different ways of informal learning on firm performance: A comparison between, classroom, web 2 and workplace training. *Computers in Human Behavior*, *51*, 812–820. doi:10.1016/j.chb.2014.11.080

Manca, S., & Ranieri, M. (2016). "Yes for sharing, no for teaching!": Social Media in academic practices. *Internet and Higher Education*, *29*, 63–74. doi:10.1016/j.iheduc.2015.12.004

Miler, R. 2016. Digital transformation requires total organizational commitment. Retrieved from https://techcrunch.com/2016/01/31/digital-transformation-requires-total-organizational-commitment/

Neomatix (2007) Neomatix company institutional web site. Retrieved from http://www.neomatix.com

Neomatix. (n.d.). [Image]. Retrieved from http://neomatix.com/wp-content/uploads/2017/05/optimus-brmp.jpg

Nookhong, J., & Wannapiroon, P. (2014). Development of collaborative learning using case-based learning via cloud technology and social media for enhancing problem-solving skills and ICT literacy within undergraduate students. *Procedia: Social and Behavioral Sciences*, *174*, 2096–2210. doi:10.1016/j.sbspro.2015.02.007

Ochian, A., Suciu, G., Fratu, O., Voicu, C., & Suciu, V. 2014. An overview of cloud middleware services for interconnection of healthcare platforms, In *10th IEEE International Conference on Communications (COMM)* (pp. 1-4). 10.1109/ICComm.2014.6866753

Petrucci, A., Kaneva, P., Lebedeva, E., Feldman, L., Fried, G., & Vassiliou, M. (2015). You Have a Message! Social Networking as a Motivator for FLS Training. *Journal of Surgical Education*, *7*(32), 542–548. doi:10.1016/j.jsurg.2014.10.007 PMID:25496795

Portugal Startups. (2016) AddVolt makes Luis Simões the first company combining solar and kinetic energy in refrigerated trucks. Retrieved from http://portugalstartups.com/2016/09/addvolt-makes-luis-simoes-first-company-combining-solar-kinetic-energy-refrigerated-trucks/

Portugal Ventures. (2017) AddVolt prepares internationalisation with funds from three investors. Retrieved from http://www.portugalventures.pt/en/article/addvolt-prepares-internationalisation-funds-three-investors

Rashid, M. A., Hossain, L., & Patrick, J. D. (2002). The Evolution of ERP Systems: A Historical Perspective. In Enterprise Resource Planning: Global Opportunities and Challenges. Hershey, PA: IGI Global.

Reddi, V., & Zhu, Y. (2017). The Red Future of Mobile Web Computing. *Communications of the ACM*, *60*(1), 52–53.

Rogers, D. (2016). *The Digital Transformation Playbook: Rethink Your Business for the Digital Age.* New York: Columbia Business School Publishing. doi:10.7312/roge17544

Romero, D., & Vernadat, F. (2016). Enterprise information systems state of the art: Past, present and future trends. *Computers in Industry*, *79*, 3–13. doi:10.1016/j.compind.2016.03.001

Salminen, L., Gustafsson, M.-L., Vilén, L., Fuster, P., Istomina, N., & Papastavrou, E. (2016). Nurse teacher candidates learned to use social media during the international teacher training course. *Nurse Education Today*, *36*, 354–359. doi:10.1016/j.nedt.2015.08.026 PMID:26358632

Solis, D. (2017). Digital Transformation – The six stages of digital transformation. Retrieved from http://www.briansolis.com/2017/01/definition-of-digital-transformation/

Sawy, O. A., Kræmmergaard, P., Amsinck, H., & Vinther, A. L. (2016). How LEGO Built the Foundations and Enterprise Capabilities for Digital Leadership. *MIS Quarterly Executive*, *15*(2), 141–166.

Thames, L., & Schaefe, D. (2016). Software-defined cloud manufacturing for Industry 4.0. *Procedia CIRP*, *52*, 12–17. doi:10.1016/j.procir.2016.07.041

Team, O. R. (2011). *Big data now: current perspectives.* OReilly Media.

Uhl, A., & Gollenia, L. (2016). *Digital Enterprise Transformation: A Business-Driven Approach to Leveraging Innovative IT.* USA: Routledge.

Xu, L. D. (2011). Enterprise Systems: State-of-the-Art and Future Trends. *IEEE Transactions on Industrial Informatics*, *4*(7), 630–640. doi:10.1109/TII.2011.2167156

ADDITIONAL READING

Balina, S., Baumgarte, D., & Salna, E. (2017). Cloud Based Cross-system Integration for Small and Medium-sized Enterprises. *Procedia Computer Science*, *104*, 127–132. doi:10.1016/j.procs.2017.01.084

Bauer, W., Hämmerle, M., Schlund, S., & Vocke, C. (2015). Transforming to a Hyper-connected Society and Economy–Towards an "Industry 4.0". *Procedia Manufacturing, 3*, 417–424. doi:10.1016/j.promfg.2015.07.200

Chen, Y. Y. K., Jaw, Y. L., & Wu, B. L. (2016). Effect of digital transformation on organisational performance of SMEs: Evidence from the Taiwanese textile industry's web portal. *Internet Research, 26*(1), 186–212. doi:10.1108/IntR-12-2013-0265

Del Rowe, S. (2016). Digital Transformation Requires a Cultural Change. *CRM Magazine., 20*(7), 15–15.

Dumitrica, D., & Wyatt, S. (2015). Digital technologies and social transformations: What role for critical theory? *Canadian Journal of Communication, 40*(4). doi:10.22230/cjc.2015v40n4a3044

Funtenberger, V. (2017). German Music Libraries in the Age of Digital Transformation—Challenges and Perspectives. *Fontes Artis Musicae, 64*(1), 86–91.

Hinchcliffe, D., & Kim, P. (2012). Social business by design: Transformative social media strategies for the connected company. John Wiley & Sons.

Hess, T., Matt, C., Benlian, A., & Wiesböck, F. (2016). Options for Formulating a Digital Transformation Strategy. *MIS Quarterly Executive, 15*(2).

Jäger, J., Schöllhammer, O., Lickefett, M., & Bauernhansl, T. (2016). Advanced Complexity Management Strategic Recommendations of Handling the "Industrie 4.0" Complexity for Small and Medium Enterprises. *Procedia CIRP, 57*, 116–121. doi:10.1016/j.procir.2016.11.021

Kaivo-oja, J., Roth, S., & Westerlund, L. (2017). Futures of robotics. Human work in digital transformation. *International Journal of Technology Management, 73*(4), 176–205. doi:10.1504/IJTM.2017.083074

Kane, G. C., Palmer, D., Phillips, A. N., Kiron, D., & Buckley, N. (2015). Strategy, not technology, drives digital transformation. MIT Sloan Management Review and Deloitte University Press.

Kans, M., & Ingwald, A. (2016). Business Model Development towards Service Management 4.0. *Procedia CIRP, 47*, 489–494. doi:10.1016/j.procir.2016.03.228

Majchrzak, A., Markus, M. L., & Wareham, J. (2016). Designing for digital transformation: Lessons for information systems research from the study of ICT and societal challenges. *Management Information Systems Quarterly, 40*(2), 267–277. doi:10.25300/MISQ/2016/40:2.03

Marcum, D. (2014). The digital transformation of information, education, and scholarship. *International Journal of Humanities and Arts Computing, 8*(Suppl.), 1–11. doi:10.3366/ijhac.2014.0095

Ozcaglayan, M., & Çelik, R. (2014). Self Expression, Exposition and Surveillance in Social Media (A Qualitative Study on the Transformation of Surveillance through The Digital Information). *Journal of Media Critiques, 1*(3).

Romero, D., & Vernadat, F. (2016). Enterprise information systems state of the art: Past, present and future trends. *Computers in Industry, 79*, 3–13. doi:10.1016/j.compind.2016.03.001

Schumacher, A., Erol, S., & Sihn, W. (2016). A maturity model for assessing Industry 4.0 readiness and maturity of manufacturing enterprises. *Procedia CIRP*, *52*, 161–166. doi:10.1016/j.procir.2016.07.040

Suarez-Guerrero, C., Lloret-Catalá, C., & Mengual-Andrés, S. (2016). Percepción docente sobre la transformación digital del aula a través de tabletas: un estudio en el contexto español [Teachers' Perceptions of the Digital Transformation of the Classroom through the Use of Tablets: A Study in Spain]. *Comunicar*, *24*(49), 81–89. doi:10.3916/C49-2016-08

Uhlemann, T. H. J., Lehmann, C., & Steinhilper, R. (2017). The Digital Twin: Realizing the Cyber-Physical Production System for Industry 4.0. *Procedia CIRP*, *61*, 335–340. doi:10.1016/j.procir.2016.11.152

Wang, Y., Wang, G., & Anderl, R. (2016). Generic procedure model to introduce Industrie 4.0 in small and medium-sized enterprises. In *Proceedings of the World Congress on Engineering and Computer Science* (Vol. 2).

KEY TERMS AND DEFINITIONS

Agility: The capability of an organization to rapidly change or adapt in response to changes in the market.

Collective Intelligence: shared knowledge jointly constructed by a group of people through their collaborations and instructions.

Digital Transformation: The definition of digital transformation is the realignment of, or new investment in, technology, business models, and processes to drive new value for customers and employees and more effectively compete in an ever-changing digital economy.

Industry 4.0: Industry 4.0 is a term often used to refer to the developmental process in the management of manufacturing and chain production. The term also refers to the fourth industrial revolution.

Innovation Accelerators: Critical technologies to business transformation that can radically expand its capabilities and applications

Nomadic Workers: Someone who works in different places while away from their office, often using a mobile phone and the internet.

Well-Being: Is when individuals have the psychological, social and physical resources they need to meet a particular psychological, social and/or physical challenge.

Chapter 3
The Role of Universities in Industry 4.0 Era:
Entrepreneurship and Innovation Perspectives

Haldun Yavuztürk
Marmara University, Turkey

Zeynep Tuğçe Kalender
Marmara University, Turkey

Ozalp Vayvay
Marmara University, Turkey

ABSTRACT

In every industrial era, one of the most important issues is the adaptation of business conditions to operational and administrative processes. Nowadays, as in every field of business-life, globalization and high competition highlighted the importance of entrepreneurship and innovation. The focal point of competition has changed to the position to innovate products at desired and required amounts, times, and quality with affordable price so taking advantage requires integrated management skills. Universities play a key role at this point thus entrepreneurs and researcher universities has emerged. In this study, after a general overview on evaluation of innovation and entrepreneurship concepts, the importance of them are explained in the view of industry 4.0. Then, a model of entrepreneurial and researcher universities is explained in detail with the analysis of several countries' models. The study reveals the role of universities in the innovation and entrepreneurship framework and to generate awareness about industry-university collaborations in the new industrial era.

DOI: 10.4018/978-1-5225-4936-9.ch003

INTRODUCTION

In today's business environment, creating and sustaining competitive advantages require an effort beyond cost reduction and operational efficiency. While globalization offers wider markets for industrial companies, improved purchasing conditions and opportunities for more comfortable production, it also presents some challenges for the industrial sector. Global competition is more intense, and actors must be constantly more flexible and efficient in order to stand out in this competitive position. According to the sector, several methodologies can be used however the key point is customer. It is a well-known fact that, today customer is the focal point of the business world and the power of the customer changes traditional management approaches. In this process, it is a necessity to offer target products in a shorter time as well as having the ability to produce new, high quality, efficient and personalized products.

Nowadays in the beginning of the industry 4.0 era, as in every field of business-life practices, globalization and high competition highlighted the importance of entrepreneurship and innovation. In the first industry era, water and steam power is used by the industry. Thus, steam powered machines were developed and used in factories to mechanize some of the work. The second industrial revolution came about by the use of electricity in production systems and the control of electrical power on assembly lines. The main application of the system was serial production lines which installed at Ford Motor Plants. This system applied by Ford Motor Plants in automobile production led to the enlargement of the production scale and thus the lowering of costs and prices. Techniques applied in these factories led to the rewriting of business management models. By the 1970s, a microprocessor-based programmable logic circuit was developed that transfers the information from the sensors to business units within a program framework. By applying this system to production systems, automation of the production system becomes possible. This development has greatly reduced human contribution and minimized the error. Thus, a new industrial revolution, the third era, started from the beginning of the 1970s to the present day. In this period, the use of computers, smartphones and the widespread use of the internet has affected and shaped the production in every way. The developments in communication and transportation, trade and industry are globalized. Overall, industry 3.0 can be defined as the reduction of human labor in production and the automation of production. At finally, the industry 4.0 era introduced as smart factories where data and knowledge integrated and cyber physical systems monitor the physical processes. It is not wrong to say that the 4th industrial revolution is building on the third and digital revolution is started. Schwab (2016) stated that the way of life and work will alter with these technological revolutions in the industry 4.0 era.

Industry 4.0 was first introduced into Germany as a concept which was a project of the German government to encourage traditional industries like manufacturing and to equip them with high technology. According to the Boston Consulting Group (BCG) analysis, Germany has begun to implement Industry 4.0, achieving productivity gains in industrial production, while reducing total production costs by 5 to 8 percent (Moment Expo, 2017). The concept of industry 4.0 integrates data and knowledge to contain several new technologies and related paradigms. Since the industry 4.0 also creates a competitive advantage with its cost advantages, countries that adapt quickly to this evolution and adapt it to production processes seem to be looking for a new economic system. New product development and optimizing existing products and services are the greatest areas of growth potential for analytics and big data using Industry 4.0 technologies. The primary goal is reorganizing the production systems with information technologies to have advantage in global competition. In this new industry era, cost and time efficiency gains much more importance and quality of the product /service becomes crucial. From this point of view, industry focuses on entrepreneurial an innovative studies in terms of gaining advantage among rivals.

Innovation has become one of the top priorities for a majority of enterprises than ever. Although innovation becomes an ever more central issue for the development of firms and world economies, processes have become increasingly complex. There is a larger variety of knowledge sources and inputs to be used by organizations and firms and there is more interdependence and division of labor among actors. The surrounding conditions of the innovation system are composed of different aspects, which can potentially influence the output (Marxt & Brunner, 2013). Today, not only understanding the customer needs and expectations but also presenting new or innovative products at the desired and required amount, time, and quality with an affordable price requires integrated management skills. Therefore, the need for well-qualified personnel grows more urgent. The increase in the quality of all economic returns, including production in industry, is supported not only by technological adaptation but also by the increase in qualified human power.

Universities play a key role at the need for well-qualified personnel thus the concept of entrepreneurial university is emerged. Entrepreneurial universities meet the qualified personnel demand but also increase the national and international collaborations and research and development (R&D) capacities. There is awareness that the definition of innovation has broadened— it is no longer restricted to R&D laboratories and to published scientific papers. For that reason, universities can be evaluated with an index system based on five dimensions, (I) Scientific and technological research competence, (ii) Intellectual property, (iii) Collaboration and interaction, (iv) Entrepreneurship and innovation culture and (v) Economic contribution and commercialization.

In this study, after a general overview about the evaluation of innovation and entrepreneurship concepts, the importance of such concepts are explained in the view of industry 4.0 era. Then, model of entrepreneurial university is explained in detail with the analysis of several countries' education models. To better understand the role of universities, entrepreneurial index system for the universities is explained with a detailed dimension analysis. Each dimension of the index system discusses to enhance the importance of the universities role. Study aimed to reveal the role of universities in the innovation and entrepreneurship framework and also tried to generate awareness about the industry-university collaborations in the new industry era.

OVERVIEW OF INNOVATION AND ENTREPRENEURSHIP

In the late 18th century the term "entrepreneur" is firstly mentioned by Richard Cantillon (Koh & Hatten, 2002). According to this definition entrepreneurs buy goods with certain prices however he/she takes the risk of selling them with uncertain prices. Say (1803) extended the Cantillon's entrepreneur definition and stated that "entrepreneur brings people together to build a productive item". In literature, dimension of risk-taking in the definition of entrepreneur firstly mention by Knight (1921). According to his theory, uncertainty considered as a factor. In 1890, Principles of Economics is published by Marshall and it is stated that a successful entrepreneur should understand the industry, have good leadership skills, foresight on demand, supply changes and willing to act in risky foresights. Mcclellands (1987) considered entrepreneurs as decision makers in times of uncertainty.

In literature, there are several definitions for the term "entrepreneur" and in decades definition has been extended. In the narrowest sense the person who captures ideas then translate these ideas into products and create an initiative to offer services and products to market (Zhao, 2006). In Table 1, different definitions of entrepreneur are listed.

Table 1. Entrepreneur definitions based on the reviewed literature

Author (Year)	Definition
Cantillon (1755)	"… is the person who undertakes the risk separately from the owner of the capital."
Say (1803)	"…is a concept that expresses a new formation that changes the current economic order"
Schumpeter (1934)	"… is someone who innovates and develops new technology."
Shapero (1975)	"… is the person who takes initiative in the social and economic functioning of the organization and assumes the risk of possible failure."
Hisrich (1985)	"… is someone who produces time and effort to produce different values, assumes various risks and eventually receives financial or moral support."
Sahlman and Stevenson (1991)	"Following up opportunities regardless of available current resources."
Drucker (1995)	"… is the person who transfers resources from low efficiency areas to high productivity areas and is able to keep them there."
Bridge et al. (1998)	"Starting a new job, having a job or growing the business"
Timmons (1999)	"… is the person who has balanced behavior and thinking with contemporary leadership, who constructs truths with honesty and sincerity."
Hisrich and Peters (2001)	"… is a value creation process with taking into account financial, physical and social risks, allocating sufficient labor and time"
Wickham (2001)	"… is the person who develops a specific project with his entrepreneurial understanding."
Brenkert (2002)	"… is the indispensable element of market economy."
Shane and Jonathan (2003)	Identification, development and operation of future goods and services

Today, entrepreneurship is considered one of the most important solution ways of unemployment thus it can be seen as a resource of innovation and economic growth. Since the concept cannot be fully described and difficult to understand, entrepreneurship should be examined in both traditional and conceptual dimensions. In traditional way, it can be defined as an individual effort that can translate a vision to an enterprise initiative, however in a conceptual way it is defined as the establishment and operation of a new business (Zhao, 2006).

In today's globalized world, the term "entrepreneurship" is changes day by day with the social, cultural and economic transformation. Over time, the old courageous entrepreneur who likes to take risk has left its position to the entrepreneur who has a knowledge-based and innovative understanding. Therefore, today the term "innovative entrepreneurship" is emerged in the rapidly changing business world where the transition from the industrial society to the information society is experienced (Er, 2013). It is not wrong to say that innovation, which is the use of new or improved products and processes, is a specific function of entrepreneurship.

Innovation which originated from the Latin word "innovates", is simply defined as "…the introduction of new methods in social, cultural and administrative settings." According to economist Schumpeter, who described innovation for the first time, everything that provides profits to entrepreneur and emerges because of technological advances is innovation (Schumpeter, 1934). The concept of innovation is both a process and a result. In literature, there are several definitions for innovation. Dosi (1988) defines innovation as the adoption and commercialization of a new product or method by exploring, testing, developing or imitating it while Fisher (2000) describes innovations as new ways of thinking, producing new ways of doing things, experimenting with production and adopting in economic and social activities related to people. According to the European Union (EU) and The Organization for Economic

Co-operation and Development (OECD) literature, innovation refers to the process of "transforming an idea into a marketable product or service, a new or improved manufacturing or distribution method, or a new method of social service" (OECD/Eurostat, 1997). This innovation involves significant changes in techniques, equipment or software and is designed to reduce unit production and delivery costs, increase quality, and produce new products (OECD, 2005: 53).

Over the last 50 years, the innovation literature has sought different factors to explain what enables successful innovation and how this can be fostered by organizations. Caird et. al (2013) stated that in the 1980s, there was an emphasis on enterprise, small firm start-ups and organizational spin-offs, with increased understanding of the innovative role of entrepreneurs, entrepreneurs and social entrepreneurs. Studies reveal that in the 1990s, there was a focus on multidisciplinary approaches, multi-functional teams and collaborative networks as a source of innovation. In the 2000s, the role of users and industry supply-chain intermediaries attracted particular attention. More recently the Internet and the availability of information and communication technologies have led to new trends in innovation (Caird et al., 2013). Business entrepreneurship features and entrepreneurial approaches make innovation easier. Innovation which can be considered as the main component of entrepreneurship is seen as the most important way of gaining sustainable competitive advantage for an organization (Porter, 1980).

In parallel with the developments in science and technology in the 21st century, the economic value of the entrepreneur and entrepreneurship and the prominence in the society have increased considerably. Companies pay much more attention to innovation than ever because of the rapidly developing technology, differentiation in the customer expectations, compelling competition in briefly because of the changing industry era. In an interview published in the magazine of Fortune magazines in 1998, Steve Jobs stated that "… innovation is not a concept of spending millions of dollars on innovation; more clearly, it is a human concept and related to how to be led and what can be gained from it" (Tidd et al., 2005).

Evaluation of new technology and analysis of its effects is important in terms of making future decisions. Technological innovation development consists of a long process. This process starts with an idea or an invention and covers all the processes until the new product or services, which arise from R&D, offered to the market and customers. In fast technological changing environments, it becomes hard to survive for companies. Today, companies are working with many different types of data in other words challenging with big data which can be defined anything from databases to photos, videos, sound recordings, written text and sensor data. Moreover, objects can connect to each other to exchange and analyze data in a giant network which is achieved by well-known concept, Internet of things (IoT). Initially, the IoT was mostly used in business especially in manufacturing, in which it is also known as machine-to-machine (M2M). However, nowadays IoT is in our daily life, with smart devices, and it is estimated that IoT market reach $267B by 2020 according to Boston Consulting Group.

In this kind of environments where competitive conditions is tough, having abilities to manufacturing/ producing technology is become more important (Akyos, 2005). Because of the rapid changes in technology, companies start to integrate their own internal abilities with external resources such as universities, research companies and etc. It is not needed that all technologic resources have to be in company, companies should only realize which external resources that they will be need and where these resources can find.

In today's business world, employers express the need for qualified graduates with a range of enterprise skills. These skills are included creativity, capacity for innovation, networking relationship management and risk taking. According to the study conducted by Gibb et al. (2009) university paradigm is changing from certain to uncertain environment so structures becomes more complex. In this uncertain and complex environment, universities should be entrepreneurial and innovation driven. With an open innovation system and interacting business world and governmental institutions universities can increase their innovative and entrepreneurial capacity instead of being a closed research institution.

ENTREPRENEURIAL UNIVERSITY MODEL

The concept of entrepreneurial university is emerged because of the direct effects of globalization, rapidly increasing competition and the inadequacy of public funds in the face of increasing demand for higher education. Changing conditions have led universities to seek new sources of funding and to take on income saving and cost reduction strategies, generate knowledge-based fundraising strategies, develop student tuition and aid schemes, and integrate commercial life into a variety of ways (Kezar & Eckel, 2002). In this context, entrepreneurial university becomes an educational pillar of entrepreneurship which is formed by innovation, creativity and change.

In literature, many researcher (Acs & Audretsch, 1991; Drucker,1985; Storey,1994) consider entrepreneurship as a key factor of economic growth of countries. The underlying reason is that entrepreneurial societies can generate different projects to different scenarios of economic challenges. Since universities are the natural incubators to create new ideas and technologies, the importance of the universities in entrepreneurial society conditions increases. As a conclusion of this changing era, universities become a bridge education and R&D to become more productive (Kirby, 2006).

Entrepreneurial university is firstly defined by Etzkowitz (1983), as "...a university which provides activities such as collaborations with private enterprises, research contracts, patents, etc. and new financial resources..." In 1995, Chrisman et al. stated that it is a creation of new initiatives by university actors (academicians, technicians, students). Clark (1998) extended concept of the definition and defined that these kinds of universities are researching several ways to enable the business world becomes more innovative and creative. In 2004, Etzkowitz stated that there are 5 common concepts in entrepreneurial universities such as (i) capitalization of knowledge, (ii) managing interdependence with industry, (iii) independent, (iv) hybrid in managing the both and (v) continuous structure renewal. In 2008, Robertson determined key factors of entrepreneurial universities such as;

- Leadership that develops entrepreneurial capacities of all university actors on campus
- To establish strong ties with external stakeholders producing added value, to increase university-industry collaboration
- Creating entrepreneurial outcomes affecting individuals and institutions
- Implement innovative learning techniques that lead to entrepreneurial action
- Removing borders to support effective cross-institutional information flow
- Multidisciplinary education approaches that focus on complex world problems and reflect real-world experiences
- To encourage entrepreneurial thinking and leadership practices

Odabaşı (2008) defined entrepreneurial universities as a pioneer in producing knowledge. These universities should be able to find sources, establish relationship between the university and industry, and contribute to the economy by evaluating the opportunities. What is more, these universities should be capable of innovating, evaluating opportunities, being team-oriented, and changing organizational structure in order to be successful (Odabaşı, 2008). It is aimed to increase the profitability and profitable advantages with the entrepreneurial universities (Özlem & Sarı, 2008).

The indispensable part of the entrepreneurial university mode is providing university-industry cooperation. At this point, common projects should be developed with the cooperation between the industrial zones and the universities. For that reason, universities should remove their educational programs' boundaries and become more project oriented. Educational curriculums should be adapted to their changing conditions in the industry 4.0 era. It is well-known fact that today, there is an awareness that the definition of innovation has broadened— it is no longer restricted to R&D laboratories and to published scientific papers. International cooperation, which began in the early 1990s, has not only become a source of income for universities, it has also become a key instrument in the development of new activities and academic programs. As well as university graduates, faculty, researchers and managers are influential in the spread of international cooperation, science and education cultures. For this reason, universities play an important role in international cooperation and outreach of the region (Kniazev, 2004). Universities are the carriers of the study capital, especially for innovative research. In this context; they are an important factor in accelerating innovation and developing new products and services.

Nowadays, universities make many improvements to become and entrepreneurial university. National and international collaborations and industry-university cooperation are increasing, project-oriented programs are developed, etc. on the other hand these universities can be evaluated with an index system based on five dimensions, (I) Scientific and technological research competence, (ii) Intellectual property, (iii) Collaboration and interaction, (iv) Entrepreneurship and innovation culture and (v) Economic contribution and commercialization. These criteria are interrelated with each are. In other words, every criterion is directly related the other one and increment in the one criterion is caused a positive impact on the other one. It can be considered as a chain, each of the rings should be powerful individually but most importantly they affected the overall strength of the university by creating an impulse to each other. To better understand the relationship, it is better to investigate each of them individually.

Scientific and Technological Research Competence

The competition of scientific and technological competence emerged between countries since technological innovations lead to changes in competition and product/ process as well as in the markets (Zerenler et al., 2007). There is a very close relationship between the level of development of a country and the scientific investments. Three basic indicators are considered to be used in the comparison of science and technology levels of countries. These are the ratio of R&D spending in GDP, the number of R&D personnel per 10 thousand working population and the scientific articles published by the countries. However, to evaluate the first dimension not only publication numbers are used but also, contributions of these publications to the literature, number of project that are supported by innovation support programs, financial support that is taken from the related programs, national and international awards, number of graduated person with a doctorate degree should also be considered.

Post-graduate education is an important educational step for the growth scientists who will advance science in various ways in the future of a country and for the advancement of technology. It is a well-known fact that education in science and technology is constantly changing and in this new industry era 4.0 with the developments made in definitions and science, the need for academicians who are educated with post-graduate education is increasing day by day. Both quantitative and qualitative aspects of educational programs constitute one of the key points that will determine the long-term success of the growth. The underlying reason is that the human power is one of the most important inputs of the system.

Scientific and technological research competence is evaluated according to the new products and scientific publications that show the effectiveness of the human power, financing, physical infrastructure.

The publication of scientific publications is an important development that enables the comparison of the education system and its development at the international level. Bibliometric investigations have an important role in the evaluation of these contributions. It is defined by Pritchard (1969) as "the application of mathematical and statistical methods to books and other communication environments," and is an area where various findings about scientific communication are obtained by analyzing certain properties of publications. The results of bibliometric surveys have led to the publication of reports reflecting the science and technology indicators of various countries. The National Science Board's report (1972) has taken its place in the literature as the first report on this subject (Aksnes, 2005). There has been a significant increase in the number of publications in bibliometric studies. When the contents of the studies are examined, it is seen that comparisons are sometimes made between countries (Braun, et al. 1995a, 1995b), sometimes the performance of a single country is evaluated (Jacobs & Ingwersen, 2000; Moed, 2002; Ortiz-Rivera et al., 2000) and sometimes the subject / area-based approach is taken as basis.

Publications generally analyzed by the international literature indexes such as Information Sciences Institute-Web of Science, SCOPUS, etc. In these indexes scientific publications statistics can be monitored in national and international level with benchmarking countries. It is stated in the literature that the main function of the reference is to establish a link between the cited person and the cited document (Smith, 1981). Today, the Science Citation Index, the Social Sciences Citation Index and the Arts and Humanities Citation Index, which are widely used, have begun to be published in the Institute for Scientific Information-ISI. Most recently, h-index which is generated by Hirsch (2005) is used to evaluate scientists. Commonly, it is concerned with how many of his/her publications refer to a certain value. There are two variable numbers that should be known when calculating the h-index, are the number of publications and the number of citations to these publications. Accordingly, each publication's citation counts however there is a must that each considered publications should be published in journals covered by citation indexes. Then publications are sorted from the most cited to the least.

The development of the ability to create new ideas, spread them and transform them into new and profitable products, processes and services, and thus developing the R&D and innovation-generating infrastructure, is one of the key factor in increasing the value-added activities. A new or significantly improved product or service is an activity with a continuity of innovation and entrepreneurship plays an important role. R&D and innovation-driven high and sustainable productivity gains are key factors in determining competitive power. From this point of view, number of projects and the financial support financial support that is taken from the related programs becomes an important indicator for universities especially entrepreneurial universities. When the business conditions are considered in the industry 4.0 era, contributions of the universities become more essential in terms of R&D.

Scientific and technological research competence has an important weight in the overall grading system in the entrepreneurial university index. Since the university tries to increase its research capacity and develop more R&D project other dimensions are directly affected. Therefore, an impulse can be created to improve overall grading in the index system.

Intellectual Property

The goal of all countries is to develop in the industry and the economy, and to take this development further by protecting it against changing conditions. The most important factor in the development of countries in industrial and economical aspects is the production of new technologies and the application of these technological developments to industrialization.

Intellectual property dimension of the index system for entrepreneurial universities considers the number of patent applications, number of patents both national and international and number of industrial designs. Many countries have adopted the principle of encouraging the inventor to be awarded so that innovations and inventions can be applied to industrialization and consequently contribute to economics. The patent system gives the inventor a privilege to use the invention for a certain period of time, thus helping the industry develop by encouraging the invention. Inventions that can be protected by patents are required to carry three basic qualities such as (i) innovation, (ii) exceeding the known state of the method and (iii) industrial applicability. The European Patent System (EPC) is a system established to create a single patent granting system in Europe with the "Convention on the Export of European Patents" signed in 1973 in Munich. The system provides for the patenting of a single center on behalf of the participating countries, granting the validity of these patents and the resolution of the disputes from a single center. Many countries develop patent application incentive and support programs in order to increase the number of national and international patent applications and to encourage individuals and institutions to make patent applications. Moreover, patent prizes are awarded to universities and businesses according to several categories such as the most number of patent applications prize. Thus, the creation of value-added projects is becoming more important in which R&D and innovation works are supported.

In the industry 4.0 era, technological improvement is rapidly changing. In these fast-moving conditions, universities collaborate with industries and opened techno parks to create more close relationships with industry. Universities try to increase their scientific and technological competence and number of patents or their applications are radically increase when the university gives much more importance to the first criterion. These collaborations not only increase the R&D capacity and the intellectual property of the university, they also provide national and international achievements. For this reason, intellectual property is a vital criterion for entrepreneurial universities since there is a direct relation between the term entrepreneur and innovation. When universities give much more importance to its intellectual property, it is clearly seen that there is a great increase in the number of patent applications.

Collaboration and Interaction

Today, access to new ideas and approaches that are important for solving tough and complex problems becomes easier with the various mechanisms for the transfer of knowledge, and using such mechanisms is a necessity for businesses. In order to increase knowledge and realize technology transfer, perhaps one of the important methods is university industry collaborations. Through these collaborations, businesses can transfer information through a variety of methods, including consulting, publicly funded R&D projects or qualified R&D personnel. Especially in the industry 4.0 era, these collaborations not only become a great advantage among rivals in the competition but also a vital necessity. For that reason, today, many businesses are cooperating with universities or research institutions especially in geographically close ones, and have the opportunity to use the human resources and infrastructure facilities of universities.

From universities view of point, they have the opportunity to turn their research results into practice and explore new research areas through collaborations with industry. In other means, cooperation facilitates the transfer of technology and the commercialization of the produced knowledge for universities, which provide scientific and technological knowledge production through basic research and applied research and qualified personnel development. From industry view of point, industries benefit from the academic human resources of the university, facilities of the university and the experiences of the industry in order to produce new technologies and to develop the research and development activities (Şahin, 2010). The importance of university-industry cooperation is better understood when the main source of scientific development is thought to be the universities and the main actors of economic development and new technologies is industry. When universities combine strong research capacities, trained human resources, scientific know-how, and time-based and profit-oriented working principles of the industry, a structure that benefits both sides are formed.

Universities can be graded according to their number of project that is achieved with international cooperation and industry collaborations. Nowadays, many countries developed special supporting programs to motivate universities and the academicians to create an impulse in collaboration and interactions. There are several international R&D support programs provided by European Cooperation in Science and Technology, European Space Agency, European Science Foundation etc. which aims to support developing products and processes that are market-oriented and can be commercialized in a short time. Moreover, there are some international cooperation platforms such as EUREKA which established to promote the research and development of advanced technologies, products and services that will increase the competition power of the world's industrial and research institutions in European countries, and the creation and execution of joint projects between countries. In 2016, 329 projects are supported by EUREKA and according to the report %15 of the participants is large companies, %23 of them is research organizations and %62 of them are SMEs.

Number of academic staff or students can also be great criterion for collaboration and interaction dimension. There are special programs such as ERASMUS which is a European Union program to encourage higher education institutions to cooperate with each other. These kinds of programs provide financial support to higher education institutions to produce joint projects with each other, short-term student and staff exchanges. It also encourages the development of relationships and cooperation between higher education institutions and workplaces in order to improve the higher education system in accordance with the needs of the business world and to increase the employability of university graduates in the business world.

Entrepreneurship and Innovation Culture

Universities have a significant effect to increase the culture of entrepreneurship and innovation. Entrepreneurship education can be carried out in the context of formal education, non-formal education (certificate programs) or various projects. In formal education, universities can rearrange their curriculum based on the new improvements in the business world. In this sense, entrepreneurship education, technology management, innovation, creation and development of business projects, management of small and medium enterprises etc. can be added as common lectures to curriculums. Entrepreneurship concept entered the lectures of higher education generally in the mid-twentieth century (Gutierrez & Baquero, 2017). Today, many universities opened specialized graduate programs in the entrepreneurship and innovation field. Thus, not only the culture of innovation is created but also qualified personnel demand of the businesses is met.

Non-formal education consists of several certificate programs provided by the authorized institution or foundations. In these programs, at the end of the education a certificate is given to successful participants to approve their training. In some governmental support programs, approved certification is needed to apply a support program for new startups. As it is mentioned above, entrepreneurship and innovation culture can be enhanced by various projects as well. These include conferences, competitions, joint academic projects etc. in which undergraduate and graduate students who have the idea of innovating in various fields are supported within the scope of the goal of raising awareness in the field of entrepreneurship and innovation.

Universities can contribute to creation and development of entrepreneurship and innovation culture through collaborations. In this sense, techno parks and technology development zones are being established in order to strengthen the technology-intensive industries where they are located to support the development of the industry in new areas, to carry out innovative studies, to increase productivity through more effective working conditions as well as to increase the international competitiveness of the field (Harmancı & Önen, 1999). According to the International Association of Science Parks (IASP, 2012), the techno park is designed to encourage the establishment and development of technology-based firms with formal ties to one or more universities or research centers.

Techno parks first appeared in the United States in 1950 with the efforts of a group of researchers seeking to commercialize their research in the pioneering of Stanford University. Today it is known as the "Silicon Valley" and this techno park has become the most famous techno park in the world (Başalp & Yazlık, 2014). The second country where techno parks are established is England. The Heriot-Watt University Techno Park in Edinburgh in 1972 and the Cambridge Techno Park in London were the first techno parks to be built in the UK. After the Second World War, the rise of oil prices, the decline of production, the economic recession forced the countries to watch and the countries started to give importance to university industry cooperation in order to get rid of these negative situations. The development of techno parks in the 1980s was the beginning of speed. At that time, the total number of techno parks reached 12 in the US, 7 in Belgium and France and 2 in the UK (Çelik, 2011). When the periods of establishment of the technoparks in the world are examined, it is observed that the greatest increase is in the year of 1990 with 48% and the numbers have started to increase quite rapidly after 2000 years.

Techno parks play an important role in the development of strategic priority sectors. These centers provide contribution with R&D-oriented activities to regional development depending on functions such as increasing the productivity and competitive power of the enterprises in the region, providing advanced technology and innovation infrastructure, transferring technology, diversifying economic activities of the region and providing new business opportunities. In these centers new entrepreneurs are also supported by the centers called as "Incubator Centers". The main aim of establishment of these centers is especially developing young and new enterprises by providing office services, equipment and management support, access to financial resources, critical business and technical support services under a single roof.

Economic Contribution and Commercialization

Today, the most important factor determining the economic power and the level of welfare of societies is the developments in science and technology. Countries that cannot adapt to these developments are no longer able to sustain their assets in a strong and respectful manner in the 21st century. In order to ensure that the country's industry is able to compete internationally by providing university-industry cooperation; it is aimed to create regions that will produce technological knowledge, commercialize the produced information, develop innovations not only increase quality and efficiency of the processes / products but also decrease the production costs, provide adaptation of the small and medium sized enterprise to new and advanced technology, provide business opportunities to researchers and accelerate the entry of foreign capital into the country.

In these specialized regions, academicians have great opportunities to develop new projects. As an example, in some countries academic staff can establish a company in these regions with the aim of commercializing the results of their research with permission of the university board of directors and can become a partner in an established company and / or take part in the management of these companies.

Economic contribution and commercialization dimension is directly related with the number of enterprises which are established by academicians in techno parks. In addition to that, number of enterprises that established by students are graduated students in 5 years. As it is mentioned before, these collaborations enable to develop new products and production methods aimed at technological production by bringing industry, researchers and universities together. Technology Transfer Offices (TTO) which is an organization that conducts activities that enable the commercialization of academic research results in an efficient and rapid manner, play a vital role in collaborations as well. They provide needed communication links between universities, research institutions and private sector in other words between researchers, entrepreneurs and investors. The basic services provided by TTO are (i) to evaluate the commercialization potential of the inventions of the academicians and the companies (ii) to provide financial and legal consultancy support for the inventions with the potential of patenting and commercialization and (iii) to make commercialization of the inventions decided to be supported.

ENTREPRENEURIAL UNIVERSITY MODELS IN THE WORLD

Entrepreneurial universities accelerate innovation and play a key role in the creation of entrepreneurs from the new establishment phase to the growth phase. As a result, there may be a lot of job opportunities and new income streams that facilitate the future investment opportunities in the economy. Guerrero et al. (2016) stated that in the world many universities are currently in a state of transition because they are expected to develop a wide range of relationships with stakeholders. Therefore, universities are changing according to governance and strategies based on their content and structure. This development process is not a process of co-creation but rather in a series of transitions in which stakeholders are continually shaping/adapting the university model.

In the world, there is an important trend to the entrepreneurship and innovation-based education. First education programs are started in Harvard Business School USA in 1947 however first academic programs are started in 1960s and become widespread after 1965s. According to Koh's (1996) study; In 1993, more than 400 US universities had entrepreneurship courses at the undergraduate or graduate level, which reached 1600 in 2003.Katz stated that (2003) today, entrepreneurship programs are provided by more than 1600 schools with more than 2200 lectures and there are 44 academic journals and several management journals in USA.

Entrepreneurship training has grown more slowly in Europe than in the United States. England and France were the first countries to start entrepreneurship education in the 1970s however education programs about entrepreneurship are become widespread in 1990s. Over the past two decades UK governments supported higher education in terms of funding, more than ever. The underlying reason is that the links between research and teaching with economic growth are getting stronger day by day. The biggest challenge to the UK engineering sector right now is people. The UK needs 200,000 engineering graduates each year (Rossi, 2016). Attention to the entrepreneurship education gains importance immediately and this impulse created awareness in the political activities.

In 2000, Etzkowitz stated that there is a structural shift at European Universities. They changed their mission from traditional education to commercialization of new knowledge. Primary goal of this shift is the economic development (Arnaut, 2010). The development of entrepreneurship in the direction of creating a more entrepreneurial Europe is among the priority policies of the European Union. Today, many governments are putting innovation at the center of their growth strategies. The EU is involved in a wide range of activities under these policies, but also monitors the development performance of the EU member states and candidate countries. Therefore, there will be incremental improvements in the high education and some examples from Europe as follows;

- In Germany, the "EXIST Program" was established by the Federal Ministry of Education and Research to develop regional cooperation between universities, technical schools, the business community and other interested parties. The program brings together interested parties to ensure a more entrepreneurial mentality in higher education and research institutions and to put the enterprise potential into practice.
- A coordinated approach to entrepreneurship has been introduced in the Netherlands and interaction has been established between various ministries, government agencies and business owners. Regions that are directly involved in the implementation of entrepreneurship policies have agreed to have joint ventures with the Ministry of Economy in important issues such as the development of entrepreneurship spirit in higher education and the provision of integrated support services. In order to create entrepreneurship spirit among high school students, "Mini Business Project" is started. Students establish a business, produce goods or provide services thus the "Mini Enterprise Award" is given once a year.
- In Austria, the activities of entrepreneurship in universities have increased and three entrepreneurship institutes have been established. In addition, programs are being implemented to ensure convergence between schools and the business world, and applications are being made by top business executives to visit schools, organize trips to businesses, and gain business experience with teachers. With the Business Start Program (Junior), which is applied to the students in the 15-19 age group, it is possible to learn from the development of the idea of starting a business to production, marketing and sales.

- Ireland has implemented a supportive package called as "Campus Company Program" to increase the number and intensity of university-based business initiatives to develop new businesses with high growth potential. This program is part of the Irish regional development strategy and includes a one-year business support program that provides practical information and management support to college graduates seeking to establish a workplace. Teaching staff in the framework of the program can receive financial support to pass research results to commercial life.

In Turkey, institutions such as Turkey Scientific and Technological Research Institution (TUBITAK, Ministry of Science, Industry and Technology, KOSGEB, EU Project Support Services, etc. support many projects on behalf of entrepreneurship in universities. TUBITAK releases an annual index of top 50 entrepreneurial and innovative universities in Turkey aiming to foster the development of entrepreneurship ecosystems within and around higher education institutions (Gür et al., In Press). In 2016, more than 400 of the 80 universities are taught in "entrepreneurial entrepreneurship" with the support of KOSGEB and the Ministry of Science, Industry and Technology.

FUTURE RESEARCH DIRECTIONS

In the social and economic development of the countries the role of universities is vital. In the technological advancement era, universities become a bridge between research and the commercialization of the related product or service. Globalization affects both developing and developed countries as the same rate of pressure. On the other hand, internal challenges and difficult conditions in developing countries generating disadvantages to them. For that reason, entrepreneurial character of university concern for the development of the both internal and external conditions of the environment. This situation also contributes to creating a good image of the university, which plays a very important role in university development. A good image brings in more students, larger number of projects, and consequently a higher income, essential for normal functioning and development of the university.

Chang et al. (2016) stated that there are rapid developments in the entrepreneurial university concept since the concept is emerged firstly in 1983 however the question of how to develop an appropriate organizational context still remains unclear. However, there is no single solution for transforming university to entrepreneurial university. The objective of building entrepreneurial university is to generate accomplishments that will allow achievement of a sustainable university, a desirable partner to the business and government. Each university must discover its own mode of transformation, which best fits the situation, resources and environment of the university.

In future studies, entrepreneurial university index should be extended in order to consider all dimensions of today's business environment. In this study main pillars are investigated in a detailed way and some common criterion for each of them explained. On the other hand, to evaluate a university there should be more detailed criterion in each pillar. From this point of view, it can create great advantage to extend the index system. According to the developed index system, it is possible to evaluate universities and benchmark them based on the main five pillars and their common sub-criteria. On the other hand, with a more detailed criteria analysis, more detailed classifications can be made.

Today indexes are used to evaluate universities and according to overall grades universities are ranked in the descending order. It is a quite practical approach and gives important ideas about the position of the related university. Especially with an extended index system it can give more clear results for the analysis. Although the index system creates awareness to the universities about their position among other universities, with an appropriate mathematical model it is possible to evaluate the constraints as well. Entrepreneurial universities can be evaluated with mathematical models such as goal programming as well. Mathematical models can be used to maximize the overall performance of the related entrepreneurial university with considering the main constraints. It is possible to generate some hybrid decision making solutions with these models. After the determination of the weights of each pillar and sub criteria model can establish. Therefore, universities can determine their improvement points and can prioritize them based on the effects to the overall performance. These models also create a proper benchmarking system among entrepreneurial universities.

CONCLUSION

Entrepreneurs or entrepreneurial organizations that have industry experience or find research opportunities in industry-developed societies can achieve positive results by dedicating themselves to further research and development. It is well-known fact that, in today's business environment conditions research and development is one of the most essential issues for countries and plays a vital role not only in competition but also in survival of companies at the target market. One of the most important places that can make this positive situation is the universities. In new industry era 4.0, like the revolutions that preceded it, industrial revolutions have the potential to raise global income levels and improve the quality of life for populations around the world. Thus, quality of international competition radically changes the role and functions of universities.

According to the initiatives about the entrepreneurial and innovative universities and the data which is given by indices, it is observed that universities that allow scientific studies around the world and the number of expert academicians directly affects the entrepreneurial spirituality and university-industry collaboration. It is a well-known fact that R&D and innovation culture contribute to the development of entrepreneurship and innovation concepts. However, before the implementation of the entrepreneurial university model, not only to the quantitative increase of universities but also to qualitatively researching, producing innovation, developing and contributing country's economy is a vital step to be successful in the international arena. In theory, universities should determine the entrepreneurship as a core strategy thus an entrepreneurship culture can be created. Only after a culture created the transformation to entrepreneurial structure is possible.

It is very important for a university to cooperate with the structures that will make the information given to students become concrete. The university, which is nested with the industry thanks to techno parks, is able to follow the current changes in technology and offers its students the opportunity to get to know the industry early. Data gathering is a business, but it is a separate business to interpret the database correctly and make accurate results. For this reason, industrial personnel and students have to acquire new qualifications in order to be ready for Industry 4.0. In this sense, university- industry collaborations should be considered as a must to be an entrepreneurial university. Thanks to these collaborations, universities provide budget for new researches, fulfill mission for public benefit, create application field for academic staff, implement employment policies such as internship or new job opportunities for students and contribute to the development of economy and technology. Besides, industry which is a stakeholder in this technology transfer method; has access to university knowledge accumulation infrastructure, increases its own research opportunities, reaches the human power and plays an active role in the development of the country. Techno parks, which are one of the best examples of university industry collaborations, are the ones having the new and advanced technology of their stakeholders, the creation of high-value-added products or processes, and the most important of them are the missions such as preparing the ground for technology transfer between the university and industry.

The most important step to contribute to the economy of the whole country is developing, producing, regenerating, controlling, and developing the technology rather than becoming just a consumer. The production infrastructure of the old world, which started with the first, second and third industrial revolutions, now open doors to a new economic perspective with Industry 4.0. Major shifts on the demand side are also occurring, as consumer engagement, and new patterns of consumer behavior force companies to adapt the way they design, market, and deliver products and services. Universities that will meet the needs of a dynamic industry will be the key element in development in Industry 4.0 era. If the spirit of entrepreneurship and innovation in universities is kept dynamic, there will be opportunities to create many products or services that can contribute to the economy of the country.

REFERENCES

Acs, Z. J., & Audretsch, A. (1991). *Innovation and small firms*. Cambridge: MIT Press.

Aksnes, D. W. (2005). Citations and their use as indicators in science policy: Studies of validity and applicability issues with a particular focus on highly cited papers. Unpublished doctoral dissertation, University of Twente, Netherlands.

Akyos, M. (2005), Firma Düzeyinde Yenilikçilik (Yenilik) ve Bilgi Yönetimi. Retrieved 10.04.2017 from http://sistems.org./know_info1.htm)

Arnaut, D. (2010). Towards an entrepreneurial university. *International Journal of Euro-Mediterranean Studies*, *3*(1), 135–152.

Başalp, A., & Yazlık, B. (2006). Türkiye'de Teknoparklar ve Sorunları. In TOBB Ekonomi ve Teknoloji Üniversitesi Türkiye'de İnternet Konferansı Bildirileri (pp. 275–278).

Braun, T., Glanzel, W., & Grupp, H. (1995a). The scientometric weight of 50 nations in 27 science areas, 1989-1993, 1. All fields combined, mathematics, engineering, chemistry and physics. *Scientometrics*, *33*(3), 263–293. doi:10.1007/BF02017332

Braun, T., Glanzel, W., & Grupp, H. (1995b). The scientometric weight of 50 nations in 27 science areas, 1989-1993. *Scientometrics*, *34*(2), 207–237. doi:10.1007/BF02020421

Brenkert, G. G. (2002). Entrepreneurship, Ethics, and the Good Society. *Ethics and Entrepreneurship*, *3*, 5–43.

Bridge, S., O'Neill, K., & Cromie, S. (1998). *Understanding enterprise, entrepreneurship & small business*. London: MacMillan Press; . doi:10.1007/978-1-349-26171-0

Caird, S., Hallett, S., & Potter, S. (2013). The Open2-Innovation Tool—A software tool for rating organisational innovation performance. *Technovation, 33*(10-11), 381–385. doi:10.1016/j.technovation.2013.06.003

Cantillon, R. (1755). *Essai sur la Nature du Commerce en General, edited with an English translation and other material by Henry Higgs, C.B. Reissued for The Royal Economic Society by Frank Cass and Co*. London: LTD.

Çelik, M. (2011). *Şirketlerin İnovasyon Yapma Eğilimlerinde Üniversite Sanayi İşbirliğinin Rolü ve ODTÜ Teknokent Örneği*. İstanbul: Marmara Üniversitesi Sosyal Bilimler Enstitüsü, Yüksek Lisans Tezi.

Chang, Y. C., Yang, P. Y., Martin, B. R., Chi, H. R., & Tsai-Lin, T. F. (2016). Entrepreneurial universities and research ambidexterity: A multilevel analysis. *Technovation, 54*, 7–21. doi:10.1016/j.technovation.2016.02.006

Chrisman, J., Hynes, T., & Fraser, S. (1995). Faculty entrepreneurship and economic development: The case of the University of Calgary. *Journal of Business Venturing, 10*(4), 267–281. doi:10.1016/0883-9026(95)00015-Z

Clark, B. R. (1998). *Creating entrepreneurial universities*. Oxford, England: Pergamon Press.

Dosi, G. (1988). Sources, Procedures and Microeconomic Effects of Innovation. *Journal of Economic Literature, 26*, 1120–1171.

Drucker, P. F. (1985). *Innovation and entrepreneurship practice and principles*. New York, NY: Harper & Row.

Drucker, P. F. (1995). *Managing in a time of great change*. Boston, MA: Harvard Business Press.

Er, P. H. (2013). Girişimcilik ve Yenilikçilik Kavramlarının İktisadi Düşüncedeki Yeri: Joseph A. Schumpeter. *Selçuk Üniversitesi Sosyal Bilimler Enstitüsü Dergisi, 29*, 75–85.

Etzkowitz, H. (1983). Entrepreneurial Scientists and Entrepreneurial Universities in American Academic Science. Minerva.

Etzkowitz, H. (2004). The evolution of the entrepreneurial university. *International Journal of Technology and Globalisation, 1*(1), 64–77. doi:10.1504/IJTG.2004.004551

Etzkowitz, H., & Leydesdorff, L. (2000). The dynamics of innovation: From national systems and 'mode 2' to a triple helix of university industry-government relations. *Research Policy, 29*(2), 109–123. doi:10.1016/S0048-7333(99)00055-4

EUREKA. (2017) Annual Report 2016, Belgium.

Eurostat. (1997). OECD Proposed Guidelines for Collecting and Interpreting Technological Innovation Data - Oslo Manual. Paris: OECD.

Fisher, F. M. (2000). *Innovation and Monopoly Leveraging. Dynamic Competition and Public Policy: Technology, Innovation, and Antitrust Issues* (J. Ellig, Ed.). Cambridge University Press.

Guerrero, M., Urbano, D., Fayolle, A., Klotfsten, M., & Mian, S. (2016). Entrepreneurial universities: Emerging models in the new social and economic landscape. *Small Business Economics*, *47*(3), 551–563. doi:10.100711187-016-9755-4

Guiterrez, J. G., & Baquero, J. E. G. (2017). New cross-proposal entrepreneurship and innovation in educational programs in third level (tertiary) education. *Contaduría y Administración*, *62*(1), 239–261. doi:10.1016/j.cya.2016.10.005

Gür, U., Oylumlu, İ. S., & Kunday, Ö. (2016). Critical assessment of entrepreneurial and innovative universities index of Turkey: Future directions. *Technological Forecasting and Social Change*.

Harmancı, M., & Önen, M. O. (1999). Dünyada ve Türkiye'de Teknopark ve Teknokent Uygulamaları. Türkiye Kalkınma Bakanlığı, GA/99-2-4.

Hirsch, J. E. (2005). An index to quantify an individual's scientific research output. *Proceedings of the National Academy of Sciences of the United States of America*, *102*(46), 16569–16572. doi:10.1073/pnas.0507655102

Hisrich, R. D. (1985). The woman entrepreneur in the United States and Puerto Rico: A comparative study. *Leadership and Organization Development Journal*, *5*(5), 3–8. doi:10.1108/eb053558

Hisrich, R. D., & Peters, M. P. (2001). *Entrepreneurship*. McGraw-Hill College.

International Association of Science Parks. (2012). A Governance Model for Science and Technology Parks. In *29th IASP World Conference*.

Jacobs, D., & Ingwersen, P. (2000). A bibliometric study of the publication patterns in the sciences of South African scholars 1981-96. *Scientometrics*, *47*(1), 75–93.

Katz, J. A. (2003). The Chronology and Intellectual Trajectory of American Entrepreneurship Education. *Journal of Business Venturing*, *18*(2), 283–30. doi:10.1016/S0883-9026(02)00098-8

Kezar, A., & Eckel, P. (2002). The effect of institutional culture on change strategies in higher education: Universal principles or culturally responsive concepts? *The Journal of Higher Education*, *73*, 435–460.

Kirby, D. A. (2006). Creating entrepreneurial universities in the UK: Applying entrepreneurship theory to practice. *The Journal of Technology Transfer*, *31*(5), 599–603. doi:10.100710961-006-9061-4

Kniazev, E. (2004). University Strategy in Regional Development. In *Entrepreneurialism and The Transformation of Russian Universities: New Trends in Higher Education*. International Institute for Educational Planning.

Knight, F. (1921). *Risk, Uncertainty and Profit*. Boston, MA: Houghton, Mifflin.

Koh, Y. K., & Hatten, S. T. (2002). The Tourism Entrepreneur. *International Journal of Hospitality & Tourism Administration*, *3*(1), 21–48. doi:10.1300/J149v03n01_02

Marshall, A. (1890). *Principles of Economics*. London, UK: Mcmillan and Co.

Marxt, C., & Brunner, C. (2013). Analyzing and improving the national innovation system of highly developed countries—The case of Switzerland. *Technological Forecasting and Social Change, 80*(6), 1035–1049. doi:10.1016/j.techfore.2012.07.008

McClelland, D. (1987). Characteristics of Successful Entrepreneurs. *The Journal of Creative Behavior, 21*(3), 219–233. doi:10.1002/j.2162-6057.1987.tb00479.x

Moed, H. F. (2002). Measuring China's research performance using the Science Citation Index. *Scientometrics, 55*(3), 281–296. doi:10.1023/A:1014812810602

Moment Expo. (2017). Dijital Sanayi Çağı: Endüstri 4.0. Retrieved from http://www.moment-expo.com/dijital-sanayi-cagi-endustri-4-0

National Science Board. (1972). Patterns and perspectives in environmental science.

Odabaşı, Y. (2008). *Meslektaş Girişimci Üniversite ve Yönetimi, Türk Üniversiteleri İçin Bir Model Önerisi.* Girişimcilik Kongresi.

Organization for Economic Co-operation and Development. (2005). *Handbook on Economic Globalization Indicators.* OECD, Paris Organization for Economic Co-operation and Development.

Ortiz-Rivera, L. A., Sanz Casado, E., & Suárez-Balseiro, C. A. (2000). Scientific production in Puerto Rico in science and technology during the period 1990 to 1998. *Scientometrics, 49*(3), 403–418. doi:10.1023/A:1010585522899

Özdem, G., & Sarı, E. (2008). *Yükseköğretimde Yeni Bakış Açılarıyla Birlikte Yeni Kurulan Üniversitelerden Beklenen İşlevler (Giresun Üniversitesi Örneği).* Üniversite ve Toplum.

Porter, M. E. (1980). *Competitive Strategy.* New York: Free Press.

Pritchard, A. (1969). Statistical Bibliography or Bibliometrics. *The Journal of Documentation, 25*, 348–349.

Robertson, I. (2008). Comment: How universities and graduates can thrive. *Independent.* Retrieved from http://www.independent.co.uk/student/career-planning/getting-job/comment-how-universities-and-graduates-can-thrive-918572.html

Rossi, B. (2016). Industry 4.0: top 10 critical trends for 2017. *Information Age.* Retrieved from http://www.information-age.com/123463810-123463810/

Şahin, E. (2010). Teknoloji Transferi Yöntemleri Bağlamında Türkiye'de Yabancı Sermaye Yatırımları ile Üniversite Sanayi İşbirliğinin Gelişimi.

Sahlman, W. A., & Stevenson, H. H. (1991). Introduction. In W. A. Sahlman & H. H. Stevenson (Eds.), *The Entrepreneurial Venture.* Boston, MA: McGraw Hill; . doi:10.2190/Y2NP-MLD5-UQKX-XMNP

Say, J. B. (1803). A Treatise on Political Economy (C.R. Prinsep, trans. and Clement C. Biddle). Library of Economics and Liberty

Schumpeter, J. A. (1934). *The Theory of Economic Development: an Inquiry into Profits, Capital, Credit, Interest, and the Business Cycle.* Cambridge, MA: Harvard University Press.

Schwab, L. (2016). The Fourth Industrial Revolution: what it means, how to respond, Global Agenda. *World Economic Forum*. Retrieved from https://www.weforum.org/agenda/2016/01/the-fourth-industrial-revolution-what-it-means-and-how-to-respond/

Shane, S., & Eckhardt, J. (2003). The individual-opportunity nexus. In Z. J. Acs & D. B. Audretsch (Eds.), *Handbook of Entrepreneurship Research* (pp. 161–194). Dordrecht, NL: Kluwer.

Shapero, A. (1975). The Displaced, Uncomfortable Entrepreneur. *Psychology Today*, *9*(6), 83–88.

Smith, L. (1981). Citation analysis. *Library Trends*, *30*, 83–106.

Storey, D. (1994). *Understanding the small business sector*. London, UK: Routledge.

Tidd, J., Bessant, J., & Pavitt, K. (2005). *Managing innovation: integrating technological, market and organizational change*. UK: John Wiley & Sons Ltd.

Timmons, J. A. (1999). *New venture creation. Entrepreneurship for the 21st century*. Singapore: Irwin McGraw-Hill.

Wickham, P. A. (2001). Strategic Entrepreneurship: A Decision-Making Approach to New Venture Creation and Management. 2nd Edition, Financial Times/Prentice Hall, London

Zerenler, M., Haslioglu, S. B., & Sezgin, M. (2008). Intellectual capital and innovation performance: Empirical evidence in the Turkish automotive supplier. *Journal of Technology Management & Innovation*, *3*(4), 31–41. doi:10.4067/S0718-27242008000200003

Zhao, F. (2006). *Entrepreneurship and Innovations in E-business: An Integrative Perspective*. Hershey, PA: IGI Global; . doi:10.4018/978-1-59140-920-5

ADDITIONAL READING

Barringer, B. R., & Ireland, R. D. (2015). Entrepreneurship: Successfully Launching New Ventures (5th ed.). UK: Pearson.

Butler, S., & Magelli, P. J. (2003). *The ABCs of Entrepreneurship: Entrepreneurial Professors From the World's Top Business Schools on the Fundamentals All Business Professionals Should Know About Entrepreneurship*. Aspatore Books.

Landström, H. (2009). *Pioneers in Entrepreneurship and Small Business Research, International Studies in Entrepreneurship*. Springer.

Schilling, M. (2012). *Strategic Management of Technological Innovation* (4th ed.). New York: McGraw-Hill.

KEY TERMS AND DEFINITIONS

Entrepreneurial University: Entrepreneurial University refers to the university, which is able to survive and adapt in highly complex and uncertain conditions of the environment in which it operates.

Entrepreneurship: The capacity and willingness to develop, organize and manage a business venture along with any of its risks in order to make a profit. The most obvious example of entrepreneurship is the starting of new businesses.

Globalization: Worldwide integration and development in technology and all related areas. Globalization creates a pressure on developing companies/countries to follow rapidly changing technology via technology management processes.

Industry 4.0: Industry 4.0 refers to the fourth Industrial Revolution. After the steam, electricity and IT, this time is Internet that is going to totally change the production / service systems.

Innovation: Something new or different introduced to the business environment to make radical changes and incremental improvements in technology. The process of translating an idea or invention into a good or service that creates value or for which customers will pay.

Intellectual Capital: Collective knowledge of the individuals in an organization or society. This knowledge can be used to produce wealth, multiply output of physical assets, gain competitive advantage, and/or to enhance value of other types of capital. Intellectual capital includes customer capital, human capital, intellectual property, and structural capital.

Strategic Management: Strategic management consists of the analysis, decisions, and actions related to R&D and technological improvements and organizations undertake in order to create and sustain competitive advantages among its rivals in terms of strategic management of technology.

Technology Transfer: The movement of new technology from its creator to a user in other terms from developed companies/countries to developing ones in several methods. In other words, assignment of technological intellectual property developed and generated in one place, to another through legal means such as technology licensing or franchising.

Chapter 4
Technologies for Industry 4.0 Data Solutions

Ander García
Vicomtech, Spain

Ander Arbelaiz
Vicomtech, Spain

Javier Franco
Vicomtech, Spain

Xabier Oregui
Vicomtech, Spain

Bruno Simões
Vicomtech, Spain

Zelmar Etxegoien
Gaindu, Spain

Andoni Bilbao
Gaindu, Spain

ABSTRACT

One of the main objectives of Industry 4.0 (I4.0) is to generate new opportunities based on the convergence of traditionally isolated technologies such as industrial control systems (ICSs) and information and communication technology (ICT). This presents new opportunities to take advantage of ICT technologies to develop new applications and services related to industrial processes. However, there are a variety of requirements and constraints that must be addressed for the attainment of this purpose. Moreover, the large amount of existing technologies and tools that can cope with these requirements makes the definition and selection of a solution a cumbersome task for traditional industrial workers with a non-ICT focused background. This chapter analyses and describes the main requirements and technologies required to provide a data-based Industry 4.0 solution.

DOI: 10.4018/978-1-5225-4936-9.ch004

INTRODUCTION

Traditionally, technologies from ICS (Industrial Control System) and ICT (Information and Communication Technology) have been historically isolated research fields. However, in Industry 4.0 these two domains have been experiencing a strong convergence that is enabling the develop advanced industrial applications.

Nevertheless, this new industrial paradigm is still in an early stage. Thus, knowledge and technologies from ICS and ICT are still isolated within organizations, slowing the adoption of Industry 4.0 applications. This chapter focuses on the technologies behind Industry 4.0 applications, highlighting the point of view from ICT and aiming to clarify their main requirements and related technologies.

The chapter starts introducing main technologies and architectures required to acquire data, communicate it between devices, process it, and design and deploy Industry 4.0 applications.

The chapter continues proposing a classification of the main requirements of Industry 4.0 data applications that directly affect their design and the resources required for their development. These requirements include whether data acquisition and process should be performed in magnitudes of milliseconds; if data volumes generated by the application require a Big Data architecture; if access to the application is remote or local; and if a cloud infrastructure would be available. Then, a classification of industrial application is presented according to the previous requirements.

Finally, an Industry 4.0 application developed for Gaindu Automation is described. The objective of this application is to manage and visualize the state of the stations and machines of production lines, consolidating the concepts and technologies highlighted during the chapter.

MAIN TECHNOLOGIES RELATED TO INDUSTRY 4.0 APPLICATIONS

The revolution of Industry 4.0 is directly related to several ICS and ICT technologies. However, only when all the different actors involved on the industrial manufacturing process have achieved certain degree of evolution on these technologies, it will be possible to push industry standards one step forward.

This section introduces the most important technologies related to the Industry 4.0.

Communication

Different elements and devices operating in industrial environments communicate among them to exchange information and to guarantee their correct working. Thus, adapted communication protocols that enable components to share information have played a fundamental role in the development of productive plants.

Traditionally, communication protocols in ICS technologies were proprietary, based on undisclosed specifications and not interoperable with the devices of any other manufacturer. Therefore, enterprises were locked to solutions and families of hardware of specific manufacturers. Back in those days, security was not an important issue, since the ICS were completely isolated from other networks, such as Internet.

However, in the last years different needs have been identified, such as the interoperability between different devices, remote access, and security risks in the communications. Thus, some standardization attempts have been accomplished to provide open solutions for these needs. Thanks to these attempts, the Industry 4.0 paradigm has a solid base in terms of interoperability and communications.

Among these attempts, two main initiatives have been proposed: OPC-UA, the evolution of the classic OPC, and DDS:

OPC-UA (Open Platform Communications Unified Architecture)

It was created and is currently supported by the OPC Foundation (OPC Foundation, 2017), an international industrial consortium with the objective of specifying a standard way of communication between industrial devices. It is an evolution of its predecessor, OPC (OLE for Process Control).

OPC is a communication standard in control and supervision of industrial processes. OPC was released in 1996 as a solution against the classic problem of proprietary drivers, and nowadays most control system manufacturers include OPC in their products. It is based on Microsoft DCOM technology and offers a common communication interface, which allows individual software components to share and interact with data (Mahnke, Leitner, & Damm, 2009).

OPC communication follows a client-server architecture. The OPC Server is the data source, like a device in an industrial environment, and any number of OPC applications can access to read or write any variable of the server. OPC includes different modules and specifications such as OPC Data Access, Historical Data Access, Alarms & Events and Security.

Unfortunately, OPC was based on Microsoft DCOM, which restricted the solution only to a subset of devices (e.g. devices running Windows). Moreover, it added several vulnerabilities, and limited its performance in distributed environments. On top of that, OPC DA (the module responsible for the instant access to published data) does not have security mechanisms to protect the access to the information.

OPC-UA was created to overcome the mentioned weaknesses of OPC. OPC-UA extended the successful OPC protocol for the acquisition of data, data modelling and the communication between the industrial environment and industrial applications in a secure and trustworthy way (García, Irisarri, Pérez, & Elisabet, 2017). It removed the dependency with the DCOM technology, offering a lot of advantages. Instead of on DCOM, OPC-UA is based in SOA (Service Oriented Architecture), which is much more interoperable and less restricted.

Table 1 represents the differences between classic OPC and OPC-UA

Its main characteristics are the following ones:

- It is a neutral platform, it can work in any operative system
- It covers the requirements demanded nowadays in terms of security and interoperability. However, it is also able to communicate with legacy systems
- It is easy to configure and maintain.
- It is a service-oriented technology.

Table 1. Comparison between OPC Classic and OPC-UA

	OPC Classic (DA)	**OPC-UA**
Architecture	Client/Server based in DCOM	Client/Server Based in SOA
Access Control Mechanisms	No	Yes
Platform	Only Windows	Any Platform
Configuration in restricted environments	Difficult and Restricted	Easy

- It has high performance
- It provides security mechanisms to protect the data

OPC-UA is based on a client server architecture, as the previous OPC DA Classic (Pfrommer, 2017). In this architecture (Figure 1), servers publish information using internal variables, while clients connect to them and can read and write (if they are granted the necessary access rights) the variables they are interested in. To sum up, there is a point to point connection between clients and servers. This discovery service is not automatic, servers must be identified a priori to enable the possibility of establishing a connection, making the configuration of new devices more cumbersome, since programming work is required.

In terms of Data Modelling, OPC-UA is Object Oriented. Therefore, each server maintains a hierarchical data address inside them, which models the information the server is publishing. This makes possible to browse all the information in the address space of a server if it allows it.

The main idea is to map the signal from the sensors, the controls and other values with variables in the address space. Apart from data, OPC-UA can offer services based in calculations over the published data. For example, the access to the data is made using requests to the server, which is responsible for the answering this request. It is also possible to subscribe to events of a server or to data changes, in order to read the updated values of the variables from the clients.

In terms of Security, OPC-UA includes in its specification a security model to secure the communication between clients and servers. It provides the way to build a secure channel between extremes and mechanisms for authentication of the clients and servers. OPC-UA is usually built over TCP-IP messages. It provides good results with a low latency considering the restrictions of TCP-IP and its default payload.

Recently, OPC-UA has been recommended by the Plattform Industrie 4.0, sponsored by the German Government, as the communication protocol within the RAMI 4.0 (Reference Architecture for Industry 4.0) architecture for industrial applications (Schleipen, Gilani, Bischoff, & Pfrommer, 2016).

Figure 1. Client/Server architecture OPC-UA

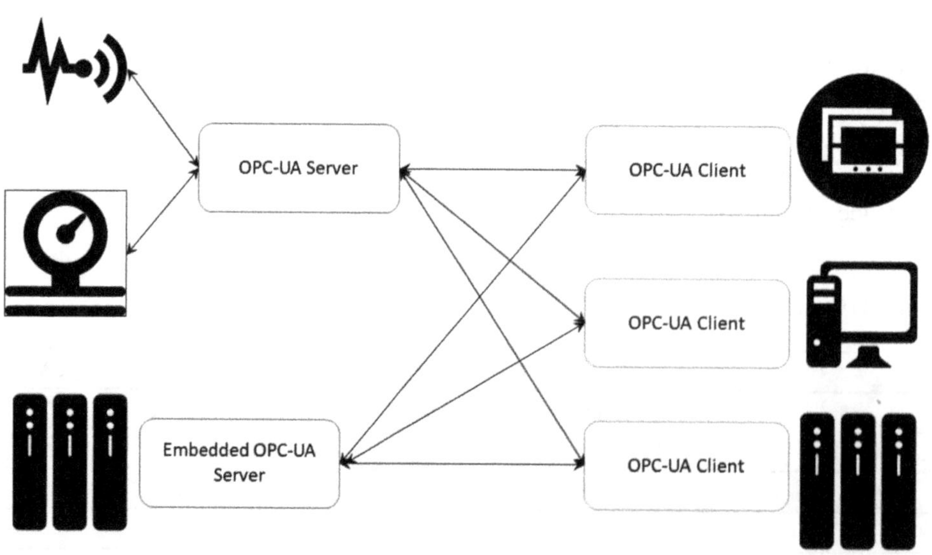

DDS (Data-Distribution Service for Real-Time Systems)

DDS is the specification of a public/subscribe middleware in distributed computation. It has been developed and sponsored by the OMG (Object Management Group) (DDS, 2017). It has been devised for its performance in the IoT (Internet of Things) world, using the data centricity as a principle across a network infrastructure (Köksal & Tekinerdogan, 2017).

As a middleware, it can work in any platform or operative system and it can be implemented in any language. Unlike OPC-UA it is not based in a client server architecture, but in a public/subscribe architecture with data centricity as its main principle. DDS works in a similar way to existing messaging approaches such as MQTT (MQTT, 2017).

This means that components taking part in a classic DDS architecture are going to be Writers, Readers and topics inside a network infrastructure (Figure 2). From this point of view, the DDS Writers are responsible of publishing data and injecting it to the network using a topic, which categorizes the data. In the other extreme, there are DDS Readers. These readers can subscribe to the topics in which they are interested, and they will receive the information when new data is available. With DDS, it is not necessary to manage a point to point connection. Moreover, DDS makes possible to discover new applications and topics in an automatic way, adding new plug & play elements of data without making changes in the configuration (Al-Madani & Ali, 2017).

The data centricity architecture provides several advantages:

- **Writer and Readers Are Decoupled:** Components in both extremes can be added or removed without significant changes in the configuration.
- **It Is Easy to Integrate:** It offers a discovery interface and there is no need of infrastructure knowledge to integrate new devices.
- **It Is Decentralized:** There is no need to manage point to point connections.

Figure 2. DDS architecture based in data centricity

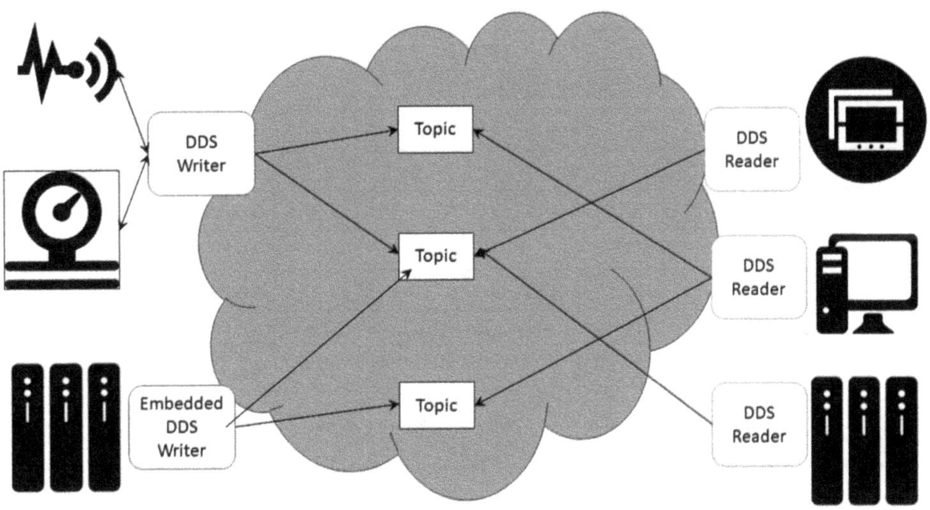

Moreover, DDS offers a very low latency and real time QoS, per benchmarks performance by RTI is in the order of magnitude of sub-milliseconds (RTI, 2017). This is possible thanks to its very light protocol, similar to an extended messaging protocol.

The data modelling in DDS, unlike OPC-UA, is relational. Thus, it is possible to map the published data in DDS to an ER (Entity Relationship) diagram. Moreover, this also makes integrations with traditional relational databases easier or compatible with SQL style queries to correlate the information.

In terms of security, in the last versions DDS includes the add-on DDS Security, which offers information centric security for the data in motion as well as data in rest. In this case, this module includes support for authentication of the readers/writers, cryptography to make de communication confidential, as well as access control.

DDS and OPC-UA

OPC-UA and DDS are two valid options that are already being deployed to apply the Industry 4.0 paradigm in productive environments. Although they are very different, both offer a very good performance, flexibility to define and publish information, interoperability, robustness, security and are easily configured (DDS, 2017).

It can be said that OPC-UA and DDS are different approaches to the same issue (DDS and OPC UA, 2017). Table 2 highlights the main relevant differences between them, which have been described by the previous paragraphs. However, main bodies supporting OPC-UA and DDS are working together to increase their interoperability.

Data Acquisition

Data acquisition is another one of the key components of the Industry 4.0. Each I4.0 application starts having access to data related to the production process. The information about how the machine is physically behaving and the different status of the machines that belong to a production process must

Table 2. Comparison between OPC-UA and DDS

	OPC-UA	DDS
Architecture	Centralized, based in Client-Server	Data Centricity
Data Model	Object Orientation, Possible to Browse	Relational Data Model
Services	Yes	No
QoS	No	Yes
Discover Service	Manual	Automatic
Performance	High (over TCP-IP)	Very High
Security	Access Control, Authentication and confidentiality (secure channels) include natively	Using an add-on: it provides Access Control, Confidentiality and Authentication
Publish-Subscribe	Possible to subscribe to data changes and events of the server	Natively included

be detected and controlled. In order to do so, both devices that can turn the physical phenomena into an electric signal that can be measured (sensors); and the information regarding the status of a machine, usually defined by the input and the output of the PLC (Programmable Logic Controller), are required.

However, retrieving this information directly from the sensors and the controllers is not optimal. These devices do not have the capability to make this information easily readable to other components. Thus, devices that can gather the information and transmit it to other devices performing further processing operations are required.

On the hazardous industrial environments, durable devices that can be integrated near the measured machine and perform different set of tasks are required. These devices are known as CPS, (Cyber-Physical Systems), or CPPS (Cyber-Physical Production Systems) within the industrial environment, and have become the main resource for data acquisition on very different kinds of environments. This section focuses on the CPSs.

CPSs (Cyber-Physical Systems)

The Cyber-Physical Systems (CPS) are new generation devices of intelligent ICT embedded systems that are interconnected, interdependent, collaboratives, autonomous and provide computational, communication, monitorization, process and cyber-physical control capacity. They are defined as transformative technologies for managing interconnected systems between its physical assets and computational capabilities. However, one of the most important feature provided by this kind of component it's their use as a high speed acquisition devices (Baheti & Gill, 2011; Yang & Zhou, 2013).

CPSs allow to integrate computation and physical processes, which generates a big number of possibilities in multiple sectors (manufacturing, automation, health, construction, aeronautical, railroad, energy, smart cities…). By definition, they can perform several tasks as integrators, as information generators or as actuators. They can be applied to generate intelligent processes and products, designed and manufactured in an intelligent way. Based on CPSs, the service they offer and the analysis of the generated data, it is possible to optimize the resources consumed during the production process, or develop new products or variations of existing products.

Thus, CPSs are considered as a Key Enabling Technology, engine to the innovation that can lead to the appearance of new productive mechanism, smarter products and improvements on the quality and the efficiency of the production, the birth of new business models or the aperture of new markets.

It is necessary to differentiate the value proposal through the innovation and technology, with a philosophy of advanced manufacturing, speeding up the cycles of technology transference regarding significant advances on R&D (Research and Development). There are global tendencies, in economically advanced countries with high industrial GDP such as Germany, EEUU or Japan, that pay more attention to productive processes and to the integration in a global and narrow way the advances in production technologies and in ICT technologies. Nowadays, the main part of this global tendency is the introduction of new possibilities for the CPS.

The introduction of this new technology is giving place to, what in some places such as the strategic German project Industry 4.0 is beginning to be known as, the fourth industrial revolution. According to this theory, the first three revolution came as the result of the mechanization, the electricity and the information technologies.

The application of the cyber physical systems in the manufacturing industry is known as the Cyber-Physical Production Systems (CPPS) (Lee, Bagheri, & Kao, 2015). The CPPS include:

- Smart machines
- Storing systems able to manage a massive load of data
- Installations a production plants that have a digital representation in a natural way and that include complete integration (end to end) based on ICTs and that are able to exchange information, trigger actions and control each other in an autonomous way.

Contrary to traditional embedded systems, formed by isolated devices, a CPSS system is designed as a network formed by interconnected interactive elements, with physical inputs and outputs. CPPSs interact with the physical world and must operate in a safe, efficient and, usually, on a real-time way. Moreover, they allow saving and processing the process information in real-time and detect tendencies and useful patterns in order to reconfigure the production. Not only they optimize existing processes based on TICs, but they can take advantage of the potential of the information obtained through a differentiated traceability, both in the detailed processes and the global scale effects on the production. This was previously impossible to store and manage in a connected and a cooperative way.

CPSs, and CPPSs in a more specific way, provide a solution that can solve most of the requirements of an industrial application. They need to be designed, developed and installed in a customized way depending on each situation, plant or machine. Thus, there are several alternatives with different functionalities and target prices, from low-cost devices such as an eWon (Ewon, 2017) to devices such as Compact RIO systems (Compact RIO, 2017), which provide powerful real-time calculation power at a much higher price tag.

Architecture Patterns

Regarding Industry 4.0 application architecture, as user requirements strongly diverse at each enterprise and scenario, it is difficult to define a "one size fits all" architecture. Thus, this section describes not only traditional architectures based on monolithic applications, but also real-time Big Data architectures, and referential architectures for Industrial applications.

The objective is to introduce main concepts related both to a Big Data Architecture, which can be considered transversal to the sector and are directly to the ICT world; and to an Industrial Reference Architecture, which is directly related with the industrial sector and is focused in the convergence between ICS and ICT worlds. This section summarises the most relevant aspects of these two architectures.

Real-Time Big Data Architectures

In the last few years the Big Data concept has come on the scene to approach the acquisition, processing and storage of large volumes of data. However, it is a complex concept which can have several interpretations.

The first definition of Big Data was given by the META Group (Currently Gartner), and it identifies the opportunities behind the increasing volume of data characterizing this data using the 3Vs: Volume, variety and velocity. This data also presents the need of new processing patterns to address the overload of data for the improvement in the decision-making process, the discover of new visions of the information and the process optimization (Laney, 2001). Later, the same group identified the need of the fourth V: Veracity, showing the need of trust in the data as an explicit requirement (Beyer & Laney, 2012). Recently new Vs are being added such as visualization or value, that are more related with the final functionality of the big data than the characteristics of the processing performed with the data.

While the way in which big data processing is executed has been varying during the years, three main tendencies have been identified:

- **Batch Processing:** It is the first approach for big data processing. It is based on the processing of static batches of great volumes of data. This model addresses mainly the problem related with the volume of the data. The technologies used to build this model are scalable, distributed, based in parallel computation. Their main disadvantage is that they present a high latency.
- **Stream Processing:** Given that the batch processing is limited (processing of finite datasets and high latency), in 2010 the real time Big Data processing model was proposed. Unlike the previous case, this model addresses mainly the velocity. The technologies which supported this model allow to process an incessant stream of data in real-time, with a very low latency and based in a distributed and fault tolerant architecture.
- **Hybrid Computational Model:** It is based in the combination of the two previous models.

Related to these tendencies, some reference architectures emerged to apply big data technologies in the best possible way. Mainly, there are two reference architectures which are nowadays being dominating the market: Lambda architecture and Kappa architecture.

Lambda Architecture

Lambda has been the reference for Big Data processing, becoming the de facto Big Data architecture for years. Lambda (Marz, 2012) was defined as a generic, scalable and fault tolerant architecture for Big Data processing based in the hybrid computational model.

Lambda architecture (Figure 3) has several layers to separate and differentiate the different process models applied to data. In this case the acquired new data is processed by a speed layer (based in stream processing and which enables the access to a real-time view of the data); and by a batch layer in which static datasets of data are stored. This centralized combination allows the users to access the data in different states using queries able to access to precomputed views of the data. The stages identified by the Lambda architecture are the following ones:

1. Acquired data is sent to both the batch layer and the speed layer.
2. The batch layer manages a master dataset with raw data in which new data is being added to allow its use in new analysis and the precomputation of batch views.
3. The serving layer indexes the batch views in such a way that they are accessible using queries with low latency.
4. The speed layer complements the high latency of update in the batch views with recent data (processed in real-time).
5. All the data is accessible using ad-hoc queries.

Figure 3. Lambda architecture

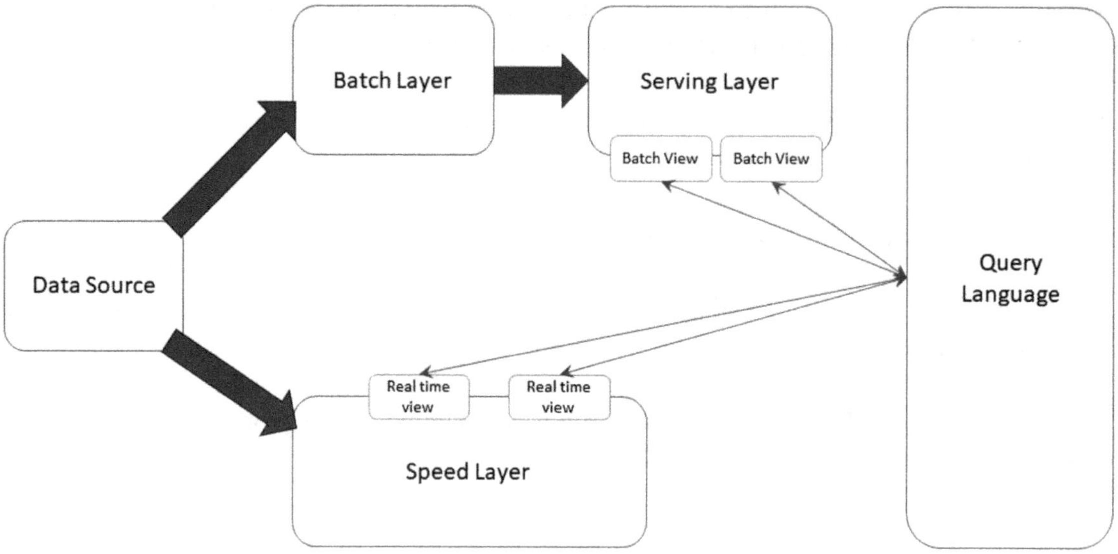

Kappa Architecture

In 2014, some doubts and supposed weaknesses about the Lambda architecture and hybrid processing model are posed (Kreps, 2014):

- The code maintaining become complicated since the developers need to maintain two complex infrastructures in parallel. The test and depuration of the code of the architecture is hard and tricky.
- To avoid this complexity, some abstraction tools have appeared with the objective of coding once for the two models of computation (batch and streaming). Nevertheless, this programming paradigms are opposed and they are in constant evolution. This fact makes the adaptation of this abstraction codes very difficult for all the environments.
- It is argued that lambda architecture surpasses the CAP theorem, which says that storage systems only can fulfil 2 of 3 characteristics between Consistency, Availability and Partition Tolerance. However, lambda is an asynchronous processing architecture, and for example the data which is being computing in batch is not consistent with the rest immediately.
- Streaming processing near to real-time is as powerful and precise as batch processing.

Based on these statements, Kreps proposed an alternative to lambda architecture, he called it Kappa Architecture (Figure 4). Kappa proposed to avoid completely the batch processing layer, executing all the processing using real-time processing models (streaming). He argues that it's not true that the batch processing is more efficient for big volumes of historical data and thinks that these opinions are biased or based in the limitations of the tested systems or in the fact that the systems were badly scaled.

So, what Kappa Architecture offers are the following points:

Figure 4 Kappa architecture

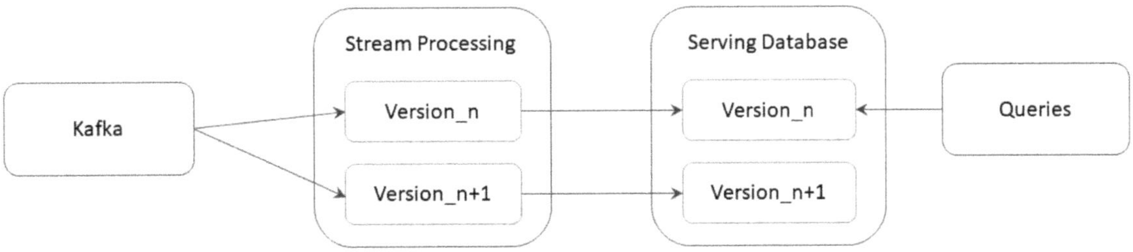

- Use Kafka or other technology which allows to retain the completeness in case of need reprocessing for the distribution of the data, allowing multiple subscriptions.
- In case of need reprocessing, start a second instance of streaming processing from the beginning with the retained data.
- Change the processing to the new instance when it reaches the desired results, changing the external data access to the new results.
- Stop the previous processing instance and delete old entries.

Referential Architectures for Industrial Applications

The rise of the concept of Industry 4.0 has launched a new paradigm for industrial environments, making possible to generate value from analysis of the data generated by productive processes. The convergence of ICS and ICT technologies is a requirement that must be met to fulfil this purpose.

Given that Industry 4.0 responds to a strategy devised by the German Government, and since it implies a lot of changes in the value chain and in the communications of the different stakeholders, it is necessary to agree on a common basis. The objective is to ensure the alignment of the different agents and stakeholders in the process and to define several concepts and ideas which will be very helpful to implement the strategy.

In this context, various Referential Architectures have been proposed to fulfil this requirement. The two main proposals that are being developed are RAMI 4.0 in Europe and IIRA in United States.

RAMI 4.0

RAMI was launched by the "Platfform Industrie 4.0" in Germany and was presented in the Hannover Messe of the year 2015 (Hankel & Rexroth, 2015). The main objective is to connect production environments and communication technology and take advantage of the connection using the data from the industrial processes. Furthermore, the development of RAMI has been driven by new devices that allow the communication between machines without human intervention, making this convergence easier. RAMI 4.0 is basically focused in Manufacturing Sector and it has been devised for it.

In the case of RAMI 4.0 it has been drawn by defining three axes of integration: hierarchy, life cycle and architecture layers (Plattform Induistrie 4.0, 2016).

Hierarchy Axis

Hierarchy axis can be considered an improvement of the automation pyramid defined by ISA 95, in which five different levels (Production, Sensors and actuators and Control, Manufacturing Operations, Business Planning and Enterprise Planning) exist (Figure 5).

In this case, and with the aim of including the Industry 4.0 key concepts, two more levels have been added to this pyramid. Firstly, below the lower level of the ISA 95 pyramid a "Product" level has been included. This level is very important in the Industry 4.0 paradigm since a product can also be smart and connected.

Secondly, above the Enterprise planning level the Connected World level appears. it marks that the infrastructure of an automation production environment will not be isolated anymore, allowing the connection with the cyberspace and allowing the creation of new services related with it.

Life Cycle and Value Stream Axis

This axis represents the life cycle of facilities and products. It is based on IEC 62890 for lifecycle management, and it makes the distinction between "types" and "instances". For RAMI "type" represents a prototype and it becomes an "instance" when the design of the type has been finished and the product is in manufacturing process) (Figure 6).

Figure 5. ISA 95 automation pyramid vs hierarchy Axis in RAMI 4.0

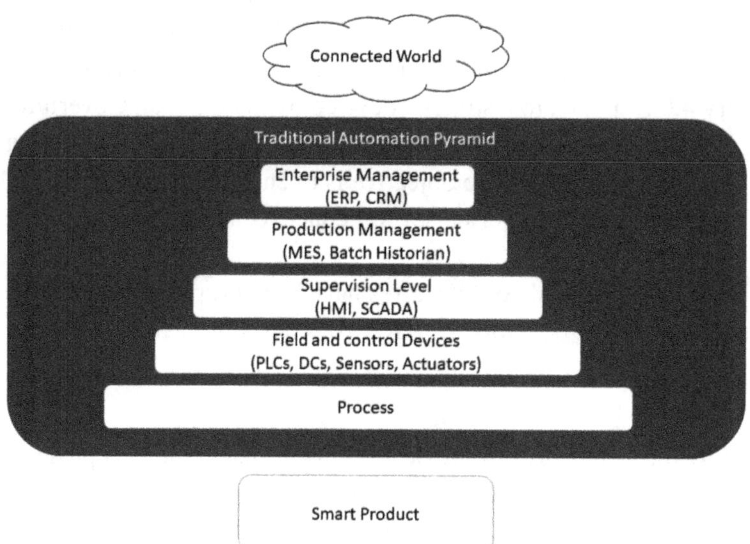

Figure 6. Product life cycle on RAMI 4.0

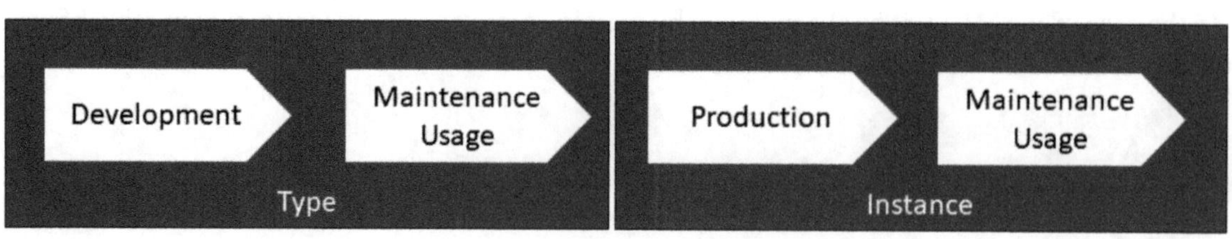

The first stage of a new idea is the development of the type. This stage includes development, construction, computer simulation and finally the prototype of the product. During this process a type maintenance has to be done. This maintenance includes software updates, instruction manuals or maintenance cycles, among others.

On the other hand, when a prototype becomes a regular product the type happens to be an instance, and a production stage must be accomplished. The production stage approaches the traceability of the product and collect data from it, identifying it with a serial number. In the case of an instance, it is also necessary to look at a maintenance and usage stage. Within this stage issues such as measurement, usage, service, maintenance, recycling or scrapping should be addressed.

Architecture Layer Axis

These layers represent the vertical axis and the different architecture layers from things (in this case represented by physical devices) and business process which is facilitated by the Industry 4.0 environment.

The different layers are the following ones:

- **Asset:** It represents the physical things and devices which exists in real world.
- **Transition:** Is the layer responsible from manage the transition from real to digital world, enabling the data from the asset
- **Communication:** It manages the access to the information and the communication between the different modules and elements. In the case of RAMI 4.0, OPC-UA is the unique standard defined as a communication protocol.
- **Information:** Represents the raw data extracted from the asset.
- **Functional:** This module is responsible for the functionalities offered by the asset. It can be used to servitization inside the industrial environments.
- **Business:** This module represents the organization and business processes inside the Industry 4.0 architecture.

IIRA

The IIRA or Industrial Internet Reference Architecture (IIC, 2015) is the Reference Architecture of the IIC (Industrial Internet Consortium) which is a consortium set up by several Large Multinationals (AT&T, Cisco, General Electric, IBM, and Intel) and it is considered as the reference Architecture in USA. Unlike RAMI 4.0, IIRA has a multisector orientation (Energy, Healthcare, Manufacturing, Public Domain or Transportation are the key sectors).

In the case of IIRA there is a distinction between the different viewpoints identified in the process (from low level to high level):

- Implementation viewpoint
- Functional viewpoint
- Usage viewpoint
- Business viewpoint

Each of these viewpoints is associated with a category. The processes and analysis in each viewpoint should be adjusted to it, existing a very close relation between them. While the design of the industrial applications build should be guided from business viewpoint to lower viewpoints; on the other hand, from the implementation viewpoint, the validation and revisions are going to elevate from implementation viewpoint to higher levels.

On another dimension, functional domains have been defined. IIRA defines a set of functional domains which interact with a physical system, shown by Figure 7. Control is located on the next level, which includes the sense and the actuation on the physical system. In the following level the categories are operations, information, application, and finally the business level. Between this functional domain, there is a constant flow of information in order to apply an interconnected and efficient industrial architecture.

IIRA also introduced the following concepts of cross-cutting functions:

- Connectivity
- Distributed data management
- Industrial analytics
- Intelligent and resilient control

Figure 7. IIRA functional domains

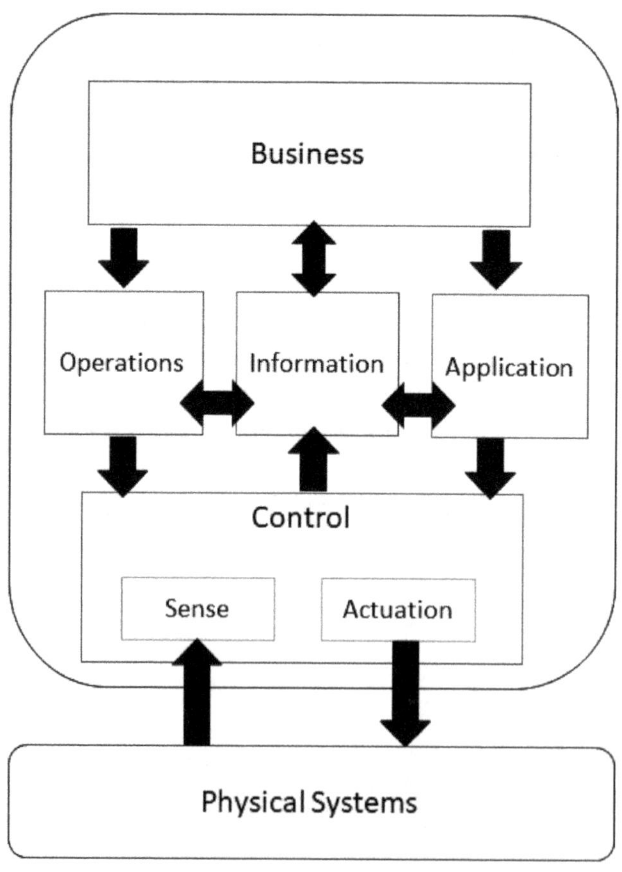

The basis of IIRA is the same one from RAMI 4.0: to build a complete architecture capable of offering added value from the data. Thus, connectivity is fundamental to guarantee the access to the generated information.

Since the data volume generated by industrial applications is overseen to exceed the capabilities of a single computer, it is important to have a ready-for-service infrastructure based in Big Data and Distributed Data Management. This will allow to build a scalable and reliant system, given that the fall of one or the systems will not provoke the fall of the general infrastructure and the optimization of this kind of systems.

The collection of all the data generated in the industrial environment will enable its exploitation. This information could be exploited with the objective of improving the industrial process using industrial analytics. The production aspects to be improved can cover the productivity, the predictive maintenance or the optimization of resources among others. These analytics can be distributed in three major categories, such as descriptive analytics, predictive analytics, and prescriptive analytics.

Finally, IIRA includes the intelligent and resilient control. The objective is to control the system not only based in a fixed control, but considering several parameters which can be variable with the context. This implies to have a complete vision of the world from the model, and to consider deterministic or probabilistic models depending on the situation. In conclusion, to build an adaptable control system depending on the situation and on the recollected parameters.

On the other hand, IIRA sets several system characteristics to be trustworthy, resilient and scalable that should be fulfilled by industrial applications:

- Safety
- Security
- Resilience
- Reliability
- Privacy
- Scalability

Local vs. Cloud Infrastructure

Last but not least, the manufacturing industries will require compute and storage capacity for processing and storing the huge quantity of data they are generating. Hence, they will be collaborating with external resources that can perform a considerable part of the data processing tasks.

The design of strategies for deploying computing and storage infrastructures that enable the manufacturing industry to process and store the huge quantity of data they generate is not self-evident. On the one hand, production lines are built with components that live at the edge of network and are basic in terms of hardware. Additionally, the physical ecosystem encapsulates operational machines that raise restrictive constraints, such as limited power energy, low power communications, unreliable communications, and are further constrained by requirements like factories spanning over a vast geographical area, arranged in various configurations, and cover a plethora of distinct use cases. On the other hand, deciding the location of computational and storage resources that can handle data processing tasks poses many non-trivial questions to the designers of the platform because of stringent applicational requirements, such as high throughput during short time periods, and the need for very low latency and real-time analytics.

Solutions based on cloud computing have proved to facilitate the flow within ecosystems characterized by a very high heterogeneity of devices, technologies, and protocols, as well as to enable rapid setup and integration of new things, while maintaining low costs for deployment and for complex data processing (Rao, Saluia, Sharma, Mittal, & Sharma, 2012). The benefits of adopting cloud extend also to the possibility of analyzing data of unprecedented complexity and volume (Dash, Mohapatra, & Pattnaik, 2010; Parwekar, 2011) at low cost, thus delivering means for increasing revenues and reduced risks (Zaslavsky, Perera, & Georgakopoulos). The architecture of cloud can be split into four layers: data centers (hardware), infrastructure, platform, and application (Zhang, Cheng, & Boutaba, 2011). Layers are designed to function as a service for the layer above and as a consumer for the layer below. However, this type of architecture might not be suitable for latency-sensitive applications, which require nodes in the vicinity to meet their delay requirements. In addition to the problem of geo-distribution and low latency, the situation worsens when these "things" have mobility capabilities.

Edge computing, previously named as fog computing is "a highly-virtualized platform that provides storage, compute, and networking services between end devices and traditional cloud computing data centers, typically, but not exclusively located at the edge of network." (Bonomi, Connected vehicles, the internet of things, and fog computing, 2011). Edge and fog computing enable applications to run closer to sensed data by pushing intelligence strategies down to the local network architecture, and by performing data processing within the IoT gateway. Edge computing can push these limits even further by bringing the intelligence to the processing power and communication capabilities of an edge gateway or directly into devices like programmable automation controllers (PACs). The above computing architectures provide data, computation, storage and application services to end-users, but the former can be distinguished from the later by its proximity to end-users, and by its support for mobility, while later focus on providing global centralization (Bonomi, Milito, Zhu, & Addepalli, 2012).

Future industrial applications might require both fog and edge support for localization and mobility, and cloud for globalization, particularly when analytics and multi-latency systems are involved. Finding an optimal balance between cloud, fog and local resources is still an open challenge. For further details, see surveys (Mao, You, Zhang, & Huang, 2017; Yi, Li, & Li, 2015; Mahmud, Kotagiri, & Buyya, 2018) that provide an exhaustive overview of existent applications of Edge, Fog and Cloud computing.

INDUSTRIAL REQUIREMENTS AND SCENARIOS FOR DATA APPLICATIONS

This section presents a classification proposal for industrial data applications. This classification is based on three scenarios related to the main requirements of these type of industrial applications. These requirements are focused on real-time constraints, data volume, final application constraints, and underlying IT infrastructure.

Real-Time Constrains

While some scenarios require to acquire data from several sensors in real-time at a millisecond frequency, others only require data being updated few times per second or minute. Furthermore, in some scenarios, data should be processed in real-time to be relevant, whereas in others, data can be analyzed offline to generate relevant information about a process.

Real-time constraints may be grouped within three general categories for industrial applications: milliseconds for real-time applications; seconds or few minutes for near real-time applications; and offline (several minutes, hours, days), for the rest of the applications.

Data acquisition constraints at rates within milliseconds magnitude are mainly related to sensors of production machinery, such as vibrations or speed of engines. In this case, time must be synchronized across the network using specialized protocols such as TSN (Time-Sensitive Networking, IEEE 802.1), which guarantee that all devices that are participating in real-time communication have a common understanding of time.

Lower rates, within seconds or few minutes of magnitude, are mainly related to physical or production variables changing at a similar rate, such as temperature, measurement of parts. Finally, data acquired at lower rates (hourly, daily, weekly, etc.) is mainly generated from existing data to summary or report the status of the production processes.

In these cases, data process time constraints could be related both to milliseconds, for example to react to dangerous situations; seconds, for example to adjusts production speeds of different machines to optimize the production flow, or lower rates, for example to be used within advanced predictive maintenance algorithms.

Data process could also be divided in several steps, starting for example with a fast real-time acquisition, storing data on a database and performing a fast preprocess to detect and react to unwanted values. Then, this real-time data could be aggregated at a lower rate, for example counting the times an event is triggered, to feed different data processes (generate reports, predictive maintenance, etc.).

Data Volume

The volume of data generated by I4.0 applications is directly related to three main factors. Firstly, the number of variables acquired from industrial processes. Secondly, the acquisition rate of these variables. Thirdly, the rate at which historical data is purged or aggregated.

These three factors have to reach a compromise and have to be aligned with the requirements of the industrial application to be developed. Data volume has a direct impact not only on the cost of the final solution (more storage space, faster networks, etc.), but also on the scalability and response times of the applications. Purging historical data, or aggregating data at lower rates for older data, is a compromise solution that is compatible with the requirements of most industrial use cases.

Data volume constraints may be grouped within two general categories for industrial applications. Even if there is not a defined frontier, up to several Terabytes data volume constraints should fit on one computer´s disk. Higher volume constraints should be considered as big data applications and require a distributed architecture.

Application Constrains

The constraints imposed by industrial applications should lead the design of an I4.0 solution. For example, within an industrial process where PLCs already react in real-time to key sensor data, even if the data acquisition may be performed in real-time, probably the data process would not have such a constraint. Moreover, if the application is targeting reporting or tracing functionalities, sensor data could be aggregated at lower rates without affecting the final functionality.

Within these group of constraints, remote access to the application is a relevant constraint that directly impacts its design and cost. When final users may connect to the application from outside the internal network of the plant or the company, further functionalities and investments are required from the IT infrastructure to not affect the security of the company or the response time of the application.

IT Infrastructure

Even if usually constraints imposed by the IT infrastructure can be bypassed, this may involve costs and business risks that pose challenges to companies, mainly SMEs. One of the first constraints is related to the speed of the Internet connection. Depending on the location of the company, faster connections may not be available at affordable prices. When this is the case, or even when there is no Internet connection at all, the application itself has to guarantee the connection between each component. In these cases, either all the components are within the same LAN or, more commonly, the application is self-contained in a computer located close to the production machines.

A critical infrastructure constraint is related to the security of the network and the application. A properly configured IT environment should isolate the different networks of a company (operations, engineering…) and provide the means to limit and supervise the connections made between networks, reducing the risk of a cyber-attack.

Finally, the use of traditional local servers or the availability of a cloud (either public, private or a fog), impose constraints that affect other aspects such as the volume of data that could be stored and accessed, or the rate at which data can be acquired and processed.

As the security is a must have requirement to be fulfilled for any I4.0 application, and connectivity is widespread in the majority of locations where I4.0 are being deployed, within this section applications would be categorized in two groups, ones running on any type of cloud, and one running on local servers.

Classification and Categorization

Following the previous requirements, Industry 4.0 data applications could be categorized in three main groups (basic, medium and advanced), as shown by Table 3.

Basic applications, which comprise most of the I4.0 applications that are being developed nowadays, are not targeting neither real-time data acquisition nor process, which are performed by traditional industrial equipment (PLCs, SCADAs…). These applications, which manage a volume of data fitting on a server, are not hosted on a cloud (public or private) and do not provide a direct remote access.

Table 3. Classification of Industry 4.0 applications

Group\ Requirement	RT Acquisition	RT Process	Big Data	Remote	Cloud
Basic	No	No	No	No	No
Medium	Yes	No	No	Yes	No
Advanced	Yes	Yes	Yes	Yes	Yes

Medium applications, which are already being developed by some companies, include real-time data acquisition and allow users to connect remotely without compromising the security or the response time of the application.

Finally, advanced industrial applications, mainly found at big industrial enterprises, take full advantage of I4.0 technologies and fulfill the previous constraints.

Each group of application does not require the integration of all technologies introduced on the previous Section. Thus, Table 4 presents a general guideline of the technologies that are more suitable for each of the application groups. This table could be considered as a starting guide for a technology decision process, it does not enforce any technology. Its purpose is to provide a general reference, but it does not enforce a specific technology.

REAL USE CASE: GAINDU AUTOMATION

Introduction

Gaindu S.L. is composed by professionals dedicated to the demanding field of industrial automation. Terms like concept development and design make up GAINDU's identity and together with its experience are a guarantee for the integration of assembly technologies and turnkey installations of industrial automation systems.

Gaindu has the three main areas of its manufacturing process:

- Complete assembly machines and assembly lines of automotive components.
- Robotized cells for loading / unloading components into machines.
- Automation of complete machining lines.

Table 4. Relation between technologies and application groups

Technology\Group		Basic	Medium	Advanced
Com.	**OPC UA**	No	Yes	Yes
	DDS	No	No	Yes
CPSs		No	No	Yes
Big Data Architectures	**Lambda**	No	Yes	Yes
	Kappa	No	No	Yes
Ref Architectures	**RAMI 4.0**	No	No	Yes
	IIRA	No	No	Yes
Infrastructure	**Local**	Yes	Yes	No
	Cloud	No	Yes	Yes
	Edge	No	No	Yes

The foundations for its activity lie within the areas of engineering, manufacture/integration and assembly/installation of industrial automation systems on a turnkey basis.

The most representative sectors of Gaindu activity are:

- **Machine Tool:** Automation of machine loading / unloading and realization of ancillary operations.
- **Renewable Energies:** Definition and development of the process required for manufacturing energy generators.
- **Aeronautics:** Cutting machines for composite materials and manufacture of structural components in light materials.
- **Electromedicine:** Machining and assembly of parts for machines used in medicine.
- **Nuclear:** Specific machines for handling and control of nuclear fuel.

Traditionally Gaindu has provide solutions based on ICSs, where customs applications where developed with traceability purposes for each of its installations. This working solution presents relevant drawbacks related to scalability, code manageability, and reusability. Moreover, adapting this solution for each installation is a time-consuming task.

Thus, following an Industry 4.0 approach, Gaindu has decided to develop an updated application to monitor, visualize and manage the state of production lines. Since this application has been the first attempt to include I4.0 technologies, its requirements are suited with basic applications defined earlier: acquisition and process times in near real-time, data volume manageable by one computer, no direct remote access to the application, and deployment on a self-contained computer. Table 5 summarises the main technologies integrated on this application. This solution runs on an industrial PC assembled with the production line. This PC has a limited amount of resources (CPU, RAM and hard disk), which have influence the selection of software components described on the following sections.

The scenario defined by Gaindu is aligned with the assembly lines it currently designs, develops and deploys. The objective of the application is to manage, store and visualize the state of production lines. The application targets four different user roles: production manager, line manager, machine manager and station manager.

Table 5. Technologies integrated on the prototype

Technology		Gaindu
Com.	OPC UA	Yes
	DDS	No
CPSs		No
Big Data Architectures	Lambda	No
	Kappa	No
Ref Architectures	RAMI 4.0	No
	IIRA	No
Infrastructure	Local	Yes
	Cloud	No
	Edge	No

The production manager has access to information related both to any station, machine or line; and to the production process as a whole. Line managers have access to information related to the lines they manage and their stations or machines. Finally, machine managers have access to information of their machines and their stations; and station managers can only access information about their stations.

The application must fulfil the following requirements:

- Visualize information on a control panel located next to a station, machine or line
- Launch predefined actions base on rules analysing data related to the production line. For example, generating an alarm or sending an email when a parameter goes beyond a threshold
- Store both (i) relevant data and aggregated data on a database, and (ii) no directly relevant data, for historical and analysis purposes
- Communicate with external systems such as MES, cloud systems…

Description of the Application

The architecture of the application, presented by Figure 8, has been designed to be flexible and configurable to adapt to specific characteristics of each client of Gaindu. Although this application has been categorized as a basic application, the architecture integrates several components that could be distributed and could also be valid for medium applications, and even advanced applications with real-time constraints updating at rates as high as hundreds of milliseconds.

The main components of the application and the architecture are the following:

- **OPC-UA Server (SIMATIC.NET):** The PLC directly communicates with the SIMATIC.NET (Siemens, 2017) software of Siemens running on an industrial PC. This software has access to the variables of the PLC and creates an OPC-UA server that publishes them. Both each state of the stations, machines and lines; and events and alarms generated by the PLC, are published by this OPC-UA server.

Figure 8. Architecture of the application

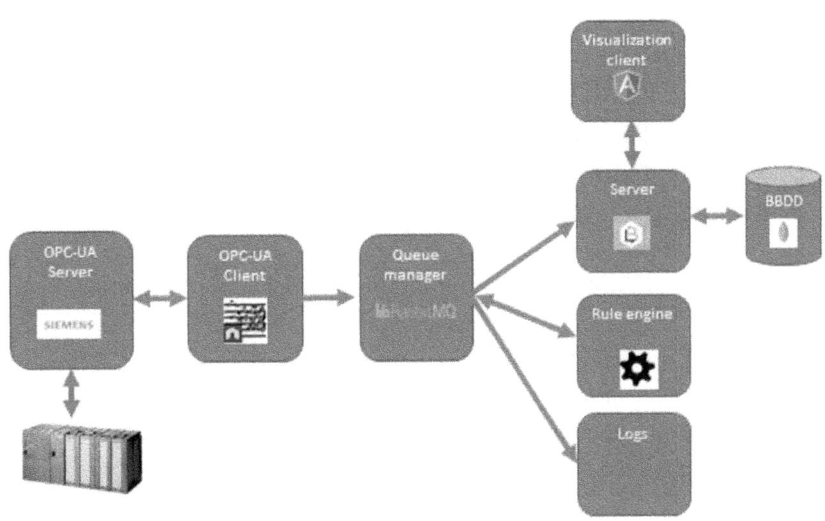

- **OPC UA Client (Nodejs):** The OPC-UA client connects to the OPC-UA server and subscribes to the previous variables. Each time this client receives an update of a variable, it wrappers it as a message that is send to the queue manager. This OPC-UA client has been developed using an Open Source OPC-UA node.js library, Node OPCUA (Node OPC-UA, 2017). However, any other OPC-UA library or programming language could be used, and only this component should have to be changed.
- **Queue Manager (RabbitMQ):** The queue manager, also called message broker, is responsible of the coordination between the rest of the components. Producer components send messages to the manager. Then, consumer components receive these messages and process them, performing the corresponding action and generating new messages when required. RabbitMQ, the most widely deployed open source message broker, has been selected as the queue manager of the application due to its versatility and proved reliability (RabbitMQ, 2017).
- **Rule Engine (json-rules):** This component receives updates of the states, events and alarms of the stations, machines and lines. Then, it applies rules codified as AND and OR conditions on the values of the updates (Json rules, 2017). For each rule that is activated with the update, the rule engine performs the associated action. For example, when state changes of a machine are received, the rule engine calculates the state of the line based on a logic defined by Gaindu. If this line state is updated, it is send as a new message to the queue manager.

The actions already integrated with the application can be combined and they are triggered producing new messages that are consumed with specialized consumers that perform the associated action. Actions of the application are focused on these tasks:

- Generate new messages
- Generate HTTP POST requests to an external URL API endpoint, for example to notify a failure to the MES of the company.
- Store logs to perform further analysis offline
- Store information on the database
- Send an email

The rule engine of the application is based on json-rules, a lightweight node library that store rules as JSON documents.

- **Database (MongoDB):** The database is responsible of the persistence of the information. The NoSQL database MongoDB has been selected as the database engine due to its flexibility and scalability (MongoDB, 2017). MongoDB stores information as JSON documents with no predefined structure, avoiding to define new schemas for each enterprise the application is deployed. Moreover, its scalability assures a good performance both on standalone and distributed deployments.
- **Application Server (Loopback.io):** The application server (Loopback, 2017) performs the tasks associated with traditional server:
    - Receives POST requests from other components as messages, and stores their information on the database

- Publishes a REST API to give access to the visualization client to the data of the database and the service it requires to properly visualize the information of the stations, lines or machines

The server has been developed using Loopback.io. This library, supported by IBM, is focused on the generation of node.js REST APIs and application backends.

- **Visualization Client (Angular):** This final component generates the Web application that allow final users to interact with the information of the database. This Web application loads the information store on the database using the REST API of the server, and present and visualize it to users.

The client has been developed using Angular (Angular, 2017), an updated version of a Web library supported by Google to develop Web applications.

One of the most relevant component of the architecture is the queue manager. This manager acts as a gateway between components, making them independent and lousy coupled. Thus, any component could be changed or updated without affecting the rest of them. The queue manager integrated on the architecture has been RabbitMQ. RabbitMQ supports several communication patterns to connect message producers to message consumers (workers). A basic patter consists on an intermediate queue that protects against possible failures or overloads. In more advanced patterns a producer can send a message to several queues, where several consumers receive and process them. Thus, the system can be adapted to the special characteristics of each client of Gaindu, and scale to meet their real-time and data volume requirements increasing the available computational power adding more server or more resources from the cloud. For example, when a worker is being continually overloaded and cannot cope with the message production rate, a new worker can be launched, nearly doubling the process rate of the messages.

Recently, OPC-UA has included a Publish /Subscribe message pattern to its specification. However, this functionality is still being implemented by most OPC-UA libraries and has not been included on this application, integrating RabbitMQ. Once the functionality is included on the OPC-UA library, the application will be migrated to exchange RabbitMQ with the corresponding OPC-UA implementation.

Graphical Interface of the Application

This section describes the Web application for the visualization of information regarding the system. The time interval that affects the information to be visualized can be customized to predefined intervals (last hour, shift, day, week, month, year) or to user-defined time ranges.

Figure 9 shows the general overview of the lines accessed by production managers. For each line, the current state of each of its machines, and the existing warnings and faults, is visualized. Moreover, the application also shows the total number of parts produced on the time interval, the evolution of the production against the production objective values, and the number of warning and faults of the line.

After this overview, the detailed view of a machine presents the most recent states of the machines, and details about its faults and warnings (Figure 10).

Besides previous visualization, oriented towards a real-time monitoring of a production line, the application also presents and offline visualization for a more detailed analysis based on historical data. In this offline mode, presented by Figure 11, for each line the system represents the time spent on each state by each machine.

Figure 9. Line overview

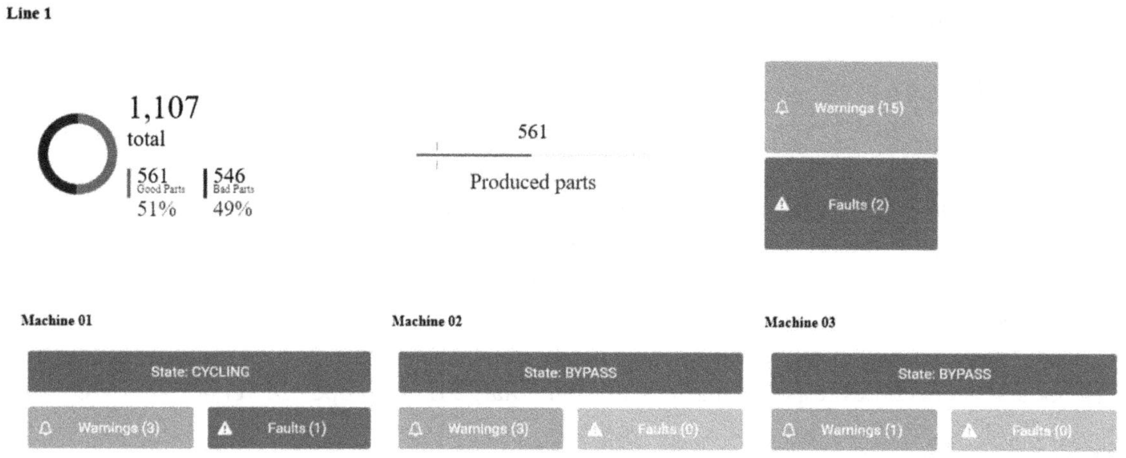

Figure 10. Machine overview

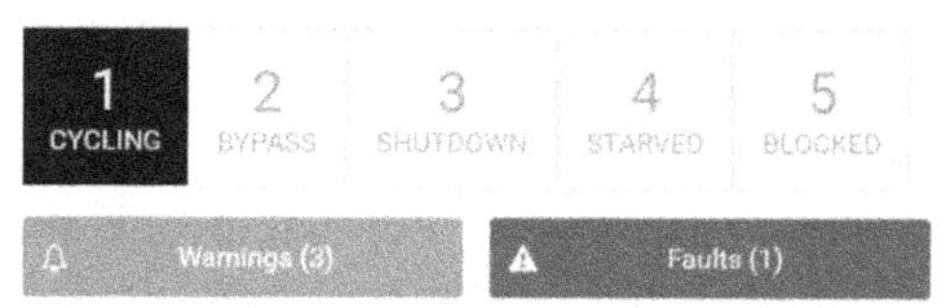

MACHINE FAULTS

Faults	Time
Machine Trimming Open Request, Stop Automatic Cycle	2017-06-09T08:58:41.217Z

MACHINE WARNINGS

Warnings	Time
PLC Battery Bad Or UPS Alarm	2017-06-09T08:43:41.199Z
Energy Reduction Shutdown Phase 2	2017-06-09T09:13:41.235Z
Error Proofing Disabled	2017-06-09T09:28:41.251Z

Then, for each machine the application presents the states of the PLC connected to it (Figure 12). First, the time spent at each state by the PLC is visualized as a percentage of the time of the period under analysis.

Finally, in order to ease the analysis of this information, the chart of Figure 13 shows the temporal evolution of the previous states. Colours represent if a state was active at a certain time point.

CONCLUSION

The Industry 4.0 paradigm requires a convergence of traditionally isolated technologies, mainly ICS (Industrial Control System) and ICT (Information and Communication Technology). However, due to this traditional isolation between them, most of the ICT requirements and technologies required to develop Industry 4.0 application are not deeply understood by enterprises.

Figure 11. States of machines of a line

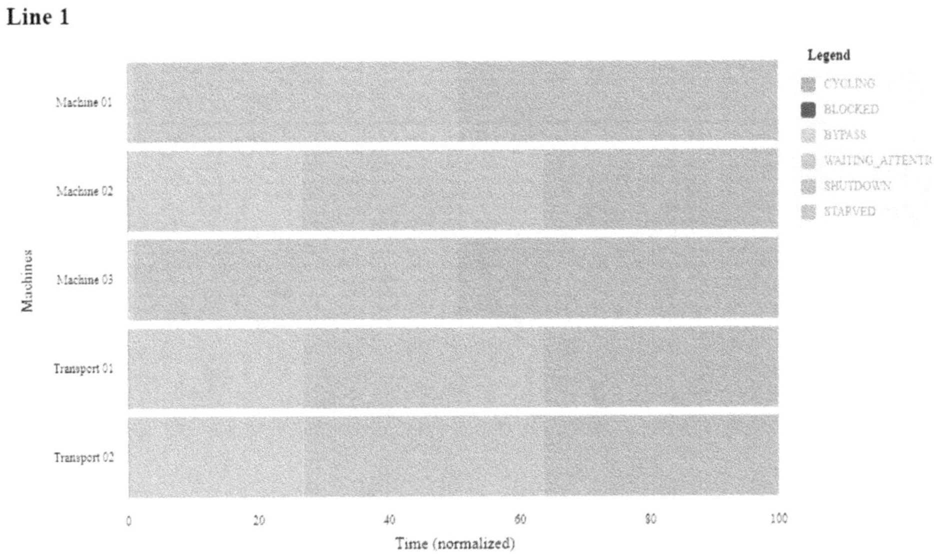

Figure 12. Details about the states of a PLC of a machine

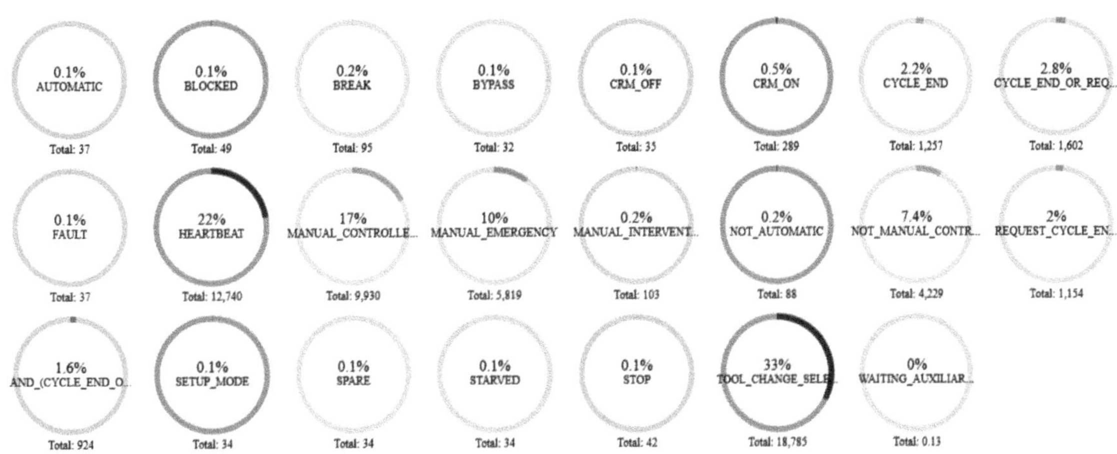

Nowadays, the complexity of underlying technologies; the isolation between the existing ICS and ICT knowledge within companies; and the low quantity of Industry 4.0 applications outside big corporations; is a drawback to the implantation of these applications, mainly among SMEs. This chapter focuses on these drawbacks with the objective of promoting the Industry 4.0 paradigm.

Firstly, this chapter has described main technologies related to the development of Industry 4.0 data applications. Secondly, an analysis of the main requirements of these data applications has been presented. Based on these requirements, a classification for data applications has been proposed, according to their complexity. This complexity has been reflected as a combination of several requirements (real-time data acquisition, real-time data process, Big Data nature, remote access, and cloud infrastructure).

Figure 13. Sequence of the states of a machine

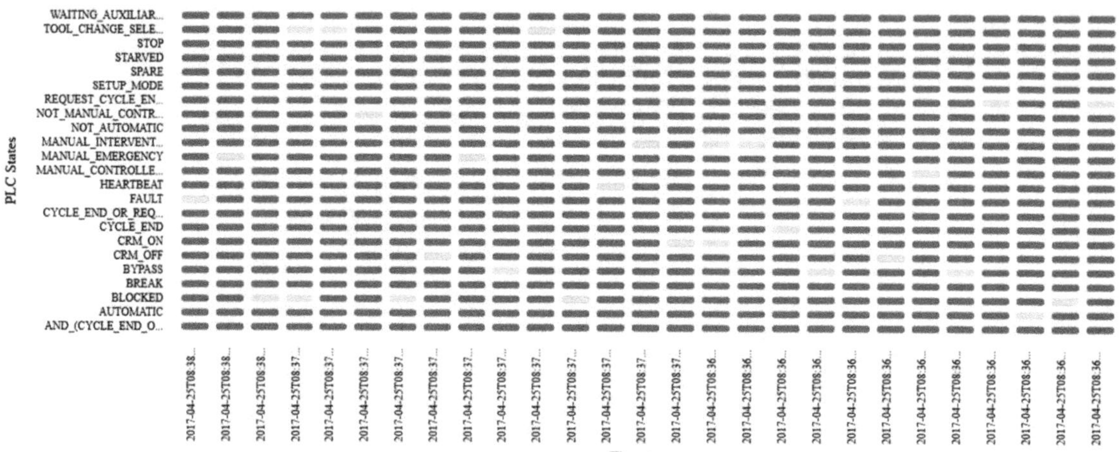

Finally, a basic Industry 4.0 application developed for Gaindu Automation has been described, focusing on its architecture to foster the future development of Industry 4.0 applications.

A deeper understanding of main technologies related to Industry 4.0 applications, and examples of applications suited to the requirements of different companies (from SMEs to big corporations), still presents a gap that needs to be closed to strengthen and spread the Industry 4.0 paradigm. The analysis of technologies, classification proposal and example presented in this chapter is oriented to help people inside companies entering the Industry 4.0 path.

ACKNOWLEDGMENT

This research was partially supported by the Centre for the Development of Industrial Technology (CDTI) and the Spanish Ministry of Economy and Competitiveness (IDI-20150643).

REFERENCES

Al-Madani, B., & Ali, H. (2017). Data Distribution Service (DDS) based implementation of Smart grid devices using ANSI C12.19 standard. *Procedia Computer Science*, *110*, 394–401. doi:10.1016/j.procs.2017.06.082

Angular. (n.d.). Retrieved from https://angular.io/

Baheti, R., & Gill, H. (2011). Cyber-physical systems. The Impact of Control Technology, 12, 161-166.

Beyer, M., & Laney, D. (2012). *The importance of big data: A definition*. Stamford, CT: Gartner.

Bonomi, F. (2011). Connected vehicles, the internet of things, and fog computing. In *The Eighth ACM International Workshop on Vehicular Inter-Networking (VANET)*, Las Vegas, NV (pp. 13-15).

Bonomi, F., Milito, R., Zhu, J., & Addepalli, S. (2012). Fog computing and its role in the internet of things. In *Proceedings of the first edition of the MCC workshop on Mobile cloud computing* (pp. 13-16). 10.1145/2342509.2342513

Boyer, S. A. (2009). *SCADA: supervisory control and data acquisition*. International Society of Automation.

Compact RIO. (n.d.). Retrieved from http://www.ni.com/compactrio/esa/

Dash, S. K., Mohapatra, S., & Pattnaik, P. K. (2010). A survey on applications of wireless sensor network using cloud computing. *International Journal of Computer Science & Emerging Technologies*, *1*(4), 50–55.

Dastjerdi, A. V., Gupta, H., Calheiros, R. N., Ghosh, S. K., & Buyya, R. (2016). Fog computing: Principles, architectures, and applications. arXiv:1601.02752

DDS. (n.d.). Retrieved from http://portals.omg.org/dds/

Adlink. (n.d.). DDS and OPC UA. Retrieved from http://www.prismtech.com/vortex/technologies/dds-and-opc-ua

Ewon. (n.d.). Ewon official web page. Retrieved from https://ewon.biz/

García, M. V., Irisarri, E., Pérez, F., & Elisabet, E. (2017). An Open CPPS Automation Architecture based on IEC-61499 over OPC-UA for flexible manufacturing in Oil & Gas Industry. *IFAC-Papers OnLine*, *50*(1), 1231–1238. doi:10.1016/j.ifacol.2017.08.347

Hankel, M., & Rexroth, B. (2015). *The reference architectural model industrie 4.0 (rami 4.0)*. ZVEI.

Hong, K., Lillethun, D., Ramachandran, U., Ottenwälder, B., & Koldehofe, B. (2013). Mobile fog: A programming model for large-scale applications on the internet of things. In *Proceedings of the second ACM SIGCOMM workshop on Mobile cloud computing*, (pp. 15-20). 10.1145/2491266.2491270

IIC. (2015). *Industrial Internet Reference Architecture* [technical report]. Industrial Internet Consortium.

Ghithub. (n.d.). Json rules. Retrieved from https://github.com/cachecontrol/json-rules-engine

Köksal, Ö., & Tekinerdogan, B. (2017). Obstacles in Data Distribution Service Middleware: A Systematic Review. *Systems. Future Generation Computer*, *68*, 191–2210. doi:10.1016/j.future.2016.09.020

Kreps, J. (2014). Questioning the lambda architecture.

Laney, D. (2001). 3D data management: Controlling data volume, velocity and variety. *META Group Research Note*, *6*(70).

Lee, J., Bagheri, B., & Kao, H.-A. (2015). A cyber-physical systems architecture for industry 4.0-based manufacturing systems. *Manufacturing Letters*, *3*, 18–23. doi:10.1016/j.mfglet.2014.12.001

Loopback. (n.d.). Retrieved from http://loopback.io/

Madsen, H., Burtschy, B., Albeanu, G., & Popentiu-Vladicescu, F. L. (2013). Reliability in the utility computing era: Towards reliable fog computing. In *2013 20th International Conference on Systems, Signals and Image Processing (IWSSIP)* (pp. 43-46).

Mahnke, W., Leitner, S., & Damm, M. (2009). *OPC unified architecture*. Springer Science & Business Media. doi:10.1007/978-3-540-68899-0

Mao, Y., You, C., Zhang, J., Huang, K., & Letaief, K. B. (2017). A Survey on Mobile Edge Computing: The Communication Perspective. *IEEE Communications Surveys and Tutorials, 19*(4), 2322–2358. doi:10.1109/COMST.2017.2745201

Marz, N. (2012). Big data lambda architecture.

MongoDB. (n.d.). Retrieved from https://www.mongodb.com

NI. (2017). National instruments official web page. Retrieved from http://www.ni.com/en-us/shop.html

Nishio, T., Shinkuma, R., Takahashi, T., & Mandayam, N. B. (2013). Service-oriented heterogeneous resource sharing for optimizing service latency in mobile cloud. In *Proceedings of the first international workshop on Mobile cloud computing \& networking* (pp. 19-26). 10.1145/2492348.2492354

Node OPC-UA. (n.d.). Retrieved from http://node-opcua.github.io/

OPC Foundation. (n.d.). Retrieved from https://opcfoundation.org/

Parwekar, P. (2011). From Internet of Things towards cloud of things. In *2011 2nd International Conference on Computer and Communication Technology (ICCCT)* (pp. 329-333).

Pfrommer, J. (2017). *Semantic Interoperability at Big-Data Scale with the open62541 OPC UA Implementation. In InterOSS-IoT 2016: Interoperability and Open-Source Solutions for the Internet of Things* (pp. 173–185). Springer.

Plattform Induistrie 4.0. (2016). Reference Architectural Model Industrie 4.0 (RAMI4.0) - An Introduction.

RabbitMQ. (n.d.). Retrieved from https://www.rabbitmq.com/

Rao, B. B., Saluia, P., Sharma, N., Mittal, A., & Sharma, S. V. (2012). Cloud computing for Internet of Things & sensing based applications. In *2012 Sixth International Conference on Sensing Technology (ICST)* (pp. 374-380).

Rao, B., Saluia, P., Sharma, N., Mittal, A., & Sharma, S. (2012). Cloud computing for Internet of Things & sensing based applications. In *2012 Sixth International Conference on Sensing Technology (ICST)* (pp. 374-380). IEEE.

Rezai, A., Keshavarzi, P., & Moravej, Z. (2017). Key management issue in SCADA networks: A review. *Engineering Science and Technology, an International Journal, 20*, 354-363.

RTI. (2017). *Context DDS Professional*. Retrieved from https://www.rti.com/products/dds/benchmarks

Corsaro, A. (2016, August 15). A Tale of Two Industrial IoT Standards: DDS and OPC-UA. *RTInsigihts*. Retrieved from https://www.rtinsights.com/dds-opc-ua-industrial-iot-standards/

Schleipen, M., Gilani, S.-S., Bischoff, T., & Pfrommer, J. (2016). OPC UA & Industrie 4.0 - Enabling Technology with High Diversity and Variability. *Procedia CIRP, 57*, 315–320. doi:10.1016/j.procir.2016.11.055

Siemens. (n.d.). Industrial Ethernet: the basis for efficient automation. Retrieved from http://w3.siemens.com/mcms/automation/en/industrial-communications/industrial-ethernet/Pages/industrial-ethernet.aspx

Stojmenovic, I. (2014). Fog computing: A cloud to the ground support for smart things and machine-to-machine networks. In *Telecommunication Networks and Applications Conference (ATNAC), 2014 Australasian,* (pp. 117-122). 10.1109/ATNAC.2014.7020884

Yang, G., & Zhou, X. (2013). Cyber-physical systems.

Zaslavsky, A., Perera, C., & Georgakopoulos, D. (n.d.). Sensing as a service and big data. arXiv:1301.0159

Zhang, Q., Cheng, L., & Boutaba, R. (2011). Cloud computing: state-of-the-art and research challenges. *Journal of internet services and applications, 1*(1), 7-18.

KEY TERMS AND DEFINITIONS

CPS: Cyber-Physical Systems.
CPSS: Cyber-Physical Production Systems.
DCOM: Distributed Component Object Model.
DDS: Data-Distribution Service for Real-Time Systems.
I4.0: Industry 4.0.
ICS: Industrial Control System.
ICT: Information and Communication Technology.
IIRA: Industrial Internet Reference Architecture.
LAN: Local Area Network.
OPC-UA: Open Platform Communications Unified Architecture.
PLC: Programmable Logic Controller.
QoS: Quality of Service.
RAMI 4.0: Reference Architectural Model Industrie 4.0.
SCADA: Supervisory Control and Data Acquisition.
SME: Small and Medium Enterprise.
SOA: Service Oriented Architecture.
TSN: Time-Sensitive Networking.
WAN: Wide Area Network.

Chapter 5
Smart DC Microgrid:
A Cyber-Physical System Perspective

Brijendra Pratap Singh
MNNIT Allahabad, India

M M Gore
MNNIT Allahabad, India

ABSTRACT

The objective of this chapter is to elucidate on microgrid technologies, a comparison of direct current (DC) microgrid technology and alternating current (AC) microgrid technology, the role of the information and communication technology, demand response programs, and the evolution of Industry 4.0 in detail. The microgrid is a cyber-physical system. ICT is used for computing control algorithms and sending control information to actuators for physical processes. In a cyber-physical system, the physical processes, which are governed by the laws of physics, are controlled by computers. The computers are used for computing or executing the algorithms (i.e., the control logic) and the result is sent to the actuators in the form of control signal for actual control. In a microgrid, a consumer can act as a producer also, which is termed as the prosumer. This chapter explains the maximum power point tracking algorithm, software-defined battery, the operation of parallel converters, the working of prosumer, the demand response program, communication technologies, and the (industrial) Internet of Things.

INTRODUCTION

A cyber-physical system is a mechanism in which physical phenomena are controlled by algorithmic logic through computers. Physical phenomena are governed by the laws of physics. When these physical phenomena are controlled by algorithmic logic executed in computers, are called cyber-physical system. In a cyber-physical system, algorithms and physical phenomena are tightly coupled. Computers or communication units execute the algorithmic logic, and the outcome is used by the actuator to perform the physical phenomena. The term "cyber-physical system" is in mass consumption since 2007.

DOI: 10.4018/978-1-5225-4936-9.ch005

To understand algorithmic logic and physical phenomena coupling consider working of VLC media player (a program to play video or audio file) as shown in Figure 1. The VLC media player is a program which processes the media file (here mp3 file), and the VLC media program and data (mp3 file) are processed by the processor. A processor can do operations, such as addition, subtraction, multiplication, division, logical operations (e.g., and, not, or, shift, etc.). During these operations, the processor takes data and instructions from memory and again puts it into the memory.

This is the stored program and data computing model. The processor is capable to do some arithmetic and logical operations. So, the question is how these arithmetic and logical operations affect the physical world. Actually, when processor processes, it generates output voltage signal. Here, suppose a speaker is connected to the computer using a signal amplifier. VLC media player executes the mp3 file. It generates voltage level for the speaker, which is actually the audio sound (i.e., dialog, song, etc.). When the voltage is applied to the speaker, current flows through the wire connected to the speaker. Speaker is a device which has a coil and a magnet when current flows through the coil, it generates a magnetic field, and this magnetic field is opposite to the fixed magnet's magnetic field. The magnetic field generated by the coil depends on the amount of current flowing through it. By this variable current, the diaphragm in the speaker moves and generates longitudinal waves which are the sound waves. This way the VLC media program controls the movement of the diaphragm, whose movements generate controlled sound waves i.e., dialog or song. And human ear listens to that dialog or song. This is how an algorithmic logic affects the physical phenomena. When this algorithmic logic affects biological phenomena, it leads to the cyber biological system.

Above example is not directly related to the smart DC microgrid. But the simplicity of this example fascinates to be included in this chapter for the explanation of the concept of cyber-physical system as it spans across a very large area, such as energy, healthcare, manufacturing, transportation, aerospace, chemical processes, Quantum Information Science and Technology (QIST), etc. The next example also

Figure 1. VLC media player program execution to generate sound

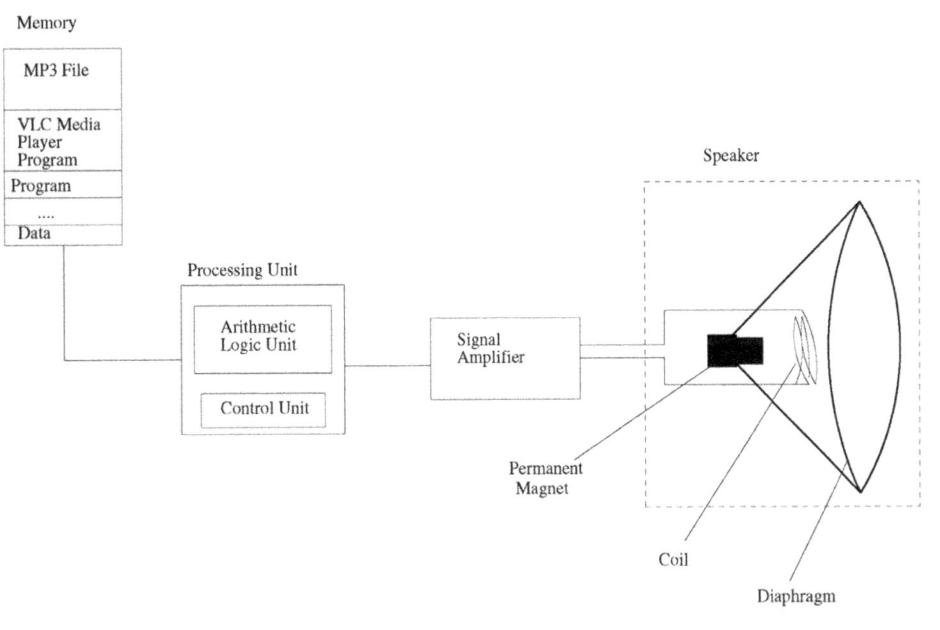

explains the concept of cyber-physical system in different flavor in the context of the smart DC microgrid and also gives the global overview of the smart DC microgrid.

Figure 2 depicted a smart DC microgrid in which voltage of the distribution line is 380 volt. The distribution line voltage varies between 360 volts to 400 volts. This willful variation is just for sharing power among different energy resources (especially voltage droop technique). DC/DC converter is a component by which the DC microgrid can be realized. DC/DC converter is a power electronic device by which voltage can be up (using boost converter) and down (using buck converter). Power electronic converter is the device which has potential to replace the transformers from the electricity distribution system. In Figure 2, renewable energy resources, such as solar power (generates power at 48 volt and with the help of boost converter, the voltage is stepping up to 380 volt), wind energy (generates power at 600 volt and with help of AC/DC converter, the voltage is down to 380 volt and converts AC to DC), energy storage (battery storage which stores energy at 12 volt), electric vehicle (also uses battery and stores power from the DC microgrid with the help of buck converter), a smart building, and a smart home constitute a smart DC microgrid. The Information and Communication Technology (ICT) elements are not shown here that will be explained later in this chapter. ICT is necessary for the implementation of demand response program which is one of the major concerns of the smart grid.

Let us take an example of DC/DC buck converter to understand the how the logic or algorithm used is for creating desired variable voltage level. The example will explain the concept in the top-down approach. In, a DC/DC buck converter is shown. Buck converter is used for stepping down the voltage. The realization of the switch is made through the semiconductor devices such as MOSFET (Metal Oxide Semiconductor Field Effect Transistor), IGBT (Insulated Gate Bipolar Transistor), BJT (Bipolar Junction Transistor), and thyristors. Switching frequencies vary from 1 kHz to 1 MHz. When switch S is ON, the power is supplied from V_{in}, and when S is OFF, the power is not supplied from the V_{in}, only stored power in inductor and capacitor is supplied to the load. So, the average output voltage is decreased

Figure 2. A smart DC microgrid

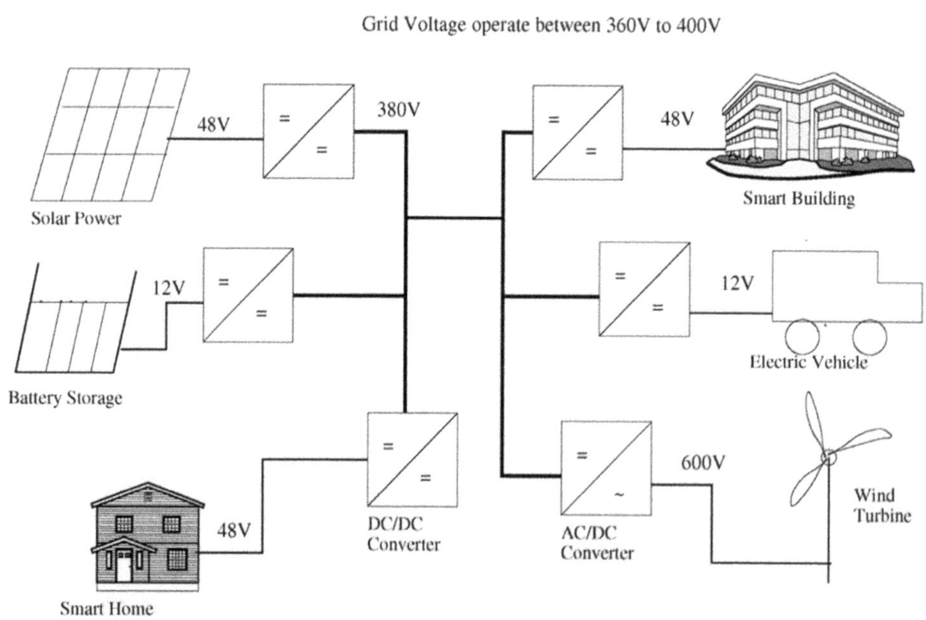

than the input voltage (V_{in}). Switching frequencies and duration of ON and OFF time in a period control the output voltage. Bigger the ON time of the switch higher the voltage, lower the ON time or bigger the OFF time of the switch, lower the voltage. So the voltage can be controlled in the range of 0 volt to maximum V_{in} volt. To generate the desired ON time or greater OFF time in a period to control the output voltage, a microcontroller is used. Here to explain the concept, 8051 microcontroller is assumed. The microcontroller normally has timers. These timers are used for counting. Timer counts from 0 to 255.

When timer overflows, it generates interrupt and at this interrupt, toggling from ON to OFF or from OFF to ON for the switch is performed. With the help of program which executes in the microcontroller, it would be possible to control the ON and OFF time of the switch, consequently output voltage of the converter.

The smart grid is a cyber-physical system. The basic functionalities of the smart grid are generating electrical energy, transmit energy, and control this energy. Generation, transmission, and distribution of the energy are governed by the laws of physics. Appropriate algorithms are used in control systems to ensure the stability of grid, demand and & supply of the energy, and fault protection of the grid. Algorithms are written to control the smart grid as per the laws of physics. A microgrid is an independent small power system. A microgrid can be applicable for communities/neighborhoods or villages, academic/university campuses, corporate buildings, home buildings, etc. A microgrid can work independently (islanded mode) or with the smart grid (in grid-connected mode). This chapter will describe about the smart DC microgrid, which includes maximum power point tracking algorithm, software-defined battery, sharing of power among multiple distributed energy resources, working of a prosumer, demand response program, and communication technologies.

In the traditional electrical grid, power is generated in bulk and then transmitted to the end users (industries and residential consumers) through transmission lines and distribution lines. Distribution substation connects with the transmission system. Step-down transformers lower the voltage and power is transmitted to the end users. Circuit breakers, switches, and transformers are used for transmitting and distributing power. But there are several issues such as blackout (e.g., some recent blackouts are listed as follows (i). 30th and 31st July 2012 in India (700 million persons are affected), (ii). 18th August 2005

Figure 3. DC/DC buck converter

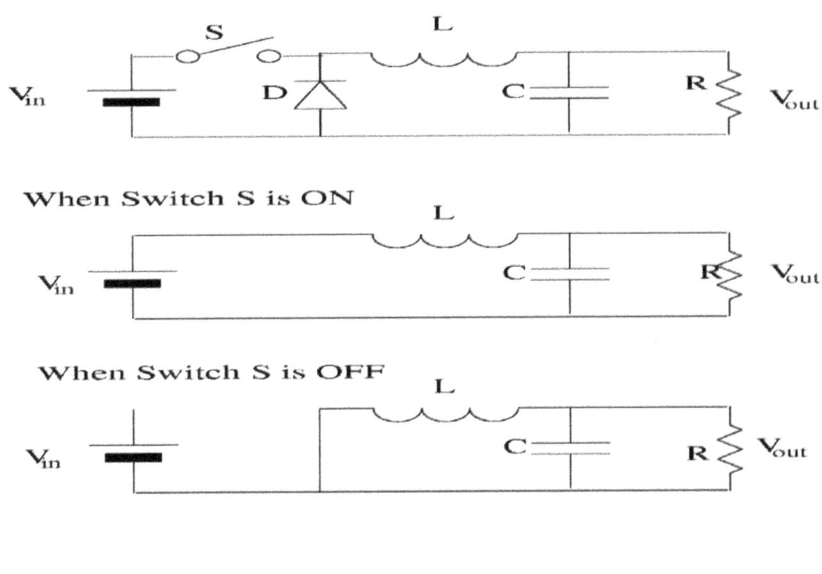

in Java and Bali island of Indonesia (120 million persons are affected), and (iii). 2nd January 2001 in India (230 million persons are affected), etc. (Duddu, 2015)), the reliability of electricity network, power quality, safety, etc. in the traditional electrical grid. Supervisory control and data acquisition (SCADA) is used for remote monitoring and control in power system. SCADA uses real-time programmable logic controllers (also called remote terminal unit), communication system, and the operator workstation. SCADA paves the way for automation of distribution system. SCADA improves the reliability of distribution system. But the advancement in the power electronics, phasor measurement unit, information and communication technology, renewable energy resources pave the way for the development of the smart grid. A smart grid is actually an enhancement of the existing electrical grid which makes use of intelligent electronic devices and communication technologies. The smart grid has several features, such as two-way power supply (i.e., a consumer can be a producer), digital communication, distributed renewable energy resources, smart metering, smart appliances, demand response program, etc.

Before diving into the technical aspects of the smart grid, it would be better to know some other related facts. Energy is required for living. Biological creatures take foods for energy. The human being (also a biological creature) requires the huge amount of energy for living life. Humans require energy for things such as hot water, fan, air conditioner, transportation (bus, rail, etc.), lighting, for machines, agriculture, etc. Therefore, the production of energy is one of the major concerns. So, the humans need a very efficient way of energy production, energy transportation (the electrical grid is a good way to transport energy i.e., the electrical grid is a transport system for electrical energy), and energy usage. There are many forms of energy in the nature, such as chemical energy, gravitational energy, mechanical energy, thermal energy, nuclear energy, electrical energy, etc. One of the easy ways to produce energy is to burn fossil fuel, such as coal, gas, petroleum, etc. But the fossil fuels are limited resources on the earth. The earth is running out of these resources. Another consequence of burning fossil fuel is the creation of carbon dioxide (CO_2) in huge amount. A huge increment of CO_2 consequences the global warming. Global warming creates problems, such as flood, drought, the rise of sea level, and affects the creature's life.

The Key World Energy Statistics 2017 (Produces by International Agency) (Biro, 2017) shows that world total primary energy supply in 1973 was 6101 Mtoe (Million tonnes oil equivalent). In 2015, world total primary energy supply was 13647 Mtoe. It shows huge increment in demand for the energy. In 2015, China alone had 21.9% share while whole Asia has 47.7% of all world primary energy supply. Total crude oil production of the world in the year 2016 was 4321 million tonnes. Saudi Arabia produced 13.5% of the total being the highest producer. Natural gas production of the world in the year 2016 was 3613 billion cubic meters. The highest producer was United States (produced 20.7%). Coal production of the world was 7269 million tonnes, and the highest producer was China (produced 44.6%).

PBL Netherlands Environmental Assessment Agency provides a report in December 2017 entitled "Trends in Global CO_2 and Total Greenhouse Gas Emission" (Olivier, Schure, & Peters, 2017). In 2016, total greenhouse gas emission was 49.3 giga tonnes in CO_2 equivalent. Emission consists of carbon dioxide (CO_2), methane (CH_4), nitrous oxide (N_2O), and fluorinated gases. The combined share of methane, nitrous oxide, and fluorinated gas is approximately 28% of total greenhouse gas. CO_2 is approximately 72%. 25% of methane emission is accounted for coal mining, natural gas production, and gas distribution. Other 75% of methane emission is accounted for cattle (23%), rice cultivation (10%), and others. Only 25% of nitrous oxide emission is energy-related emission. Other 75% is accounted for agriculture, animal, and fertilizers.

Paris Climate change conference held November 2015 – December 2015 (Nations, 2015) resulted in the "Paris Agreement." This agreement states about the reduction of greenhouse gas emission to reduce the temperature to reduce the risk of climate change. Article 2 of the Paris Agreement states that "Holding the increase in the global average temperature to well below 2°C above pre-industrial levels and pursuing efforts to limit the temperature increase to 1.5°C above pre-industrial levels, recognizing that this would significantly reduce the risks and impacts of climate change".

International Energy Outlook 2017 (produced by U.S. Energy Information Administration) (Administration, 2017) reports that world energy consumption increases by 28% between the year 2015 to the year 2040. And more than half of the increment is due to India and China.

All the above data clearly indicate that the need for a clean and renewable energy to fulfill increasing demand and to make the earth a better place to live. The smart grid is a technology (which is still in developing phase) to harvest energy from renewable energy resources, such as solar, wind, tidal, etc. and optimize the usage of energy in efficient manner to make our earth sustainable for life, not only for humans but also for other creatures. "Smart Grid Projects Outlook 2017: fact, figures, and trends in Europe" is a report published by the Joint Research Center (JRC), European Commission (Gangale, Vasiljevska, Covrig, Mengolini, & Fulli, 2017) which tells about smart grid projects in Europe. In Europe, total 950 projects in 50 countries, total investment cost€4.97 billion, 2900 organizations are involved. India Smart Grid Forum (ISGF) (Forum, 2015) reports that smart grid projects in several countries, such as Austria, Belgium, Canada, Denmark, France, Germany, India, Ireland, Italy, Japan, Mexico, Netherland, Portugal, South Korea, Spain, Sweden, United Kingdom, United States of America, and Vietnam.

Discussion of the smart grid also requires the discussion of Industry 4.0. Industrie 4.0 is a project launched by the German government in 2011 (i-SCOOP, n.d.). Industrie 4.0 aims to revolutionize the manufacturing and production and leads to smart factory (Germany Trade And Invest, n.d.). A paper entitled "Industrie 4.0: Smart Manufacturing for the Future" by the Germany Trade and Invest describes about the Industrie 4.0 which states that two technological concepts namely "Cyber-Physical Systems" and "Internet of Things". These two concepts pave the way for new technological era Industrie 4.0. The first industrial revolution introduced the mechanical production with the help of water and steam power. The second industrial revolution introduced mass production with the help of electrical energy. The third industrial revolution used the electronics, embedded system, and computer for automation. The fourth industrial revolution introduces the concepts of "Cyber-Physical System" and "Internet of Things". The association of more than 6000 members (companies) concluded a cooperation agreement for the further development of Industrie 4.0 in April 2014 known as "Plattfom Industrie 4.0" (Federal Ministry For Economic Affairs And Energy, n.d.).

Although, Industrie 4.0 is originated from Germany and focused on the production and manufacturing. Another leader also around the world such as Industrial Internet Consortium (IIC) explores the fourth industrial revolution and stretches the boundary of Industrie 4.0 beyond the smart factory. Thus, Industry 4.0 includes smart transportation and logistics, smart buildings, smart cities, smart healthcare, smart villages, smart factories, etc. IIC is an organization (founded in March 2014, by AT&T, Cisco, General Electric, IBM, and Intel) to accelerate the development of interconnected machines, devices, and intelligent analytics named as "Industrial Internet" (Wikipedia, n.d.).

In March 2016, Plattform Industrie 4.0 and Industrial Internet Consortium agree on cooperative development of their architectures namely the Reference Architecture Model for Industrie 4.0 (RAMI4.0) and the Industrial Internet Reference Architecture (IIRA), respectively (Federal Ministry For Economic Affairs And Energy, 2016).

The basic building blocks of Industry 4.0 are "Cyber-Physical Systems" and "Internet of Things". National Science Foundation (National Science Foundation, 2006) describes a cyber-physical system as "It refers to the right conjoining of and coordination between computation and physical resources". IEEE Internet of Things (IEEE Internet Of Things, 2015) defines IoT as "Internet of Things envisions a self-configuring, adaptive, complex network that interconnects 'thing' to the Internet through the use of standard communication protocols. The interconnected things have physical or virtual representation in the digital world, sensing/actuation capability, a programmability feature and are uniquely identifiable. The representation contains information including the thing's identity, status, location or any other business, social or privately relevant information. The things offer services, with or without human intervention, through the exploitation of unique identification, data capture and communication, and actuation capability. The service is exploited through the use of intelligent interfaces and is made available anywhere, anytime, and for anything taking security into consideration".

Now let's come back to the original context of this chapter that is the smart DC microgrid. Contents of this chapter are organized in the following manner. Section "SMART DC MICROGRID" elucidates about the elements of the smart DC microgrid and comparison of AC and DC microgrid. Section "LAYERED OVERVIEW OF THE SMART DC MICROGRID" presents a layered architecture of the smart DC microgrid. Section "WORKING OF SMART DC MICROGRID" elucidates the working of smart DC microgrid which includes the working of solar panel, software-defined batteries, operation of parallel converters to share power among multiple distributed energy resources, working of a smart home as a prosumer, and demand response program (which utilizes the Internet of Things), communication protocols, and Internet of Things architecture. Section "CONCLUSION" describes the conclusion.

SMART DC MICROGRID

Smart DC microgrid is an interconnected infrastructure of distributed energy resources, controllable loads, and communication technologies. The smart DC microgrid can work as a single controllable unit. The distributed energy resources and controllable loads are connected through the intelligent power electronic devices, microcontrollers, and network protocols. A smart DC microgrid system is depicted in Figure 4 in which renewable energy resources, AC/DC converters, DC/AC converters, DC/DC converters, smart home, smart building, network routers, energy management systems, smart devices, communication links, and microgrid controller are depicted. The distributed generation units and storage units in DC microgrid constitute a Virtual Power Plant (VPP). Multiple DC microgrids constitute a larger virtual power plant. Smart DC microgrid uses the communication network to create Home Area Network (HAN), Building Area Network (BAN), Neighborhood Area Network (NAN), and Field Area Network (FAN). These communication networks have many different protocols and computation devices. These computation devices and communication networks are used for controlling the Internet of Energy. Department of Energy, United States defines microgrid as "a group of interconnected loads and distributed energy resources with clearly defined electrical boundaries that acts as a single controllable entity with respect to the electric utility grid".

Elements of the Smart DC Microgrid

Smart DC microgrid constitutes distributed energy resources (e.g., photovoltaic cell, wind energy, diesel generator, micro Combined Heat and Power (μCHP), etc.), intelligent power electronic devices (e.g., DC/DC converter, DC/AC converter, AC/DC converter with communication and computation capabilities), energy storage systems, controllable loads, algorithms, and protocols.

Distributed Energy Resources

Distributed Energy Resource (DER) in the smart DC microgrid increases resilience. In the case of a fault, most of the load can have the power supply due to DER. One of the prominent renewable energy resources is solar power. Photovoltaic cell converts solar energy to electrical energy. But photovoltaic cell has a very low efficiency (between 12% to 20%). Theoretical efficiency limit of the silicon-based photovoltaic cell is 29.1%. Photovoltaic cell efficiency is improved to 26.3% (Yoshikawa, et al., 2017). Increasing the efficiency of solar cell is a hot research area. Extracting maximum power from photovoltaic cell requires maximum power point tracking algorithm. Photovoltaic cells produce DC power that is also a strong factor to move from to AC system to DC system. Parallel use of multiple distributed energy resources requires power sharing and control algorithms. In the smart DC microgrid, one DER can communicate with the other DER for better power sharing.

Intelligent Power Electronic Device

Intelligent Power electronic devices are the key component for power system automation. Intelligent power electronic devices have microprocessor-based controllers, which can do computation and communicate over the communication network. These devices can control the power system components, such as circuit breaker, transformer, capacitor bank, and distributed energy resource. The DC/DC converter, DC/AC converter, and AC/DC converter are the key component of the smart DC microgrid system. These converters are of very small size. These converters are alternate of the transformer for stepping up and stepping down voltage in the smart DC microgrid.

Figure 4. Smart DC microgrid

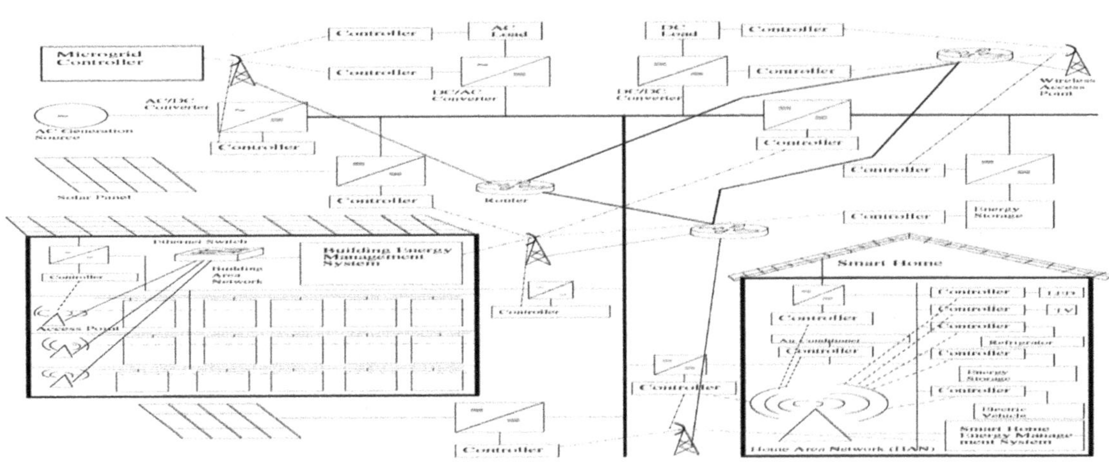

Energy Storage

Energy storage technologies are mechanical energy storage, electromagnetic energy storage, and electrochemical energy storage. Mechanical energy storage is pumped hydro storage, flywheels, compressed air storage, etc. Electromagnetic energy storage system stores energy in a superconducting coil. Electrochemical storage includes a lead-acid battery, lithium-ion battery, etc. Lithium-ion, Sodium-sulfur, and advanced lead-acid battery technologies are compared on the characteristics, namely (i) size, (ii) discharge rate, (iii) efficiency, (iv) cost per kWh, (v) lifetime, (vi) environment, (vii) calendar life, (viii) power to energy ratio (kW/kWh), (ix) maintenance cost, (x) commercial maturity, and (xi) security. The authors showed that lithium-ion and advanced lead-acid battery would be an appropriate energy storage technology for an electrical grid (Saez-de-Ibarra, et al., 2013; Ammar, Hafsa, & Hammami, 2013).

Controllable Load

Controllable loads are classified in three categories namely (i) passive controllable load, (ii) active controllable load, and (iii) broad controllable load (Shen, Jiang, & Li, 2015). Passive controllable load loads are residential load, such as refrigerator, washing machine, lights, air conditioner, etc. These devices can be interrupted or shifted to smooth the load curve. This type of load cannot inject energy into the electrical grid. Active controllable loads are the electric vehicles, battery storage, etc. These loads can inject energy into the electrical grid. Broad controllable loads are microgrid itself as one controllable unit. The passive and active controllable loads have the computation and communication capabilities. They can have a unique address (e.g., ipv6 (Internet protocol version 6) address). These loads are controllable over the Internet.

Algorithms and Communication Protocols to Control Smart DC Microgrid

Algorithms are required for maximum power extraction from photovoltaic cell (e.g., perturb and observe algorithm for maximum power point tracking), power sharing among distributed energy resources (e.g., voltage droop mechanism), controlling power converters to ensure stability (e.g., feedback control techniques (proportional-integral-derivative controller)), and smoothing peak load curve (e.g., demand response programs). Communication protocols are used for communication among distributed energy resources, controllable loads, and smart DC microgrid controller. Communication protocols, such as TCP/IP, HTTP, IEEE 802.11, Zigbee, power line communication protocol, etc. are used in the smart grid.

AC Microgrid vs. DC Microgrid

AC and DC microgrid are compared on parameters such as economy and technology.

Economy

Modular design promotes effective capital investment and better planning for power system installation. A well-designed DC microgrid saves money regarding hardware installation and its operation than AC system. DC distribution line is more efficient than AC distribution line due to less energy loss. DC distribution line requires fewer conductors than AC distribution line. DC distribution line has not reactance, therefore, more power transmission. DC distribution does not require frequency monitoring.

Technology

There is not any requirement to control frequency and phase in DC microgrid. Therefore, the DC microgrid is simpler to control. Most of the distributed energy generation resources and loads are DC in nature, e.g., solar cell, battery, LED (Light Emitting Diode), electronic devices (cell phone, computer), etc. DC microgrid is more reliable than AC microgrid (Ikebe, 2003). Energy efficiency in DC distribution system is higher than AC distribution system (Liu & Li, 2014). 860 watt equivalent coal power is needed to power a 100 watt load at data center. DC microgrid is 5% to 7% more efficient. DC microgrid is highly modular and scalable than AC microgrid because DC converters are easy to control and easy in parallel operation. DC microgrid components are smaller. The modular design of DC microgrid provides more flexibility and scalability. DC microgrid stability is simpler than AC microgrid. In AC microgrid voltage and frequency both need to be regulated through active and reactive power. In DC microgrid only voltage regulation is required. Smart DC microgrid is expected to have following functionalities:

1. On-demand integration and disintegration of distributed energy resources
2. Real-time demand response program
3. Automatic energy regulation
4. A secured information and communication infrastructure

LAYERED OVERVIEW OF THE SMART DC MICROGRID

Smart DC microgrid system can be thought of as six layer system as in self-explanatory Figure 5. Layer 1 to Layer 4 constitutes the Internet of Things. Although a smart DC microgrid consists of several small cyber-physical systems. Layer 1 to Layer 5 constitutes a broader cyber-physical system as a whole. On the top of all these, business/market mechanism can be thought of as Layer 6.

WORKING OF SMART DC MICROGRID

Microgrid controller strategies are required for connecting multiple power converters and distributed energy resources (Majumder, Ghosh, Ledwich, & Zare, 2010). The nature of DER units also imposes constraints on the control strategy. For example, the solar cell uses the maximum power point tracking algorithm to extract maximum power. Microgrid controller controls power flow and voltage of the DER units in response to load change and disturbance. The control strategy in DC microgrid is usually for voltage regulation, power flow, stability, and fault detection & protection (Balog & Krein, 2011). Algorithms are required for smooth functioning of DC microgrid in grid-connected and islanded mode. Power sharing among DER units varies from centralized control scheme to decentralized control scheme. DC microgrid system is simpler because it does not have reactive power. Only voltage is available for control. Power sharing among the DER units can be determined by droop characteristic. The improved

Figure 5. Layered overview of the smart DC microgrid

		Business/Market mechanism	Layer 6
Cyber Physical System (CPS)		Energy management system (home, office, academic building, etc.), demand response mechanism, microgrid controller mechanism, etc.	Layer 5
Cyber Physical System (CPS)	Internet of Things (IoT)	Machine to machine communication and network protocols and security mechanism	Layer 4
Cyber Physical System (CPS)	Internet of Things (IoT)	Operating system for intelligent devices, device drivers, maximum power point tracking algorithm, control algorithms, etc.	Layer 3
Cyber Physical System (CPS)	Internet of Things (IoT)	Hardware connections: Smart DC microgrid infrastructure	Layer 2
Cyber Physical System (CPS)	Internet of Things (IoT)	Hardware: Solar cells, wind energy farm, batteries, intelligent power electronic devices, controllable loads, routers, access points, wires, smart meters, etc.	Layer 1

droop characteristics have been researched by several researchers for better power sharing and dynamic response (Vandoorn, Meersman, Kooning, & Vandevelde, 2012). Smart DC microgrid increases power system resilience because it has the provision of having small distributed generation sources and energy storage elements to supply power locally during faults. DC microgrid uses a huge amount of renewable energy resources (solar cell) mounted on the homes and the buildings which also provide environmental benefits. Use of renewable energy reduces greenhouse gas emission. In smart DC microgrid system, each generation unit and controllable load need a controller. There is a central controller which acts as a coordinator for other controllers to give the global view of the microgrid. The information and communication technology assists for communication and processing. A DC/AC grid tied inverter is used for a grid-connected mode of the microgrid. The DC/AC inverter is fully controlled by the central controller to decide whether the DC microgrid will work in grid-connected mode or islanded mode.

Maximum Power Point Tracking (MPPT) Algorithms for Photovoltaic Cell

The photovoltaic cell is a semiconductor device (made with semiconductor materials e.g., silicon), which directly absorbs the solar energy and converts to the electrical energy. A French physicist Edmond Becquerel observed the phenomena of the photovoltaic effect in 1839. Albert Einstein explained the phenomena of photovoltaic effect, in 1905, on which photovoltaic cell technology is built. Albert Einstein got the Nobel Prize in physics for the explanation of the photoelectric effect (Phillips, 2005). In 1954, Bell Laboratories developed first photovoltaic cell. The photoelectric effect is a phenomenon in which an electron can emit from the surface of the metal when light incident on the metal. When light incident on the photovoltaic cell some of the photons are absorbed, and the electrons get energy for conduction. Figure 6 depicts single junction and multi-junction photovoltaic cell. In the single junction photovoltaic cell, only those photons are useful whose energy is equal or greater than the band gap energy of absorbing material. Photons which have lower energy than band energy is not useful. To use these lower energy photons, multi-junction photovoltaic cell is created. Multi-junction photovoltaic cell is constructed of different semiconductor materials and has multiple p-n junctions.

Open circuit voltage of the photovoltaic cell is approximately 0.5 volt to 0.6 volt. Therefore, photovoltaic cells are connected in series for desired voltage and connected in parallel for the desired current. Figure 7 (RS-Data-Sheet, 1999) shows the I-V curve for irradiance and temperature (the Figure 7 is taken from the RS Data Sheet, for more details follow the reference). The curve shows that greater the sunlight irradiance greater the power. The electric power from the photovoltaic cell is highly dependent on illuminance of light and environment temperature. To extract the maximum power from the photovoltaic panel or solar panel, the impedance seen by the solar panel needs to be adjusted dynamically. The MPPT algorithm is implemented in a microcontroller. The MPPT algorithm adjusts the duty ratio of the DC/DC converter to extract maximum power from the solar panel.

Let us take an example to understand the requirement of the MPPT algorithm. Let there is a battery of 12 volt. The actual voltage of a battery depends on the actual state of charge; therefore, the voltage of the battery is approximately between 10.5 volt to 12.7 volt. For full charging, most batteries need voltage between 13.2 to 14.4 volt. Let there is a solar panel of 176 watt rated at the 17.6 volt and 10 amp (17.6 volt x 10 amp =176 watt). So, the solar panel gives 10 amp current at 17.6 volt voltage. If the battery is charged with the above specified solar panel. A 12 volt voltage battery can extract 12 volt x 10 amp = 120 watt power. So, the 176-120 =56 watt power is not used. The 56 watt power is actually not being produced by the solar panel due to poor match of generation and consumption.

If the battery voltage level is at 10.5 volt, then the power extraction from the solar panel is 10.5 volt x 10 amp =105 watt. Power utilization is decreased further. Here, MPPT algorithm is implemented in microcontroller to control the duty cycle of a buck converter. At the solar panel terminal, the power is extracted at 17.6 volt and 10 amp, at the battery terminal buck converter changes it to the 12 volt and 14.6 amp (which is almost 176 watt). So, the maximum power from the solar panel is extracted using MPPT algorithm. MPPT algorithm controls the voltage for maximum power point on power-voltage (P-V) curve.

Question is how an MPPT algorithm affects the solar cell's voltage and current? The solar cell is a semiconductor device which is governed by the laws of physics. So, how an algorithm is able to change the voltage and current?

Figure 6. Single junction and multi-junction solar cell

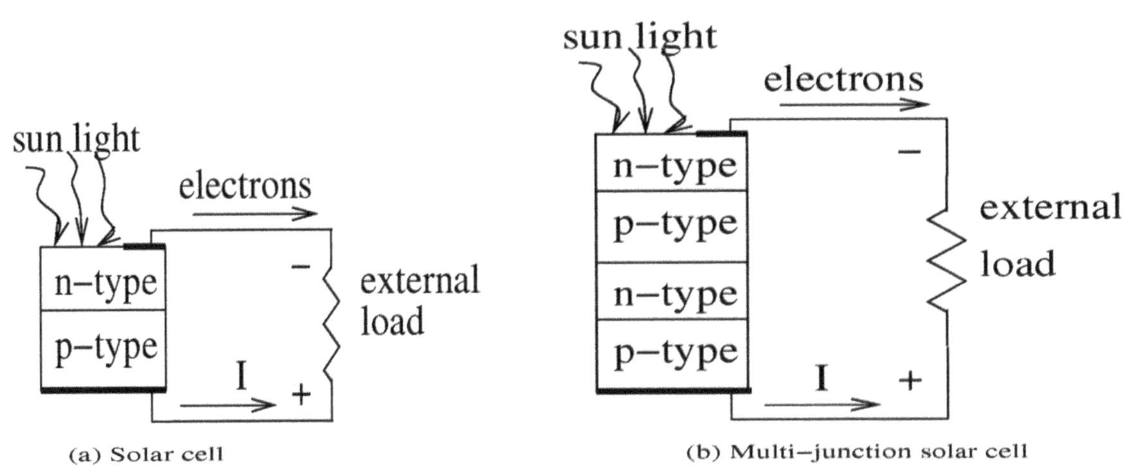

The answer is that when the current of solar panel increases the voltage eventually begins to decrease and when the voltage of the solar increases the current eventually decreases. This is the natural phenomena of the solar panel. With the help of the DC/DC converter, when duty cycle is increased, the current allowed to pass from the solar panel is increased. This increased current cause eventually decrement in the voltage of the solar panel. This way the operating point of the solar panel moves from its one operating point to another operating point. So the operating point of the solar panel can be changed by changing the duty cycle of the DC/DC converter.

Therefore, an algorithm can be written to control the duty cycle of the DC/DC converter by which the solar panel operating point can be controlled. This way the MPPT algorithm affects the solar panel's voltage and current. In Figure 8 an arrangement for the implementation of an MPPT algorithm.

There are several algorithms (such as, perturb and observe (P&O) algorithm, incremental conductance algorithm, fuzzy logic, neural network, and particle swarm optimization-based algorithm, etc.) for the maximum power point tracking (Saravanan & Babu, 2016). Here, perturb and observe algorithm (most simple) is elucidated to illustrate the concept of maximum power extraction from the solar cell. Figure 9 shows the flowchart of the P&O algorithm. In the P&O algorithm voltage and current is measured and power is calculated. The calculated power $P\left(k\right)$ is compared with the previous power $P\left(k-1\right)$. If the power is increased $(P\left(k\right) > P\left(k-1\right))$ and voltage is also increased $(V\left(k\right) > V\left(k-1\right))$ then the voltage of the PV panel again going to increase

$$\left(V\left(k\right)_{\_ref} = V\left(k-1\right)_{\_ref} + C\right)$$

Figure 7. I-V curve of solar panel

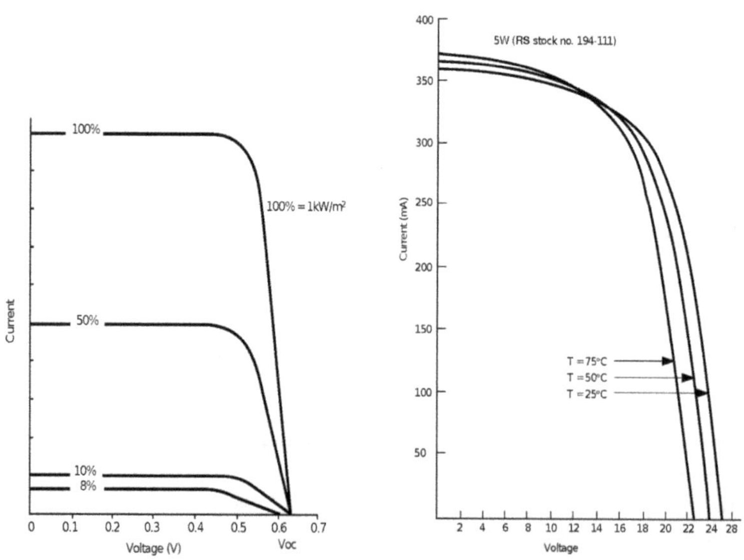

(a) I–V Curve for various irradiance

(b) I–V Curve for various temperature

Figure 8. An arrangement for implementation of an MPPT algorithm

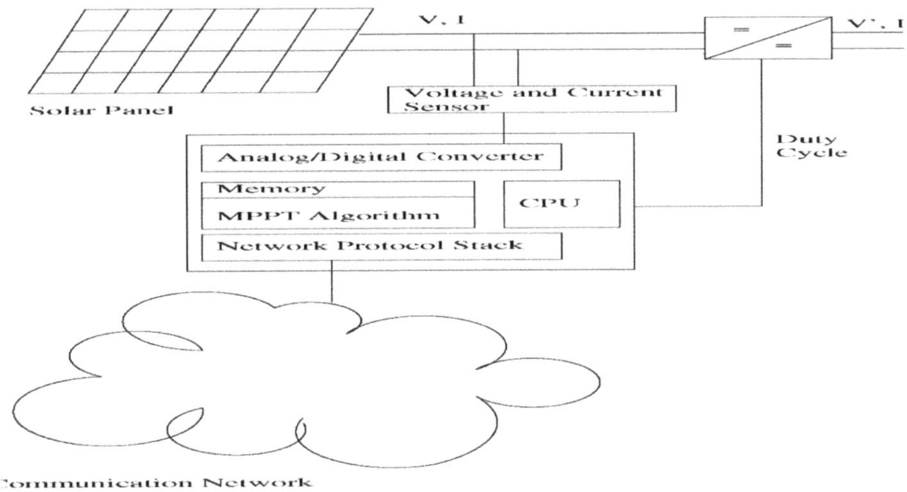

(i.e., duty cycle is going to decrease). If the voltage is decreased $(V(k) < V(k-1))$ then voltage is again going to decrease

$$\left(V(k)_{-ref} = V(k-1)_{-ref} - C\right)$$

(i.e., duty cycle is going to increase). If the power is decreased $\left(P(k) < P(k-1)\right)$ and voltage is increased $(V(k) > V(k-1))$, then the solar panel voltage is going to be decreased

$$\left(V(k)_{-ref} = V(k-1)_{-ref} - C\right)$$

(i.e., duty cycle is going to increase). If the power is decreased $\left(P(k) < P(k-1)\right)$ and voltage is also decreased $\left(V(k) < V(k-1)\right)$, then the solar panel voltage is going to be increased

$$\left(V(k)_{\_ref} = V(k-1)_{\_ref} + C\right)$$

(i.e., duty cycle is going to decrease). The basic analogy of P&O algorithm is that if the power is increased by increasing voltage, it further checks power by increasing voltage of solar panel (i.e., solar panel is operating at left side of maximum power point).

As shown in Figure 10, suppose solar panel works at voltage V_1 (corresponding power P_1). P&O algorithm increase voltage V_1 to V_2 (corresponding power P_2). P_2 is greater than P_1. P&O algorithm again increases voltage V_2 to V_3 (corresponding power P_3). P_3 is greater than P_2. This only happens

Figure 9. Flow chart of perturb and observe algorithm

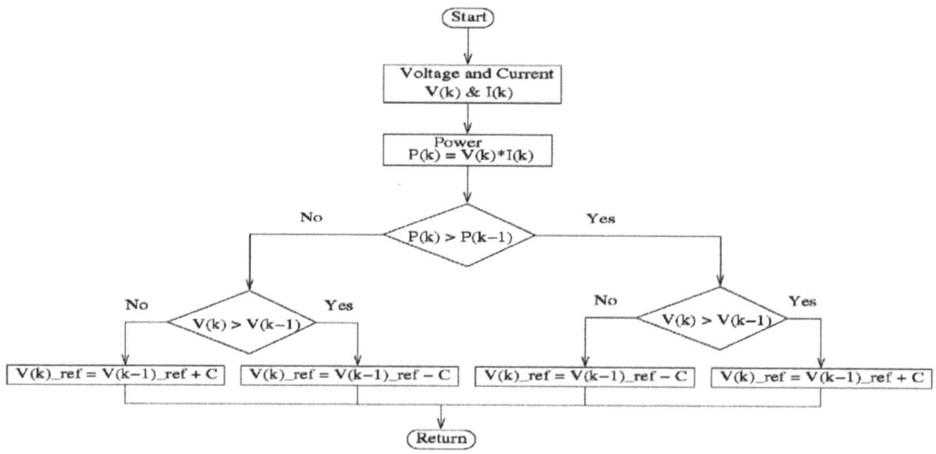

when the operating point is left side of maximum power point. If the power is decreased by increasing voltage, that means solar panel is operating at right side of maximum power point, so that algorithm check for power by decreasing voltage of the solar panel. Suppose solar panel works at voltage V_4 (corresponding power P_4). P&O algorithm increase voltage V_4 to V_5 (corresponding power P_5). P_5 is less than P_4. Here the operating point is right side of the maximum power point. Suppose solar panel works at voltage V_6 (corresponding power P_6). P&O algorithm decreases voltage V_6 to V_5 (corresponding power P_5). P_5 is greater than P_6. So, P&O algorithm again decreases voltage V_5 to V_4 (corresponding power P_4). P_4 is greater than P_5. This way P&O algorithm searches for maximum power point. The P&O algorithm is very easy to implement. But the limitation of this algorithm is that it shows oscillation around the maximum power point and less response speed.

Figure 10. P-V graph for a solar panel

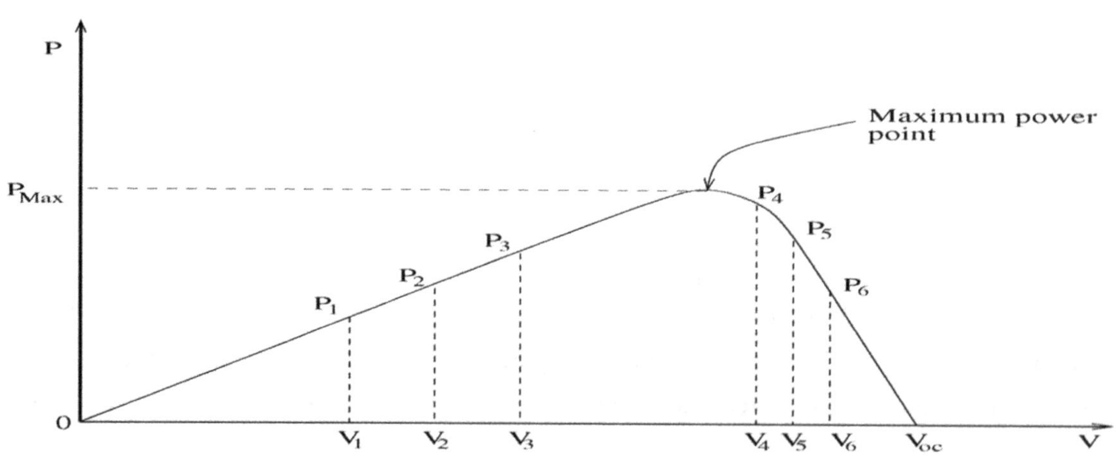

Software Defined Battery

Software Defined Battery (SDB) has software interface to control the batteries operations, such as charging and discharging (Badam, et al., 2016). Batteries technologies perform differently on different characteristics, such as energy density, cost per kWh, recharge time, life time, discharge rate, power to energy ratio (kW/kWh), and maintenance cost. SDB is a hardware-software system, in which many different battery technologies can be interconnected to get the multiple features of the battery characteristics. Some properties of batteries are not cooperative with each other. For example, batteries having high power densities are likely to have low volumetric (Wh/l) and gravimetric (Wh/kg) energy densities vice versa. In a single charge-discharge cycle, energy capacity of the battery is dependent on the rate of current draining (discharge rate) from the battery. If the current draining rate is high, then the energy delivered by battery is low (battery energy capacity decreases). This behavior shows by the battery due to energy loss inside the battery is proportional to the square of the current. The energy capacity and rate of current delivery should not be increased simultaneously. The longevity of a battery highly depends on the charge and discharge rate. Creation of the fissures in electrode speeds up in the case of high current which reduces the energy capacity of the battery.

The modern devices require the characteristics, such as fast charging, low cost, high capacity, less volume, high discharge rate, best longevity, and less heating. A single battery technology is not able to provide all these characteristics. SDB technology interconnects heterogeneous battery technologies to provide the best characteristic of each battery technology. SDB comprises of hardware and software system. Hardware system implements the fine-grained control of charging and discharging (amount of current is controlled for charging and discharging) using switching circuitry.

Software system implements the policies, such as from which battery how much current should be drained, what should be the charge and discharge rate.

Figure 11. SDB system overview

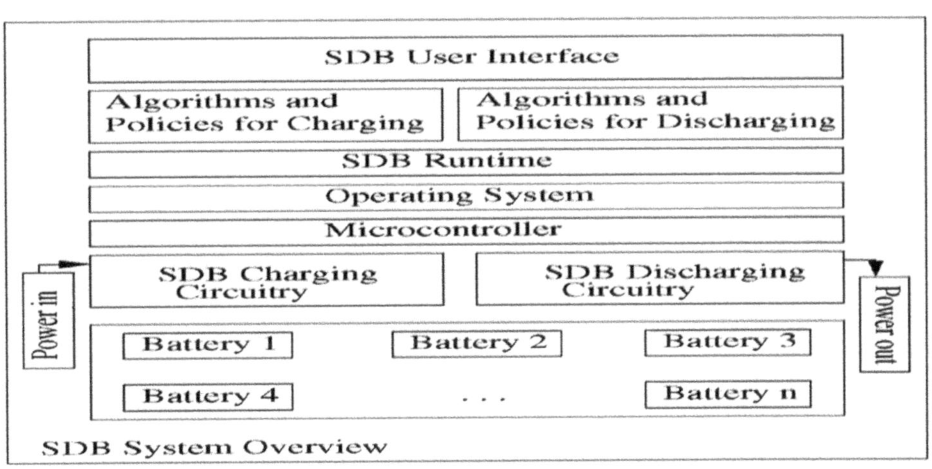

Figure 12. SBD charging/discharging circuitry

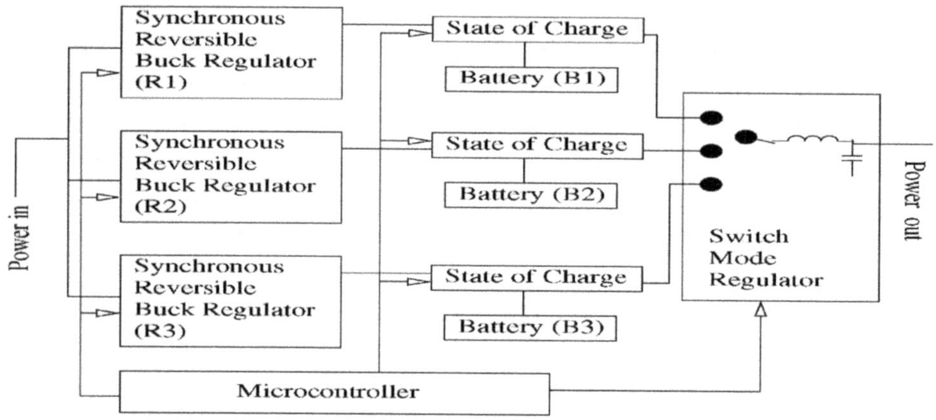

SDB System Overview

SDB system consists of three layers: (i) the batteries and their chemistry, (ii) battery charging/discharging circuit, and (iii) microcontroller & software as shown in Figure 11.

SDB Hardware Components

SDB hardware components provide two facilities: (i) coarse grain switching of the load across multiple batteries, (ii) fine-grain sharing of load by which load is shared among each battery. In Figure 12 the charging/discharging circuit design for three batteries are shown. State of charge monitors are used for indicating the state of charge of the battery. Synchronous reversible buck regulators are used for charging the battery. When external power supply is available, R1, R2, and R3 are configured by the microcontroller in buck mode to charge the batteries. In the above circuitry, a battery can also charge other battery, for example, when battery B1 is charging from B2, then R2 works in reverse buck mode and R1 works in buck mode to charge battery B1. The discharging circuitry drains current from multiple batteries in weighted round robin fashion. The capacitor is used for smoothing the output load current variations.

SDB Software Components

SDB Runtime takes decision for charging and discharging of the batteries. SDB Runtime communicates with the SDB controller for charging and discharging. A SDB system with N batteries has two N-tuples $(c_1, c_2, \ldots, c_N)$ and $(d_1, d_2, \ldots, d_N)$. Value of c_i and d_i are non-negative number, where $1 \geq i \leq N$ and $c_1 + c_2 + \ldots + c_N = 1 \,\&\, d_1 + d_2 + \ldots + d_N = 1$. The numbers represent the fraction of energy that must go in and out from each of the N batteries. There are four interfaces through the SDB Runtime communicate with SDB controller. The four interfaces are:

1. $Charge(c_1, c_2, \ldots, c_N)$: Charge batteries in the given proportion
2. $Discharge(d_1, d_2, \ldots, d_N)$: Discharge batteries in the given proportion
3. $ChargeOneFromAnother(X, Y, W, T)$: Charge battery Y from battery X with power W for time T
4. $QueryBatteryStatus()$: Return state of charge of the battery, terminal voltage, and cycle count for each battery

Charging and discharging algorithms are implemented in the SDB Runtime. By the algorithms, battery characteristics are optimized. SDB system is a cyber-physical system, where algorithmic computations can provide benefit for the diverse battery characteristics requirements.

Operation of Parallel Converters

In smart DC microgrid, there are several distributed energy resources, which are connected through the converters. DC loads and AC loads are also connected through the converters as shown in Figure 4. There are strategies such as centralized control and distributed control for operating converters in parallel for sharing power from DERs. To elucidate the concept of power sharing among the DERs, the droop control method is explained. Droop control method is a decentralized technique and does not require any communication among the control units (Ferreira, Braga, Ferreira, & Barbosa, 2012). The droop control method is easy to implement, but it has poor current sharing and poor voltage regulation. The principle of droop control method is that, for increment in power, the voltage is going to decrease linearly. Let there are three converters connected to DC bus as shown in Figure 13 (a). As shown in Figure 13 (b) at voltage V the power is $P_1 = V*I_1 + V*I_2 + V*I_3$ and at the voltage V' the power is $P_2 = V'*I_1' + V'*I_2' + V'*I_3'$. To illustrate the concept, the constraints are derived here for three converters. Let the slope of the three lines are m and the equation of the lines are (the lines are drawn in Figure 13 (c))

$$y = m*x + c_1 \tag{1}$$

$$y = m*x + c_2 \tag{2}$$

$$y = m*x + c_3 \tag{3}$$

From equation 1, 2, and 3, respectively, x can be calculated

$$x = \frac{y - c_1}{m} \tag{4}$$

Figure 13. Operation of parallel converters

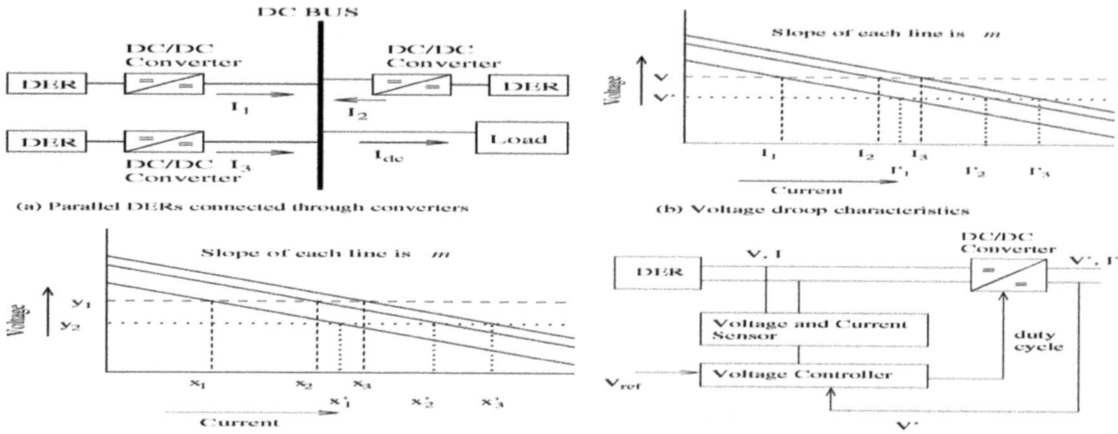

(a) Parallel DERs connected through converters
(b) Voltage droop characteristics
(c) Three linear line corresponding to three sources
(d) Voltage droop control circuit

Let the voltage is y_1, then corresponding current values would be $x_1, x_2, and\ x_3$. Therefore, power P_1 would be $P_1 = y_1 * (x_1 + x_2 + x_3)$ i.e.,

$$P_1 = y_1 * \left(\frac{y_1 - c_1}{m} + \frac{y_1 - c_2}{m} + \frac{y_1 - c_3}{m} \right)$$

or

$$P_1 = \frac{y_1}{m} * \left(3 * y_1 - (c_1 + c_2 + c_3)\right).$$

Similarly, when voltage is y_2, then power would be

$$P_2 = \frac{y_2}{m} * \left(3 * y_2 - (c_1 + c_2 + c_3)\right).$$

For $P_2 > P_1$, i.e.,

$$\frac{y_2}{m} * \left(3 * y_2 - (c_1 + c_2 + c_3)\right) > y_1 * \left(\frac{y_1 - c_1}{m} + \frac{y_1 - c_2}{m} + \frac{y_1 - c_3}{m} \right),$$

if $c_1 + c_2 + c_3 = c$ then the relation would be $3 * (y_1 + y_2) > c$. This means that the power P_2 will be greater than P_1, if $c < 3 * (y_1 + y_2)$. For example, consider the equation

$y = -x + 14, y = -x + 17, and\ y = -x + 12.$

Therefore, $c = 14+17+12 = 43$. Let $y_1 = 10$, then $x_1 = 4, x_2 = 7, and\ x_3 = 2$.

Power $P_1 = 10*(4+7+2) = 130$ and let $y_2 = 8$, then $x'_1 = 6, x'_2 = 9, x'_3 = 4$ and $P_2 = 8*(6+9+4) = 152$. Here $P_2 > P_1$ and also satisfy the condition $c < 3*(y_1 + y_2)$ i.e., 43 < 3*(10+8). Let another value $y_1 = 4\ and\ y_2 = 3$, then

$x_1 = 10, x_2 = 13, x_3 = 8\ and\ x_1^{'} = 11, x_2^{'} = 14, x_3^{'} = 9.$

Therefore,

$P_1 = 4*(10+13+8) = 124\ and\ P_2 = 3*(11+14+9) = 102.$

Here $P_2 < P_1$ and also not satisfy the condition $c < 3*(y_1 + y_2)$ i.e., 43 > 3*(4+3).

Voltage droop technique decreases/increases the voltage for power sharing among multiple DERs. The voltage decrement/increment of the converter (which is associated with DER) is controlled through the voltage control circuit (as shown in Figure 13 (d)) which generate duty cycle for the converter. Voltage controller circuit can be proportional controller, proportional-integral controller, etc. Each controller has its benefits and drawbacks. Proportional controller exhibits faster response. Proportional-integral exhibits zero steady-state error and better power regulation.

Working of Smart Home as a Prosumer

A prosumer can act as energy supplier as well as energy consumer. Here the bidirectional flow of energy from consumer to electrical grid and from electrical grid to consumer is supported (Alagoz, Kaygusuz, & Karabiber, 2012). At any instance of time, prosumer acts either as consumer or producer. Energy storage acts as energy buffer. As shown in Figure 14, smart home has "smart home energy management system", which controls the smart home either as producer or consumer. When home has surplus energy, it "OFF" the charger (which is connected to DC microgrid) and "ON" the DC/DC converter (which is connected to DC microgrid) to supply energy to the DC microgrid. When home needs energy, it drains energy from DC microgrid and acts as a consumer. Smart home energy management system implements the demand response program to control appliances on the basis of price signal. Smart home energy management system uses the communication technology, such as IEEE 802.15.4 (personal area network), Zigbee, etc. Communication infrastructure enables the communication between the energy management system and controllable load.

Demand Response Program

Demand response strategies are applicable for both AC and DC microgrid. It is something which is higher level of control for smart devices regardless of AC or DC infrastructure. Department of energy, United States, describes demand response as "…changes in electric usage by end user customer from their normal consumption patterns in response to changes in the price of electricity over time, or incentive payments designed to include lower electricity use at times of high wholesale market prices or when system reliability is jeopardized…"

With the help of Information and Communication Technology (ICT), load (i.e., consumer) can participate to balance energy supply from the electricity grid. In case of low energy, load can reduce its consumption to balance the power supply. In case of high energy, load can increase its consumption to balance the power supply. ICT assists controllable loads to take active part in demand and supply. Use of ICT in electrical grid introduces the market concept, day ahead pricing scheme, real-time pricing scheme, etc. Demand response program can be used for peak load reduction, emergency condition, etc. Power system shows non-linear characteristic. In case of high load, losses are also very high due to non-linear I^2R (where I = current and R = resistance) losses. Demand response program reduces the peak load, by which, I^2R losses reduce. A good demand response strategy reduces peak load, which reduces power loss and also reduces the infrastructure cost of electricity grid installation. Demand response program is implemented either in centralized or distributed fashion. In centralized implementation, all the appliances in the smart grid are controlled by the utility provider (control center). Appliances send request to the control center. Control center schedules the appliances for operation. Control center processes the requests for the optimization of generation and distribution. In distributed implementation, appliances requests are processed locally on the basis of price or incentive signal. Consumers schedule their appliances having goal of minimizing electricity bill and maximizing comfort (or satisfaction). There are several algorithms proposed in the literature for real-time electricity price prediction by which the consumers schedule their appliance to reduce their electricity bill and increase their comfort. There are several demand response approaches, which are modeled and solved using linear programming, game theory, particle swarm optimization, dynamic programming, evolutionary algorithm, stochastic optimization, etc. Centralized implementation of demand response program is only suitable for small size smart grid. For large scale smart grid, the distributed implementation of demand response program is preferable (Vardakas, Zorba, & Verikoukis, 2015; Zhou, Rodrigues, & Oliveira, 2012).

A distributed real-time demand response approach, heuristic algorithm for demand response program, direct load control algorithm, decentralized demand response method for temperature constrained devices while satisfying temperature requirements, multi-objective optimization model for optimizing the energy usage and comfort of users, and an adaptive real-time scheduling for demand response program analogous to multi-processor real-time scheduling algorithm are some recent enhancement in the

Figure 14. Smart home acts as a prosumer

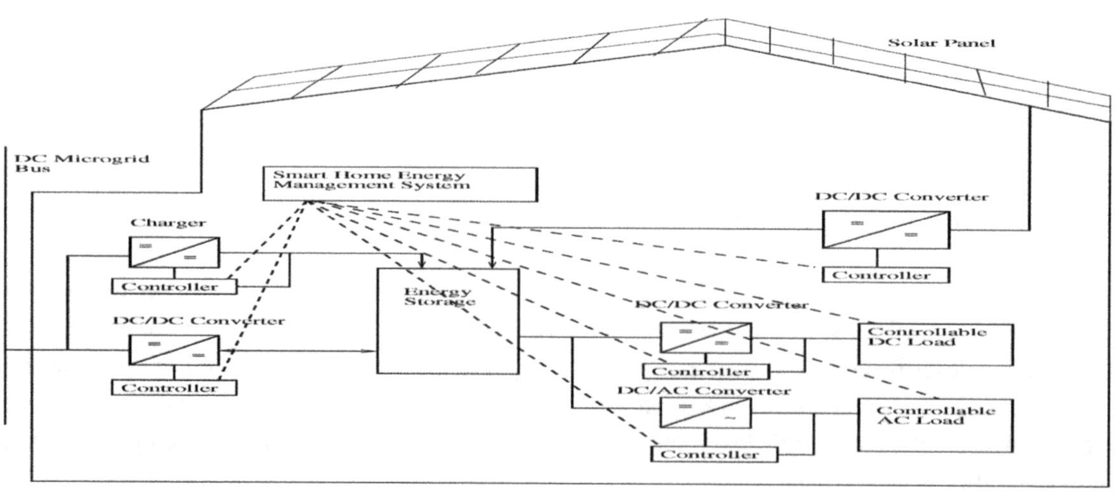

demand response strategies, given by the researchers (Srikantha & Kundur, 2017; Park, Jang, Cho, & Kim, 2017; Mortaj, Ow, Moghavvemi, & Almurib, 2017; Duong, Maillé, Ram, & Toutain, 2017; Ren, Liu, Talari, Shafie-khah, & Catalão, 2017; Martini, Benetti, Vedova, & Facchinetti, 2017), respectively.

To explain the concept of demand response program, an example (which is a demand response scheme proposed by (Mohsenian-Rad & Leon-Garcia, 2010)) is elucidated. This example explains the demand response problem formulation. The system model is as follows. Let A is set of appliances. A includes dish washer, refrigerator, PHEV (Plug-In Hybrid Electric Vehicle), etc. For each appliance $a\varepsilon A$, an energy consumption scheduling vector X_a as follows: $X_a = \left[x_a^1, x_a^2, \ldots, x_a^h, \ldots, x_a^H\right]$ where $H \geq 1$ is the scheduling horizon. For example,

$$H = 24 \ or \ H = 48, h\varepsilon\mathcal{H} = \left\{1, 2, \ldots, H\right\}.$$

Let E_a denotes the total energy needed for the operation of an appliance $a\varepsilon A$. For example, in case of PHEV, total $E_a = 16kWh$ is needed to charge the battery. For washing machine, $E_a = 3.6kWh$ energy required per load. Each appliance $a\varepsilon A$ has beginning and end time interval $\alpha_a, \beta_a\varepsilon\mathcal{H}$. For example, dishwasher should finish its work before dinner (e.g., $\alpha_a = 2PM, \beta_a = 8PM$), for PHEV ($\alpha_a = 10PM, \beta_a = 7AM$). For the given parameters E_a, α_a, and β_a, energy requirement for each appliance $a\varepsilon A$ within the time interval $[\alpha_a, \beta_a]$ satisfies the relation $\sum_{h=\alpha_a}^{\beta_a} x_a^h = E_a$. For each appliance, $a\varepsilon A$, maximum power consumption denoted by $\gamma_a^{\max}$. For example, PHEV may charge only up to $\gamma_a^{max} = 3.3kWh$. For each appliance, $a\varepsilon A$, minimum stand-by power consumption denoted by $\gamma_a^{\min}$. Therefore, for a particular time interval for an appliance the relation $\gamma_a^{\min} \leq x_a^h \leq \gamma_a^{\max}, \forall h\varepsilon\left[\alpha_a, \beta_a\right]$ is satisfied. There is usually a limit on the total energy consumption at each residential unit at each hour denoted by E^{max}. This limit poses the constraint on energy scheduling to satisfy the relation $\sum_{a\epsilon A} x_a^h \leq E^{max}$, $\forall h\varepsilon\mathcal{H}$. The constraint in the relations $\sum_{h=\alpha_a}^{\beta_a} x_a^h = E_a$, $\gamma_a^{\min} \leq x_a^h \leq \gamma_a^{\max}, \forall h\varepsilon\left[\alpha_a, \beta_a\right]$, and $\sum_{a\int A} x_a^h \leq E^{\max}$, $\forall h\varepsilon\mathcal{H}$ determine all the valid choice for the energy consumption scheduling vectors.

A feasible scheduling set χ for all possible energy consumption scheduling vectors as

$$\chi = \{x| \sum_{h=\alpha_a}^{\beta_a} x_a^h = E_a,$$

$$\gamma_a^{\min} \leq x_a^h \leq \gamma_a^{\max}, \forall h\varepsilon\left[\alpha_a, \beta_a\right], x_a^h = 0, \forall a\varepsilon A, h\varepsilon\mathcal{H} \setminus \left[\alpha_a, \beta_a\right], \sum_{a\int A} x_a^h \leq E^{\max}, \ \forall h\varepsilon\mathcal{H},$$

where x denotes the vector of energy consumption scheduling variable for all appliances. Real–time pricing with inclining block rates is as follows for this example. Let $l^h = \sum_{a\int A} x_a^h$ denotes the total hourly energy consumption. Inclining block rate is calculated with the relation

$$p^h\left(l^h\right)=\begin{cases}a^h, if\ 0\le l^h\le c^h\\ b^h, if\ l^h>c^h\end{cases} \quad where\ p^h\left(l^h\right)$$

depend on three parameters $a^h, b^h, c^h \geq 0$. Problem formulation for this example can done as following. Electricity payment for all appliances modeled as $\sum_{h=1}^{H} p^h(\sum_{a\epsilon A} x_a^h)\times\left(\sum_{a\epsilon A} x_a^h\right)$ and cost of waiting modeled as $\sum_{h=1}^{H}\sum_{a\varepsilon A}\rho_a^h x_a^h$ where ρ_a^h is a waiting parameter. The waiting parameter for each appliance $a\varepsilon A$, modeled as $\rho_a^h=\frac{\left(\delta_a\right)^{\beta_a-h}}{E_a}$, $\forall a\varepsilon A, h\varepsilon\left[\alpha_a,\beta_a\right]$ where $\delta_a \geq 1$ is an adjustable control parameter. Here, energy consumption scheduling problem modeled as following optimization problem

$$\text{minimise}\sum_{h=1}^{H} p^h(\sum_{a\varepsilon A} x_a^h)\times\left(\sum_{a\int A} x_a^h\right)+\lambda_{wait}\sum_{h=1}^{H}\sum_{a\varepsilon A}\frac{\left(\delta_a\right)^{\beta_a-h}}{E_a}x_a^h$$

where λ_{wait} is used for controlling the importance of waiting cost. The linear programming technique is used for solving the above energy consumption scheduling problem. In this example, the constraints are based on three criteria namely (i) energy requirement of each appliance in the specified time interval, minimum standby power requirement and maximum power requirement of each appliance, and energy consumption limit of each residential house, (ii) pricing with inclining block rate, and (iii) waiting parameters for each appliance. The goal is to make use of each appliance to meet the user requirement & satisfaction and reduction in the electricity bill. The above optimization problem finds out when to use particular appliance to optimize the requirement, satisfaction, and electricity bill. An interested reader can find fine-grained detail of this example in the (Mohsenian-Rad & Leon-Garcia, 2010).

Communication Protocols

Communication infrastructure is the essential component in the realization of the smart grid. Communication infrastructure enables the consumers to take active participation in the smart grid energy market. Smart grid communication infrastructure includes the Home Area Network (HAN), Building Area Network (BAN), Neighborhood Area Network (NAN), and Field Area Network (FAN). Smart grid communication infrastructure constitutes multiple communication technologies, such as fiber optics, power line communication, copper wireline communication, and wireless communication technologies (e.g., GPRS, GSM, 3G, 4G, WiMax, LTE, WLAN, etc.) (Yan, Qian, Sharif, & Tipper, 2013). Smart grid constitutes billion of smart devices which will communicate over communication infrastructure. The intelligent smart devices are embedded with the network hardware and protocol stack as shown in the Figure 15 (Wang, Xu, & Khanna, 2011).

Figure 15. Smart device having network capability

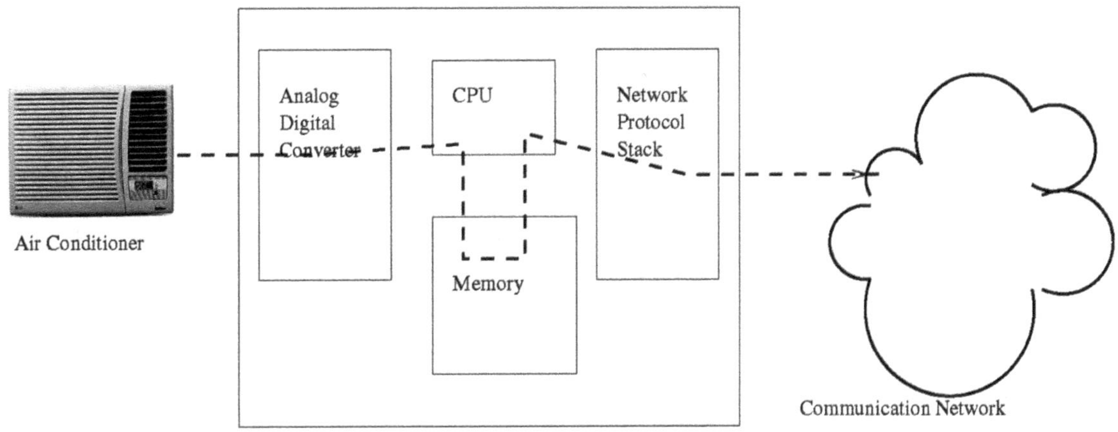

Smart grid communication infrastructure needs to have energy efficient wireless, wireline, and optical network (Erol-Kantarci & Mouftah, 2015). Energy efficient wireless communication technologies are 3G, 4G, WiMax, LTE-A, etc. (these are well suited for FAN and NAN). Energy efficient wireless communication technologies, such as Zigbee and Wi-Fi are well suited for BAN and HAN. Energy efficient wireline communication technologies, such as 802.3az (energy efficient Ethernet) and PLC (Power Line Communication) are good for HAN and BAN. Fiber-Wireless (Fi-Wi) network is also suitable for BAN and NAN. Billions of smart devices will be connected through the Internet. Each of them is uniquely identifiable. These billions of interconnected smart heterogeneous devices help the realization of the smart grid. Therefore, it required flexible layered architecture for communication and service. Among the several proposed architectures, three layer architecture and five layer architecture (Al-Fuqaha, Guizani, Mohammadi, Aledhari, & Ayyash, 2015; Sethi & Sarangi, 2017) will be explained here. The reader who wants to know more about the IoT architecture can refer the references, which include cloud and fog-based architecture also. IoT architecture stack is shown in the Figure 16.

Figure 16.IoT architecture (three layer and five layer)

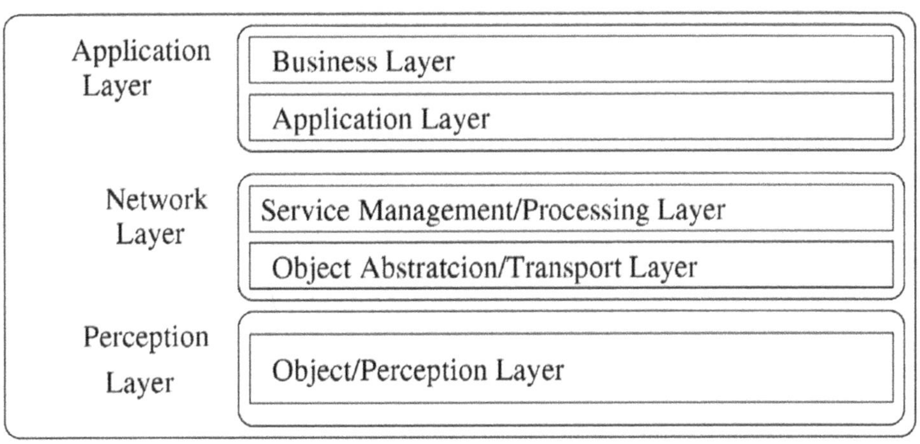

- **Object Layer/Perception Layer:** The first layer represents the physical sensors and actuators which perform the functionalities, such as measuring temperature, weight, humidity, motion, vibration, etc. This layer also digitizes and transfer data to Object Abstraction Layer/Transport Layer. There are several kinds of sensors and actuators, for example, neural sensors, medical sensors, environmental sensor, chemical sensor, bio sensor, magnetometer, location sensor, movement sensor, camera, light sensor, etc. available to make our earth smart.
- **Object Abstraction Layer/Transport Layer:** This layer transfer object layer data to Service Management Layer through one of the communication technologies, such as 3G, 4G, 5G, GSM, Wi-Fi, Zigbee, Visible Light Communication, Power Line Communication, etc.
- **Service Management Layer/Processing Layer:** This layer processes the data received by Object Abstraction Layer and delivers the required services. Service management can be database oriented, semantic-based, event-based, application specific, etc.
- **Application Layer:** Application layer is used for providing services for the customers. Application layer explores the application of several areas, such as energy conservation, smart agriculture, smart supply chain and logistics, smart healthcare, smart home, smart transportation, smart city, smart environment, etc.
- **Business Layer:** Responsibility of this layer is to create business model based on the data received from the application layer.

Smart devices help in the realization of the smart city. Smart grid uses Internet of Things to connect smart meters used for collecting energy consumption data, controlling & monitoring. There are also several issues of cyber security in the smart grid. Network security threats in a smart grid are denial-of-service attack on integrity and confidentiality.

CONCLUSION

This article elucidates the evolution of Industry 4.0. "Cyber-Physical Systems" and "Internet of Things" (both are the key technologies for the Industry 4.0) are discussed with very fine details. This article presents a layered architecture of cyber-physical system (taking example as smart DC microgrid). Smart DC microgrid components, such as distributed energy resources, intelligent electronic devices, energy storage, controllable load, and comparison of AC and DC microgrid are discussed. Working principles of extracting maximum power from solar panel, software-defined batteries, operation of parallel converters, working of a prosumer, demand response program, and IoT architecture are elucidated. Realization of the smart DC microgrid technology still needs more research and development. In the smart grid, algorithmic logic is written to control the smart grid as per the laws of physics. To realize a secure and stable smart grid, the correctness proof of the algorithmic logic is required. Any fault in algorithmic logic can lead to huge loss and failure of the smart grid. Smart grid needs to be carefully analyzed for the cases of communication failure of control signals.

REFERENCES

Administration, U. E. (2017). *International Energy Outlook 2017.* Retrieved December 17, 17, from U.S. Energy Information Administration: https://www.eia.gov/outlooks/ieo/pdf/0484/%282017/%29.pdf

Al-Fuqaha, A., Guizani, M., Mohammadi, M., Aledhari, M., & Ayyash, M. (2015). Internet of Things: A Survey on Enabling Technologies, Protocols, and Applications. *IEEE Communications Surveys and Tutorials*, *17*(4), 2347–2376. doi:10.1109/COMST.2015.2444095

Alagoz, B., Kaygusuz, A., & Karabiber, A. (2012). A user-mode distributed energy management architecture for smart grid applications. *Energy*, *44*(1), 167–177. doi:10.1016/j.energy.2012.06.051

Ammar, F. B., Hafsa, I. H., & Hammami, F. (2013). Analytic hierarchy process selection for batteries storage technologies. In *2013 International Conference on Electrical Engineering and Software Applications* (pp. 1-6).

Badam, A., Chandra, R., Dutra, J., Ferrese, A., Hodges, S., & Hu, P. … & Skiani, E. (2016). Software-defined Batteries. *Communications of the ACM*, *59*(12), 111–119. doi:10.1145/3007179

Balog, R. S., & Krein, P. T. (2011). Bus Selection in Multibus DC Microgrids. *IEEE Transactions on Power Electronics*, *26*(3), 860–867. doi:10.1109/TPEL.2010.2094208

Biro, D. F. (2017). Key world energy statistics. *International Energy Agency*. Retrieved December 17, 2017, from https://www.iea.org/publications/freepublications/publication/KeyWorld2017.pdf

Duddu, P. (2015). *The 10 worst blackouts of the last 50 years*. Retrieved December 17, 2017, from http://www.power-technology.com/features/featurethe-10-worst-blackouts-in-the-last-50-years-4486990/

Duong, N. H., Maillé, P., Ram, A. K., & Toutain, L. (2017). Decentralized demand response for temperature-constrained appliances. *IEEE Transactions on Smart Grid*, 1. doi:10.1109/TSG.2017.2778225

Erol-Kantarci, M., & Mouftah, H. T. (2015). Energy-efficient information and communication infrastructures in the smart grid: A survey on interactions and open issues. *IEEE Communications Surveys and Tutorials*, *17*(1), 179–197. doi:10.1109/COMST.2014.2341600

Federal Ministry For Economic Affairs And Energy. (2016). *Plattform Industrie 4.0.* Retrieved December 17, 2017, from http://www.plattform-i40.de/I40/Redaktion/EN/PressReleases/2016/2016-03-02-kooperation-iic.html

Federal Ministry For Economic Affairs And Energy. (n.d.). *The background to Plattform Industrie 4.0.* Retrieved December 17, 2017, from http://www.plattform-i40.de/I40/Navigation/EN/ThePlatform/PlattformIndustrie40/plattform-industrie-40.html

Ferreira, R. A., Braga, H. A., Ferreira, A. A., & Barbosa, P. G. (2012). Analysis of voltage droop control method for dc microgrids with Simulink: Modelling and simulation. In *2012 10th IEEE/IAS International Conference on Industry Applications* (pp. 1-6).

Gangale, F., Vasiljevska, J., Covrig, C. F., Mengolini, A., & Fulli, G. (2017). JRC Publications Repository. *Smart Grid Project Outlook 2017.* Retrieved December 17, 2017, from http://publications.jrc.ec.europa.eu/repository/bitstream/JRC106796/sgp/outlook/2017-online.pdf

Germany Trade And Invest. (n.d.). *INDUSTRIE 4.0.* Retrieved December 17, 2017, from https://www.gtai.de/GTAI/Navigation/EN/Invest/Industries/Industrie-4-0/Industrie-4-0/industrie-4-0-what-is-it.html

i-SCOOP. (n.d.). *Industry 4.0: the fourth industrial revolution – guide to Industrie 4.0.* Retrieved December 17, 2017 from https://www.i-scoop.eu/industry-4-0/

IEEE. (2015). *Towards definition of the Internet of Things.* Retrieved December 17, 2017, from https://iot.ieee.org/images/files/pdf/IEEE_IoT_Towards_Definition_Internet_of_Things_Revision1_27MAY15.pdf

Ikebe, H. (2003). Power systems for telecommunications in the IT age. In *The 25th International Telecommunications Energy Conference INTELEC '03* (pp. 1-8).

India Smart Grid Forum. (2015). *Smart grid project book - A global snapshot April 2015.* Retrieved December 17, 2017 from http://www.indiasmartgrid.org/document/ISGF-Smart%20Grid%20Project%20Book%20-%20A%20Global%20Snapshot.pdf

Liu, Z., & Li, M. (2014). Research on energy efficiency of dc distribution system. In *2nd AASRI Conference on Power and Energy Systems* (pp. 68-74). 10.1016/j.aasri.2014.05.031

Majumder, R., Ghosh, A., Ledwich, G., & Zare, F. (2010). Power management and power flow control with back-to-back converters in a utility connected microgrid. *IEEE Transactions on Power Systems, 25*(2), 821–834. doi:10.1109/TPWRS.2009.2034666

Martini, D. D., Benetti, G., Vedova, M. L., & Facchinetti, T. (2017). Adaptive Real-Time Scheduling of Cyber-Physical Energy Systems. *ACM Transactions on Cyber-Physical Systems, 1*(4), 20–25.

Mohsenian-Rad, A.-H., & Leon-Garcia, A. (2010). Optimal residential load control with price prediction in real-time electricity pricing environments. *IEEE Transactions on Smart Grid, 1*(2), 120–133. doi:10.1109/TSG.2010.2055903

Mortaj, H., Ow, S. H., Moghavvemi, M., & Almurib, H. A. (2017). Load shedding and smart-direct load control using internet of things in smart grid demand response management. *IEEE Transactions on Industry Applications, 53*(6), 5155–5163. doi:10.1109/TIA.2017.2740832

National Science Foundation. (2006). *Cyber-Physical Systems.* Retrieved December 17, 2017 from https://www.nsf.gov/publications/pub/summ.jsp?ods/key=nsf12520/&org=NSF

United Nations. (2015). *Paris Agreement.* Retrieved December 17, 2017, from https://unfccc.int/files/essential/background/convention/application/pdf/english/paris/agreement.pdf

Olivier, J., Schure, K., & Peters, J. (2017). Trends in Global CO2 and total greenhouse gas emissions. *PBL Netherlands Environmental Assessment Agency*. Retrieved December 17, 17, from http://www.pbl.nl/sites/default/files/cms/publicaties/pbl-2017-trends-in-global-co2-and-total-greenhouse-gas-emissons-2017-report/_2674.pdf

Park, L., Jang, Y., Cho, S., & Kim, J. (2017). Residential Demand Response for Renewable Energy Resources in Smart Grid Systems. *IEEE Transactions on Industrial Informatics*, *13*(6), 3165–3173. doi:10.1109/TII.2017.2704282

Phillips, D. T. (2005). Was Einstein a Space Alien. *NASA*. Retrieved December 17, 2017, from https://www.nasa.gov/vision/universe/starsgalaxies/23mar_spacealien.html

Ren, F. W., Liu, X., Talari, S., Shafie-khah, M., & Catalão, J. P. (2017). Multi-objective Optimization Model of Source-Load-Storage Synergetic Dispatch for Building Energy System Based on TOU Price Demand Response. *IEEE Transactions on Industry Applications*.

RS-Data-Sheet. (1999). Solar Panels, RS Data Sheet: RS stock numbers 194-082, 194-105, 194-111, 194-127,194-133, 194-149, 194-161, 194-183.

Saez-de-Ibarra, A., Milo, A., Gaztañaga, H., Etxeberria-Otadui, I., Rodríguez, P., Bacha, S., ... (2013). Analysis and comparison of battery energy storage technologies for grid applications. In 2013 IEEE Grenoble Conference (pp. 1-6).

Saravanan, S., & Babu, N. R. (2016). Maximum power point tracking algorithms for photovoltaic system – A review. *Renewable & Sustainable Energy Reviews*, *57*, 192–204. doi:10.1016/j.rser.2015.12.105

Sethi, P., & Sarangi, S. R. (2017). Internet of Things: Architectures, Protocols, and Applications. *Journal of Electrical and Computer Engineering*.

Shen, J., Jiang, C., & Li, B. (2015). Controllable Load Management Approaches in Smart Grids. *Energies*, *8*(10), 11187–11202. doi:10.3390/en81011187

Srikantha, P., & Kundur, D. (2017). Resilient Distributed Real-Time Demand Response via Population Games. *IEEE Transactions on Smart Grid*, *8*(6), 2532–2543. doi:10.1109/TSG.2016.2526651

Vandoorn, T. L., Meersman, B., Kooning, J. D., & Vandevelde, L. (2012). Analogy Between Conventional Grid Control and Islanded Microgrid Control Based on a Global DC-Link Voltage Droop. *IEEE Transactions on Power Delivery*, *27*(3), 1405–1414. doi:10.1109/TPWRD.2012.2193904

Vardakas, J. S., Zorba, N., & Verikoukis, C. V. (2015). A Survey on Demand Response Programs in Smart Grids: Pricing Methods and Optimization Algorithms. *IEEE Communications Surveys and Tutorials*, *17*(1), 152–178. doi:10.1109/COMST.2014.2341586

Wang, W., Xu, Y., & Khanna, M. (2011). A survey on the communication architectures in smart grid. *Computer Networks*, *55*(15), 3604–3629. doi:10.1016/j.comnet.2011.07.010

Wikipedia. (n.d.). *Industrial Internet Consortium.* Retrieved December 17, 2017, from https://en.wikipedia.org/wiki/Industrial/Internet/Consortium

Yan, Y., Qian, Y., Sharif, H., & Tipper, D. (2013). A Survey on Smart Grid Communication Infrastructures: Motivations, Requirements and Challenges. *IEEE Communications Surveys and Tutorials*, *15*(1), 5–20. doi:10.1109/SURV.2012.021312.00034

Yoshikawa, K., Kawasak, H., Yoshida, W., Irie, T., Konishi, K., & Nakano, K., … & Yamamoto, K. (2017). Silicon heterojunction solar cell with interdigitated back contacts for a photoconversion efficiency over 26%. *Nature Energy*, 2.

Zhou, L., Rodrigues, J., & Oliveira, L. (2012). QoE-driven power scheduling in smart grid: Architecture, strategy, and methodology. *IEEE Communications Magazine*, *50*(5), 136–141. doi:10.1109/MCOM.2012.6194394

Chapter 6
Cyber-Physical Systems Framework for Smart Built-Environments

Yaseen Srewil
Technische Universität Dresden, Germany

Raimar J. Scherer
Technische Universität Dresden, Germany

ABSTRACT

Cyber-physical systems (CPS) can be seen as digitally-driven technological fusions of physical, spatial and virtual models. Thereby, the construction entities, locations and processes are tightly linked to the digital building models (BIM) using automated data capturing (ADC) solutions. This chapter focuses on the challenge of modelling a cyber–physical systems approach for constructing domains in order to effectively exploit the computation, control and communication capabilities. Pairing BIM and ADC technologies offers an adequate solution bridging the information gaps and promoting the collaboration among digital, spatial and physical construction entities. The results are cyber-physical objects (CPOs) at a level of "smartness" that provides enhancement of digital capabilities and the ability of context-awareness. These hybrid objects of digital equivalents and physical entities can be embedded in the construction and logistics processes in order to track progress and monitor the construction site activities and control flow close to real-time.

INTRODUCTION

Investigation of a digitally-driven production strategy is important to enable bridging the information gaps between physical construction and the digital models based on sensing and emerging technologies. A breakthrough has been achieved using Building Information Modelling (BIM) technology that can extrapolate into a digital fabrication strategy. Nevertheless, the granularity of the relationships and the interconnectivity between the digital models and physical fields has to be carefully examined in order to establish a cyber-physical coordination and integration in the construction.

DOI: 10.4018/978-1-5225-4936-9.ch006

Cyber-Physical Systems (CPS) can provide for tight interaction between the physical objects and the virtual "cyber" models. Current Automated Data Capture (ADC) and BIM technologies are at a level of maturity that facilitates the integration of physical construction of spaces and objects into virtual information models which allows a bi-directional data flow in a near real-time. A cyber-physical system is about the conjunction between the physical world of things and the digital "cyber" world of software and computing. Like other industry 4.0 initiatives, a CPS approach can be seen as promising paradigms for a digitally driven technological fusion. Thereby the construction entities and site processes can be interrelated and linked directly to the digital building models using advanced ADC, i.e. QR-codes, RFID, GPS, RTLs, laser scanning, etc.

This work describes a CPS framework to close the information loop between the digital models and the real physical objects on-site by providing "near" real-time feedbacks about the actual state of physical things and their related activities.

This chapter focuses on design a cyber–physical approach in construction domain. A definition of CPS, state-of-the-art are reviewed. The current data flow problems between physical and virtual models on-site due to traditional data handling is briefly discussed. Next the chapter presents the importance, impacts and barriers of modelling a CPS for construction.

A generic framework enables the integration of physical construction entities with cyber models is addressed. The work introduces the CPOs as a basic unit in the CPS framework in order to have a complete digital footprint of the construction entities the during project lifetime. A digital construction site approach is developed to recognize the CPOs and maintain consistent relationships among the spatial, virtual and physical of construction.

A scenario for modelling of the CPOs based on object-oriented approach is developed in order to ensure a timely update in objects' status. Finally, the work proposes an approach for embedding these situational-aware objects in construction process models to support tracking activities and facilitate process monitoring and control close to real-time.

BACKGROUND

The rapidly progressing in ICT and sensing technologies open up the opportunity to achieve a bi-directional coordination and integration between virtual information models and physical construction in real-time. These functionalities have been introduced as features of the recent appearing of the CPS term, which is about the intersection of the physical and the cyber world (Lee, 2015).

CPS was originally developed in computer science as the next wave of innovation in ICT enabling the new generation of "smart systems". It has drawn great attention to academia, governments and industries due to its environmental and economic impact (NIST, 2013a). Beyond the embedded system, CPS is open, capable of multi-scale and designed as a network of interactive components more than standalone devices (Talcott, 2008; Lai, Ma, Chang, Chao, & Huang, 2011). Such smart systems encompass computing (virtual space) and physical components, seamlessly integrated and closely interacting to sense the changing state of the real world (NIST, 2013b). While CPS is still in its infancy, there are a few design principles, system architectures and prototypes -mainly in scientific works- for different application domains such as manufacturing, energy, transportation and civil infrastructure and medical systems (Sanislav & Miclea, 2012).

Recently, more research efforts are carried out in the framework of the industry 4.0 initiative. The concept has been promoted in Germany to highlight the work of more efficient, intelligent and sensor-rich manufacturing facilities. CPS is the backbone of this revolutionary cycle. For instance, in the smart factory stand for the efficient and optimized manufacturing industry, the CPS is adopted for tracking and monitoring a product throughout its life cycle (Kagermann, Wahlster & Helbig, 2013)

The cyber-physical systems approach can also be a paradigm shift in the construction and infrastructure engineering sectors. The collaboration of current cutting-edge technologies, namely, BIM and ADC technologies can establish a robust digitally-driven environment. As in Industry 4.0, CPS in the construction domain can be defined as a tight conjoining of virtual information models and physical construction to enable bi-directional data exchange, integration and coordination.

Generally, the physical construction entities have cyber capabilities. The interaction between the physical reality and cyber world is based on event/ action (or behaviour) relations. Consequently, if a specific change in one system part takes place, the impact in other system parts is reflected in a time-sensitive and/or a spatial-sensitive way.

The Need for Interconnected Systems in Construction

Unlike manufacturing, a construction site is typically in an open air and has limited production space. The building products are one-of-a-kind in a hugely fragmented industry (Akinsola, Dawood, & Hobbs, 2000). A permanent interruption in the schedule and changes are accordingly common. Currently, project's actors on-site are mostly confronted with accessing of the information to support decision making on time (Chen & Kamara, 2008). Although BIM technology as a digital form for entire building lifecycle is rapidly evolved and the digital models are available and can be exchanged, yet implementation of BIM methodology on site is at the early phase and is not often used to transmit the construction plan to production process during the construction phase. One reason can be the lack of the bi-directional coordination between the virtual models and site activities because of manual and paper-based format for information exchange. Actually, the manual method for monitoring activities and tracking assets are still dominant in many fields of practices. The result is that the information of actual construction processes is incomplete labour intensive, prone to errors and delays (Gehbauer, Zülch & Börkircher, 2007).

Adoption of new technological innovations of digitally driven as BIM and interconnected system like CPS are going to change a lot of issues facing the construction today. They will help a digital transformation in the construction industry that can drive down costs and reduce time while increasing the quality, reliability and performance of an asset through its life. The digital transformation, being about building and maintaining assets in a smarter way, is partly a matter of coupling of automated data capture solutions and data processing (collection, store, transfer, etc.).

The possible implementation of advanced sensing technologies in construction highlights the need for tight integration between physical entities and BIM models in order to close the information loops and allow an “automatic„ data exchange between both virtual and real construction.

Several approaches are recently proposed to integrate sensing technologies with BIM technology, especially for assets tracking, logistics, safety on site and monitoring of project progress as in Azimi, Lee, Abourizk and Alvanchi, (2011); Teizer, Cheng and Fang (2013); Motamedi, Setayeshgar, Soltani and Hammad (2013). Others focus on the use of BIM models in the design of mixed-reality systems based on wearable computers (HoloLens from Microsoft, google Lens, etc.) for coordination purposes (Ho, 2017).

Few research efforts propose bi-directional links to bridge the gap between virtual information models and construction components for information sharing and building operation management (Akanmu, Anumba, & Messener. 2013; Sørensen, Christiansson & Svidt, 2010). Summing-up, the solutions were in some cases a one direction relationship for tracking project progress while in others promised approaches were scenarios based on expensive IT solutions for bi-directional coordination. These hinder a large scale deployment and adoption CPS in the fragmented construction industry.

CPS in Construction: Promise and Challenges

The CPS role in construction is to close the information loop by equipping the information models with a timely feedback about the actual physical production processes (Figure 1). Here, the physical construction entities and information are concurred during logistics and construction. Hence, the differences between information and physical entities narrow down, since the products are inextricably linked to "their" information.

The CPS serves several purposes especially for monitoring and control applications, e.g. at construction and operation phases respectively:

1. Site information is used to describe the processes on-site AS-IS at a real time or close to. In turn, a continuous monitoring of AS-IS against AS-Planned performances supports project progress tracking and process simulation models with real inputs. Moreover, the analysis of deviations helps exceptions anticipation and alternative process planning (Srewil, Kadolsky, Scherer, 2012).
2. The data are collected using sensor-based networks (e.g. building integrated temperature and humidity sensors, sensors for surface crack detection for structural health monitoring, etc.). They support advanced computing platforms and enable a frequency and dynamic analysis for sensing elements.

Figure 1. Merging and interaction of information and physical production in built environments domain according to CPS approach after (Björk, 2002).

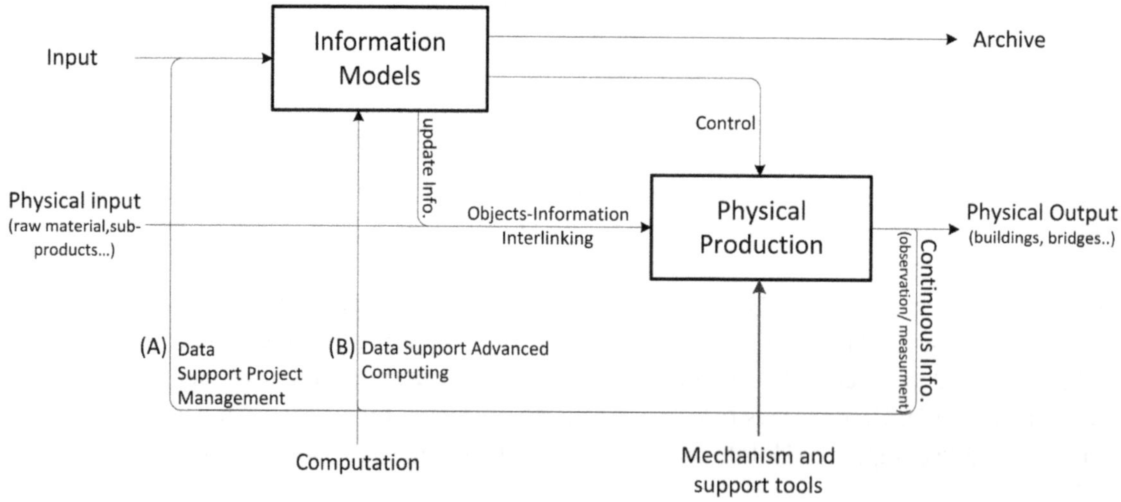

The adopted CPS scenario in the built environments offer several potential benefits by improving monitoring of construction activities, safety management, permanents inspection of stability and structural health monitoring at operation phase and enable automated services for buildings and infrastructures. This means an increase of effective project management and elimination hazards.

However, developing an approach based on CPS principles for construction industry is a challenging task due to several reasons:

3. The digital project models are heterogeneous with huge intensive and diverse information. This can hinder correct coordination and integration with real physical construction.
4. The level of integration: Understanding at what point the physical reality should conjunctively coexist with BIM, one needs to define the level of details and information placed on BIM models. The level at which the field data are retrieved. These aggregated data shall be also added upwards into estimates of the higher level of a process and the whole project.
5. The complexity of CPS approach itself includes not only the physical layer and probably people's behaviour and human/machine interaction at lower level, also the network layer and computing capabilities (Wang & Zhou, 2014).
6. Security and reliability: Cyber-attacks are more common nowadays. Such attacks on CPS may disclose business and/ or personal information. Also, unexpected failure in the CPS may occur due to errors it any system's parts lead to system (Cardenas, Amin & Sastry, 2008).

Aforementioned challenges imply the importance of development an adequate paradigm at various levels of abstraction to solve CPS-related problem in the construction domain.

CPS Framework for Construction

The chapter addresses an integrated system based on CPS concept for smoothly intersection of the physical construction entities and virtual information models controlled by semantic rules. The dataflow, rules and relationships among systems subcomponents maintain a consistent integration within the framework.

Figure 2 depicts a solution bundles different aspects and facets into the framework's conceptual model. Abstractly, it encompasses virtual "cyber" models, physically real construction, data handling (computing, analysis and management) and finally quasi real time services. The framework provides a closed loop for information interoperability among diverse subsystems. This extended concept allows the individual entities of diverse subsystems to work together toward a scalable system with new capabilities. Furthermore, the entities "objects", which are embedded in framework of CPS, are at a level of "smartness" with enhanced digital capabilities and context-aware content.

Cyber Construction

It is a classified and semantic enrichment of BIM models beyond the design phase with a tight connectivity to the real construction products and/or process. Simply it represents the digital equivalent of things and has enough detailed information allowing all objects in the model to be constructed and/or installed on-site. Several methods can be used to leverage the BIM models and create virtual equivalent of physical construction entities manually or in (semi-) automatic manner such as:

Figure 2. a generic cyber-physical integration framework describes the relations and interaction among the four main aspects in construction domain

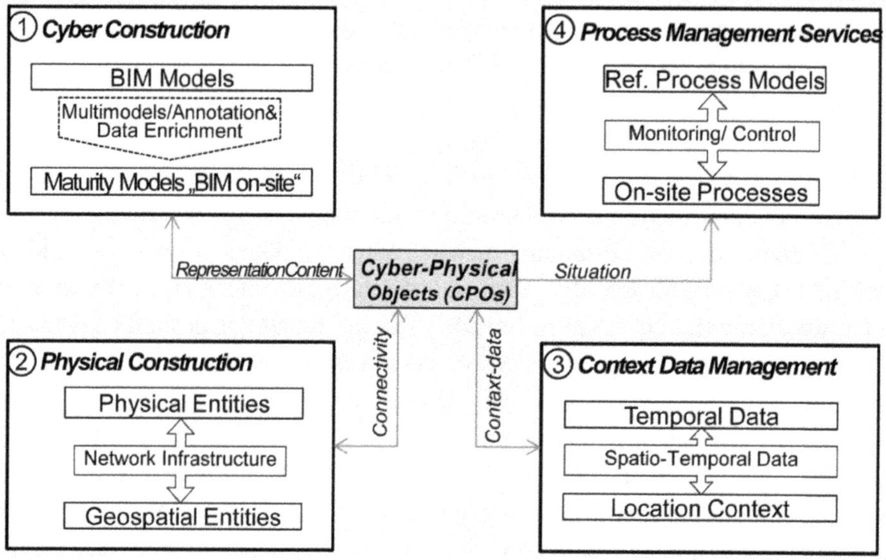

- Semantic enrichment and annotation of the building models add necessary data and classification that is not yet carried in virtual models, keep consistent relationships with the site physical entities. A web-based tool like BIM-Annotator based on Industry foundation classes (IFC) standard (ISO 16739:2013) use to add semantic data to the model (Ismail, Srewil, Scherer & Mansperger, 2016).
- Multimodels method is a method for creation, exchange and access of cross-domain information (Fuchs & Scherer, 2017). Thereby, the elementary and domains models are loosely coupled such as building models, schedule plans, bill of quantities, etc.

The results are virtual construction models (BIM onsite models) that can provide access to information on construction site for supporting constructing, and on/ off-site logistics, etc.

Physical Construction

It is the real construction environment and resources in which the construction activities are accomplished. This means, the physical construction is dynamic and evolves constantly over time. According, the physical construction entities include the physical places and resources in suitable data structures:

1. Physical places or (job-site spatial entities) which roughly describe the construction site boundary, neighbourhoods and work zones, as well as their functionality and constraint. Precisely, it loosely determines the construction site's layout model.
2. Physical assets encompass basically the tractable entities such as engineered materials, modular elements and prefabricated components), equipment, machines and personnel on-site.
3. Network infrastructure includes the hardware and devices (e.g. tracking devices, sensors…).

Context Data Management

Changes in the physical construction take place rapidly which are reflected in a stream of intensive and diverse data. The project's data are heterogynous and huge that may take up several gigabytes even for a small building project. The spatial-temporal data in our case coexist with the objects and can be enrich during their lifetime to identify continuously the state of the objects.

This deluge of information has to be considered by using several manipulations. Data analysis, algorithms and computational procedures process the raw facts of the physical construction into meaningful information. While users and software applications are usually interested in specific data that should be delivered to meet one's own needs and purposes. These data are designed to play an actuation role within the system, especially in a business process context. For instance, at process level, a specific action can release as soon as a well described event occurs, which can derive based on the changing of the CPO's statues

Process Management Services

This facet is an access point, where the CPOs interact to deal with real-world problems. Thereby the bi-directional coordination between the real world and digital models are accomplished by monitoring (physical→cyber) and control (cyber→physical). The mutual-influence is established as services in the CPS's framework by exploring the ongoing changes in the real flow and doing necessary adjustments in the digital models and vice versa based on:

1. Monitoring and control of site processes: A continuous monitoring of AS-IS against AS-Planed performances supports process progress tracking. Moreover, the analysis of deviations helps exceptions anticipation and on time alternative process planning (Srewil et al., 2012). Thereby, the site data are analysed and correlated. Next, semantic business events will be prepared in order to set up or shut-down a specific process fragment in the process models.
2. Process based construction simulation: The real field data serve the construction simulation model in two ways. First, they update the simulation model frequently to consider the changes during project construction and keep the correctness of the results. Second, the real data can be used to improve the quality of simulation parameters which increased the reliability of simulation results (Ismail, Srewil & Scherer, 2017).

In this framework, the raw facts regarding the construction site are processed to derive meaningful information (actionable insights) that supports on-site activities and makes sound decisions.

The results are hybrid objects at a level of "smartness" with enhanced digital capabilities and ability of context-awareness. These Cyber-Physical Objects (CPOs) have a centric role in the framework, in which physical objects and construction processes are mutually-influencing. Since objects and information are inextricably linked, there are no longer differences between information and physical items especially in context of flow. These key system's components are supposed to work together towards a higher level of functioning and collaboration using sophisticated coupling technologies (e.g. RFID, GPS).

Merging Methods for a Consistent CPS

An integration approach to create a kind of a closed-loop information flow between virtual and physical construction is necessary to meet the interlinking requirements within the CPS framework. It encompasses two levels:

- The real physical construction entities identification and continuous discovery their locations which link directly to the digital equivalents. At this level of integration, the focus is on the global object position and semantically association with a well-defined construction site. Each physical construction element identified and connected to its temporal location using RFID reader/ transponders communication over air protocol interfaces and GPS-unit mounted on the readers.
- The virtual construction models of physical entities and their location context are bound together. Here, a method to pair the RFID system with BIM models is needed to maintain interlinking between digital and physical environment. Also tracked entities have to associate with pre-defined construction site layout data in order to determine their location context. Here, the site layout model is a spatial, functional, semantic structure and is used to formalize the structure of construction site.

Accordingly, the spatial and temporal data of identified objects are combined and linked to the virtual equivalent (i.e. the model element) in order to maintain a consistent relationship between the physical construction cyber models. The results are cyber-physical objects (CPOs) in a digitalized construction site which are recognized throughout the following methods:

1. A cyber-physical product link establishment: It is to guarantee that the physical construction objects and digital equivalents are uniquely interlinked during the project lifetime.
2. Objects spatial-temporal data collection: It includes the identification and approximate temporary location of an object at a specific time interval which is generated by a reading-node. A passive RFID system is used to track physical entities on construction site.
3. Context of location information association: Extracting the semantic location of the objects is necessary to conquer the limitation of the data acquisition system (i.e. passive RFID system), by leveraging the semantic content of captured data.

Here, an approach for digital construction site includes the physical construction, IT infrastructure and spaces' functionalities, is necessary to describe spatial-temporal attributes of CPOs. In CPS framework CPOs have a fundamental role. They promise richer models and are able to integrate smoothly in the process models. They provide an important insight of process performance and yield more realistic and holistic measurements.

Cyber-Physical Objects, (CPOs)

The discussed CPS framework inducts consequently embedded construction objects that have specific characteristics namely: digital representation, context-aware information and connectivity. Those three fundamental properties encapsulate the definition of Cyber-Physical Objects (CPOs) in our approach and make them different from other conventional objects (Figure 3).

Figure 3. Cyber physical-object, CPO concept

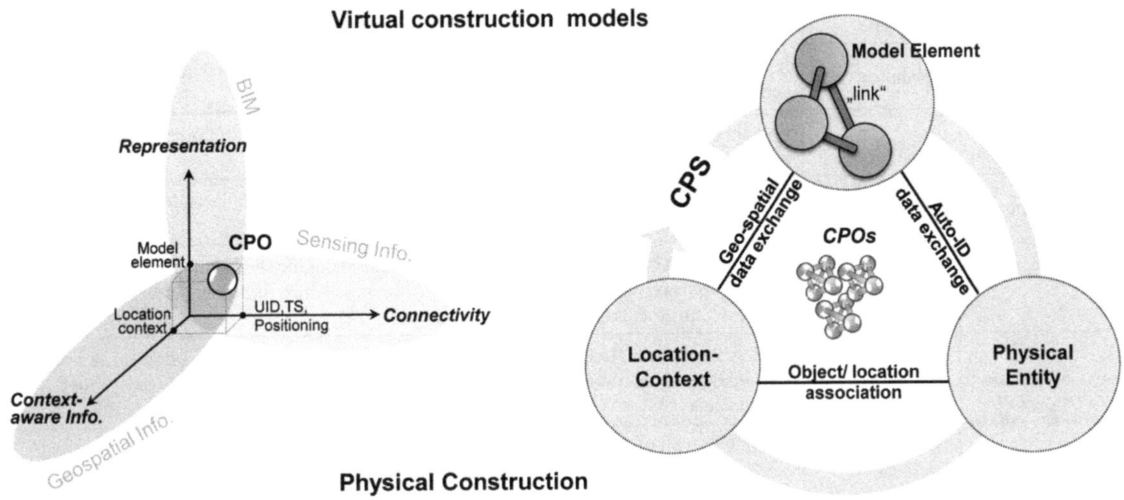

Cyber-Physical Object (CPO) is an object at a certain level of smartness with digital ability including identification communication and data storage and capable to enrich with sensing input technologies. In other word it's a digital-twin of a physical entity.

Therefore, CPOs are types of smart things which can be seen from multiple view, e.g. from an engineering point of view they are a part in the production processes.

Generally, smart objects in the era of CPS and internet of things can understand and react to their environment (Kortuem, Kawsar, Sundramoorthy & Fitton, 2010). They are shifting the interconnectivity from human-to-content and human-to-device toward content-to-content and device-to-device levels, e.g. GPS unit updates the content automatically based on object's location.

The representation of CPOs indicates that each element has a size, function and properties, and a semantic relationship within the virtual information models namely the element model. This representation can be extracted and filtered out from BIM models.

The context-aware property characterizes the relative location where objects are processed in which time window and what state/ behavior they have (i.e. the object status). In turn, state/ behavior information provides an interpretation of events and activities with respect to predefined processes (reference model).

The connectivity refers particularly to the ability of unique identification (via: IPv6, EPC code, serial no., etc.) of the individual object. The objects can communicate with each other directly or their digital-equivalent indirectly by establishing peer-to-peer (P2P) connections. This shall allow bi-directional interaction among the construction components and with their CPOs.

A CPO is defined by a set attributes, capability methods, rules and/or procedures.

A formal description of the CPO capabilities is given in table 1: The mandatory attributes refer to the simplest CPO with simple sensing structures, though optional, one may need advanced sensing capabilities like an active RFID system and/ or an object with multiple sensors to measure its temperature, humidly, vibration, etc.

Table 1. CPO capabilities

CPO Property	Type	Description
Identification	Mandatory	It is to sustain a uniquely bi-directional identification of the object and its presence in a digital context
Data storage	Mandatory	It refers to the ability of an object to store information about itself or the environment (at least its unique identity)
Communication	Mandatory	It is an object's ability to exchange the information with devices, objects, users, etc. In a simple case, passive RFID tags need readers to exchange the information.
Logging	Optional	It represents an object's ability of registering events about the surrounding environment or itself. In advance the object capable of Self-Awareness which means an object is able to know its status and structure…
context-Awareness	Optional	It refers in particular to the ability of gathering information from the environment and the surrounding objects in order to improve knowledge as well as discover and manage the environment
Process-Awareness	Optional	CPOs are involved in the business process. They can determine the start/ end of a task and assess process compliance using trace logged by CPOs to identify exceptions or data flow violations.

Construction Site Digitalization

Construction site is usually a temporary limited manufacturing place for related buildings. Moreover, it is beyond the location where the construction processes are taking place, an interface among all project parties. Leveraging the content of site's collected data in order to enrich the CPOs with spatial-temporal data, an approach for digital construction site. Figure 4 presents a layers model, in which the construction site decomposes vertically into three characteristic layers to simplify the identification and tracking of complex construction site entities (Srewil & Scherer, 2013), which are a:

- Physical event layer, it represents the real construction site where the processes take place.
- Networking layer, it is a kernel in which the two other layers is fused.
- geoSite (geo information site layer), it brings the semantics to the location where site data are captured.

According a physical entity is identified and will be pinned to the semantic location, which is given by the geoSite layer, by mapping object's location to a specific functional work zone on site. The results are objects with a context-aware content.

The geoSite Model

According to the layered model above, the geoSite is the lowest level that provides the necessary semantics location for tracked objects (i.e. reference- or geo-locations). The geoSite is a breaking down of a construction site horizontally into a set of functional work-zones like gates, storages, facility places, etc. The geoSite is identified by:

1. Extracting basic spatial data of construction site related spaces,

Figure 4. Virtual construction site based on a layer model

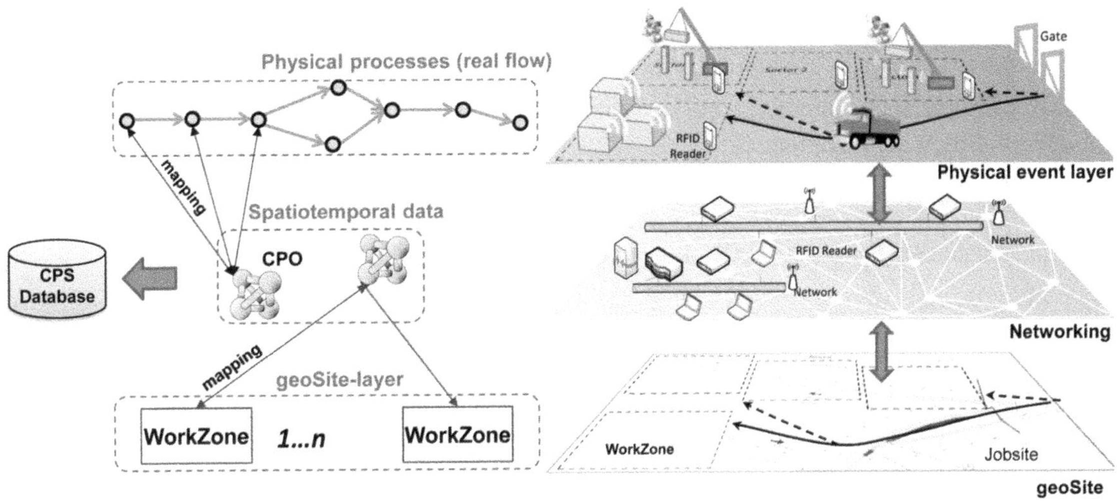

2. Adding location context using an explicit relation between the functions and geometry representation of construction site spaces over time interval.

Linking site layout plan (namely site schedule) and site geometry are crucial in order to deal with site dynamic features. In addition, designed constraints shall control the relations among work zones and the whole site with neighbourhoods. Here, generating site spatial data sets are limited to the level of details that provide site boundary and work areas as conceptual model (i.e. horizontal 2D footprint or 3D block).

Actually, longitude, latitude and altitude cannot alone interpret the semantic of locations. Therefore, an association between the global location data and functional information of a construction site is necessary to create the location context of tracked physical entities. The geoSite approach changes a complex construction site into a set of functional oriented spaces. The geoSite model is formulated by:

- LocationContext: It is a labelling class representing the semantic category (*Ci*) for a construction site which plans the functionality of diverse locations, e.g. "storage" where elements are stored.
- TempWorkZone is a 2D polygon that sizes a temporary work area with specific function over specific time window, for example "Laydown-Block XYZ", "Gate" and so forth.
- Building refers generally a stationary facility that has unchangeable location in project lifetime and represented by the footprints.
- The ConstructionSite encapsulates all TempWorkZones and Building, and it is the destination of all off-site shipments.

Mapping of geoSite and Network Layer

A mathematical algorithm is used to map the ReadingNodes (i.e. RFID readers) of network layer into geoSite (G_S) layer an envelope of all work zones and building and given as following:

$G_S = \sum_{i=1}^{k} A_{Fi}$.; i =1, 2, 3…k, and A_{Fi} is a Work zone on the construction site.

Each work area has a specific location-context (C_i). The location context of a ReadingNode (n) is formulated in following equation:

Let n (R_{POS},t): is a reading-node in network N at a position (R_{POS}) and time (t):

$$\{\forall n \in N \mid n \text{ LocatedAt } (R_{POS},t) \in A_{Fi} \wedge A_{Fi} \text{ Is-a}(P_i,C_i,t_1,t_2)\} \rightarrow n \text{ Has}(C_i,t);\ t_1 \leq t \leq t_2$$

where:

- Pi is set of point that represents a 2D polygon.
- $[t_1,t_2]$ represent the time interval that the work zone A_{Fi} has a location-context (C_i).

Furthermore, n is a reading node in a network N, it belongs to a specific work area A_{Fi} when:

$$\forall n \in N : \exists n \in A_{Fi} \subset G_S \leftrightarrow \begin{cases} PIP_n = 1 & \Rightarrow n\, is\, inside\, A_{Fi} \\ PIP_n = 0 & \Rightarrow n\, is\, directly\, on\, A_{Fi} \\ PIP_n = -1 & \Rightarrow n\, is\, outside\, A_{Fi} \end{cases}.$$

where, PIP_n is a well-known Point-In-Polygon test of a point n and polygon A_{Fi}. This is a simplified mathematical mapping function. All objects in rang n inherent the same location context as approximate and acceptable justifiable assumption.

Digital Models Interlinking

Effectively, the CPOs are established by interconnecting of the BIM component models and construction physical entities during project life time. The IFC is an open standard used in BIM, intended to describe

Figure 5. Mapping of RFID system and semantical site layout model

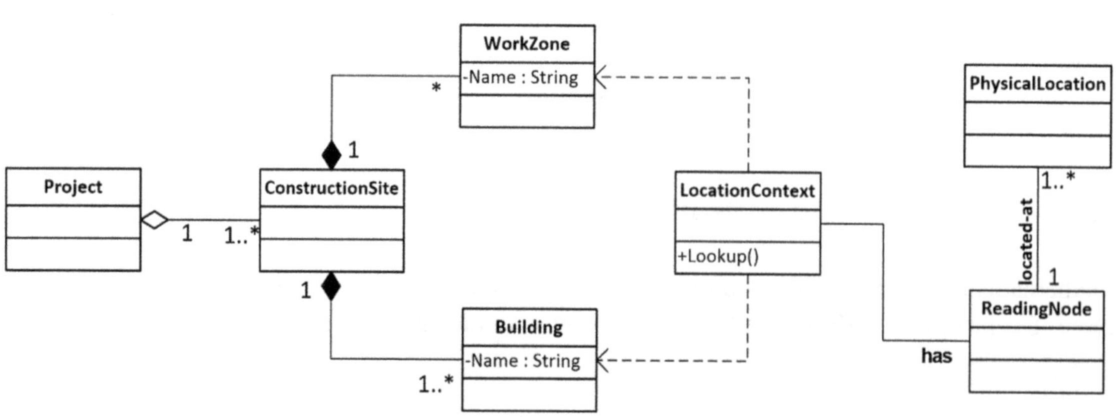

building and construction data and organized into an object-oriented inheritance hierarchy. Pairing RFID and IFC element is necessary to interlink physical and digital models in order to create CPOs.

Here, RFID data are added as user-defined properties set to IFC objects based on a standard schema to create a consistent link between the physical component and its model element. Thereby, every model element (IFC object) can be equipped with RFID Tag data like unique serial no, EPCgloble (RFIDTagID, GlobalTradeItemNumber, etc.). The new property sets are stored in a reference database for annotation. A web-based tool for annotation like "BIM-Annotator" in Ifcwebserver.org is used to enrich the BIM models specifically:

- IFC objects with correct attributes and user-defined properties, and
- Chang the model element granularity, e.g. grouping of several elements to create a component with new properties reflects the real case of physical entity in logistics and installation on site.

The reference database for classification and data enrichment is developed based on the XSD schema of Property Set Definition (PSD) of buildingSMART (Ismail et al., 2016).

CPO Tracking Model for Assembly and On-Site Logistic

CPOs should keep track to become aware of their status which is an important aspect of on-site construction logistics whether a construction component is delivered or a machine is temporary not available.

In our scenario, a ReadingNode class represents a RFID reader which can provide IoT as interface and is able to interact with the labelled physical entities. Therefore, it is an observable point sense (object ID, timestamp, and positioning information as latitude, longitude). This node has a location context and passes this semantic information to observer objects in this case they are CPOs. Extracting and linking location context information of site to reading nodes and objects are discussed in the next sections.

The work uses a design patterns modelling approach from object–oriented programming OOP, namely an observer design pattern for implementing publish-subscribe structure (Gamma, Helm, Hohnson & Vlissides, 2004). This solution is suitable for CPS applications where objects might directly communicate with each other for updating and tracking the current status. Here, a CPO fulfils the same role as objects in OOP paradigm. The data about a CPO are stored in the cyber model and the access to these data is using unique and globule identities as references for product items.

The well-known observer pattern has a significant behavioral aspect. Hence, it is adopted to define a link between objects. As an object's status changes, all dependent objects statuses are automatically updated. This type of patterns is usually used for problems addressing the frequency and timing of status update. The observer pattern is a reusable pattern allowing for communication between objects in a loosely coupled manner.

The CPO class is an implementation for CPO-Observer interface as "Object," likewise; a ReadingNode class is an implementation of ReadingNodeObservable interface as "Subject". In this way, context data (status) of a concrete reading node pass automatically to all objects of interest.

Figure 6 shows a concrete example for tracking prefabricated elements on site and the timely update of the objects status is based on the reading node location context. A RFID reader with built-in GPS unit senses its current location, and reads the objects in range. The status of the reader is derived according to the "getStatus()" method based on the location context like a laydown area which is refers to as "is-stored" status. The reader, which here is a "publisher" sends updates based on a push model to

Figure 6. A case scenario for deign patterns for tracking objects at different work zones

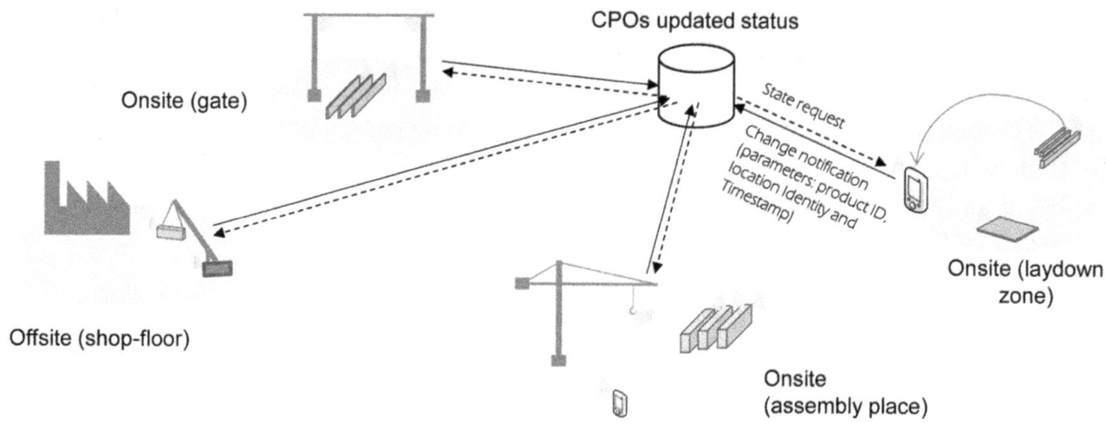

the observer (CPO), which is, in this case, a "subscriber". Whereas, the push-based notification method assumes subjects know about their observers, which are the prefabricated elements in the reader range.

Figure 7 illustrates a detailed UML model of CPO data structure based on observer pattern. While observer pattern defines and maintains a dependency among objects, it is responsible for maintaining the state information in the system. A status class is needed to be polymorphic since different things will be "statusable." The Status is defined in status name, ID, type and date.

The virtual information models (i.e. cyber construction), construction site layout model and network infrastructure are discussed in details in next sections. Figure 10 illustrates a detailed UML model of CPO data structure based on observer pattern. While observer pattern defines and maintains a dependency among objects, it is responsible for maintaining the state information in the system. The Status is defined in status name, ID, type and date.

The virtual information models (i.e. cyber construction) and construction site lay out model and network infrastructure are discussed in detail in next sections.

Linking CPS and Business Process Modeling

CPSs are designed to support and facilitate collaboration among physical entities and software services on a complex task. The business process management discipline offers a powerful representation of collaborative activities (Weske, 2007).

The CPSs help business processing by connecting the real objects with the virtual world of software and services. Consequently; there is no longer difference between information and physical items especially in context of flow, because products will be inextricably linked to "their" information (Löffler & Tschiesner, 2013). It means process and device will be inseparable; physical things become part of the process. In turn, business process will certainly become more dynamic, and capable of responding to the emerging technologies. Therefore, the objects or things being connected with smart chips (RFID, GPS, etc.) in framework of CPS can impact all the phases of the business process management lifecycle, i.e. analysis and design, implementation, enactment and monitoring, and evaluation (Mendling, 2008) (Figure 8). Especially a high impact is expected on the monitoring and enactment, and evaluation phases where data have a dominant role.

Figure 7. CPO model components

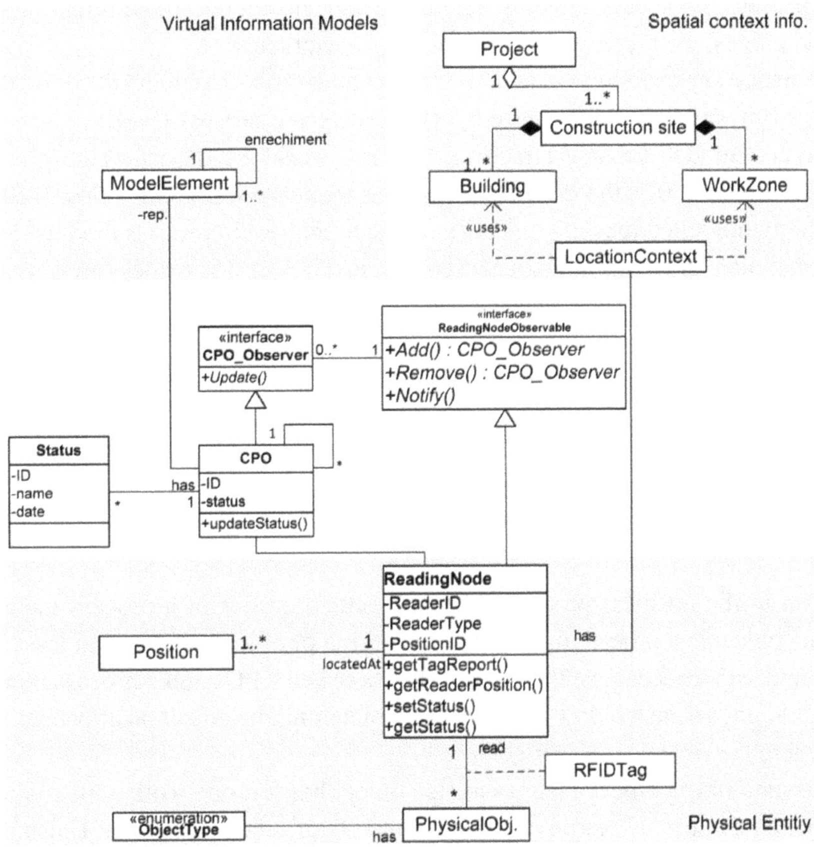

Figure 8. A possible impact of CPS on business process management lifecycle

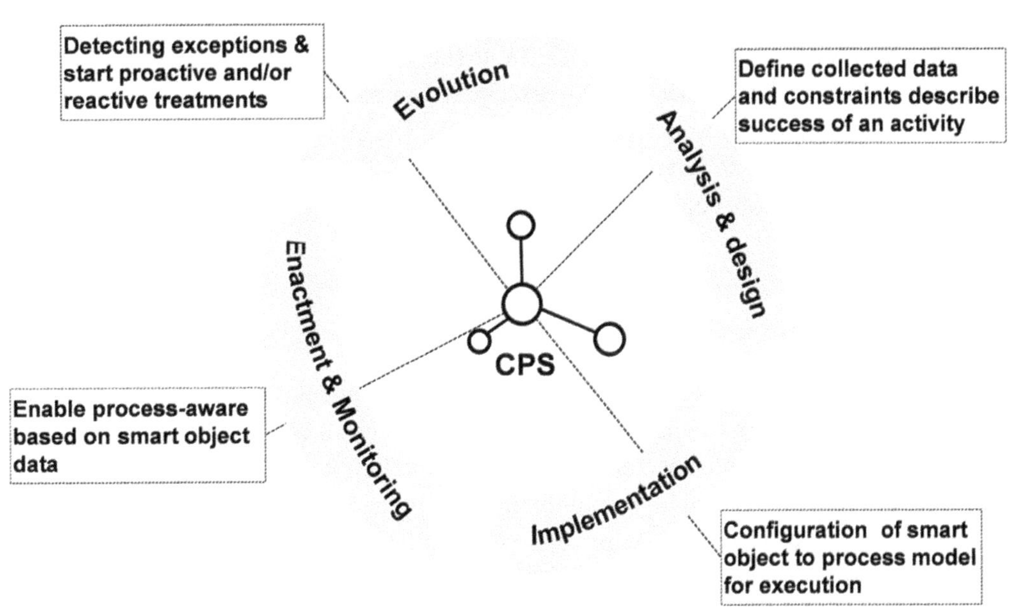

CPOs are intended to integrate in order to invoke a business process execution. In this case, the executed business process represents the work actually being carried out. In addition, they are capable of monitoring of the process correctness and controlling the workflow.

However, it is necessary to address a method allows an integration of CPO data into a process model, i.e. a modeller needs a way to consider these hybrid objects at a process level.

Although a process models are an abstraction of the real world that describes the logical and temporal order of activities (Weske 2007), the majority of process models cover control flow (i.e. activity-centric paradigm) and rarely modelled the data object (i.e. object-centric) in process models (Meyer & Weske, 2013). Recently mapping and integrating of sensed physical objects to business process models has been gained a special attention as regards to industry 4.0, smart factory and IoT applications (Meyer, Ruppen, & Hilty, 2015). However, there is still a lack modelling concepts represents CPS aspects in a business process model some approaches propose an extended standard modelling technique such as (Sungur, Spiess, Oertel & Kopp, 2013), other use the capabilities of process modelling techniques like BPMN (Business Process Modelling Annotation) in integration.

CPO Integration in a Process Model

CPOs are involved in the business process being the actual entities of a process model. They can be modelled as a data object in a process model. The definition of object expresses at least a unique ID and a status S_{Obj} attribute. Figure 9 depicts the relationship between CPOs and a process model. The current status of an object S_{Obj} at a time window t is derived by matching the object position and location context of a work area where the object is identified.

Therefore, the state of the object changes as this object has left one work zone to the next one. Let: a task T in a process, it starts by reading the status and ID of related CPO and finishes as soon as the object has a new status, e.g. an installation of prefabricated element starts when the state of this element changes from (be-stored) into (is-in-installing) and finishes as soon as the next element or next task has a new status then the predecessor element gets its final state (be-installed). It means, a task starts when one of the objects changes its status to execute this task. It finishes as soon as all related objects have the new status and so forth. Consequently, a task status (i.e. start, in progress, finish, etc.) is defined by the states of it resources which are necessary to execute this task. An aggregation of all tasks and and/or sub-processes statuses is used to derive a process state. The status attributes can use to monitoring of a process progress. Here, a continuous comparison of AS-PLANNED against AS-IS performance is used determine any deviation, conflicts or exceptions at run-time like delays in the ongoing process.

CONCLUSION

Although the construction is a conservative industry, the implementation of interconnected systems such as CPS promises to improve construction and operation phases of a project using automated data capture solutions toward more informed and precise construction.

The work suggests a CPS concept for smoothly integrating the real physical construction entities into virtual information models based on semantic rules. A generic framework relies on CPS approach for the

Figure 9. Integrated CPOs in a process model

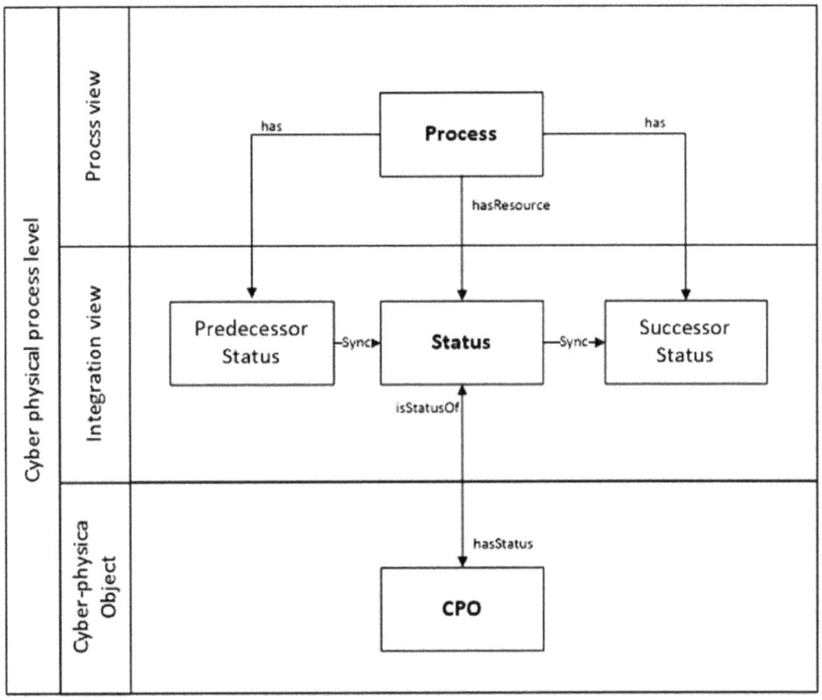

construction domain is introduced to accomplish the integration and bi-directional coordination between virtual "cyber" construction and real construction field. The framework's design enables individual entities to work together in order to form complex systems with new capacities. Tight integration methods are proposed to allow construction object recognition and ensure a high functionality and collaboration and ultimate interoperability inside the system. Consequently, cyber-physical objects are embedded in the framework with a level of "smartness", which is a digital capability, context-aware content and specified behaviour or action. The site data capture method is based on a cost-effective passive RFID system convenient for tracking the construction entities.

The layered model supports the digitalization of construction site and the underlying mathematical techniques are to facilitate deriving of the CPOs' status. For this purpose, the content of captured data has to enrich by mapping the data capture system to the semantic locations, namely functional work zones. The geoSite concept simplifies the determination of site's work zones and makes the timely changes in their functionality tractable. It shifts a complex construction site into a set of functional oriented spaces. The interoperability between BIM models and geospatial information is used to derive the geoSite model.

Pairing RFID and BIM objects is necessary to interlink physical and digital models in order to create CPOs and allow tracking the changes in the logistics and construction phase of a project.

Finally, the integration of the CPOs in construction process models supports deriving actual states of ongoing activities in order to track project progress. The analysis of actual against planned performance beyond the scope of this work can anticipate delays, errors or resources shortage during the construction phase.

Two application scenarios for future work are to discuss:

- Validation of the process simulation model and improve simulation input data and reference process models with real parameters (Ismail, et al. 2017),
- Process control in a (semi-) automated manner using configuration templates techniques can be used in order to offer alternative processes on time. According suitable treatments and recommended actions which are modeled as process fragments in the reference model. These fragments are configured in the ongoing process as reactive or proactive treatments for specific or anticipate risk expected during the process execution (Scherer & Sharmak, 2011).

REFERENCES

Akanmu, A., Anumba, C. J., & Messener, J. (2013). Scenarios for cyber-physical systems integration in construction. *ITCon*, *18*, 240–260.

Akinsola, A., Dawood, N., & Hobbs, B. (2000). Construction planning process improvement using information technology tool. *Proceedings of Construction Information Technology*, *2000*, 40–51.

Azimi, R., Lee, S., Abourizk, S., & Alvanchi, A. (2011). A Framework for Automated and Integrated Project Monitoring and Control System for Steel Fabrication Projects. *Automation in Construction*, *20*(1), 88–97. doi:10.1016/j.autcon.2010.07.001

Björk, B. (2002). A formalised model of the information and materials handling activities in the construction processnull. *Construction Innovation*, *2*(3), 133–149. doi:10.1108/14714170210814739

Cardenas, A. A., Amin, S., & Sastry, S. (2008). Secure control: Towards survivable cyber-physical systems. In *Proceedings of 28th International Conference* Distributed Computing Systems Workshops ICDCS '08. doi:10.1109/ICDCS

Chen, Y., & Kamara, J. (2008). The mechanisms of information communication on construction sites. *FORUM Ejournal*, *8*(1), 1–32.

Fuchs, S., & Scherer, R. J. (2017). Multimodels — Instant nD-modeling using original data. *Automation in Construction*, *75*, 22–32.

Gamma, E., Helm, R., Hohnson, R., & Vlissides, J. (2004). *Design patterns: Elements of reusable object-oriented software*. Addison-Wesley.

Gehbauer, F., Zülch, G., & Börkircher, M. (2007). Simulation-based analysis of disturbances in construction operations. In *Proc. of the 15th Annual Conference, International Group for Lean Construction (IGLC)* (Vol. 571, p. 579).

Ho, V. (2017). Design revolution: Microsoft HoloLens and mixed reality are changing how architects see the world. *Microsoft*. Retrieved from https://news.microsoft.com/transform/design-revolution-microsoft-hololens-mixed-reality-changing-architects-world/

Ismail, A., Srewil, Y., & Scherer, R. J. (2017). Integrated and collaborative process-based simulation framework for construction project planning. *International Journal of Simulation and Process Modelling, 12*(1), 42–53. doi:10.1504/IJSPM.2017.082789

Ismail, A., Srewil, Y., Scherer, R. J., & Mansperger, T. (2016). Semantic Enrichment and Multimodel Data Exchange Approach for CFD Analysis of Bridges. In *23rd International Workshop of the European Group for Intelligent Computing in Engineering, EG-ICE,* June 29-July 1.

International Organization for Standardization. (2013). ISO 16739:2013 Industry Foundation Classes (IFC) for data sharing in the construction and facility management industries. Geneva: International Organization for Standardization.

Kagermann, H., Wahlster, W., & Helbig, J. (2013). Recommendations for implementing the strategic initiative INDUSTRIE 4.0. (No. Final Report). Berlin, Germany: acatech – National Academy of Science and Engineering.

Kortuem, G., Kawsar, F., Sundramoorthy, V., & Fitton, D. (2010). Smart objects as building blocks for the internet of things. *IEEE Internet Computing, 14*(1), 44–51. doi:10.1109/MIC.2009.143

Lai, C., Ma, Y., Chang, S., Chao, H., & Huang, Y. (2011). OSGi-based services architecture for cyber-physical home control systems. *Computer Communications, 34*(2), 184–191. doi:10.1016/j.comcom.2010.03.034

Lee, E. A. (2015). The past, present and future of cyber-physical systems: A focus on models. *Sensors (Basel), 15*(3), 4837–4869. doi:10.3390150304837 PMID:25730486

Löffler, M., & Tschiesner, R. (2013). The Internet of Things and the future of manufacturing. *McKinsey.* Retrieved from http://www.mckinsey.com/business-functions

Mendling, J. (2008). *Metrics for Process Models: Empirical Foundations of Verification, Error Prediction, and Guidelines for Correctness.* Springer. doi:10.1007/978-3-540-89224-3

Meyer, A., & Weske, M. (2013). Extracting data objects and their states from process models. In *Proceeding of 17th IEEE International Enterprise Distributed Object Computing Conference* (pp. 27-36). 10.1109/EDOC.2013.13

Meyer, S., Ruppen, A., & Hilty, L. (2015). The Things of the Internet of Things in BPMN, In J. Barjis, A. Gupta & A. Meshkat (Eds.), Advanced Information Systems Engineering Workshops (pp. 42-56). Springer. doi:10.1007/978-3-319-19243-7_27

Motamedi, A., Setayeshgar, S., Soltani, M. M., & Hammad, A. (2013). Extending BIM to incorporate information of RFID tags attached to building assets. In *Proceeding of the 4th construction specialty conference,* Montreal Canada, May 29-June 1.

National Institute for Standard Technology. (2013a). *Strategic vision and business drivers for 21st century cyber-physical systems.* USA: NIST.

National Institute for Standard Technology. (2013b). *Foundations for innovation in cyber-physical systems.* USA: NIST.

Sanislav, T., & Miclea, L. (2012). Cyber-physical systems-concept, challenges and research areas. *Journal of Control Engineering and Applied Informatics*, *14*(2), 28–33.

Scherer, R., & Sharmak, W. (2011). Process Risk Management Using Configurable Process Models. In L.M. Camarinha-Matos, A. Pereira-Klen & H. Afsarmanesh (Eds.), *Adaptation and Value Creating Collaborative Networks: 12th IFIP WG 5.5 Working Conference on Virtual Enterprises PRO-VE 2011* (pp. 341-348). Springer.

Sørensen, K. B., Christiansson, P., & Svidt, K. (2010). Ontologies to support RFID-based link between virtual models and construction components. *Computer-Aided Civil and Infrastructure Engineering*, *25*(4), 285–302. doi:10.1111/j.1467-8667.2009.00638.x

Srewil, Y., Kadolsky, M., & Scherer, R. J. (2012). Integrating RFID technology in dynamic construction process planning. In *Proceeding of Joint 29th CIB W78 the Beirut* (pp. 384–393). Lebonon.

Srewil, Y., & Scherer, R. J. (2013). Effective construction process monitoring and control through a collaborative cyber-physical approach. In L. M. Camarinha-Matos, & R. J. Scherer (Eds.), *Collaborative systems for reindustrialization: 14th IFIP WG 5.5 working conference on virtual enterprises, PRO-VE 2013*, Dresden, Germany, September 30-October 2 (pp. 172-179). Springer. 10.1007/978-3-642-40543-3_19

Sungur, C. T., Spiess, P., Oertel, N., & Kopp, O. (2013): Extending bpmn for wireless sensor networks. In *2013 IEEE 15th Conference on Business Informatics (CBI)* (pp. 109–116). IEEE. 10.1109/CBI.2013.24

Talcott, C. (2008). Cyber-physical systems and events. In M. Wirsing, J. Banâtre, M. Hölzl, & A. Rauschmayer (Eds.), *Software-intensive systems and new computing paradigms* (pp. 101–115). Springer. doi:10.1007/978-3-540-89437-7_6

Teizer, J., Cheng, T., & Fang, Y. (2013). Location tracking and data visualization technology to advance construction ironworkers' education and training in safety and productivity. *Automation in Construction*, *35*(0), 53–68. doi:10.1016/j.autcon.2013.03.004

Weske, M. (2007). *Business Process Management: Concepts, Languages, Architectures*. Germany: Springer.

Chapter 7
Smart Gateways for IOT-Factory Integration:
Trends and Use Case

Eva Masero Rubio
Allbesmart, Lda, Portugal

Pedro Miguel Baptista Torres
Instituto Politécnico de Castelo Branco, Portugal

Rogério Pais Dionísio
Instituto Politécnico de Castelo Branco, Portugal

ABSTRACT

This book chapter proposes a description of smart gateways and cyber-physical systems (CPS) for the industrial internet of things (I-IOT). It also presents a case study where a smart gateway is developed to be used in different types of industrial equipment for the shop floor. The case study is developed under the specifications of different industries in the region of Castelo Branco. It is a proof that the 4th industrial revolution will be the engine for SME innovation, independence of the regions and their financial strength. It is also proof that the cooperation between universities, industries and startups can evolve to break barriers and add value in the improvement of regional industries competitiveness. Topics that will be addressed on the chapter can be used for developers, students, researchers and enthusiasts to learn topics related to I-IOT, such as data acquisitions systems, wired and wireless communication devices and protocols, OPC servers and LabVIEW programming.

INTRODUCTION

Factories are continually changing due to emerging technologies and different ways to process information, from computing and automation systems towards big data and a smart world, to digitising the industry, and create an intelligent and an autonomously network which provides: an increase in the efficiency of manufacturing with a reduction of time, efforts during set-up, re-configuration, a reduction

DOI: 10.4018/978-1-5225-4936-9.ch007

of production costs and an increase in production quality. In this field, Industry 4.0 is the current trend of automation and data exchange in manufacturing technologies. It was initially developed by the German government to create a coherent policy framework to maintain industrial competitiveness. Related terms used internationally include Internet of Things, Internet of Services, Industrial Internet, Advanced Manufacturing and Smart Factory (Smit et al., 2016). Now, the introduction of the Internet of Things (IoT) and Services into the manufacturing environment is ushering in a fourth industrial revolution (Kagermann et al., 2013).

This chapter examines challenges to the adoption of the Internet of Things in factory – or Industrial Internet of Things (I-IoT)– to implement new applications and services in more efficient manufacturing systems. Several studies, projects and companies have focused their attention on this new paradigm. The authors present the necessary background by introducing the fourth industrial revolution. Consequently, they provide a survey of the state of the art related to I-IoT and cyber-physical production systems. Some industrial solutions for IoT are also presented –hardware and software solutions– from companies from power systems and automation technology areas. Readers interested in this field could learn which are the benefits of theses factories of the future, and meet the developments of the latest technologies for smart production with Co-Factor –an action supported by the European Commission under its Research & Innovation Program Horizon 2020 (Co-FACTOR project, 2017) –whereby the researchers introduce several projects focus on a new system architecture. Then, this chapter presents development platforms for I-IoT solutions. Moreover, a practical case is described. The main purpose this case –study is to measure vibration on electrical motors from machine factories, and implement predictive maintenance to assess anomalies. To implement the practical case, the authors will use several platforms and technologies: LabVIEW and NI DAQ hardware, Arduino, OPC server, machine learning algorithms and EtherCAT communication protocol. Finally, the authors highlight improvements and some possible directions for future research.

This chapter is organized as follow. First, a background of industrial revolution over the years and how is going toward Industry 4.0. The second section presents a general vision about Cyber-Physical Systems (CPS) and Smart Gateways. The third section describes several available platforms that are useful to implement smart networks in factories and after that, some current European projects towards digitizing industry and diverse development platforms are shown. One case study is also exposed and afterwards the benefits obtained and possible improvements. Finally, future research direction and conclusions are pointed out.

BACKGROUND

Industry has changed since the beginning of the Industrial Revolution and it can be described as the fourth stage of industrialisation or Industry 4.0. Over the years, the industrialisation concept has been changing simultaneously with technological advances.

Industrialisation began with the introduction of mechanical manufacturing equipment at the end of the eighteenth century when water and steam-powered machines like the mechanical loom revolutionised the way goods were made. This first industrial revolution was followed by a second one that began around the turn of the twentieth century and involved electrically-powered mass production of goods based on the division of labour. This was in turn superseded by the third industrial revolution that started during the early 1970's and has continued right up to the present day. This third revolution employed electronics and information technology (IT) to achieve increased automation of manufacturing processes with PLCs, as machines took over not only a substantial proportion of the manual labour but also some of the brainwork. The fourth stage of the Industrial Revolution is based on cyber-physical systems, where real objects and virtual processes are interlinked, leading to the connected Industry and the spread of the Internet of Things and Services into the manufacturing environment. Industry 4.0 is the name given to the German strategic initiative and the main aim is to set a paradigm shift from centralised to decentralised smart manufacturing and production (Kagermann et al., 2013).

Industry 4.0 describes the organisation of production processes based on technology and devices autonomously communicating with each other along the value chain: a model of the smart factory of the future where computer-driven systems monitor physical processes, create a virtual copy of the physical world and make decentralised decisions based on self-organisation mechanisms. According to Smit et al. (2016), these are the main factors that are involved in Industry 4.0:

- **Interoperability:** Cyber-Physical Systems (work-piece carriers, assembly stations and products) allow humans and smart factories to connect and communicate with each other.
- **Virtualisation:** A virtual copy of the Smart Factory is created by linking sensor data with virtual plant models and simulation models.
- **Decentralisation:** Ability of Cyber-Physical Systems to make decisions of their own and to produce locally thanks to technologies such as 3D printing.
- **Real-Time Capability:** The capability to collect and analyse data and provide the derived insights immediately.
- **Service Orientation.**
- **Modularity:** Flexible adaptation of smart factories to changing requirements by replacing or expanding individual modules.

STATE-OF-THE-ART OF INDUSTRY 4.0

Cyber-Physical Systems

In the Factories of the future, the physical and virtual world are merging. These two worlds are linked, through Cyber-Physical Systems (CPS), into an Internet of Things, Data and Services. They are physical and engineered systems whose operations are monitored, coordinated, controlled and integrated by a computing and communication core (Li et al., 2011).

Cyber-Physical Systems are being developed as part of a globally networked future world, in which products, equipment and objects interact with embedded hardware and software beyond the limits of single applications. With the help of sensors, these systems process data from the physical world and make it available for network-based services, which in turn can have a direct effect on processes in the physical world using actuators (Geisberger et al., 2011).

A CPS can work in three different modes: passive, passively active and active (Ray, 2017):

1. **Passive Mode:** A passive CPS only monitors an environment and collects and processes data.

2. **Passively Active Mode:** In this mode, a CPS monitors the environment, and executes an indirect actuation by changing their own behaviour to satisfy some pre-determined criteria of the physical process.
3. **Active Mode:** An active CPS executes a direct actuation by modifying the behaviour of the physical process to satisfy some pre-determined criteria of that process.

For CPS, the role of Smart Gateways simplifies greatly the process to export data and functionality of the physical devices to be provided to end users on the web (Li et al., 2011). Smart gateway must manage various aspects of underlying IoTs. Smart gateway performs several tasks, for instance, collecting data and performing pre-processing, filtering data and reconstructing it into more useful form, uploading only necessary data to the Cloud Computing, keeping check on IoT objects and sensors' activities, keeping check on energy consumption of power constrained nodes of IoTs, security and privacy of the data, and overall service monitoring and management (Aazam, Hung, & Huh, 2014).

The strategic initiative Industry 4.0 will give rise to novel CPS platforms geared towards supporting collaborative industrial business processes and the associated business networks for all aspects of smart factories and smart product life cycles (Kagermann et al., 2013). Nowadays, there are several I-IoT solutions –from companies from power systems and automation technology areas– which are useful ways to implement smart networks in factories.

Solutions for I-IoT

Several companies are currently developing software and hardware solutions for I-IoT. In this section, the authors expose some platforms from companies from power systems and automation technology areas.

- **CompactRIO (From National Instruments):** This platform can process and analyse acquired data in real time to make informed decisions fast. CompactRIO Controllers are high-performance embedded controllers with industrial I/O modules, extreme ruggedness, industry-standard certifications, and integrated vision, motion, industrial communication, and human machine interface (HMI) capabilities. Its controller combines a real-time processor (Intel and ARM), a user-programmable FPGA (to implement high-speed control and custom timing and triggering directly in hardware) and industrial I/O, giving users the performance and flexibility to meet their application requirements National Instruments (2016).
- **Nanobox and Nanopanel SIMATIC IPC227E (From Siemens):** The Nanobox PCs are compact and flexible embedded industry PC and display, suitable for measuring, controlling, regulating, and verifying process and machine data for industrial image processing or for decentralized visualization in combination with SIMATIC panels. The Nanobox PC and Nanopanel (SIMATIC, 2017).
- **MindSphere, MindApp and MindConect (From Siemens):** MindSphere is the open, cloud-based IoT operating system from Siemens that lets users connect their machines and physical infrastructure to the digital world. It lets harness big data from billions of intelligent devices, enabling to uncover transformational insights across their entire business. It is also an open interface for development of customer specific apps (MindApps). Moreover, MindConnect offers numerous

possibilities for connecting machines, plants and worldwide fleets to MindSphere, regardless of the manufacturer. There are devices as MindConnect Nano and MindConnect IoT2040 for transferring data and allow connectivity to MindSphere (MindSphere, 2017).

- **AC 800M Controller (From ABB):** ABB offers to implement Internet of Things, Services and People (IoTSP) in industrial environment based on 800xA control system. The AC 800M controller is a family of rail-mounted modules, consisting of CPUs, communication modules, power supply modules and various accessories. Several CPU modules are available that vary in terms of processing power, memory size, SIL-rating, and redundancy support (AC 800M, 2017).
- **LimeNET (From Lime Microsystems):** LimeNET is a software defined radio (SDR) high capacity network in a box for mobile and IoT applications, based on an Intel i7 processor and the open source LimeSDR PCIe card (LimeNET, 2017). LimeNET base stations hold the potential to completely transform the way telco networks run by shifting the emphasis and value away from proprietary hardware to open hardware with app stores on top (C, 2017).
- **Kuhnke FIO Controller WL (From Kendrion Kuhnke Automation):** Kuhnke FIO Controller WL is a PLC with integrated WIFI data interface. It offers scalability and modular expandability. It can connect to any available Wi-Fi infrastructure, but can also be set up as an access point with DHCP (Malente, 2016).
- **LioN-Power (From Lumberg Automation):** Lumberg Automation presents new LioN-Power energy and data transmission elements for I-IoT. The LioN-Power Active I/O modules meet Ethernet / IP and PROFINET protocol specifications and detect both input and output data directly in the machine (LioN-Power, 2017).
- **IoT Gateway (From Bosch Rexroth):** Bosch Rexroth has created a IoT gateway that connects new and existing machines with the IoT without intervening in the automation logic. The IoT Gateway cost-efficiently makes new and existing plants ready for the Internet of Things. It collects sensor and process data and transfers it to higher-level IT-based solutions such as MES, cloud applications, systems for local monitoring of machine states or the analysis of process data (Boschrexroth, 2016).
- **FPGA Based Gateways and CPS (eWon and SOCe):** eWON as a low cost industrial CPS, eWON flexy, that can fit the requirements of basic I4.0 application (eWON, 2018). SOCe provides affordable programmable CPS such as CPPS-Gate40 Sensor with increased performance, for more advanced applications (SOCe, 2018).

European Research on Industry 4.0

There is a current European initiative called Co-FACTOR (Cooperate, Communicate and Connect to boost smart Components for tomorrows Industry) which aims to create a community about smart components in the industry of the future. These smart components are included in cyber physical systems or these are CPS in themselves (Co-FACTOR project, 2017).

This current initiative is driven by six European projects located in the future industry (T-REX, INTEFIX, Power-OM, ReBorn, I-Ramp3 and SelSus) and allows participating organisations to cooperate with experts in this area and to get results from the projects. In this section, the authors focus on three projects toward digitising industry: SelSus, ReBorn and I-Ramp3.

SelSus (SELf-SUStaining Manufacturing System) is a reference architecture which aims to enable the provisioning of diagnostic and prognostic capabilities in manufacturing systems that utilize the notions of smart automation devices. The overall strategy of the architecture is based on the concept of independent smart automation components which can be integrated bottom-up into a wider automation system. Furthermore, SelSus architecture support three levels of communication which can be roughly categorized into data, information and knowledge. The main procedures are: read data, statistics analyses and contextualize information. The vision of SelSus is to create a new paradigm for highly effective, self-healing production resources and systems to maximise their performance over longer life times through highly targeted and timely repair, renovation and up-grading (Sayed, Lohse, & Madsen, 2015; SelSus, 2017).

ReBorn started with the idea to transmit the general ReBorn concept of innovative re-use in manufacturing and the technologies that enabled such a re-use. The vision is to demonstrate strategies and technologies that support a new paradigm for re-use of production equipment in old, renewed and new factories, by maximizing the efficiency of this re-usage and making the factory design process straightforward, shortening ramp-up times and increasing production efficiency and flexibility. This paradigm will give new life to decommissioned production systems and equipment, making it possible to be 'reborn' in new production lines (Pinto et al., 2016).

I-Ramp3 (Intelligent Reconfigurable Machines for Smart Plug&Produce Production) is an ongoing project under the wing of the Seventh Framework Programme of the European Commission. The objective is to create solutions to improve the European Industry competitiveness with innovative concepts towards smart manufacturing systems, converting the conventional product equipment into intelligent agent-based production devices known as NETDEVs (NETwork-enabled DEVices). The main responsibility of these components lies in equipping the manufacturing equipment –both complex machines, as industrial PCs or PLC, and sensors & actuators– with standardized communication skills, along with intelligent functionalities for inter-device negotiation and process optimization. Hence, the main goals of the project are to reduce the production costs and increase the manufacturing efficiency, by enabling fast and optimized ramp-up making use of the plug&produce concept (Gonçalves et al., 2014).

Other European initiatives are developing towards implementing IoT solutions. FIWARE platform is a standardized approach to provide a set of APIs (Application Programming Interfaces) that ease the development of Smart Applications in multiple vertical sectors (FIWARE, 2017). VIRTUS is a middleware layer employed to enforce secure seamless integration of devices and data within the same information network (Brizzi et al., 2011). Under the SPIRE initiative (Sustainable Process Industry through Resource and Energy Efficiency), other projects are developing solutions on industrial manufacturing or predictive maintenance, such as Maestri (MAESTRI, 2018) or Monsoon (MONSOON, 2018).

Development Platforms for I-IoT

In this section, the authors describe some of the available platforms to develop I-IoT solutions.

- **Predix:** Predix is an industrial platform that helps to develop apps on cloud infrastructure that connect industrial machines with people through data and analytics for better business outcome (deployed from the edge to the cloud). General Electric (GE) created Predix to guide industrial companies through this complex technology and business transition, putting them in charge of their I-IoT journey. By using this comprehensive platform, businesses can create innovative apps on Predix that turn real-time operational data into actionable insights. (Predix, 2016).

- **ThingSpeak™:** ThingSpeak is an Internet of thing platform that lets you collect and store sensor data in the cloud and develop IoT applications. ThingSpeak service is operated by MathWork and it lets you analyse and visualise your data in MATLAB®, and then act on the data. Sensor data can be sent to ThingSpeak from Arduino®, Raspberry Pi™, BeagleBone Black, and other hardware. (Thingspeak, 2017)
- **ThingWorx & Axeda Machine Cloud™:** This end-to-end technology platforms from PT Comunicações (PTC) is designed specifically for the Internet of Things. Combining ThingWorx and Axeda Machine Cloud™ technology empowers developers to connect, create, and deploy breakthrough, enterprise-ready IoT applications, solutions, and experiences, (Thingworx, 2017).
- **EcoStruxure™ Platform (Schneider Electric):** Schneider Electric launched its EcoStruxure™ architecture and platform, designed to deliver IoT-enabled solutions at scale for building, infrastructure, industry and data centre customers. The platform is designed to meet customer challenges associated with implementing IoT solutions seamlessly, cost-effectively and at scale. Essential components include: Embedded connectivity and intelligence that enables smart sensing, embedded computing, IP networking and edge analytics; a foundation for smart operations to serve as the building blocks for control, management, automaton and optimization; and the infrastructure for cloud-connected digital services that enables the delivery of cloud-based connected apps, analytics, services, control and monitoring (Rueil-Malmaison, 2016). As shown, nowadays, several companies, and start-ups promote IoT innovation, and they will certainly reap the benefits of that expansion. This point brings European initiatives and development platform which aim to create a community about smart components in the industry of the future.

According to these I-IoT solutions, the next step is to implement smart networks in factories. Due the importance of smart gateways –an intelligent bridge between virtual and real world, the following section shows an example for digitally enhanced electrification and automation, using two different gateways to collect data and control process.

In addition to the gateways, predictive maintenance on electric motor is addressed. Anticipating anomalies before occurring is synonymous with a proactive and intelligent system, fundamental to increase the productivity of the factories. Furthermore, a classification methodology will be presented to identify the anomalies and their level of severity, according to the legislation.

CASE STUDY

Predictive Maintenance on Electrical Motors

Vibration monitoring and analysis in rotating machineries offer very important information about anomalies formed internal structure of the machinery. The information gained by vibration analysis enable to plan a maintenance action (Orhan, Aktürk, & Celik, 2006).

In this use study, the authors study and analyses vibrations on an electrical motor to achieve predictive maintenance mechanisms supported by IoT. Firstly, vibration signals are collected by means of the vibration analyser equipped with a sensor in the time domain. These signals are converted into frequency domain by processing an FFT, and the information gained from the vibration signals can be used to predict catastrophic failures, to reduce forced outages, to maximize utilization of available assets, to increase the life of machinery, and to reduce maintenance costs related to health of machinery (Orhan et al., 2006). The predictive maintenance by vibration analysis is the best tool for this purpose. The vibration analysis is a philosophy or attitude that, simply stated, uses the actual operating condition of plant equipment and systems to optimize total plant operation. A comprehensive predictive maintenance management program uses the most cost-effective tools (e.g., vibration monitoring, thermography, tribology) to obtain the actual operating condition of critical plant systems and based on this actual data, schedules all maintenance activities on an as-needed basis. Including predictive maintenance in a comprehensive maintenance management program optimizes the availability of process machinery and greatly reduces the cost of maintenance. It also improves the product quality, productivity, and profitability of manufacturing and production plants (Mobley, 2002).

The transducer which, nowadays, is more-or-less universally used for vibration measurements is the piezoelectric accelerometer. It exhibits better all-round characteristics than any other type of vibration transducer. It has very wide frequency and dynamic ranges with good linearity throughout the ranges. It is relatively robust and reliable so that its characteristics remain stable over a long period (Brüel & Kjær, 1982). Accelerometers are typically placed at key locations on the motor and fan bearings. Vibration sensors should be placed in the radial (vertical & horizontal) and axial locations on the motor and fan bearings. This will provide the best detection of all vibration components including: bearing vibration, unbalanced, misalignment, electrical faults, blade pass (aerodynamic disturbances), belt frequencies (Connection Technology Center, 2017).

A good routine vibration program using a portable data collector, or permanent on-line vibration monitoring, will become the foundation of any predictive maintenance initiative. Also, it is very important to select an appropriated sensor to measure. The three parameters representing motion detected by vibration monitors are displacement, velocity, and acceleration. These parameters can be measured by a variety of motion sensors and are mathematically related. Selection of a sensor proportional to displacement, velocity or acceleration depends on the frequencies of interest and the signal levels involved (SKF, 1999).

In this case study, vibrations data are detected by the piezoelectric accelerometer placed in the structure of a motor. The authors show two different architectures to collect the vibration data. The first one, using the NI DAQ hardware and the second one, using Arduino over EtherCAT protocol. An OPC Server and LabVIEW software are used to store and visualise data, respectively.

1. Vibration Data Acquisition

In the Figure 1, there are two block diagrams of the architectures implemented.

As shown, both architectures are formed by an electric motor, an accelerometer, a signal conditioner, a LabVIEW application and an OPC Server. The principal difference between them is the device used to process vibration data. In the A architecture, the data acquisition board is a NI DAQ 6008, and in the B architecture, is an Arduino UNO board with the shield EasyCAT to do the communication over EtherCAT.

The specific equipment that is used in these architectures:

Figure 1. Block diagrams for data acquisition with: A) NI-DAQ B) Arduino over EtherCAT

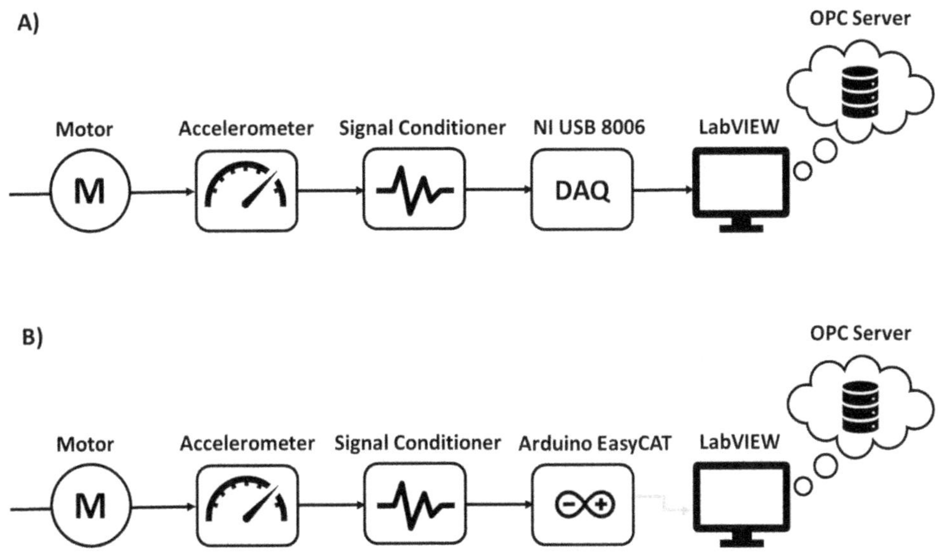

- **Electric Motor:** A motor from Industrie Elektronik (typ. LGHA 63) which characteristics are: power input P = 0,2 kW; motor speed n_1 = 3000; voltages U_a/U_f = 160/200 V; currents I_a / I_f = 1,6 / 0,22 A; F_f = 1.0; I_p = 44.
- **Piezoelectric Accelerometer:** It is the one-axis accelerometer –DYTRAN model 3134D– which operates with a sensitivity of 500 mV/g.
- **Signal Conditioner:** It is the model 482A21 Line Powered Signal Conditioner. This unit provides constant current excitation to the built-in transducer amplifier and decouples the signal from the DC bias voltage.
- **NI LabVIEW:** The authors use this platform in 32 bits –from National Instruments– to develop measurement and control systems.
- **OPC Server:** The OPC (OLE for Process Control) specifications, a series of open interface standard for the communication between automation/control applications, field systems/devices, and business/office applications in the process control industry, are established by OPC foundation (a non-profit corporation sponsored by Intellution, Microsoft and so on). Based upon Microsoft's OLE/COM/DCOM technology and characterized by client/server structure, OPC gets independent of programming language and supports applications' integration and interoperability in the heterogeneous or distributed environment. In the sense of data transfer, the OPC serve acts as a "Software Gateway" reading data from field various kind of devices and transforming those varied data to uniform OPC data for the integration of different systems and devices (Ling, Chen, & Yu, 2004) Concretely, in this study case, MatrikonOPC server is used. It provides equipment data connectivity products based on the OPC standard. Users have an OPC Server and one or more OPC Clients that communicate with the server to write or read data.
- **NI DAQ 6008:** It is a DAQ device for connection to USB port from National Instruments. It acts as an interface between the computer and the outside world. Its function is to digitises incoming analog signals so that the computer can interpret them. To use the block DAQ acquisition in LabVIEW, the NI-DAQmx driver is needed (NI-DAQmx, 2017).

- **Arduino UNO + Shield EasyCAT:** The shield EasyCAT Arduino from AB&T Tecnologie Informatiche is needed to do the communication over EtherCAT (Bausano, 2017).

EtherCAT is an open industrial real-time and deterministic Ethernet Fieldbus standard that is widely regarded as the best industrial network to date. It is orders of magnitudes faster than many well-known networks such as Profibus, DeviceNet, and ModbusTCP. Automation equipment vendors can utilize EtherCAT on their own device implementations to improve performance and flexibility. End users and automation system designers can also implement their own EtherCAT compliant devices for specific needs (EtherCAT, 2017).

The Table 1 indicates differences between NI-DAQ and Arduino with shield EasyCAT. The main advantage of using USB is its transmission rate and the easy configuration in LabVIEW, by contrast, the main advantage of using EtherCAT is its cost and the effective distance between I/O stations.

These devices are connected to form the networks that are shown in the Figure 2. From left to right there are: a frequency converter for regulating the motor, an electrical motor, the piezoelectric accelerometer which is attached to this motor structure, and connected to the input port of the signal conditioner. The output signal of the signal conditioner is an analogic and differential signal which goes through an analog input of the NI DAQ device (architecture A) or through an analog input of the Shield EasyCAT Arduino (architecture B). The data acquisition device is connected to a PC via USB (architecture A) or via Ethernet (architecture B).

Once all the previous connection is done, a LabVIEW application is needed to acquire data from NI DAQ 6008 and Arduino. In both cases, the authors must configure the acquire signal as an analog input (voltage). In the architecture A, the NI DAQ device provides a voltage signal that need to be converted in acceleration –the accelerometer has a sensitivity of 500 mV/g– using the following conversion:

$$V_{input} \cdot \frac{1g}{0,5V} \cdot \frac{9,8m / s^2}{1g} = Acceleration\left(m / s^2\right)$$

Table 1. Differences between NI-DAQ and Arduino over EtherCAT

Device	NI-DAQ	Shield EasyCAT Arduino
Technology	USB 2.0	EtherCAT
Distance between I/O stations	5 meters	100 meters
LabVIEW connection	Easy configuration with a block DAQ Assist	Complex configuration with EtherCAT Acquisition blocks
Software requirements	NI-DAQmx driver	Library 'EasyMaster' for LabVIEW in 32 bits Library 'EasyCAT.h' for Arduino Configuration file 'EasyCAT.xml' (ESI)
Transmission Rate	480 Mbit/s	2 x 100 Mbit/s (Fast Ethernet, Full-Duplex)
Cost	288 €	61 €
Additional cost	—	Arduino board

Figure 2. Networks for vibration data acquisition with: A) NI DAQ B) Arduino board + shield EasyCAT

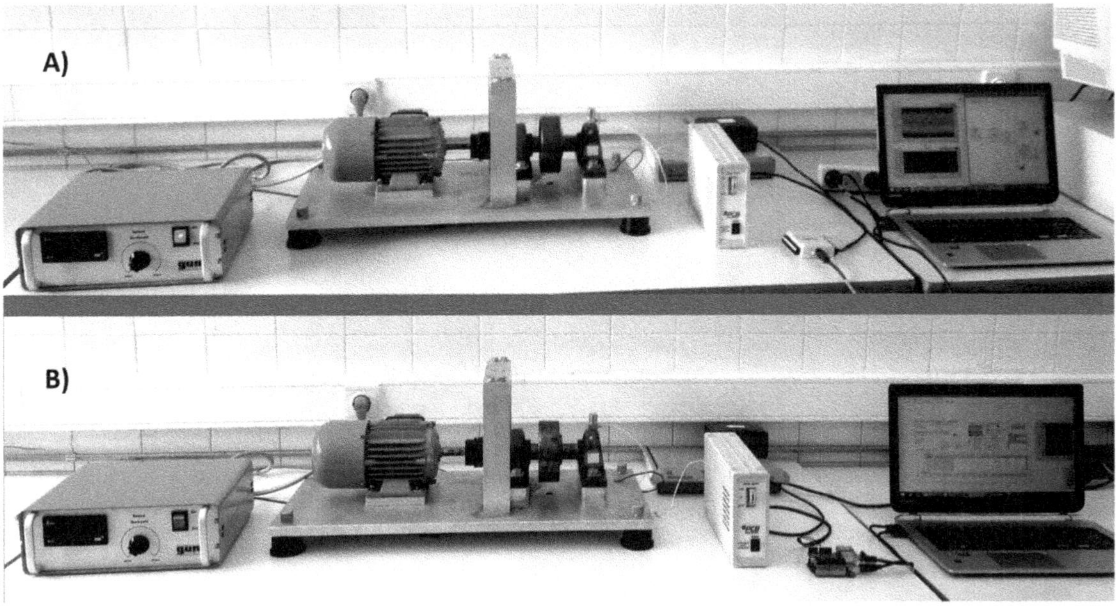

In the architecture B, the analog inputs in shield EasyCAT Arduino will be converted in 8 bits and in a range between 0 V and 5V, so first, it is important to do a scaling to convert units to voltage and the, to acceleration:

$$X\,units \cdot \frac{5V}{255\,units} \cdot \frac{1g}{0,5V} \cdot \frac{9,8m/s^2}{1g} = Acceleration\left(m/s^2\right)$$

Afterwards, the acceleration output is integrated to give velocity proportional signal and it is filtered by a bandpass filter with a bandwidth from 10 Hz to 1000 Hz. Why this range is chosen? Where a single, wide frequency band vibration measurement is made, the choice of parameter is important if the signal has components at many frequencies. Measurement of displacement will give the low frequency components most weight and conversely acceleration measurements will weigh the level towards the high frequency components. Experience has shown that the overall RMS value of vibration velocity measured over the range 10 to 1000 Hz gives the best indication of a vibration's severity. A probable explanation is that a given velocity level corresponds to a given energy level so that vibration at low and high frequencies are equally weighted from a vibration energy point of view. In practice, many machines have a reasonably flat velocity spectrum (Brüel & Kjær, 1982).

The velocity vibration, in a motor, is normally measured in three directions: horizontal, vertical and axial, as they are indicated in the Figure 3. Taking readings in these three directions give more information that helps in analysis as some defects comes with predominant vibrations in a direction. Axial is the direction parallel to the centreline of a shaft or turning axis of a rotating part. Vertical is the direction toward the centre of rotation of a shaft or rotor. Horizontal is the direction perpendicular to the centre of rotation. (Soliman, 2015)

Figure 3. Measuring directions

Once data is acquired, information is converted to be analysed. There are some types of analysis:

- **Vibration Analysis in the Time Domain:** Time-domain analysis is directly based on the time waveform itself. Traditional time-domain analysis calculates characteristic features from time waveform signals as descriptive statistics such as mean, amplitude, peak, peak-to-peak value and RMS (Wang & Chen, 2016).
- **Vibration Analysis in the Frequency Domain FFT:** Frequency-domain analysis is based on the transformed signal in frequency domain. The advantage of frequency-domain analysis over time-domain analysis is its ability to easily identify and isolate certain frequency components of interest. The most widely used conventional analysis is the spectrum analysis by means of fast Fourier transform (FFT). The main idea of spectrum analysis is to either look at the whole spectrum or look closely at certain frequency components of interest and thus extract features from the signal (Wang & Chen, 2016).

The result is vibration amplitude as a function of frequency, which lets us perform analysis in the frequency domain (or spectrum) to gain a deeper understanding of our vibration profile. Most vibration analysis will typically be done in the frequency domain (Hanly, 2016).

The vertical velocity vibration is shown in the Figure 4. The graph below shows the FFT waveform where fundamental frequency (1x) is approximately at 15 Hz, and there are harmonics at frequencies 2x, 3x, 4x, etc., which amplitudes are decreasing. As the vertical measurement's 1x amplitude is high and harmonics are less than 15% of the 1x, there may be unbalance (Mais, 2002). Moreover, the 2x indicates there may be misalignment.

Figure 4. Vertical measuring direction

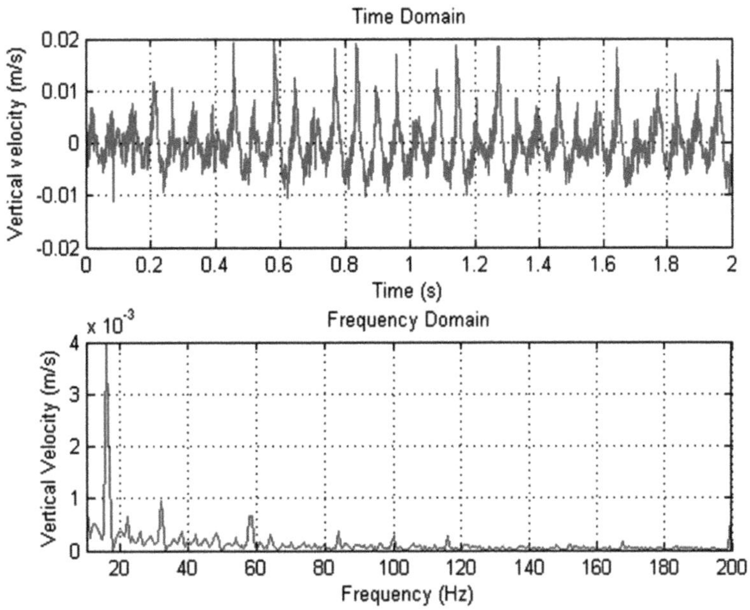

The horizontal velocity vibration is shown in Figure 5. The graph below shows the FFT waveform in horizontal direction. As there are a series of three or more synchronous or 1/2 synchronous multiples of running speed (range 2x to 10x), and their magnitudes are greater than 20% of the 1x, there may be mechanical looseness. With severe misalignment, the spectrum may contain multiple harmonics from 2x to 10x of running speed. As vibration amplitude in the horizontal plane is increased two or three times, misalignment is indicated again (Mais, 2002).

The axial velocity vibration is shown in the Figure 6. Upper graph shows the velocity time signal obtained, which ranges between -0,005 and 0,005 m/s. The graph below shows the FFT waveform in axial direction. It is an enveloped spectrum with inner race defect and bearing frequency overlays. This spectrum indicates a defect is present on the bearing's inner race (Mais, 2002).

2. Predictive Maintenance

Appropriate maintenance actions delay the possible deterioration and minimize/avoids the machining system stoppage time that leads to lower productivity and higher production cost. Moreover, measuring and monitoring machine tool condition has become increasingly important due to the introduction of agile production, increased accuracy requirements for products and customers' requirements for quality assurance (Rastegari, Archenti, & Mobin, 2017).

Condition based maintenance (CMB) is a maintenance strategy that allows optimizing maintenance operations based on e.g. measured wear for balancing production availability with costs related to spare parts and production downtime. The challenge is to get accurate measured data from many monitored components using diverse sensors which typically provide data that needs further processing and analysis to be useful. Another challenge is to integrate this heterogeneous information from sensors and systems that are distributed in different networks before it can be used to support maintenance operations (Hästbacka et al., 2016).

Figure 5. Horizontal measuring direction

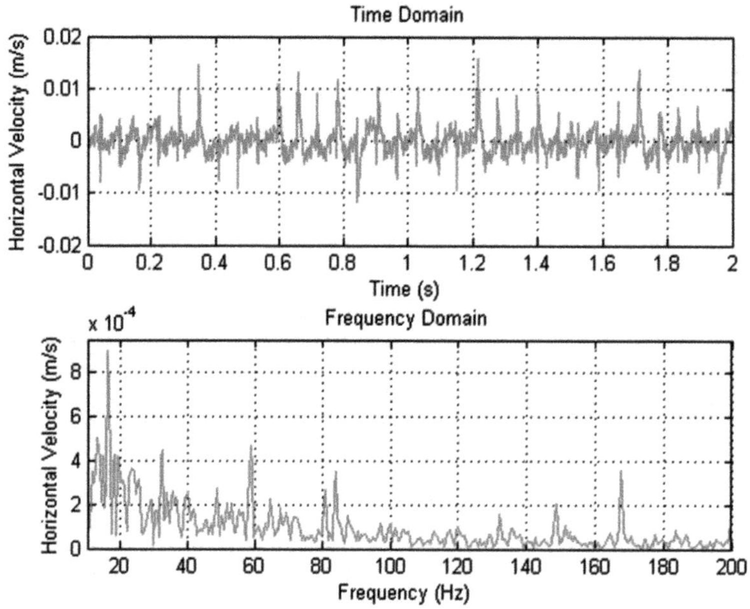

Figure 6. Axial measuring direction

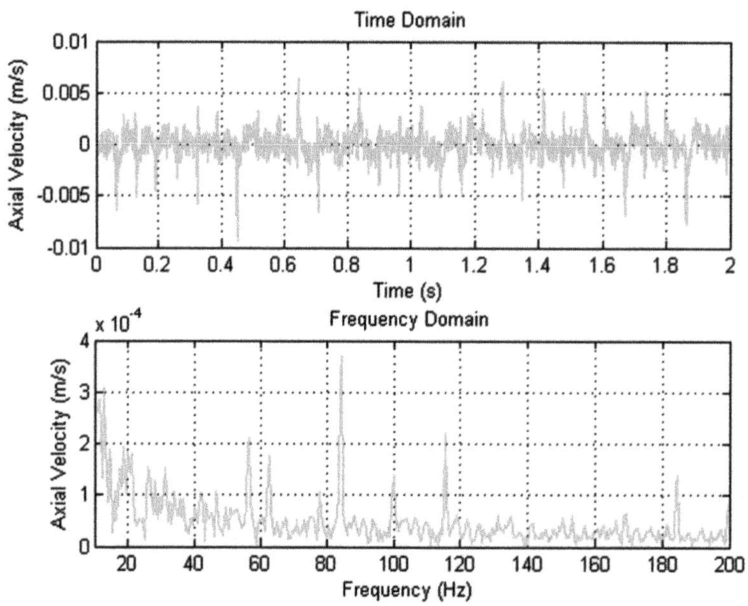

To make a fault diagnosis in this case study, the authors use a predictive modelling approaches used in Machine Learning. The model is called Decision tree classifier (DTC) and the main objectives of DTCs are (Safavian, & Landgrebe, 1991):

1. Classify correctly as much of the training sample as possible.
2. Generalize beyond the training sample so that unseen samples could be classified with as high of an accuracy as possible.
3. Be easy to update as more training becomes available
4. Have as simple a structure as possible.

The tree structure called ID3 (Interative Dichotomiser 3) is chosen to generate a decision tree from a dataset. ID3 uses entropy and information gain to construct a decision tree. ID3 algorithm uses entropy to calculate the homogeneity of a sample, and information gain approach is generally used to determine suitable property for each node of a generated decision tree (Saedsayad, 2017)

Set S is set including s number of data samples whose type attribute can take m potential different values corresponding to m different types of C_i *i(1,2,3,…,m)*. Assume that S_i is the sample number of C_i. Then, the required amount of information to classify a given data is (Jin, De-Lin, & Fen-Xiang, 2009).:

$$\boldsymbol{I}\left(\boldsymbol{S}\right)=\sum_{i=1}^{m}-\boldsymbol{p}_i\cdot\boldsymbol{log}_2\boldsymbol{p}_i$$

where $p_{i\,=}\,S_{ij}\,/\,|S_j|$ is the probability that any subset of data samples belonging to categories C_i.

Suppose that *A* is a property which has *v* different values $\{a_1, a_2, \ldots, a_v\}$. Using the property of *A*, *S* can be divided into *v* number of subsets $\{S_1, S_2, \ldots, S_v\}$, in which S_j contains data samples whose attribute *A* are equal a_j in *S* set. If property *A* is selected as the property for test, that is, used to make partitions for current sample set, suppose that S_{ij} is a sample set of type C_i in subset S_i, the required information entropy is (Jin, De-Lin, & Fen-Xiang, 2009):

$$E\left(A\right)=\sum_{j=1}^{v}\frac{s_{1j}+s_{2j}+\ldots+s_{mj}}{s}\cdot I\left(S\right)$$

Such use of property *A* on the current branch node corresponding set partitioning samples obtained information gain is (Jin, De-Lin, & Fen-Xiang, 2009):

$$Gain\left(A\right)=I\left(S\right)-E\left(A\right)$$

ID3 algorithm traverses possible decision-making space using top-down greedy search strategy, and never trace back and reconsider previous selections. Information gain is exactly the metrics for selecting the best attribute in each step of the growth tree in ID3 algorithm (Jin, De-Lin, & Fen-Xiang, 2009).

To do a classification vibration's severity of the motor with this algorithm, the authors use a table from ISO 2372 which indicates a suggested classification, according to vibration's severity range limits (velocity) and machines classes, in four groups:

1. A = Good
2. B = Satisfactory
3. C = Unsatisfactory
4. D = Unacceptable

The classification model is trained considering 4000 samples cases, where occurred evidences of the four classes. Before classification it is need identify events that evidence occurrences. For the example of vertical vibration, it is identified an upper peak of vibration velocity, depicted in Figure 7. Based on the magnitude of the peaks the system can automatically classify the vibrations.

This way, it is possible to do a classification vibration's severity in one of these four groups. As the value of the peak is 3,9 mm/s and the class of motor is Class I (< 20 HP), the vibration's severity is the Class C, that is, the vibrations are 'Unsatisfactory' –according to ISO 2372 such as it was indicated in the previous data analysis.

Figure 7. Upper peak of vertical vibration

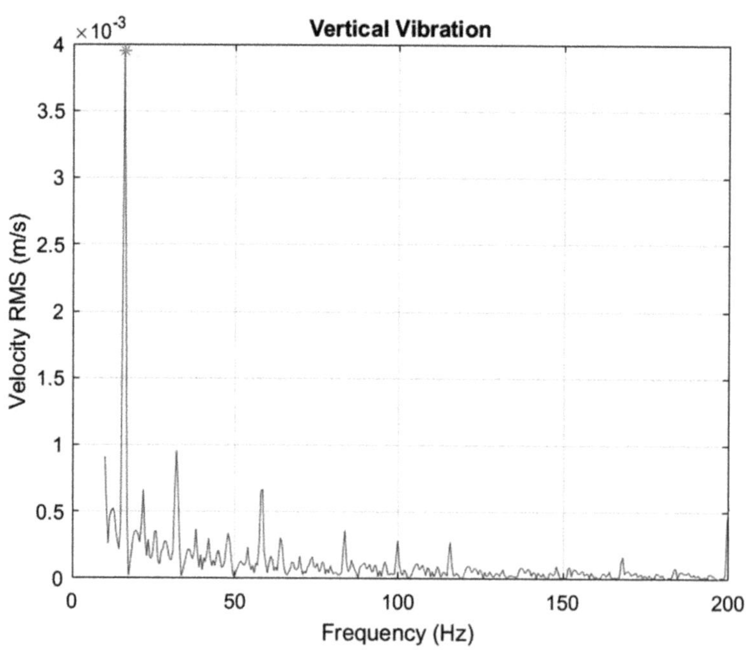

In the field of machine learning, and specifically the problem of statistical classification, a confusion matrix, is a specific table layout that allows visualization of the performance of an algorithm, typically a supervised learning one. Each column of the matrix represents the instances in a predicted class while each row represents the instances in an actual class. The name stems from the fact that it makes it easy to see if the system is confusing two classes (Confusion Matrix, 2017).

The Figure 8 shows the vibration confusion matrix. The number of elements of Class C (or Class 3) classified in the Class C are 998 and the number of elements of Class C classified in the Class B (or Class 2) are two, this means, during the training of vibration data, the 99.9% of accuracy was obtained for Class C. In this way, it is correct to ensure, vibrations of the electric motor are unsatisfactory.

FUTURE RESEARCH DIRECTIONS

The Industry 4.0 is expanding towards a Smart Manufacturing. The application of new technology based on Cyber-Physical Systems and internet leads to significant improvements, such as increase of automation and shorten period between the development of a new product and its market launch. Industry 4.0 is a model that shows how industrial production follows the latest developments and changes over time (Vuksanović, Ugarak, & Korčok, 2016).

There are four building blocks on which Smart Manufacturing revolution is working (Kevin, 2016):

Figure 8. Vibrations Confusion Matrix

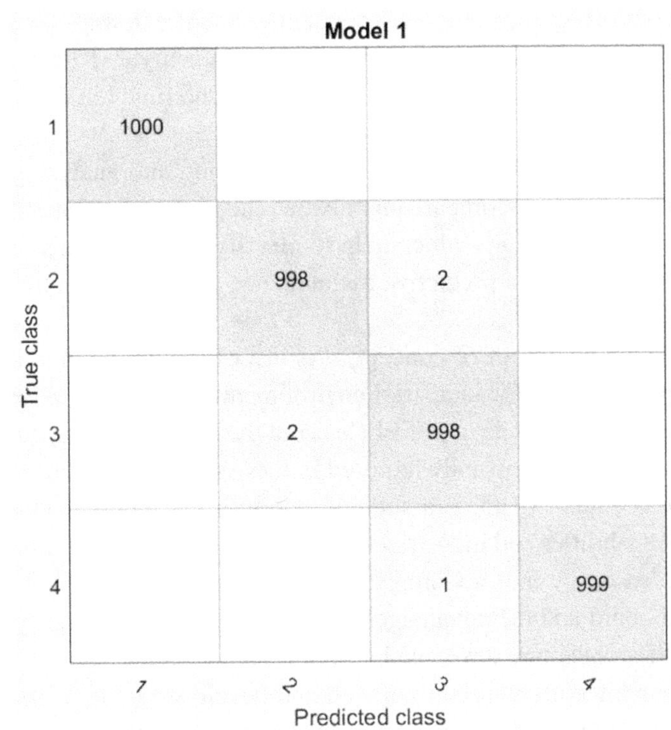

- **Smart Machines and Advanced Robotics:** These machines recognize product configurations and diagnostic information, and make decisions and solve problems without human intervention.
- **I-IoT:** Devices with network and internet connectivity that are active participants in event-driven, self-healing manufacturing processes integrated with open standards that support connectivity.
- **Cloud Services:** On-demand information technology services that can be rapidly provisioned and released with minimal management effort or service provider interaction.
- **Enterprise Integration Platforms:** Platforms that can receive data broadcast from equipment via secure open standards. These applications analyze and aggregate the data, and trigger process controls, history recording, and work flows that enable business processes across value chain systems that can then be integrated via application-to-application (A2A) and business-to-business (B2B) open standards.

These systems generate, process, and exchange vast amounts of security-critical and privacy-sensitive data (Big Data), which makes them attractive targets of attacks. Cyberattacks on IoT systems are very critical since they may cause physical damage and even threaten human lives. The complexity of these systems and the potential impact of cyberattacks bring upon new threats. This is an important issue to focus and that is why possible solutions towards a holistic security framework for Industrial IoT systems are being investigated (Sadeghi, Wachsmann, & Waidner, 2015).

CONCLUSION

The Industry 4.0 is coming with Internet of Things, Cyber-Physical Systems, Cloud Computing and intelligent networking based on wireless, mobile and sensor devices autonomously exchanging information, triggering actions and controlling each other independently. Smart gateways for I-IoT are implemented in the 4th industrial revolution due to their capabilities for data analytics, diagnostics, prognostics and self-awareness, and simplify greatly the process to export data and functionality of the physical devices to end users on the web

In the study case, the authors use two gateways for obtaining and analysing the vibration data of an electrical motor. Table 1 shows a comparison between the protocols (EtherCAT and USB) used in both gateways. They give similar results which help to identify the vibration's severity –in this case, it is unsatisfactory– and to implement a predictive maintenance using decision tree classifiers according to machine learning fundamentals.

The case study presented is a proof of concept, it is clear that the Arduino and the NI acquisition board do not have the prerequisites for industrial environments, however, it allows to demonstrate the concept of Cyber-Physical systems. The use of PLCs is not the most adequate, due to the high sampling frequencies and real-time processing usually required in these scenarios. A Programmable Automation Controller (PAC), such as compactRIO, is a suitable option in the implementation of these systems, due to the multitasking capabilities and industrial certification requirements. Our application developed in LabView can be loaded easily into a CompactRIO with minimum configuration changes. This is the reason to use the NI board and the advantages of using NI platforms. The systems upgrade can be implemented keeping the architecture developed.

As future work, our architecture will be improved and developed under a compactRIO platform in order to test and implement the smart gateway in real industrial scenarios.

REFERENCES

Aazam, M., Hung, P. P., & Huh, E. N. (2014, April). Smart gateway based communication for cloud of things. In *2014 IEEE Ninth International Conference on Intelligent Sensors, Sensor Networks and Information Processing (ISSNIP)* (pp. 1-6). IEEE.

ABB. (2017). AC 800M Controllers. Retrieved May 19, 2017, from http://new.abb.com/control-systems/system-800xa/800xa-dcs/hardware-controllers-io/ac800m-controllers

Bausano. (2017). Arduino EasyCAT. Retrieved June 14, 2017, from http://www.bausano.net/en/hardware/ethercat-e-arduino/easycat.html

Boschrexroth. (2016, November 21). Industry 4.0 for new and existing machines. Retrieved June 20, 2017, from https://www.boschrexroth.com/en/xc/company/press/index2-25280

Brizzi, P., Lotito, A., Ferrera, E., Conzon, D., Tomasi, R., & Spirito, M. (2011, December). Enhancing traceability and industrial process automation through the VIRTUS middleware. In *Proceedings of the Middleware 2011 Industry Track Workshop* (p. 2). ACM.

Brüel & Kjær. S. (1982). *Measuring Vibration.* Retrieved June 1, 2017, from https://www.bksv.com/media/doc/br0094.pdf

SOCe. (2018). Industrial Networking Modules and Boards. Retrieved May 22, 2017, from http://soc-e.com/products/

Co-FACTOR project. (2017). Retrieved May 19, 2017, from http://www.cofactor-eu-project.org/

Condition Monitoring, S. K. F. (1999). *Vibration Sensors.* Retrieved May 22, 2017, from http://www.exvalos.cz/soubory/File/SKF/SNIMACE_VIBRACI.pdf

Confusion Matrix. (2017, May 14). In *Wikipedia, The Free Encyclopedia.* Retrieved 15:06, June 20, 2017, from https://en.wikipedia.org/wiki/Confusion_matrix

Connection Technology Center. (2017). *Measuring Fan and Motor Vibration.* Retrieved May 22, 2017 from https://www.ctconline.com/pdf/pubTechPapers/23Measuring%20Fan%20and%20Motor%20Vibration.pdf

EtherCAT. (2017). What's EtherCAT? Retrieved June 13, 2017, from https://www.automation.com/automation-news/article/whats-ethercat

eWON (2018). eWON Flexy - Industrial M2M Router and Data Gateway. Retrieved January 6, 2018, from https://ewon.biz/products/ewon-flexy

Geisberger, E., Cengarle, M., Keil, P., Niehaus, J., Thiel, C., & Thönnißen-Fries, H. J. (2011). Cyber-Physical Systems-Driving force for innovation in mobility, health, energy and production. Acatech-Deutsche Akademie der Technikwissenschaften.

Gonçalves, G., Reis, J., Pinto, R., Alves, M., & Correia, J. (2014, September). A step forward on Intelligent Factories: A Smart Sensor-oriented approach. In 2014 IEEE Emerging Technology and Factory Automation (ETFA) (pp. 1-8). IEEE. doi:10.1109/ETFA.2014.7005227

Hanly, S. (2016, June 16). Vibration Analysis: FFT, PSD, and Spectrogram Basics. *Mide*. Retrieved June 6, 2017, from https://blog.mide.com/vibration-analysis-fft-psd-and-spectrogram

Hästbacka, D., Jantunen, E., Karaila, M., & Barna, L. (2016, October). Service-based condition monitoring for cloud-enabled maintenance operations. In *42nd Annual Conference of the IEEE Industrial Electronics Society, IECON 2016* (pp. 5289-5295). IEEE. 10.1109/IECON.2016.7793470

Jin, C., De-Lin, L., & Fen-Xiang, M. (2009, July). An improved ID3 decision tree algorithm. In *4th International Conference on Computer Science & Education ICCSE'09* (pp. 127-130). IEEE

Kagermann, H., Helbig, J., Hellinger, A., & Wahlster, W. (2013). *Recommendations for Implementing the strategic initiative INDUSTRIE 4.0: securing the future of German manufacturing industry; final report of the Industrie 4.0 working group*. Forschungsunion.

Kevin, L. J. (2016, November 27). *Smart Manufacturing Is Cloud Computing*. Retrieved June 8, 2017, from http://kevinljackson.blogspot.pt/2016/11/smart-manufacturing-is-cloud-computing.html

Li, Q., Qin, W., Han, B., Wang, R., & Sun, L. (2011). A case study on REST-style architecture for cyber-physical systems: Restful smart gateway. *Computer Science and Information Systems*, *8*(4), 1317–1329. doi:10.2298/CSIS110310062L

LimeNET. (2017). Network-in-a-Box. Retrieved May 18, 2017, from https://www.crowdsupply.com/lime-micro/limenet

Ling, Z., Chen, W., & Yu, J. (2004, June). Research and implementation of OPC server based on data access specification. In *Fifth World Congress on Intelligent Control and Automation WCICA 2004* (Vol. 2, pp. 1475-1478). IEEE.

LioN-Power. (2017). Active I/O modules. Retrieved May 18, 2017, from http://www.lumberg-automation.com/en/Lumberg_Automation/news/lion-power/index.phtml

MAESTRI. (2018). Resource and Energy efficiency for Process Industries. Retrieved December 28, 2017, from https://maestri-spire.eu

Mais, J. (2002, May). Spectrum Analysis. The key features of analysing spectra. *SKF*. Retrieved June 8, 2017, from http://www.skf.com/binary/tcm:12-113997/CM5118%20EN%20Spectrum%20Analysis.pdf

Malente. (2016, October 20). *Kuhnke FIO Controller WL - IoT compatible Controller.* Retrieved May 19, 2017, from http://kuhnke.kendrion.com/industrial/ics/en/press-releases/iot-compatible-controller-plc-with-integrated-wifi-data-interface.html

MindSphere (2017). Retrieved June 12, 2017, from https://www.plm.automation.siemens.com/store/en-ie/mindsphere/index.html

Mobley, R. K. (2002). *An introduction to predictive maintenance*. Butterworth-Heinemann.

MONSOON. (2018). Model-based control framework for Site-wide Optimization of data-intensive processes. Retrieved December 28, 2017, from https://www.spire2030.eu/monsoon

National Instruments (2016, March 24). The CompactRIO, Retrieved May 18, 2017, from http://www.ni.com/white-paper/52251/en/

NI-DAQmx (2017). Software. Retrieved June 14, 2017, from https://www.ni.com/dataacquisition/nidaqmx.htm

Orhan, S., Aktürk, N., & Celik, V. (2006). Vibration monitoring for defect diagnosis of rolling element bearings as a predictive maintenance tool: Comprehensive case studies. *NDT & E International*, *39*(4), 293–298. doi:10.1016/j.ndteint.2005.08.008

Pinto, R., Gonçalves, J., Cardoso, H. L., Oliveira, E., Gonçalves, G., & Carvalho, B. (2016, December). A Facility Layout Planner Tool Based on Genetic Algorithms. In *2016 IEEE Symposium Series on Computational Intelligence (SSCI)* (pp. 1-8). IEEE. 10.1109/SSCI.2016.7850280

Predix. (2016, November). *The Industrial Internet Platform.* Retrieved May 19, 2017, from: https://www.predix.com/

Rastegari, A., Archenti, A., & Mobin, M. (2017, January). Condition based maintenance of machine tools: Vibration monitoring of spindle units. In *2017 Annual Reliability and Maintainability Symposium (RAMS)* (pp. 1-6). IEEE. 10.1109/RAM.2017.7889683

Ray, L. (2017). Cyber-Physical Systems. In Handbook of Research on Applied Cybernetics and Systems Science (p. 335).

Rueil-Malmaison. (2016, November 30). Schneider Electric delivers IOT solutions platform, education. *Waterworld.* Retrieved May 19, 2017, from http://www.waterworld.com/articles/iww/2016/11/schneider-electric-delivers-iot-solutions-platform-education.html

Sadeghi, A. R., Wachsmann, C., & Waidner, M. (2015, June). Security and privacy challenges in industrial internet of things. In *Design Automation Conference (DAC), 2015 52nd ACM/EDAC/IEEE* (pp. 1-6). IEEE. 10.1145/2744769.2747942

Saedsayad. (2017). Decision Tree - Classification. Retrieved June 20, 2017, from http://www.saedsayad.com/decision_tree.htm

Safavian, S. R., & Landgrebe, D. (1991). A survey of decision tree classifier methodology. *IEEE Transactions on Systems, Man, and Cybernetics*, *21*(3), 660–674. doi:10.1109/21.97458

Sayed, M. S., Lohse, N., & Madsen, A. L. (2015, July). SelSus: Towards a reference architecture for diagnostics and predictive maintenance using smart manufacturing devices. In *2015 IEEE 13th International Conference on Industrial Informatics (INDIN)* (pp. 1700-1705). IEEE.

SelSus. (2017, April 26). Retrieved May 19, 2017, from http://www.selsus.eu/project/

Siemens. (2017). The Internet of Things is speaking. Retrieved May 19, 2017, from https://www.siemens.com/global/en/home/products/software/mindsphere.html

SIMATIC. (2017) Box PC. Retrieved May 18, 2017, from http://w3.siemens.com/mcms/pc-based-automation/en/industrial-pc/box-pc/Pages/Default.aspx

Smit, J., Kreutzer, S., Moeller, C., & Carlberg, M. (2016). *Industry 4.0 by Policy Department A.* Retrieved May 22, 2017, from http://www.europarl.europa.eu/studies

Soliman, M. H. A. (2015, March 12). *Machine Vibration Analysis*. Retrieved June 8, 2017, from https://es.slideshare.net/MohammedHamedAhmedSo/machine-vibration-analysis

Thingspeak. (2017). Retrieved May 19, 2017, from https://thingspeak.com/

Thingworx. (2017). Retrieved May 22, 2017, from https://www.thingworx.com

Ubuntu Blog. (2017, February 16). Lime Microsystems and Canonical announce LimeNET crowdfunding. Retrieved June 20, 2017, from https://insights.ubuntu.com/2017/02/16/lime-microsystems-and-canonical-announce-limenet-crowdfunding/

Vuksanović, D., Ugarak, J., & Korčok, D. (2016). Industry 4.0: The future concepts and new vision of factory of the future development. *Paper presented at International Scientific Conference on ICT and e-business related research*, Singidunum University. 10.15308/Sinteza-2016-293-298

Wang, R., & Chen, N. (2016, December). A survey of condition-based maintenance modeling of multi-component systems. In *2016 IEEE International Conference on Industrial Engineering and Engineering Management (IEEM)* (pp. 1664-1668). IEEE. 10.1109/IEEM.2016.7798160

KEY TERMS AND DEFINITIONS

Big Data: This term emerged to solve the treatment and storage massive data that has specific defined features as speed, volume and variety.

Cloud Computing: A service online which permits users to store information and access it from any part of world via web.

Cyber-Physical System (CPS): Any device which integrates computing, storage and communication capacity to manage and interact with a physical process. Through CPS, physical world is linked with virtual world or cyberspace.

EtherCAT (Ethernet For Control Automation Technology): An Ethernet fieldbus standard for control and system engineers, especially for automation purposes.

Industrial Internet of Things: It is the concept of Internet of Things –devices interconnect in a smart network via internet– applied to Industrial factories.

Industry 4.0: A new type of industrialisation, the 4th industrial revolution, based on Cyber-Physical Systems in the manufacturing environment.

LabVIEW (Laboratory Virtual Instrument Engineering Workbench): A platform from National Instrument to develop measurement and control systems.

Machine Learning: A scientific discipline within Artificial Intelligent which creates automatic learning systems.

OPC Server: A software application conforms to OPC specifications. In this architecture OPC server is the slave and OPC Client is the master.

Smart Objects (SO): Autonomous and decision-making objects interconnected within the framework of the factory.

Chapter 8
Cloud-Based Manufacturing (CBM) Interoperability in Industry 4.0

István Mezgár
Institute of Computer Science and Control of the Hungarian Academy of Sciences, Hungary

Gianfranco Pedone
Institute of Computer Science and Control of the Hungarian Academy of Sciences, Hungary

ABSTRACT

Cloud computing (CC) is generating new computing and business models because of its service-based nature, which enables collaboration and data exchange at higher level, more flexibility with better efficiency and parallel decreasing costs. Manufacturing environments can also benefit from cloud technology and follow fast changes in market demands. In these new scenarios, interoperability has vital importance in the operation and in the interaction among industrial realizations of cyber-physical systems. The chapter introduces the different cloud models and the interoperability issues concerning connected enterprise information systems. Various standardization frameworks have been developed for homogeneous integration of IT models in industrial environments: the IIRA and the RAMI 4.0 are the best-known ones. The chapter introduces both architectures; their methodological approach to industrial integration efforts and how integration can be realized through the OPC Unified Architecture. Finally, the authors propose a basic conceptual model for cloud manufacturing.

INTRODUCTION

Information and communications technologies (ICTs) have an extremely fast evolution tendency, which leads to creating new "digital" economy. Nowadays this evolution has reached a point where it can be called a revolution. The first industrial revolution was when mechanical production facilities were powered by water and steam (1784), the second one was the introduction of mass production with the help of electrical energy (1870), the third big turning point was the application of electronic and IT systems to further automate production worldwide (1969 first PLC).

DOI: 10.4018/978-1-5225-4936-9.ch008

Now the world is already in the Forth Industrial Revolution phase where in Cyber Physical Systems (CPS) real and virtual objects and processes are interlinked. This phase can be featured as a deep interdisciplinary integration of technologies in the digital, physical, and biological world (Schwab, 2015).

Also manufacturing industry belongs to those sectors which are changing basically in all of their component systems (control, ITC, fabrication) and governments are launching initiatives, global national (framework) research projects to support the research and implementation activities in the most important ITC fields (European Commission, 2016; NSF Program Announcements, 2017).

The high-tech ICT-based strategic program of the German government, called "Industry 4.0", essentially focuses on manufacturing and describes the up-to-date automation and data exchange in manufacturing technologies. Industry 4.0 includes CPSs, the Internet of Things and cloud computing. In the industrial sector the application of cloud computing is constantly growing. According to statistics from US (RightScale, 2017) 95% of firms use cloud computing technology, while 25% call security as a significant challenge. Based on a former opinion of industrial experts in cloud environments interoperability is a bigger problem than security. "The greatest challenge facing longer-term adoption of cloud computing services is not security, but rather cloud interoperability and data portability" say cloud computing experts from IEEE (Weissberger, 2011). According to the above examples it can be stated that interoperability and cloud-based manufacturing is really in the focus of the actual research topics.

Cloud-based manufacturing (CBM) make use of the cloud technology in the industry mirroring the service orientation approach of it by applying diverse cloud service- and deployment models that can easily convert and map manufacturing processes and assets into services.

The chapter, after a short overview on cloud manufacturing, provides a comparison of the most relevant features between traditional manufacturing IT systems and new Industry 4.0, cloud-oriented ones. Cloud technology is introduced in the next section with focusing on key fields of such architectures and cloud models, their combinations, and cloud interoperability. The forth section contains an overview on cloud manufacturing, highlighting its main characteristics, the different types, and a short description of interoperability challenges and how to convert traditional manufacturing to CBM through virtualization. A basic conceptual model for cloud manufacturing (CMCM) is also proposed by the authors. IIRA and RAMI 4.0 are two of the most known standardization frameworks for Industrial Internet environments: the aim of this chapter is also to present the two architectures and highlight their integration compatibility and functional interoperability. As cloud architectures become the basis of most innovative and competitive industrial IT systems, the future role of CBM and IoT in Cyber Physical Production Systems (CPPS) as part of the Industry 4.0 (or Smart Factory) has been discussed in the last section of this chapter.

MANUFACTURING ORGANIZATIONS, TECHNOLOGIES

In order to fulfil the actual market demands, production and manufacturing systems have been optimized in their structure, costs and fabrication technology. To be able to compare the different manufacturing systems a short overview is given, starting from traditional automated manufacturing system, towards the FMS (Flexible Manufacturing System), the networked, reconfigurable manufacturing systems (Virtual Enterprises), the CBM and finally the Industry 4.0 domain.

A short summary of the comparison showing the differences and the evolution of these manufacturing systems is presented in Figure 1. The qualifications of the categories are not absolute but are constructive in comparison with other manufacturing system categories.

Figure 1. Comparison of the different production and manufacturing systems

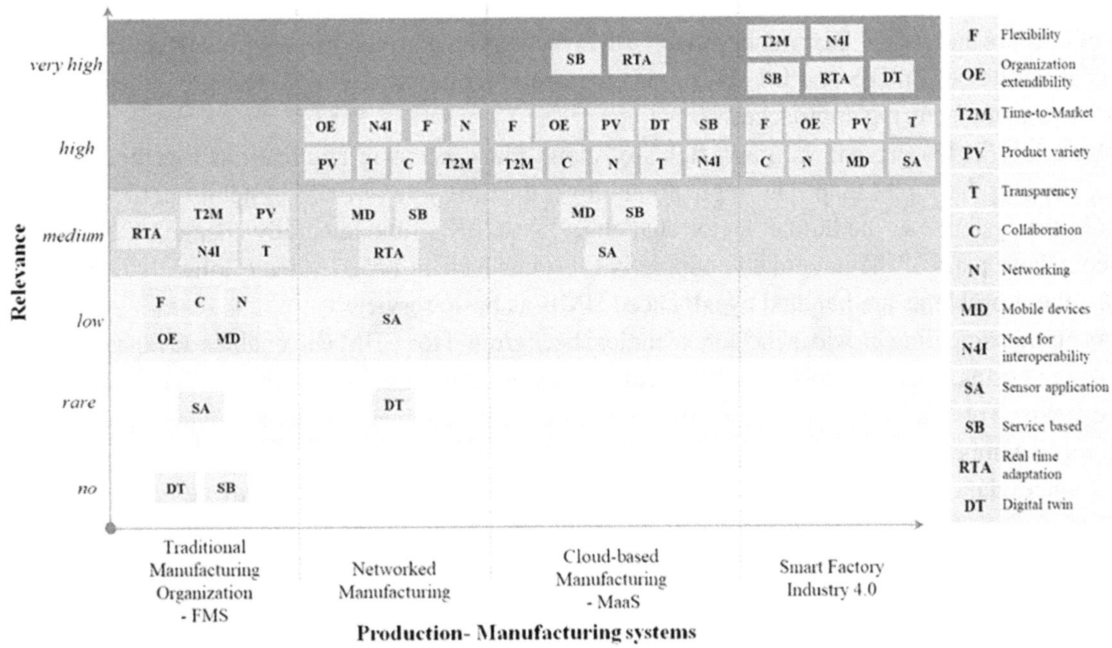

The FMS is an integrated manufacturing system that has machine tools automatized on different levels, an automated material handling system and some type of automated store. The FMS is controlled by computers and can contain beside the control SW CAD/CAM and other CAX packages. The volume of production is medium, and the variety of parts is high. Because of the flexibly applicable optimization algorithms e.g. production efficiency, machine utilization could be improved while inventory, throughput time, waste could be reduced. Different types of standards have been applied, but in many cases the manufacturing units were connected with proprietary, "home-made" protocols.

The next step in the evolution was the distributed or networked manufacturing systems. The development of network technology made it possible to connect machine tools, complex manufacturing and assembly systems both inside and outside a factory. Information technology (collaborative, agent, holonic, artificial intelligence) made it possible to raise the level of collaboration, the efficiency, to reduce time-to-market and costs. The collaboration and cooperation are main characteristics of networked enterprises, so the contacts among the users, the machines and production units have outstanding importance.

The collaborative network paradigm has been developed and described in Camarinha-Matos and Afsarmanesh (2005) that covers the main characteristic of all different networked units that are autonomous, geographically distributed, and heterogeneous considering their operating goals, environment providing a framework to describe these organizations.

A collaborative network (CN) is a network consisting of different entities (e.g. organization units and humans), social capital and culture. The collaboration is supported by computer network and makes possible to achieve common or compatible goals easier, thus generating joint value. The organizations participating in the network can be called as collaborative networked organizations (CNOs) that are introduced in Camarinha-Matos et al., (2009). The reliable exchange of data and information is critical among CNOs, so interoperability and standardization have a vital role.

Making a systematic comparison between CBM and Smart Factory (instantiation of Industry 4.0 approach) clarifies in what fields SFs are more affective, and what technological advantages they have. The concept of Industry 4.0 is broader then CBM where as an addition to CBM Industry 4.0 integrates the IT technologies of CPS and IoT. The main characteristics presented in Table 1 gives the differences of the two manufacturing approaches.

Wu et al. (2013) present the main fields of cloud manufacturing applications together with their classes of user in a topics map. The applications cover the life cycle of CBM systems but there are some gaps in the system e.g. the human factor challenges. As CBMs fundamentally change work and work systems these parts of the integrated cooperative systems have to be extended (Golightly et al., 2016). In SFs these problems are handled by advanced HMIs as basic services.

As cloud computing provides the core technical background for CBM and enables advanced concepts in manufacturing, such as networked and virtual manufacturing, the limitations are in connection with these fields. The real-time handling of big amount of data obtained during a manufacturing process (e.g. force, temperature), can cause problems in the operation as they are uploaded via Internet to the cloud where data storage and analysis are performed, and the results should provide real time control for manufacturing units (Wang et al., 2015). So, response time, communication channel (data handling) bottleneck, and communication and data storage security can cause different limitations in cloud manufacturing as well.

Table 1. Comparison of the main characteristics of CBM and Smart Factory

Factors of Comparison	Cloud Based Manufacturing	CBM in Industry 4.0 – Smart Factory
Production services Flexibility	Limited to certain fields (HMIs are not general)	In all fields – general characteristics of the system (advanced HMIs)
Organization expandability	Limited by virtualized units	Easy to expand based on CPS technologies (virtual-physical objects)
Product variety	High	Very high
Time–to-Market	Short	Very short, immediate (3D print)
Networking	Partially wireless	Fully networked with all ICT technology types
Collaboration	Based on cloud deployment models	Lean structure (Included all devices, equipment, machines, SW, sensors)
Sensors/real-time adaptive management/control	Included machines, equipment and virtual services	Real time, wireless, among all devices and sensors + processes
Transparency	Transparency limited	High transparency
Service based	Basic technology	Basic technology
Virtual twins	Not existing	Basic technology
Application of mobile devices	Limited applications (e.g. communication between machines)	High level, all round applications
Need for interoperability	Many existing standards	Numerous solutions are missing (e.g. standards for M2M communication)
Security (endpoint, network)	Endpoint and communication risk can be limited by increased security	Higher risk because of M2M (IoT)
Basic technologies	Virtualization, cloud computing, service-based technologies	As in CBM + IoT, Big Data, sensors, virtual simulation

Handling Industry 4.0 approach as a base on which "Smart Factories" (SF) are developed, physical processes are monitored by cyber-physical systems and create a virtual replica of the physical world (virtual twin) and make decentralized decisions, while all devices are cooperating with each other (including humans as well) in real time using the Internet and IoT components. In SF the need for interoperability, for security are higher (keen communication among all devices), the continuous collaboration of sensors, mobile devices and humans (through intelligent interfaces) in all fields and levels is significant (Liu & Xu, 2016). In Smart Factories BigData based analysis supported by artificial intelligence techniques are applied and the results of real-time virtual simulation (based on virtual twins) are also used.

INTEROPERABILITY IN NETWORKED ENTERPRISES

Interoperability can be defined as the ability of two or more systems or applications to exchange information and to mutually use the information that has been exchanged without special effort on the part of the customer. Interoperability is realized by the implementation of standards (ETSI SR 002 761). In the context of networked enterprises, interoperability refers to the ability of interactions (exchange of information and services) between enterprise systems.

There are different levels in the enterprise where interoperations can take place. In the ATHENA EU project, for example, four layers of interoperability concerns have been defined that can be applied also to networked enterprises (ATHENA, 2007). In Table 2 these layers are extended with activities and standards applicable to each level.

There are three different main approaches to solve interoperability challenges in enterprises:

- **Integrated Approach:** There is a common format for all models. This format must be agreed by all parties to elaborate models and build systems;
- **Unified Approach:** The common format exists at meta-level and provides a means for semantic equivalence to allow mapping between models;
- **Federated Approach:** No common format exists. Partners have to solve interoperability promptly and real-time, which means they have to share an ontology to map their concepts at the semantic level.

Table 2. Layers of networked enterprises for interoperability with activities and standards

VE Layers	Description of Functions, Activities	Standards
Business	Collaborative modelling, semantic interoperability, company culture	KIF, KQML, UEML 1.0
Process	Cross organizational business process, in networked enterprises integrating different internal processes into a common one	PSL, UEML 1.0
Service, application	Flexible execution and service composition, identifying, composing and making various application functions together	UEML 1.0, APIs, STEP, EDI, HTML, XML, or eb-XML, J2EE, Java, .NET
Data	Information interoperability – to make query languages and different data models working together	UEML 1.0, XML, flat files, DB

(Mezgar & Rauschecker, 2014)

A detailed description of the above methods can be found in Chen et al. (2008), whereas trends and the issues for enterprise integration and interoperability in manufacturing systems are presented in detail in Panetto and Molina (2008).

CLOUD TECHNOLOGY, ARCHITECTURES

To date considerable efforts have been concentrated for a commonly accepted (if not definition, at least) vision of Cloud Computing (CC) and technology. Just like clouds may have different names if seen from different physical perspectives, also CC seems to fit into several definitions, derived from different application domains. It is commonly accepted that key benefits have to encapsulate both business and technological views. In general, cloud-based computing is an information technology (IT) architectural model where computing services (both hardware and software) are delivered to customers over the Internet, on-demand, in a self-service fashion, independent of device and location (Marston et al., 2011). In other words, a CC system must enable, according also to the National Institute of Standards and Technology (NIST), a ubiquitous, convenient, on-demand network access to a shared pool of configurable computing resources (e.g., networks, servers, storage, applications, and services), which can be rapidly provisioned and released with (theoretically) minimal management effort or service provider interaction (*NIST*, 2011). The cloud model embodies unique characteristics and can have various service and deployment models, as reported in the section hereafter.

Service Models

This is typically an end-user's perspective in CC industry, where different delivery models refer to different layers of the CC architecture.

The most common and used term perhaps is Software as a Service (SaaS), in which the application runs on the vendor's infrastructure and is recognized as a service. The consumer is unaware of the application provider's infrastructure and complexity.

A Platform as a Service (PaaS) facilitates the development and deployment of applications without the cost and complexity of buying and managing the underlying hardware and software layers (operating system, network, servers, etc.).

Infrastructure as a Service (IaaS) offers also storage, network and computational capabilities as a service.

Manufacturing specific Hardware as a Service (HaaS) is a standard or technique for consistently describing and serving equipment and its functionality, behaviour, structure, etc. It represents one of the major challenges in service implementation description of core physical equipment in the context of Manufacturing as a Service (MaaS).

Characteristics and Deployment Models

Essential properties of CC systems can be classified (also according to considerations reported by Xu, 2012; Zissis & Lekkas, 2012; Wang & Xu, 2013) into core-, business-, enterprise- and manufacturing specific. Resource abstraction, self-service-centricity, network access mechanisms, rapid elasticity, service measurability, multi-tenacity, load-balancing and virtualization are all core aspects in cloud computing.

Business relevant properties comprise quality-of-service (QoS), service level agreement (SLA), user experience (UX), fault-tolerance, auditability and certifiability. When entering the enterprise level, interoperability, deployment models, security and business process management augment the complexity of cloud requirements. Furthermore, CBM is expected to provide a solid support for service-oriented environments, simulations and global services.

Illustration on Figure 2 depicts common cloud deployment models with service-critical instantiation orientations.

Cloud hosting deployment models represent the category of cloud environment and are mainly distinguished by the proprietorship, size and access. They are classified into public cloud (cloud owner provides Internet-based public services on predefined rules, policies, and a pricing model), private cloud (most of the benefits of a public cloud exclusively for an organization, with decreased security concerns, thanks to the utilization of corporate firewalls), community cloud (similar requirements, concerns and policies for a number of organizations which share the CC to lower costs, divide expenses and support high security), and hybrid cloud (a combination of two or more different public, private, or community clouds with standardized functionalities for communication and interoperations).

CC stakeholders include not only traditional technology roles (consumer and provider) but also enablers, regulators, brokers, certification authorities, auditors and others (refer to Marston et al., 2011, for a comprehensive list of key players in the CC industry).

Major Recommendations

Next to the perceivable benefits of CC, there are still significant barriers and recommendations to be taken into account for its adoption as introduced in Table 3 (Zissis & Lekkas, 2012; Lee, 2008). Mission-critical services of enterprises need to be locally reinforced, so to ensure continuity in the execution and, therefore, provision of business processes. Consulting companies reported how current cloud services may not be cost-effective for larger enterprises which have achieved best efficiencies from their own

Figure 2. Cloud deployment models with property-specific service instantiation

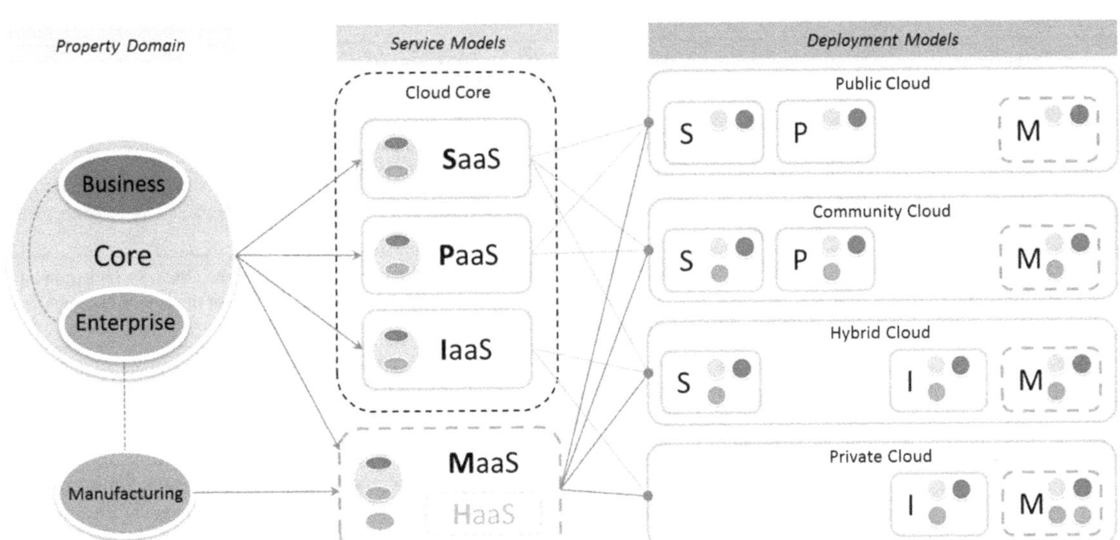

computing infrastructure (Bommadevara et al., 2016). Different is the situation for small and medium enterprises (SMEs), which usually do not have the initial capacity to set-up secure, large data centres, and for which prices and SLAs from cloud providers are far more advantageous. Unfortunately, many cloud providers still offer SLAs with rather weak user compensations on outages (measurement of service delivery, method of monitoring performance, and amendment of SLA over time). Finally, some hidden costs need to be carefully balanced, including support, disaster recovery, application modification, and data loss insurance.

CLOUD STANDARDIZATION

Standardization plays a huge impact on cloud adoption and usage as Cloud service providers usually have their own approach on interactions and Cloud APIs required to users. Complex business applications on the cloud require adequate standards, as the lack of integration between networks makes it difficult for organizations to consolidate their IT systems in the cloud and realize productivity. At present more than 15 different groups, committees and organizations have established a Wiki site for Cloud Standards Coordination (Cloud-standards, 2017), whose goal is to document activities from the various Standards Development Organizations (SDOs) and the leading technology. Table 4 gives an overview on main activities and standards on cloud-service layers.

Generally, physical hardware virtualization (ARL) follows the Open Virtualization Format, while PRL do not evidence the use of any specific standards.

Containers Technology and Serverless Computing

Containers represent a step in cloud evolution: they wrap software in a complete filesystem, containing everything needed to run (called image). This guarantees that software will always run the same,

Table 3. Barriers and issues related to cloud adoption

Issue	Relevance
Security and privacy	The very first concern CC has to adequately address: privacy regulations (Zissis & Lekkas, 2012; Wei et al., 2014) and privacy protection of individuals and business.
Connectivity	The full potential of CC depends on the availability of high-speed access to all.
Reliability	Enterprise applications are so critical that they must be reliable and available 24/7 and recovery plans must take effect smoothly in case of failures or outages.
Physical control and boundaries	Organizations justifiably wary of the loss of physical control over data on the cloud. Providers must be able to guarantee the location of a company's information on specified set of servers in a specified location (legal involvements).
Mission-criticality	Cloud providers still cannot commit to the high QoS and availability guarantees demanded in large organizations. Service Level Agreement (SLA) commitments for annual uptime have different acceptance levels for SMEs but are still insufficient for mission-critical applications in large organizations.
Interoperability	Interoperability and portability of information between private and public clouds are critical enablers for adoption of CC in enterprises: they permit integrity and consistency of company information and processes.

Table 4. Cloud service layer activities and standards

Target	Activity Description	Standards
Client	User interface and hardware mobility	Proprietary standards
Application services	Business processes, social networks, collaboration	Not cloud-specific: XML, SOAP, ebXML, OASIS SCA, SDO, SOA-RM, and BPEL
Application data	Data portability among cloud services	JSON, XML
Run-time	Application developments and testing	J2EE, .NET
Middleware	Customers software modules within the cloud	Not cloud-specific
Operating System	OS drivers transparent management in SaaS/PaaS	OCCI
Storage	Data storing	CDMI, CADF

regardless of its environment: Build Once and Run Anywhere (BORA). Docker.com containers, one of the most popular providers, are based on open standards, enabling containers to run on all major Linux distributions and Microsoft Windows, as well as on top of any infrastructure. Containers are isolated but, with respect to traditional Virtual Machines (VMs), share OS and (when appropriate) bins and libraries (Figure 3).

Serverless Computing (SC), or Function as a Service (FaaS), is the newer form of CC similar to VMs and containers running on a cloud provider. This does not mean there are no servers, but instead that the management of servers, scaling, and capacity planning are taken care by the underlying cloud provider. Before FaaS, many enterprises adopted micro-services, entirely separate mini applications in their own process/container/VM. The main benefit is modularity and separation of concerns. However, micro-service based infrastructures and operations greatly increased the complexity of continuous integration/ delivery pipelines, needed for tracking and testing. Companies like Basecamp argues that monolithic architecture can make sense for certain small companies, like start-ups.

Figure 3. Docker container vs. virtual machine architecture

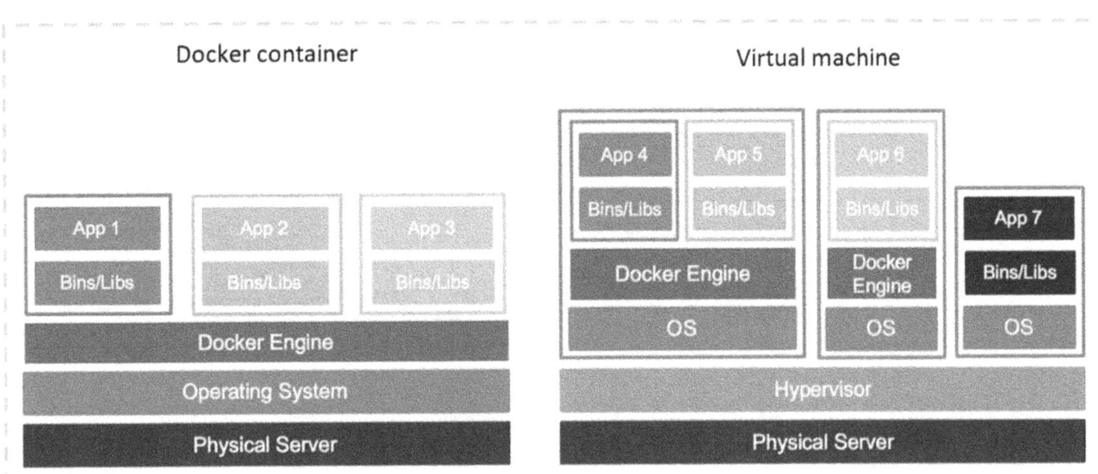

With FaaS the infrastructure, orchestration layers, and deployment become unnecessary and application developers can focus on writing business logic and functionality and leave the rest to the cloud provider.

FaaS can reduce computational costs, too: while most cloud providers will charge an hourly rate for reserving a VM, FaaS can use a consumption-based pricing model, in which there is no charge if the application is not actively using compute or memory resources.

Most FaaS applications have to adhere to an Event-Driven Architecture (EDA): this allows for a more responsive application because systems are by design asynchronous at an unpredictable scale. In terms of security, some aspects are handled by the FaaS operator, but this doesn't assure that a specific application will be free of security issues. Main advantages/disadvantages of FaaS can be briefly summarized as follows:

- **Advantages:** Management and security of infrastructure, easy deployment, scalability and reduced set-up costs;
- **Disadvantages:** Latency (FaaS adds some latency so for a high-performance application it might not be the best idea to use FaaS), limits for memory (1500MB on AWS) and execution time (300 secs on AWS, 500 secs on Google), monitoring and debugging (there are some solutions that are maturing but at this moment it is still a limitation), no local stored data (applications have to be stateless: for most of uses a good thing but it is a limit), and vendor lock-in (depending on, for example, AWS, Azure, need to a have an exit strategy if their vendor goes bankrupt in few years).

In Table 5 the major aspects are compared for the biggest three FaaS providers: AWS Lambda, Azure Functions and Google Functions.

Table 5. Comparison of biggest FaaS providers

Target	Activity Description	Standards	Google Functions
Languages support	Node.js, Java, C#, Python	Node.js, C#, F#, Python, PHP, Java	Node.js
Languages support with 3rd party	Golang by using Node.js shim	With batch files can run anything	
Monitoring	Cloudwatch, Dashbird	Azure Application Insights	Stackdriver Monitoring
Pricing	$0.20/million requests with 1 million requests per month for free	$0.000016/GBs, 400,000 GBs/month for free $0.20/million executions, with 1 million executions/month for free	$0.40/million invocations with 2 million invocations for free $0.0000025/GB-sec with 400,00 GB-sec/month for free $0.0000100/GHz-sec with 200,000 GHz-sec/month for free
Limits	Memory allocation range: Min. 128 MB / Max. 1536 MB Maximum execution duration per request: 300 seconds	Allows only 10 concurrent executions per function No limitations on max. execution time limit	Number of functions: 1000 Max function duration: 540 seconds Function calls per second: 1,000,000 per 100 seconds

CLOUD BASED MANUFACTURING

The architecture of organizations is in a recursive connection with the IC systems; the IC technology offers new possibilities for restructuring the organization (and its business processes) itself, in other cases the new demands of a business process force the development of a special IC solution. The final goal of all information systems is to provide secure data-, information-, knowledge-, or different services for the users (human beings), and for firms, enterprises (Mezgar & Rauschecker, 2014).

The idea of Cloud based manufacturing has been developed based on the cloud computing technology (Li et al., 2010). The basic idea was to apply cloud computing architecture, tasks and service structure to manufacturing systems. The result of this mirroring was a new service-oriented networked manufacturing model called Cloud Manufacturing (CMfg). Since the first basic publication numerous papers have been written with different approaches to the theme. The name of the paradigm varied from "Manufacturing Cloud - MCloud" (Zhang et al., 2010), to "Cloud Based Manufacturing - CBM" (Wu et al., 2014) but the content behind the words is the same.

The reason of the novelty of CBM lays in multi-tenancy and virtualization; these are the two main characteristics that make the difference between CBM and networked manufacturing. According to Wu (Wu et al., 2013) "the differences and similarities between CBM and web- and agent-based systems will be articulated from a number of perspectives, including (1) computing architectures, (2) data storage, (3) operational processes, (4) information and communication, (5) business models, and (6) programming models". As the differences show in every category significant positive output for CBM the closing statement of the analysis was that CBM can be considered as a new paradigm.

The definitions of cloud manufacturing also reflect how this paradigm expanded in the last years (Zhang et al., 2011). Some examples are reported as follows:

"Cloud manufacturing is a computing and service-oriented manufacturing model developed from existing advanced manufacturing models (e.g., application service providers, agile manufacturing, networked manufacturing, manufacturing grids) and enterprise information technologies under the support of cloud computing, the Internet of things (IoT), virtualization and service-oriented technologies, and advanced computing technologies" (Li et al., 2010).

"Cloud manufacturing is a model for enabling ubiquitous, convenient, on-demand network access to a shared pool of configurable manufacturing resources (e.g., manufacturing software tools, manufacturing equipment, and manufacturing capabilities) that can be rapidly provisioned and released with minimal management effort or service provider interaction" (Xu, 2012).

Qanbari et al. (2014) define "Cloud manufacturing is a distributed manufacturing execution model, where underlying resources envisaged in the internet of things are elastically exposed and utilized as cloud services, then composed and orchestrated for utilization as cloud services, then composed and orchestrated for a manufacturing task in an on demand fashion."

According to Ren and his co-authors "Cloud manufacturing is a smart networked manufacturing model that embraces cloud computing, aiming at meeting growing demands for higher product individualization, broader global cooperation, knowledge-intensive innovation and increased market-response agility" (Ren et al., 2014).

The main advantages of CBM for the manufacturers are:

- On-demand leasing/releasing of manufacturing assets,
- Flexible reconfiguration of manufacturing assets according to the needs,
- "pay-as-you-go" business model based on the measured really used service.

There are three main areas where cloud computing can be applied in manufacturing companies.

- Manufacturing software is supported as a service in the manufacturing cloud. This can be handled as the manufacturing version of cloud computing.
- Inter-factory collaboration that has an extended scope; services like supply chain visibility, transportation management, supplier/contract negotiation. Partners can create cloud computing modules to address other manufacturing issues, e.g. supply chain execution, shop floor planning, demand planning and production scheduling.
- High performance computing that use digital models to (1) virtually test the products or manufacturing system, (2) understand the business environment better through business intelligence and (3) make decisions. The models are typically highly parallelizable and fit well for a cloud environment.

Standardization and interoperability solutions have vital importance in order to effectively use cloud technology in the above applications. The control hierarchy in industry in most cases is based on the well-known ISA-95 standard (Andrew, 2015). On top-level can be found the ERP (Enterprise Resourse Planning), one level below the MES (Manufacturing Execution System), then the SCADA (Supervisory control and data acquisition) takes place. In Figure 4 it is shown how conventional manufacturing tasks/processes in ISA-95 layers are converted to cloud services.

Virtualization is the most important process of transforming a conventional manufacturing system to cloud manufacturing. Virtualization means to dynamically divide different resources into virtual units and later combine them flexibly (with other resources) to a logic unit to meet diversified demands as encapsulated services. The conversion of conventional manufacturing to cloud manufacturing has three phases; (a) identification of manufacturing resources, (b) virtualization of these resources and (c) grouping the virtualized resources into cloud manufacturing services.

The quality of virtualization determines the robustness of a cloud infrastructure. Good virtualization can effectively assist sharing of cloud facilities, managing of complex systems, and isolation of data and application.

Cloud manufacturing virtualization is more complex compared to cloud computing virtualization as both computing and manufacturing resources have to be virtualized. As manufacturing system organizations have dynamic characteristics and high uncertainty in their operations a flexible mapping strategy has to be applied. A manufacturing system can be virtualized in different levels/layers, e.g. capacity of a job shop vs. capability of a machine tool.

Different virtualization approaches have been described in details e.g. in Zhang et al. (2017) a service encapsulation and virtualization access model, in Ning and Xiaoping (2012) a cloud manufacturing virtualization framework has been described, that contains three layers: manufacturing resource layer, virtual description layer, and service encapsulation layer.

Manufacturing resources and capabilities can be shared over a manufacturing cloud platform as cloud services after having been virtualized and encapsulated.

INTEROPERABILITY IN CLOUD MANUFACTURING

Interoperability is an essential requirement for both service providers and enterprises. Services with interoperability allow applications to be ported between clouds, or to use multiple cloud infrastructures before business applications are delivered from the cloud. Interoperable solutions make easy migration and integration of applications and data between different cloud service providers (Xu, 2012).

Interoperability can be divided into two main groups in cloud-based systems/applications:

- The interoperability of the cloud systems itself (cloud layer), and
- The interoperability of the applications (Application layer – industry specific, HW/SW specific).

In the field of cloud-based manufacturing numerous solutions, systems have been developed with different focus fields. Yadgarova and Taratukhin (2016) introduced an integrated framework for building cloud-based manufacturing environment that allows to develop future production systems as the class of adaptive distributed systems with virtual cloud model and simulation. A dependency model for equipment interaction has been defined too.

Mourad et al. (2016) identified interoperability as "a key enabler for cloud manufacturing". They proposed a framework called "C-MARS" for realisation of interoperability across heterogeneous computer aided manufacturing systems (handling CAD and NC files). Using this framework, manufacturing resources can be shared by a large number of clients based on requirements and priorities.

A service-oriented, interoperable Cloud manufacturing system is proposed in Wang & Xu (2013). Service methodologies have been developed to support customer and enterprise users, along with standardized data models describing Cloud service and relevant features. By using ICMS, standardized

Figure 4. ISA-95 layers converting into cloud

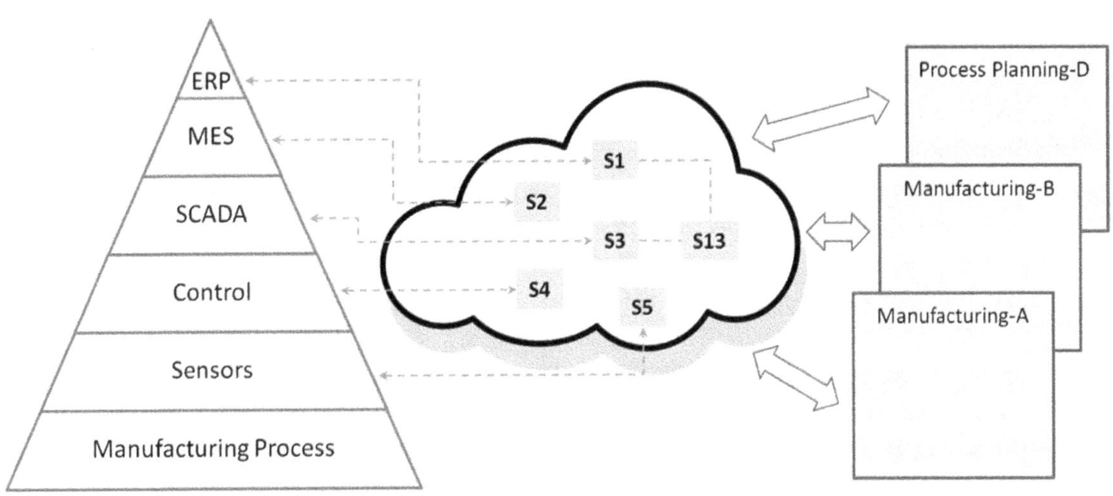

(STEP-based) communication methodologies have been deployed to support collaborative interactions in the Cloud environment.

In the next section a general cloud manufacturing conceptual model (CMCM) is proposed by the authors, that applies among others the container technology.

Cloud Manufacturing Conceptual Model

CM is the necessity for manufacturing enterprises to be described, realized, componentized, virtualized and integrated in an interoperable manufacturing cloud. As already mentioned, major research works by Wang and Xu (2013) propose service oriented architectural solutions, highlighting how the need of implementing a mechanism to organize and control cloud services at upper-level is of crucial importance. Based on research projects developed so far, the authors isolated and derived a general CM conceptual model (CMCM, illustrated on Figure 5), which encompasses the following elementary classes of actors:

- **Cloud Provider (CP):** Responsible for providing platform services both to internal components and external stakeholders. Main functionalities encompass the management of: applications patterns deployment, cloud adapters, cloud simulations workflow, privacy and security policies, virtual machines (VMs) and Docker containers;
- **Cloud Consumer (CC):** The node receiving output for service requests within the platform (i.e. digital factory model management, big data storage, applications ontology management and maintenance) through the CMCM service orchestrator;
- **Cloud Certification Authority** (CCA), **Cloud Broker** (CB) and **Cloud Auditor** (CA) have already been introduced before.

CMCM architectural back-bone should provide services through a standardized, homogeneous, application-level (i.e. HTTP) interface (C-API).

Major components of CMCM architecture are as follows:

- **Service Layer (SL):** Conceptual abstraction boundary of CMCM functionalities. It enables the creation of complex virtual services, like simulation workflows;
- **Abstract Resource Layer (ARL):** Cloud-specific back-end layer managing the dynamic allocation, bounding and routing of resources to the underlying physical level (PRL). Platform users register their applications and create VMs leveraging also supported external clouds (e.g. Amazon, CloudSigma, OpenStack or OpenNebula);
- **Physical Resource Layer (PRL):** Dynamically discovers and assigns resources to ARL. For instance, CPS elements have corresponding PRL layers running over the physical resource, eventually exploiting manufacturer-specific APIs for specific functions.

SL main components can be classified as follows:

- **Semantic Framework (SF):** Services related to the semantic description of CMCM resources with their orchestration and management, and the inferential discovery of knowledge;
- **Big Data (Big D):** noSQL-compliant data collection at CPS level. Specific mechanisms enable data processing and aggregation, reducing the volume and communication traffics;

Figure 5. Cloud manufacturing conceptual model

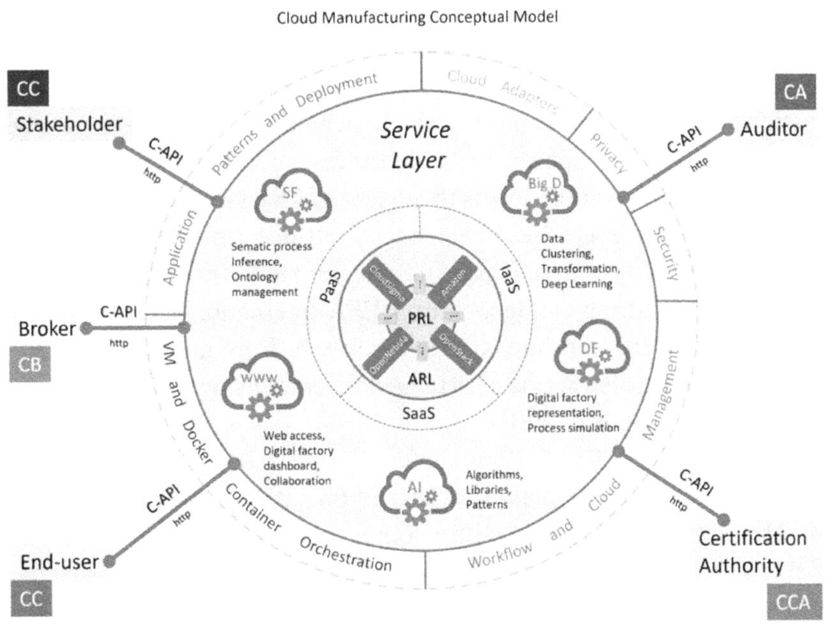

- **Web applications (WWW):** Complex functionality (like simulation workflows) provided to end-users by means of higher-level web-based solutions;
- **Cloud Application Programming Interface (C-API):** Abstracts and standardizes access to cloud-services, by means of well-defined provision technology and protocols (micro REST services, Web-services or OPC-UA).

SERVICE ARCHITECTURE INTEROPERABILITY IN INDUSTRY 4.0

The integration of new technologies produces new organizational structures in industry, requiring current frameworks, platforms, standards to be modified: the new manufacturing architectures that apply IoT, cloud, and mobile technology cannot be handled by rigid architectures. The control hierarchy in industry is fundamentally based on ISA-95 standard (developed on Purdue model) but the modern control and communication systems need a lean structure. As seen, smart connected assets communicating directly with each other require the development of new architectures, and existing standards concepts can act as the basis of new standardization evolutions. Service composition is one of the most effective approaches investigated by IT researchers and applied by cloud-technology providers so far (a comprehensive literature review on cloud-service composition can be found in Jula et al., 2014).

Selection of appropriate services from service pools, management of their composition, specification of QoS parameters, understanding of requirements dynamic and rapid changes in service-provision interface are just some of the important issues addressed to meet end-users' satisfaction. Service orchestration techniques were first applied in CC systems in 2009 (Cheng et al., 2009). Because of the exuberant growth of offered services, cloud-service brokers are facing high competition in providing quality of service. Such competition leads to difficulties in designing services in a way that their selection, orchestration, deployment and management be suitable in the cloud environment. Aspects like selection of appropriate services from service pools, composition restrictions, and substitutability of services due to emerging changes in requirements and network seem to have the higher impact on service provision. Investigations in Jula et al. (2014) demonstrate that CC service-orchestration can be categorized basically into five major groups: classic and graph-based, combinatorial, machine-based, structural and with frameworks.

With proper IT methodological approaches companies can create flexible business applications through elementary service composition: this is the essence of Service Oriented Architectures (SOAs), where services are considered as repeatable, independent, self-describing, business tasks or modules for externalized service invocation. A comprehensive introduction to SOA-applied manufacturing (SOAm) can be found in SOA Manufacturing Guidebook of MESA International and IBM (2010). This architecture allows for the creation of composite business processes from independent, self-describing, and interchangeable code modules called services, arranged together using process choreography and used via Enterprise integration Service Bus (ESB).

Based on SOA and MESA-IBM ESB (Fraunhofer IPA, 2016a), Fraunhofer IPA has developed an integration model built upon the cloud platform "Virtual Fort Knox", for networking factories (Fraunhofer IPA, 2016b). The platform is equipped with a homogeneous, service-based integration layer called Manufacturing Service Bus (MSB). MSB connects cyber-physical manufacturing systems to digital services via encrypted channels and acts as a multi-functional digital bridge between the cyber-physical environment and the digital tool. It integrates all IT-specific production control systems, including strategic and analytical business services, by means of routing, data transformation and orchestration services.

Standardization

The global networking of production resources and processes, as well as the use of globally interconnected applications crucially require the adoption of uniform standards. Two major proposals are hereafter presented, which are currently "de-facto" considered as reference guidelines in conceiving future business organizations: the Reference Architecture (IIRA) for Industrial Internet Systems (IISs) and the Reference Architecture Model for Industry 4.0 (RAMI 4.0).

IIRA

IIRA is a standard-oriented open architecture for IISs (IIC Working TWG & SWG, 2015), which aims at extending industry applicability and interoperability, and guiding technology standards development. IIRA architecture is generic-purposed and promises high level of abstraction to support broad industry applicability. IIRA abstracts common characteristics, features and patterns derived from use cases defined in the Industrial Internet Consortium (IIC). The architecture framework is based on ISO/IEC/IEEE 42010 (ISO/IEC/IEEE, 2011) standard specification, and codifies conventions and common practices for internet oriented architectural concepts, like concern (topics of interest in the system), stakeholder (entities having an interest in the system) and viewpoint (conventions framing the description and analysis of specific system concerns). Concerns of IIS are classified and grouped in four different viewpoints: business, usage, functional and implementation.

In functional viewpoint, IIS is decomposed into five sub-domains: control, operations, information, application and business. These domains represent the building blocks of an IIS and illustrate how data and control move across them. The implementation viewpoint describes the general architecture, the technological components of an IIS, and the interfaces, protocols and behaviours among them. Popular implementations are encapsulated under various architectural patterns, all including the three-tier and gateway-mediated edge connectivity patterns. IIRA is not a standard but provides guidelines on how a safe, secure and resilient IIS can help realize the vision of the Industrial Internet (II).

Figure 6 illustrates the three-tier architecture and functional domains of IIRA:

- **Edge Tier:** Responsible for collecting data from edge nodes, using the proximity network. The edge tier (comprising asset, sensors and gateways) implements the control domain;
- **Platform Tier:** Receives, processes and forwards control commands from the enterprise tier to the edge tier. It consolidates processes and analyses data flows from the edge tier and other tiers. It provides management functions, data query and analytics for devices and assets;
- **Enterprise Tier:** Implements domain-specific applications, decision support systems and provides interfaces to end-users, receiving data flows from the edge and platform tiers and originating control commands to the platform and edge tiers. It implements domain specific functionality such as MES, SCM and ERP;
- **Proximity Network:** Connects the sensors, actuators, devices, control systems and assets to a gateway that bridges to other networks, and enables data and control flow between the edge and platform tiers using typically XMPP/TLS;
- **Access Network:** Enables connectivity for data and control flows between the edge and the platform tiers (corporate network, or an overlay private network over the public Internet, or a 4G/5G network);

Figure 6. Mapping between three-tier architecture and functional domains in IIRA

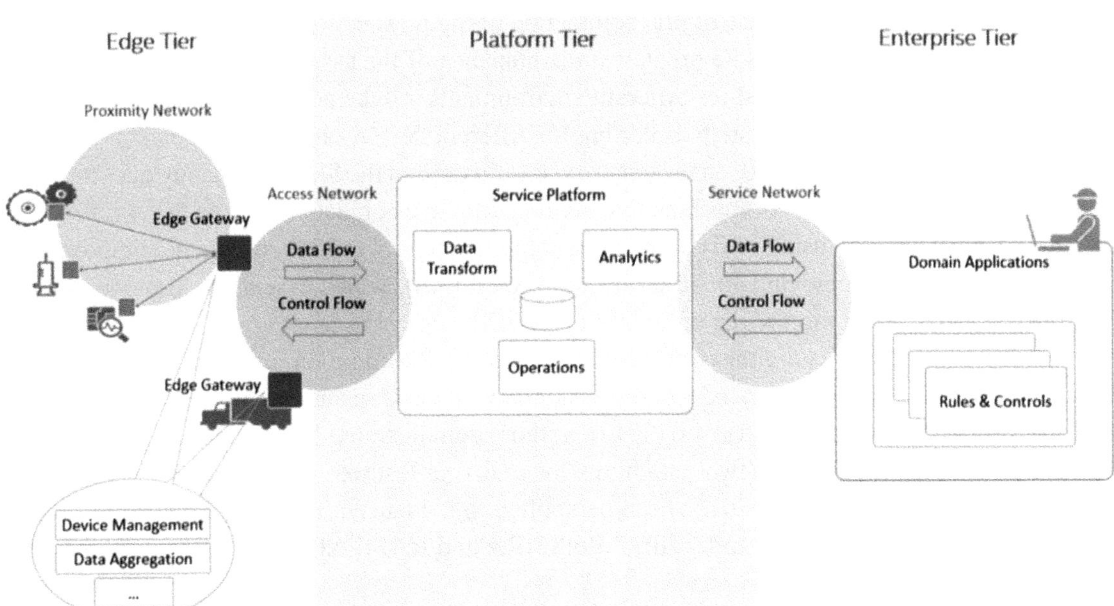

- **Service Network:** Enables connectivity (usually using TLS protocols) between services in the platform and enterprise tiers (an overlay private network over the public Internet or the Internet itself). A possible alternative may be utilize OPC UA between the various assets of the enterprise.

RAMI 4.0

The final goal of I4.0 is an interconnected factory capable of producing highly customizable products, realized through flexible mass production. Central in RAMI 4.0 is the concept of CPS – analogous to the IIS in IIRA – where autonomy is localized and participating systems make decisions on their own. The reference architecture model for RAMI 4.0 (VDI et al., 2015) is the convergence of multiple (stakeholders) visions on how I4.0 might be realized, and it is built upon existing communication standards and functional descriptions.

The six layers of the vertical axis define the nature of IT components in I4.0: business applications, functional aspects, information handling, communication and integration capability, and ability of the asset to implement I4.0 features. This layered architecture basically breaks the complexity into manageable parts. The life-cycle of products, machines, orders and factories are captured along the life cycle and value stream axis (IEC 62890), whereas the hierarchy levels represent various functions of enterprise IT and control systems (IEC 62264 and IEC 61512). An important feature of RAMI 4.0 is the identification of objects as instantiated types: an object can be a product, asset, software, machine, or even a factory.

Industrial Interoperability: IIRA vs. RAMI 4.0

Three-tier architectural pattern of IIRA shares distinctive features associated with RAMI 4.0 layered architecture, especially when considering interoperability at the functional level. Fundamental to IIC is the concept of digital twin, the computerized, often cloud-based counterpart of physical assets, which uses data from sensors, actuators, power supply and network interfaces installed on physical objects to represent their near real-time status, working condition (properties and states) or position (typical example is the use of 3D modelling to create a digital companion of the physical objects to be projected into the digital world) and efficiently schedule predictive maintenance of the asset. An important aspect of the digital twin is that it can be adapted for different environments to take advantage of data generated by other virtual machines in the ecosystem, according to different designs and requirements.

The central concept in RAMI 4.0, on the contrary, is built around the I4.0 compliancy of components, such as product, asset, software, or machine that be, referring to as objects that have the ability to communicate independently, by using I4.0 compliant communication. Non-I4.0 compliant components are made I4.0 compliant by deploying an Administration Shell (AS), which essentially provides a virtual representation and a description of the entire life-cycle of the object or asset.

It corresponds to the IIC "digital twin" of an asset in the I4.0 domain, and contains its lifecycle, technical functionality, and even procedures for sensor data integration and monitoring. In experimental results from IIC & Plattform Industrie 4.0 (2017), authors demonstrated how Industrial Digital Thread (IDT) and Asset Efficiency (AE) (two paradigms for realizing testbeds in connected industrial organizations, Infosys and IIC (2015a; 2015b) can be applied to evidence a mapping between IT layers in RAMI 4.0 and functional domains under IIRA. Both IIRA and RAMI 4.0 stipulate the need for a SOA encapsulation of functionalities into services.

Figure 7. IIRA 3-tier architecture for IIC IDT and AE testbeds, functional domains from IIRA, and communication networks (IIC & Plattform Industrie 4.0, 2017)

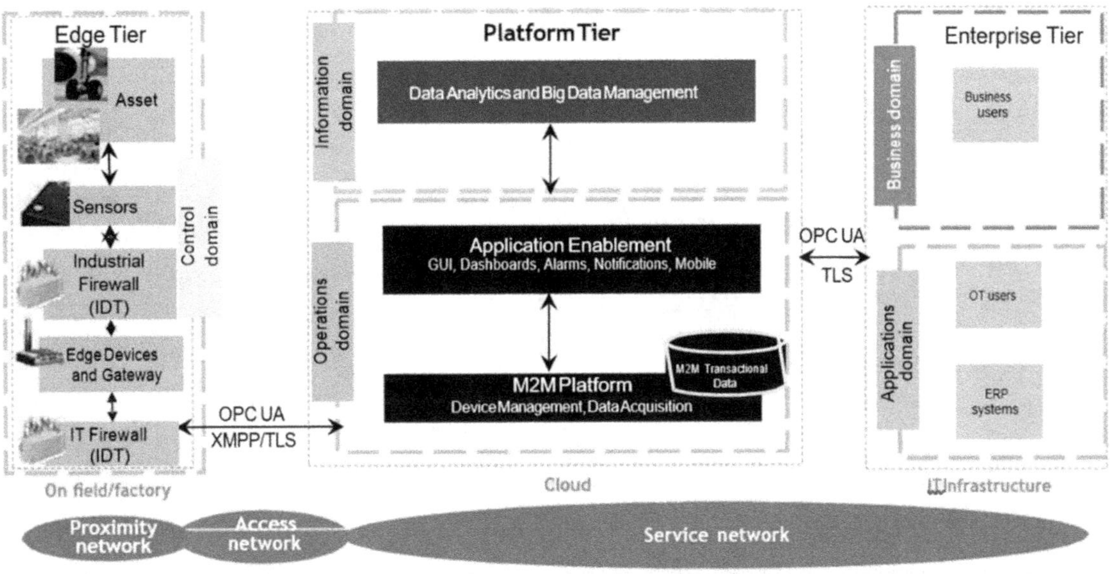

Illustration on Figure 7 shows IIRA 3-tier architecture for IIC IDT and AE testbeds, along with the functional viewpoints, the associated functional domains from IIRA, as well as all relevant communication networks associated with each tier. Mapping between the IIRA 3-tier functional viewpoints with the IT layers associated to RAMI 4.0 architecture for IDT and AE testbeds are reported on Figure 8.

IIoT (Industrial Internet of Things) solutions, such as testbeds relying on a SOA, can be considered to be semantically interoperable at a functional level between the two architectures.

OPC UNIFIED ARCHITECTURE

Both IIRA and RAMI 4.0 do not indicate any specific solution for developing the infrastructural layer underlying functionality and communication and it is here that OPC UA can play a strategic role for the realization of the service interoperability asset in industry. OPC UA is a product of the OPC Foundation (OPC UA, 2017), an industry consortium which creates and maintains standards for open connectivity of industrial automation devices and systems.

The way OPC UA enables interoperability is by mapping concepts between property sets from different physical domains and/or frameworks: the more aligned is the conceptual modelling of functions (or services) the more natural will be the mapping between referenced concepts and their direct use.

OPC UA, as a physical modelling environment and a communication standardizing architecture, can be leveraged to make different layers of both industrial architectures compliant with I4.0 guidelines, as it makes, for example, application testbeds semantically interoperable across RAMI 4.0 and IIRA architectures (Figure 9). For instance, it is interesting to note how RAMI 4.0 and OPC UA show modelling similarities in the specification of the following concepts: assets in vs. nodes; asset properties vs. object attributes and variables; sub-models vs. object methods (services) composition and orchestration.

Figure 8. Mapping between IIRA 3-tier functional viewpoints with IT layers associated to RAMI 4.0 for IDT and AE testbeds

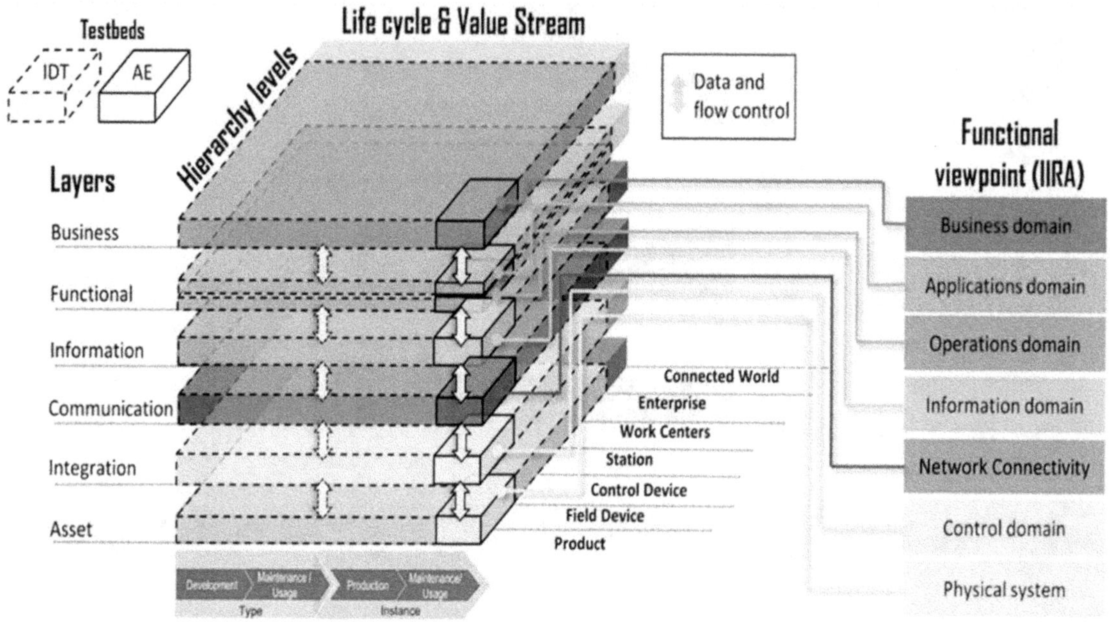

Figure 9. Example of chained OPC UA servers projected on RAMI 4.0 and IIRA layers

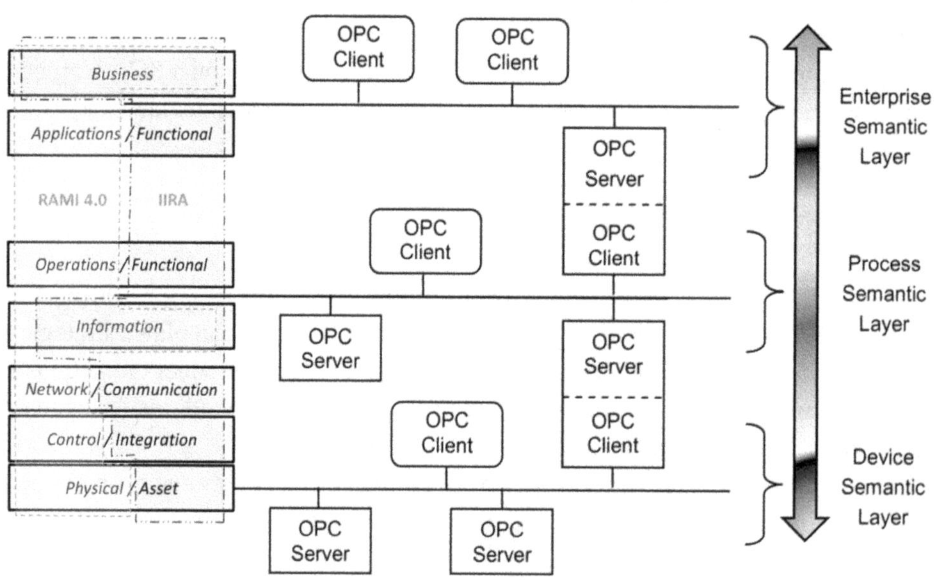

An example of effective ongoing initiative demonstrating the integration compatibility of OPC UA into RAMI 4.0 AS is the Open Asset Administration Shell, an Open Source Project (openAAS) which aims at implementing an OPC-UA-based AS (expected by the end of 2017).

OPC UA is a service-oriented, platform-independent standard through which different systems and devices can communicate by sending messages between clients and servers, over various types of networks. Figure 10 illustrates major elements of the OPC UA Server and how they relate to each other:

- **Real Objects**: Physical or software objects that are accessible by the OPC UA *Server* application;
- **OPC UA Server Application**: Code that implements the function of the *Server*. It uses OPC UA *Server* API to send and receive OPC UA *Messages* from OPC UA *Clients* and to isolate the *Server* application code from OPC UA Communication Stack;
- **OPC UA Address Space and Nodes**: The *AddressSpace* is modelled as a set of *Node*s accessible by *Clients* using OPC UA *Services* (interfaces and methods). *Node*s are used to represent real objects, their definitions and their *References* to each other;
- **Address Space Views:** Are subsets of the *AddressSpace* and are used to restrict the *Node*s that the *Server* makes visible to the *Client*. *Views* may hide some of the *Node*s or *References* in the *AddressSpace*;
- **Monitored items**: are entities in the *Server* created by the *Client* that monitor the *AddressSpace Node*s and their real-world counterparts. When they detect a data change or an event/alarm occurrence, they generate a *Notification* that is transferred to the *Client* by a *Subscription*;
- **Subscriptions**: are endpoints in the *Server* that publish *Notifications* to *Clients*. *Clients* control the rate at which publishing occurs by sending Publish *Messages*.

Figure 10. OPC UA Server architecture[1]

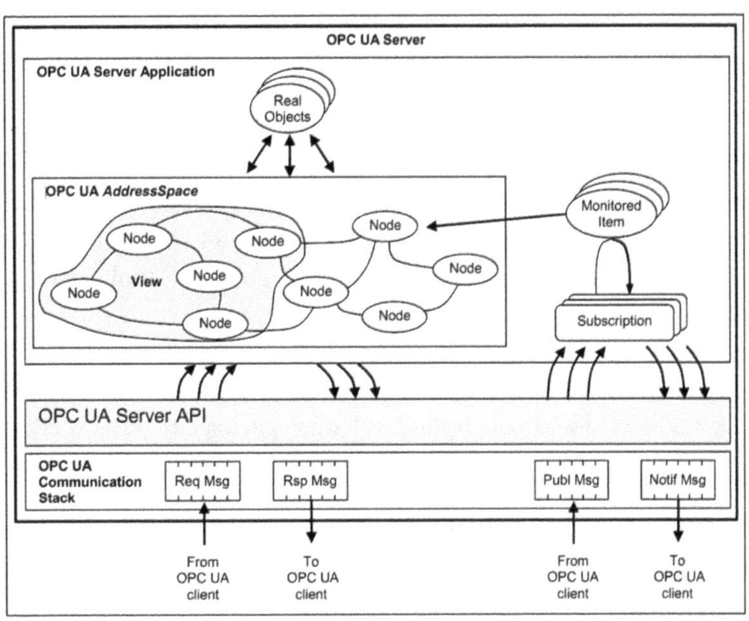

OPC UA specification is organized into several documents which relates to concepts, security model, address space model, services, information model, mappings, profiles, data access, alarms and conditions, programs, historical access, discovery and aggregates, respectively. OPC UA defines the sets of services that servers can provide, eventually mapped onto different communication protocols and in which data is encoded in various ways.

Basically, two data encodings are defined, XML/text and UA Binary; transport protocols support OPC UA TCP, SOAP/HTTP and HTTPs. Information models for a specific domain are defined on top of base specifications and organizations can build their own models on top of the UA base or on top of the OPC information model, exposing their specific information via OPC UA. Examples of standards already working on mappings to OPC UA are Field Device Integration (FDI) combining Electronic Device Description Language (EDDL) and Field Device Tool (FDT), both used to describe, configure and monitor devices and PLCopen, a standard for PLC programming languages.

FUTURE OF CLOUD MANUFACTURING IN INDUSTRY 4.0

As cloud architectures become the basis of most innovative manufacturing IT systems, the future role of cloud-technology in Cyber Physical Production Systems (CPPS) has to be properly investigated, as interoperability has vital importance in this field. The term "Industry 4.0" has nowadays totally gained the attention of scientific and economic forums. Initially conceived as a German national initiative (Wolf-Dieter, 2011), it has rapidly evolved to a broader definition which includes CPS, IoT and CC. The content of the Industrial Internet (II), for instance, is the internet of things, machines, computers and people, enabling intelligent industrial operations using advanced data analytics for transformational business outcomes.

Based on Industry 4.0 approach, so called "Smart Factories" are nowadays being developed, in which CPSs monitor physical processes, create a virtual counterpart of the physical world and make decentralized decisions, all while communicating and cooperating with each other (and with humans) in real time using the Internet and IoT components.

Besides these, IoT concept is currently transforming and redefining virtually all markets and industries in fundamental ways, and according to which any device/object (e.g. sensors, mobile equipment) and humans can be connected to the Internet (wired, wireless), also to each other, and be able to identify themselves to other devices. The significance of IoT is that if an object that can be represented itself digitally and is connected to surrounding objects forming a group of objects and/or (e.g. to a database) it becomes digitally greater than the object by itself as connected to surrounding objects and database data. Objects can be parts of machines (e.g. a touch-sensor of a machine tool, a jet engine of an airplane), wearable devices and anything that can contain a sensor, a chip.

IHS consultation firm forecasts that the IoT market will grow from an installed base of 15.4 billion devices in 2015 to 30.7 billion devices in 2020 and 75.4 billion in 2025 (Lucero, 2016). IHS also defines an IoT platform as "cloud-based and on premise software packages and related services that enable and support sophisticated IoT services". With the application of IoT platforms enterprises can manage thousands and even millions of devices and connections across multiple technologies and protocols. In the manufacturing sector IoT software enables developers to combine device and connection data with enterprise-specific customer and ERP data as well as data from third-party sources e.g. by direct using marketing data creating more valuable IoT applications. With the integration of Cloud and IoT new systems are developed, which offer effective information sharing and handling. Botta et al. (2016) call this approach as "CloudIoT" paradigm, whereas Zhou et al. (2013) developed in this sense the "CloudThings" architecture.

CONCLUSION

In industry and in business systems cloud computing and cloud-based manufacturing play nowadays a central role. Direct communication and networked collaboration among objects and human actors in manufacturing architectures are generating crucial challenges for interoperability. Establishing global interoperability in the industrial IoT is a prerequisite for its proliferation, but providing interoperability is a complex challenge. It is not a matter of agreeing on a small set of standards to rule the IIoT world, but it is about carefully orchestrating complex and partially competing protocols and standards on multiple levels (e.g. device integration, short-range and long-range wireless communication, messaging, event processing, data management, analytics, cloud operations). There is also the need to involve stakeholders and address domains of interoperability through an integrative, holistic perspective. As standards usually include both a specification and a set of industrial practices that emerge from that specification, RAMI 4.0 and IIRA are two of the most remarkable attempts which develop their efforts into this direction.

The chapter focused on the integration compatibility over the Cloud (at functional level) of their architectural layers and also indicated how OPC UA technology can play a relevant role (as a horizontal IIoT-enabling solution) in industrial service interoperability, due to its modelling semantics affinity to functional elements of RAMI 4.0 and IIRA.

To enable bidirectional data and information flows between IIoT systems, heterogeneously built on IIRA and RAMI 4.0, functional requirements driving the interoperability have to be defined. Functions and semantics that can be understood by these systems have to be specified from these requirements, with unique "understanding" for properties and the assets (parts, components and machines) and the means of communications between these systems. Therefore, identification, communication, semantic and functional mapping between IIRA and RAMI 4.0 is a fundamental step toward enabling interoperability between systems build upon these two architectures.

Interoperability is still an open issue in the current field of cloud-based Industry 4.0 implementations: the elimination of methodological incompatibilities in designing new manufacturing scenarios will surely be accelerated by the adoption of standardized solutions, while adequate attention needs to be dedicated to governance processes and partnerships. At the same time it is required to establish an open and interoperable IIoT. Achieving interoperability may require not only sound architectural principles but also dynamic and attractive negotiation between cloud providers and stakeholders of the industrial domain.

ACKNOWLEDGEMENT

This research has been supported partially by the European Commission through the EU H2020 projects EXCELL [grant number 691829] and GINOP [grant number 2.3.2-15-2016-00002].

REFERENCES

Andrew, H. (2015). *ISA-95 Unplugged: Putting MOM Things Together for IoT*. Retrieved January 06, 2018, from http://blog.lnsresearch.com/isa-95-unplugged-putting-mom-things-together-for-iot

ATHENA European project (2007). *Specification of Interoperability Framework and Profiles, Guidelines and Best Practices*. ATHENA Integrated Project (Deliverable D.A4.2).

Bommadevara, N., Kaplan, J., & Starikova, I. (2016). *Leaders and laggards in enterprise cloud infrastructure adoption*. McKinsey & Company. Retrieved January 06, 2018 from http://www.mckinsey.com/business-functions/digital-mckinsey/our-insights/leaders-and-laggards-in-enterprise-cloud-infrastructure-adoption

Botta, A., Donato, W. D., Persico, V., & Pescape, A. (2016). Integration of Cloud computing and Internet of Things: A survey. *Future Generation Computer Systems*, *56*, 684–700. doi:10.1016/j.future.2015.09.021

Camarinha-Matos, L. M., & Afsarmanesh, H. (2005). Collaborative networks: A new scientific discipline. *Journal of Intelligent Manufacturing*, *16*(4-5), 439–452. doi:10.100710845-005-1656-3

Camarinha-Matos, L. M., Afsarmanesh, H., Galeano, N., & Molina, A. (2009). Collaborative networked organizations – Concepts and practice in manufacturing enterprises. *Computers & Industrial Engineering*, *57*(1), 46–60. doi:10.1016/j.cie.2008.11.024

Chen, D., Doumeingts, G., & Vernadat, F. (2008). Architectures for enterprise integration and interoperability: Past, present and future. *Computers in Industry*, *59*(7), 647–659. doi:10.1016/j.compind.2007.12.016

Cheng, Z., Guo, X., Ou, W., & Han, D. (2009). Cloud Computing Service Composition and Search Based on Semantic. In M.Gilje Jaatun, G. Zhao & C. Rong (Eds.), *Proceedings of the 1st International Conference on Cloud Computing* (*Vol. 1*, pp. 290-300). Springer-Verlag, Berlin, Heidelberg.

Cloud-standards. (2017). *The Cloud Standards Wiki*. Cloud-standards.org. Retrieved January 06, 2018, from http://cloud-standards.org

European Commission. (2016). *European Commission sets out path to digitise European industry*. Retrieved January 06, 2018 from https://ec.europa.eu/digital-single-market/en/news/commission-sets-out-path-digitise-european-industry

Fraunhofer I.P.A. (2016a). *Manufacturing Service Bus. The Flexible Solution for Digitizing Production Processes*. Retrieved January 06, 2018, from http://www.ipa.fraunhofer.de/fileadmin/user_upload/Kompetenzen/Digitale_Werkzeuge_in_der_Produktion/Cloud-Plattformen/English_Documents/Productsheet_Manufactoring_Service_Bus.pdf

Fraunhofer I.P.A. (2016b). *Virtual Fort Knox. The Only Open It Cloud Platform For Manufacturing Companies*. Retrieved January 06, 2018, from http://www.ipa.fraunhofer.de/fileadmin/user_upload/Kompetenzen/Digitale_Werkzeuge_in_der_Produktion/Cloud-Plattformen/English_Documents/Productsheet_Virtual_Fort_Knox.pdf

Golightly, D., Sharples, S., Harshada Patel, H., & Ratchev, S. (2016). Manufacturing in the cloud: A human factors perspective. *International Journal of Industrial Ergonomics*, *55*, 12–21. doi:10.1016/j.ergon.2016.05.011

IIC & Plattform Industrie 4.0. (2017). *Industrial Internet Consortium & Platform Industrie 4.0*. Accessed January 06, 2018, from https://www.infosys.com/engineering-services/white-papers/Documents/industrial-internet-consortium-architecture.pdf

IIC Working TWG and SWG. (2015). *Industrial Internet Reference Architecture*. Industrial Internet Consortium. Retrieved January 06, 2018, from https://www.iiconsortium.org/IIRA.htm

Infosys & IIC. (2015a). *IIC - Asset Efficiency Testbed*. IIC. Retrieved January 06, 2018, from http://www.iiconsortium.org/asset-efficiency.htm

Infosys & IIC. (2015b). *IIC - Industrial Digital Thread Testbed*. IIC. January 06, 2018, from http://www.iiconsortium.org/industrial-digital-thread.htm

ISO/IEC/IEEE. (2011). *ISO/IEC/IEEE 42010:2011 Architecture description*. Retrieved January 06, 2018, from http://www.iso-architecture.org/ieee-1471/

Jula, A., Sundararajan, E., & Othman, Z. (2014). Review: Cloud computing service composition: A systematic literature review. *Expert Systems with Applications*, *41*(8), 3809–3824. doi:10.1016/j.eswa.2013.12.017

Lee, E. A. (2008). *Cyber Physical Systems: Design Challenges*. University of California. Retrieved June 30, 2017, from https://www2.eecs.berkeley.edu/Pubs/TechRpts/2008/EECS-2008-8.pdf

Li, B. H., Zhang, L., Wang, S. L., Tao, F., Cao, J. W., Jiang, X. D., ... Chai, X. D. (2010). Cloud manufacturing: A new service-oriented networked manufacturing model. *Jisuanji Jicheng Zhizao Xitong*, *16*(1), 1–7.

Liu, Y., & Xu, X. (2016). Industry 4.0 and Cloud Manufacturing: A Comparative Analysis. *Journal of Manufacturing Science and Engineering*, *139*(3), 034701. doi:10.1115/1.4034667

Lucero, S. (2016). IoT platforms: enabling the Internet of Things. *IHS*. Retrieved January 06, 2018, from https://cdn.ihs.com/www/pdf/enabling-IOT.pdf

Marston, S., Zhi, L., Bandyopadhyay, S., Zhang, J., & Ghalsasi, A. (2011). Cloud computing - The business perspective. *Decision Support Systems*, *51*(1), 176–189. doi:10.1016/j.dss.2010.12.006

MESA International & IBM. (2010). SOA Manufacturing Guidebook. Retrieved January 06, 2018, from ftp://public.dhe.ibm.com/software/plm/pdif/MESA_SOAinManufacturingGuidebook.pdf

Mezgar, I., & Rauschecker, U. (2014). The Challenge of Networked Enterprises for Cloud Computing Interoperability. *Computers in Industry*, *65*(4), 657–674. doi:10.1016/j.compind.2014.01.017

Mourad, M., Nassehi, A., & Schaefer, D. (2016). Interoperability as a Key Enabler for Manufacturing in the Cloud. In *The Sixth International Conference on Changeable, Agile, Reconfigurable and Virtual Production (CARV2016)* (pp. 30–34). Procedia CIRP. 10.1016/j.procir.2016.07.051

Ning, L., & Xiaoping, L. (2012). A Resource Virtualization Mechanism for Cloud Manufacturing Systems. In International IFIP Working Conference on Enterprise Interoperability (pp. 46-59). Springer.

NIST. (2011, October 25). The NIST Definition of Cloud Computing (NIST Special Publication 800-145). Retrieved January 8, 2018 from https://www.nist.gov/news-events/news/2011/10/final-version-nist-cloud-computing-definition-published

NSF Program Announcements. (2017). Cyber-Physical Systems (CPS) Document Number: nsf17529. December 8, 2016. Retrieved January 06, 2018, from https://www.nsf.gov/pubs/2017/nsf17529/nsf17529.pdf

OPC UAFoundation. (2017). *OPC Unified Architecture*. Retrieved January 06, 2018, from https://opc-foundation.org/developer-tools/specifications-unified-architecture

Panetto, H., & Molina, A. (2008). Enterprise integration and interoperability in manufacturing systems: Trends and issues. *Computers in Industry*, *59*(7), 641–646. doi:10.1016/j.compind.2007.12.010

Qanbari, S., Zadeh, S. M., Vedaei, S., & Dustdar, S. (2014). CloudMan: A platform for portable cloud manufacturing services. In *IEEE International Conference on Big Data* (pp. 1006-1014). IEEE. 10.1109/BigData.2014.7004334

Ren, L., Zhang, L., Wang, L., Tao, F., & Chai, X. (2017). Cloud manufacturing: Key characteristics and applications. *International Journal of Computer Integrated Manufacturing*, *30*(6), 501–515. doi:10.1080/0951192X.2014.902105

RightScale. (2017). *State of the Cloud Report*. Retrieved July 04, 2017, from http://assets.rightscale.com/uploads/pdfs/RightScale-2017-State-of-the-Cloud-Report.pdf

Schwab, K. (2015). *The Fourth Industrial Revolution: What It Means and How to Respond*. Foreign Affairs, Science & Technology. Retrieved January 06, 2018, from https://www.foreignaffairs.com/articles/2015-12-12/fourth-industrial-revolution

VDI. VDE, & ZVEI. (2015). *GMA Status Report: Reference Architecture Model Industrie 4.0 (RAMI 4.0)*. Retrieved July 04, 2017, from https://www.zvei.org/fileadmin/user_upload/Presse_und_Medien/Publikationen/2016/januar/GMA_Status_Report__Reference_Archtitecture_Model_Industrie_4.0__RAMI_4.0_/GMA-Status-Report-RAMI-40-July-2015.pdf

Wang, P., Gao, R. X., & Fan, Z. (2015). Cloud Computing for Cloud Manufacturing: Benefits and Limitations. ASME. *Journal Manufacturing Scientific Engineering*, *137*(4).

Wang, X. V., & Xu, X. W. (2013). An interoperable solution for Cloud manufacturing. *Robotics and Computer-integrated Manufacturing*, *29*(4), 232–247. doi:10.1016/j.rcim.2013.01.005

Wei, L., Zhu, H., Cao, Z., Dong, X., Jia, W., Chen, Y., & Vasilakos, A. V. (2014). Security and privacy for storage and computation in cloud computing. *Information Sciences*, *258*(10), 371–386. doi:10.1016/j.ins.2013.04.028

Weissberger, A. (2011). *IEEE CIO Says Cloud Interoperability a Bigger Problem than Security!* Retrieved February 2, 2017, from http://community.comsoc.org/blogs/alanweissberger/ieee-cio-says-cloud-interoperability-bigger-problem-security

Wolf-Dieter, L. (2011). Industrie 4.0: Mit dem Internet der Dinge auf dem Weg zur 4. industriellen Revolution. *VDI nachrichten, 13*. Retrieved July 7, 2017, from http://www.wolfgang-wahlster.de/wordpress/wp-content/uploads/Industrie_4_0_Mit_dem_Internet_der_Dinge_auf_dem_Weg_zur_vierten_industriellen_Revolution_2.pdf

Wu, D., David, W. R., Wang, L., & Schaefer, D. (2014). Cloud-Based Manufacturing: Old Wine in New Bottles? In *Proceedings of the 47th CIRP Conference on Manufacturing Systems* (Vol. 17, pp. 94-99). 10.1016/j.procir.2014.01.035

Wu, D., Greer, M., Rosen, D., & Schaefer, D. (2013). Cloud manufacturing: Strategic vision and state-of-the-art. *Journal of Manufacturing Systems*, *32*(4), 564–579. doi:10.1016/j.jmsy.2013.04.008

Xu, X. (2012). From cloud computing to cloud manufacturing. *Robotics and Computer-integrated Manufacturing*, *28*(1), 75–86. doi:10.1016/j.rcim.2011.07.002

Yadgarova, Y., & Taratukhin, V. (2016). An Interoperable Cloud Environment of Manufacturing Control System. In Mertins K., Jardim-Gonçalves R., Popplewell K., Mendonça J. (Ed.) *Enterprise Interoperability VII. Proceedings of the I-ESA Conferences (Vol. 8)*. Cham: Springer. 10.1007/978-3-319-30957-6_1

Zhang, L., Luo, Y., Tao, F., Li, B., Ren, L., Zhang, X., ... Liu, Y. (2011). Cloud manufacturing: A new manufacturing paradigm. *Enterprise Information Systems*, *8*(2), 167–187. doi:10.1080/17517575.2012.683812

Zhang, L., Luo, Y.-L., Tao, F., Ren, L., & Guo, H. (2010). Key technologies for the construction of manufacturing cloud. *Jisuanji Jicheng Zhizao Xitong*, *16*(11), 2510–2520.

Zhang, Y., Zhang, G., Liu, Y., & Hu, D. (2017). Research on services encapsulation and virtualization access model of machine for cloud manufacturing. *Journal of Intelligent Manufacturing*, *28*(5), 1109–1123. doi:10.100710845-015-1064-2

Zhou, J., Leppanen, T., Harjula, E., Ylianttila, M., Ojala, T., Yu, C., ... Yang, L.-T. (2013). CloudThings: A common architecture for integrating the Internet of Things with Cloud Computing. In *Proceedings of the 2013 IEEE 17th International Conference on Computer Supported Cooperative Work in Design* (pp. 651-657). 10.1109/CSCWD.2013.6581037

Zissis, D., & Lekkas, D. (2012). Addressing cloud computing security issues. *Future Generation Computer Systems*, *28*(3), 583–592. doi:10.1016/j.future.2010.12.006

KEY TERMS AND DEFINITIONS

Cloud Computing: In general, an information technology architecture model where computing services (both hardware and software) are delivered to customers over the Internet, on-demand, in a self-service fashion, independent of device and location.

Cloud Manufacturing: The basic idea was to apply cloud computing philosophy (architecture, tasks and service structure) to manufacturing systems. The result of this mirroring is new service-oriented networked manufacturing system.

Cyber-Physical Production System: A dual-system in which physical processes are monitored by creating a virtual counterpart of the physical world and make decentralized decisions, while communicating and cooperating with each other (and with humans) in real time using the Internet and IoT components.

Industrial Internet of Things: The IoT is a network of computers, intelligent devices, and objects that collect and share huge amounts of data. The collected data is sent to a central Cloud-based service where it is aggregated with other data and then shared with end users in a helpful way. The application of the IoT in the manufacturing industry is called the IIoT.

Industry 4.0: High-tech ICT-based strategic industrial program which focuses on manufacturing and describes the up-to-date automation and data exchange in manufacturing technologies. Industry 4.0 includes Cyber- Physical Systems, the Internet of Things and cloud computing.

ENDNOTES

1 OPC UA Part 1 - Overview and Concepts 1.04 Specification

Chapter 9
Towards Industry 4.0:
Efficient and Sustainable Manufacturing Leveraging MTEF - MTEF-MAESTRI Total Efficiency Framework

Emil Lezak
IZNAB Sp. z o.o., Poland

Enrico Ferrera
Istituto Superiore Mario Boella, Italy

Rosaria Rossini
Istituto Superiore Mario Boella, Italy

Zofia Masluszczak
Lean Enterprise Institute Poland, Poland

Malgorzata Fialkowska-Filipek
Lean Enterprise Institute Poland, Poland

Gunnar Große Hovest
ATB - Institut für angewandte Systemtechnik Bremen GmbH, Germany

Alexander Schneider
Fraunhofer FIT, Germany

Emanuel J. Lourenço
Instituto de Ciência e Inovação em Engenharia Mecânica e Engenharia Industrial, Portugal

Antonio J Baptista
Instituto de Ciência e Inovação em Engenharia Mecânica e Engenharia Industrial, Portugal

Gonçalo Cardeal
Instituto de Soldadura e Qualidade, Portugal

Marco Estrela
Instituto de Soldadura e Qualidade, Portugal

Ricardo Rato
Instituto de Soldadura e Qualidade, Portugal

Maria Holgado
University of Cambridge, UK

Steve Evans
University of Cambridge, UK

DOI: 10.4018/978-1-5225-4936-9.ch009

ABSTRACT

An overview of the work under development within the EU-funded collaborative project MAESTRI is presented in this chapter. The project provides a framework of new Industrial methodology, integrating several tools and methods, to help industries facing the fourth industrial revolution. This concept, called the MAESTRI Total Efficiency Framework (MTEF), aims to advance the sustainability of manufacturing and process industries by providing a management system in the form of a flexible and scalable platform and methodology. The MTEF is based on four pillars: a) an effective management system targeted at continuous process improvement; b) Efficiency assessment tools to support improvements, optimization strategies and decision-making support; c) Industrial Symbiosis paradigm to gain value from waste and energy exchange; d) an Internet-of-Things infrastructure to support easy integration and data exchange among shop-floor, business systems and MAESTRI tools.

INTRODUCTION

The continuous population growth has led to increased concerns regarding the protection of the environment and resource scarcity by our society and economy. In this context, the concept of sustainability was defined in 1987 in a Report of the World Commission on Environment and Development as "development that meets the needs of the present without compromising the ability of future generations to meet their own needs" (U. Nations, 1987). Since then, the interest and concern about such subjects has aroused growing interest and the sustainability concept was never as meaningful and important as it is nowadays (Fercoq et al., 2016). Likewise, the industrial sector has the concept of sustainable manufacturing, which comprehends a significant number of objectives. The most quoted definition is given by the U.S. Department of Commerce: "the creation of manufactured products that use processes that minimize negative environmental impacts, conserve energy and natural resources, are safe for employees, communities, and consumers and are economically sound" (U.S. Department, 2014).

Hence, sustainable development supported by resource efficiency - based on the principle that the efficient and effective material and energy use can reduce natural resource inputs and waste or pollutant outputs, thus avoiding environmental degradation, and eco-efficiency assessments is nowadays a clear priority for the European industry as it is largely dependent on resources imports from international markets. Moreover, European industry is also accounted for more than a quarter of total energy consumption in 2010 in Europe.

Despite the fact that the concept of sustainability might be understood intuitively, yet, its quantitative evaluation for production systems is a complex task and not intuitive. Therefore, in order to improve sustainability and resource efficiency, the appropriate sustainability assessment and performance evolution should be carried out.

Due to the complexity related with sustainability assessment, this has become a rapidly developing topic with a growing number of concepts and tools being developed during the last decades. This has been particularly relevant for manufacturing industries, main consumers of natural resources (Garetti, 2012). The response to these challenges must encompass suitable approaches enabling industries to be supported on capabilities of the novel approaches to improve production processes and to ensure resource efficiency in the delivery of high-value-added products, while maintaining a superior economic and sustainable performance.

However, it is not enough to have proper tools and to assess resource efficiency and eco-efficiency on an off-line analysis basis, as such assessments monitored in an on-line way enable an overview of the current status and support continuous analysis capabilities. So, in order to improve resources efficiency and eco-efficiency with an "on the spot" information-decision, rather than after the major inefficiencies and costs have taken place, one first has to be able monitor data in real time in order to analyse relevant results for the resource efficiency assessment.

Additionally, the process itself needs to be known in detail and to be somehow documented. One way is to model business processes with a specific language, for instance the Business Process Modelling and Notation (BPMN). BPMN help as a means of communication between technical and non-technical personnel, the models can also be executed in process engines and it is one of the standards used today in industry. In order to be able to analyse data in a much deeper way, such data coming from sensors and production systems need to be mapped to process steps and annotated automatically. This way, and in the scope of the new industrial paradigm named of "Industry 4.0", the process can be monitored in real-time and the data can be used for a more detailed analysis. In this sense, advances towards Industry 4.0 and overall control of manufacturing systems will lead to a significant increase of available information. Industry 4.0 leverages on the concept of Internet-of-Things (IoT), which incorporates several opportunities related to tracking resources, through IoT system can reveal hidden inefficiencies, costs, and ultimately provide sufficient information, with higher quality, to set goals and make informed decisions to enhance resource efficiency and eco-efficiency performance. Smart systems tend to be better than humans in capturing and communicating data, since errors related to human errors registering and managing the data may be avoided.

Sustainability may benefit a lot from IoT systems, as the lack of clear understanding of the processes and reliable data may be surpassed, since companies are able to assess in real time the performance of the process and have high level of confidence on the results. IoT's possibilities are thrilling for sustainability assessments, yet some challenges suggest that a more measured, targeted approach may be very important.

Despite all the available means to monitor in real-time and assess production systems general performance, to successfully assess and evaluate sustainability in manufacturing and process industries, suitable methods and tools continue to be necessary and more improved. Having this in mind, the main goal of this chapter is to outline the concept of the MAESTRI project, which consists of the development of a flexible and holistic integrated framework, the MAESTRI Total Efficiency Framework (MTEF). This chapter builds on and extends a previous overview of the MAESTRI approach presented in the Sustainable Design and Manufacturing conference in April 2017 (Ferrera et al., 2017). This combines different assessment methods and tools which use data from an IoT platform, as the overall purpose is to support the decision-making process and improvement on a continuous basis and increase eco-competitiveness by fostering sustainability in routine operations.

The MTEF is based on four main pillars to overcome the current barriers and promote sustainable improvements: an Effective Management System, Efficiency Assessment tools, an Industrial Symbiosis (IS) methodology, and an IoT Platform. The framework consists of a flexible and modular toolbox, which comprises eco-efficiency and efficiency performance assessment methods and tools - ecoPROSYS and MSM. The integration of these originates the Efficiency Framework, which enables standalone assessments or an integrated analysis. Moreover, this framework also integrates a Lean Management System perspective and an IS approach, i.e. resource sharing between different production processes (Holgado et al., 2016). The IoT platform acts as a major enabler regarding data collection, mainly for the tools in the form of software, especially those that aim to support the efficiency and eco-efficiency assessment.

In the following sections of this chapter an overview of the related work for each of the pillars is presented, followed by a detailed description of the MAESTRI Total Efficiency Framework and the four pillars. The final sections of this chapter focus on the strategy for implementation of the MTEF in industry, ultimately the final considerations are presented.

BACKGROUND AND RELATED WORK

The four pillars enable the total efficiency concept, encompassing the different aspects related to industrial energy and resource efficiency described above. In the following paragraphs each pillar is described along with its related work.

Effective Management System

The effective management system is introduced to boost eco-efficiency in process industry and it is based on Lean Management concept. The term 'Lean' was introduced worldwide to describe Toyota's business during the late 1980s by a research team headed by Jim Womack, at MIT's International Motor Vehicle Program (IMVP). The results of that research revealed that Toyota's management style is substantially different from the conventional management style used in the Western world. Conventional management style is frequently associated with Henry Ford, mass production, economies of scale. Lean philosophy brings a shift of paradigm. (Womack, Jones, & Roos, 1990)

Lean Management recommends the companies to maximize customer value while minimizing waste. The core of the concept was formulated in the form of five lean thinking principles (Womack & Jones, 1996):

- **Principle 1:** Precisely specify value for specific product/service;
- **Principle 2:** Identify the value stream for each product/service;
- **Principle 3:** Make value flow without interruptions;
- **Principle 4:** Let the customer pull value;
- **Principle 5:** Pursue perfection.

The fifth principle indicates striving for perfection and operational excellence, which Toyota made its strategic advantage. According to James Franz, Lean Thinking is an approach that allow involving all the people in the company (from operators to senior managers) into continuous innovation process based on daily problem solving and organizational improvements (*kaizens*) (Franz, 2013). The objective of MAESTRI is to use *kaizens* to improve also ecological performance of companies. Franz defines Lean company as "…a company that utilizes PDCA (Plan-Do-Check-Act cycle) continually at all levels of the organization and all functions to achieve challenging targets…" That approach brings significant competitive advantage.

Lean Thinking requires from leaders and employees team work to achieve common improvement objectives which results also in development of their manufacturing and business competences. A pre-condition for building a proper continuous improvement culture in organization is a constant care for people's motivation. Cheryl Jekiel (2011) identifies several motivating factors in Lean companies: bringing sense of meaning to the people's work, giving people a sense of accomplishment, engagement; removing waste from people's work, providing satisfaction from teamwork and problem solving, recognizing personal achievements, fostering a sense of ownership. These criteria are fulfilled by introducing two mechanisms: Hoshin Kanri and Lean Management System. Hoshin Kanri approach combines strategic management and operational management by linking the achievement of top management goals with daily management at the operational level. The whole Hoshin Kanri process is usually done in an annual cycle. So the strategy deployment takes place every year, and in between the annual strategy sessions the regular Hoshin Kanri board review meetings take place. Simultaneously, David Mann remarks that companies focus on improving organization of production process, while they often tend to overlook the critical element needed to sustain changes. This element is a Lean Management System, which contains four principal elements:

1. **Leader Standard Work**: daily checklists for line production leaders, team leaders, supervisors, and value stream manager, that shows expectations for what it means to focus on the process.
2. **Visual Controls**: tracking charts and other visual tools that present authentic performance output compared with planned performance.
3. **Daily Accountability Process**: brief, structured, tiered meetings focused on performance with visual action assignments and follow up to close gaps between actual results vs. expected performance.
4. **Leadership Discipline:** leaders themselves consistently following the process that defines the first three elements and following up on others' adherence to the same process. (Mann, 2015, p. xxv).

The Lean Thinking approach, Lean Management System and Hoshin Kanri make a baseline for the effective management system proposed in MAESTRI as well as for the tools developed to support its implementation and the efficiency assessment.

Efficiency Assessment

To have an exhaustive efficiency assessment of the production system, it is important to consider approaches for evaluating both operational efficiency and eco-efficiency perspective, encompassing the basic principles of eco-efficiency (Hermann, Blume, Kurle, Schmidt, & Thiede, 2015; Hauschild, 2015; Growth Within: a circular economy vision for a competitive Europe, 2015). The result of that is efficiency to support decision-making and enable the maximization of product / processes value creation and minimization of environmental burdens. Efficiency Framework which is materialized through the integration of concepts and results provided by 1) the eco-efficiency methodology ecoPROSYS – an "Integrated methodology which allows the evaluation and assessment of eco-efficiency performance" and 2) the efficiency assessment methodology MSM© – Multi-Layer Stream Mapping - a lean based method. This Efficiency Framework aims to facilitate the overall efficiency performance assessment of complex systems (such as for production systems) by an integrated multi-dimensional analysis.

The Eco-Efficiency Integrated Methodology for Production Systems (ecoPROSYS) approach relies on the use of a systematized and organized set of indicators easy to understand/analyse, aiming to promote continuous improvement and a more efficient use of resources and energy. The goal is to assess eco-efficiency performance in order to support decision-making and enable the maximization of product / processes value creation and minimization of environmental burdens.

Eco-efficiency, the base concept of ecoPROSYS, measures the relationship between environmental and economic development of activities as sustainability aspects that evidence more value from lower inputs of material and energy and with reduced emissions. Eco-efficiency is commonly expressed by the ratio between value and environmental influence.

According to the WBCSD (Michelsen et al., 2006) the two most common goals of eco-efficiency assessments are: (i) measuring progress and (ii) internal and external communication of economic and environmental performance. In order to improve overall performance, the WBCSD established seven principles (Lehni et al., 2000):

- Reduce material intensity
- Reduce energy intensity
- Reduce dispersion of toxic substances
- Enhance recyclability
- Maximize use of renewable resources
- Extend product durability
- Increase service intensity

From a conceptual point of view, in ecoPROSYS methodology the indicators are generated by a combination of three components: (1) Environmental performance evaluation (2) Life Cycle Assessment, and (3) Cost and Value Assessment. The interaction between the different modules leads to the decision support indicators and to the environmental, value and eco-efficiency profiles.

In addition, by connecting environmental influence with the inventory data and the goals defined by the organization for eco-efficiency principles, the ecoPROSYS methodology enables also the simulation of alternative scenarios and the evaluation of these goals and objectives. For the cost assessment, any change made on the production cost is reflected in the accounting indicators towards alternatives analysis.

The MSM is a novel lean based resource efficiency assessment methodology, that takes into account the base design elements from the Value Stream Mapping (VSM), namely the value streams, in order to identify and quantify, at each stage of the process system, all "value adding" (VA) and "non-value adding" (NVA) actions, as well as, all types of waste and inefficiencies along the production system (Rother & Shook, 1999). Therefore, the basic principle of the MSM relates to Lean Principles i.e. clear definition between value and waste and to pursue perfection (focus on continuous and sustainable improvement).

The MSM approach is intended to encourage achieving maximum efficiency, (i.e. 100%) and continuous improvement mind set. Moreover, unlike the VSM, that focuses only on the added value and non-added value of the time dimension, the goal of the innovative approach of MSM is to assess the overall performance, taking into account the efficiency of each process parameters/variable (e.g. time, energy, water, raw material), which are associated to one or more processing units and variable dimensions "layers", hence the "Multi-Layer Stream Mapping" and efficiency integration analysis; identify inefficiencies in a very direct and visual manner; supporting decision making; and help prioritize the implementation of improvement actions (Lourenco, Baptista, Pereira, & Dias-Ferreira, 2013).

The MSM methodology resembles a matrix (m × n), where "m" is the number of process parameters/variables evaluated and "n" the number of steps of the production system (i.e. processing units). In order to apply the MSM, the following steps should be carried out:

- Identification of the system boundaries
- Identification of the processing unit(s)
- Identification of all relevant process variables and parameters
- Definition of the associated KPI to each variable by the form of dimensionless ratios, always to be maximized and with values ranging between [0-100%]
- Analysis of the results and identification of the process parameters and processing units with lower efficiency results
- Study and prioritization the improvement actions
- Implementation of improvement actions and assessment of the efficiency gains evolution and cost reductions

One of the cornerstones of the methodology involves the systematic non-dimensionalization of the variables that characterize the production system, with the basis ratio form between the portion of the "variable that adds value" to the product and the "total of the variable that enters the unit process". Such particularity of the method, enables consecutive aggregation of the efficiency ratios along production system, sectors, or even plants, adopting a bottom-up analysis.

Overall, the MSM is intended to be used, not only for analytical evaluation, but also to support the decision-making process, namely for greenfield design or online systems monitoring, related with:

- The identification of the most critical resource or process parameters
- The identification and quantification of inefficiencies of a given production system and unit process
- The quantification of resource and operational efficiency, and overall production system performance and costs
- The implementation of improvement actions and optimization actions
- The evaluation of efficiency progress and to incite for continuous improvement sustainability within organizations

The ecoPROSYS and MSM methods and tools make an essential baseline for the efficiency assessment in MAESTRI and consequently for the Efficiency Framework.

Industrial Symbiosis

IS fosters the sharing of resources (energy, water, residues and recycled materials etc.) between at least two different companies. Thus, IS enables companies to reduce raw material and energy consumption, and waste disposal costs, promoting the efficient use of resources in manufacturing processes. The symbiotic relationship is established between processes when looked at it in terms of resource exchanges and between companies when considering the perspective of the value exchanges (Holgado et al., 2016). It, therefore, provides the foundation to think about possible connections of waste-producing facilities into a network that minimizes the amount of waste that ends up in landfill or disposal sinks or is lost in intermediate processes (Gibbs & Deutz, 2007). The definition of IS has been extended in scope to consider those cases in which IS may occur within a single company or industry as well as cases involving multiple companies from different industries (Short et al., 2014). IS exchanges can happen in five different ways, according to Chertow (2000): exchange within a factory or an organization; exchange among companies located in the same industrial area; exchange among companies not located in the same industrial area; and exchange among companies across a broader region. In this broader sense, IS covers both, the cases in which IS opportunities are realized by a single company itself alone and those realized in partnership with other companies, and happens either in a local industrial area or in a larger region that can include more than one industrial area. In the specific case of region-wide attempts to implement IS are mainly driven by facilitated activities and programmes that can support the IS networks formation. General phases for facilitated IS are conversation, where regions' resources are studies and firms' engagement and interactions are enabled; connection, where linkages among firms are introduced and encourage through project definitions; and co-creation, where there is a focus on replicating high value exchanges and developing the necessary capacity around processing key regional resource streams (Paquin & Howard-Grenville, 2012). Cases of implementation of IS between companies from different countries has been suggested with still limited evidence of practical applications (Holgado et al., 2016).

The benefits of applying IS have been reported as multi-dimensional and multi-stakeholder, having an impact on economic, environmental and social aspects and on several stakeholders, not only on the companies participating, but also in local communities, regulation bodies and the industrial system. IS enables a cyclic circulation of waste by promoting cooperation between companies, which show that the collective benefits of engaging in IS are greater than the benefits a single entity could achieve by itself (Chertow, 2000). Therefore, it represents a direct positive effect on the efficiency of the whole system and enables the decoupling between the value created through material processing or products' manufacturing, and the environmental impact associated to natural resource intake and waste/pollutants assimilation. Its main benefits reported for local communities refer to creating new jobs, securing existing jobs, improving the local ecosystems or the creation of a cleaner and safer environment (Mirata, 2004).

Nevertheless, the successful application of IS in a particular case is highly contextual. Solutions working well in one case may not work at all on other cases due to a wide range of reasons. IS opportunities are depend on context specific characteristics that include but are not limited to; company size and production processes, geographical landscape and regional industrialization as well as country-specific trade regulations and policy (Holgado et al, 2016). The factors influencing IS development and its operational characteristics are related to multiple aspects, such as technical, political, economic and financial, informational, organisational and motivational (Mirata, 2004). It has been recognized that a balance between different organisational capabilities, regarding the aspects mentioned before, is seen as a success factor in the implementation of IS (Sakr et al., 2011).

Apart from facilitated activities and programmes to support IS in industrial parks and networks, tools and methods to support the creation of IS exchanges without intervention is still limited (Holgado et al., 2016). Reviews of tools and approaches focusing on their potential application in industrial parks, ICT supported tools or knowledge repositories can be found in literature (Chertow, 2000; Kastner et al., 2015; Grant et al., 2010; Benedetti et al., 2017). The implementation gaps and challenges of different approaches have been analysed by Holgado et al. (2018). These gaps and challenges have been taken into consideration to develop the IS methodology proposed within the MTEF.

Internet of Things (IoT)

IoT technologies provide an ICT infrastructure with middleware functionality, which facilitates the interoperation of heterogeneous hardware and software components, which may be already operating inside the company or not. The aim is to extend pre-existing ICT infrastructure into a broader and more scalable network of generic "things" (i.e. shop-floor machinery, sensors, actuators, controllers, smart objects, mobile devices, servers, ERPs, MESs, third-party systems, cloud services, etc.). In the industrial domain, data sources will be hardware and software components in the shop-floor as well as business systems of the industrial companies. The ubiquitous use of sensors, the expansion of wireless communication and networks, the deployment of decision support systems has the potential to transform the way goods are manufactured. Many observers believe that IoT is the key technology for taking Europe to a new industrial revolution, considered to be the fourth and hence labelled "Industry 4.0". Industry 4.0 concept (Lee, 2013) is a tentative to answer questions, such as: "How can we exploit the opportunities that digital technologies present for industry, administration, society and political participation, and How to deal with these challenges?" The goal is for good ideas to be translated quickly into innovative products and services. This is because innovative solutions are the factors that drive our prosperity and support our quality of life. And they make it possible to find creative answers to the urgent challenges of our time – including challenges in such areas as sustainable urban development, environmentally friendly energy, individualized medicine and the digital society. Industry 4.0 paradigm is considered a step forward towards industrial future of production, i.e. with highly flexible production environments, early-stage integration of business partners within value-creation processes, also providing support to decisions.

The consensus between researchers suggests that the industrial revisions require a long-time period of development and cover the following aspects, considered as the future manufacturing visions:

- **Factory:** Future factory is going to involve all manufacturing resources (sensors, actuators, machines, robots, conveyors, etc.) are connected and exchange information automatically in purpose of predict and maintain the machines, control the production process, and to manage the factory system – the paradigm known as a Smart Factory (Lucke, Constantinescu & Westkämper, 2008) ;
- **Business:** Future business is going to involve a complete communication network between companies, factories, suppliers, logistics, resources, customers, influencing one to other section in purpose of achieving a self-organizing status and provide the real-time response (Kagermann, Helbig, Hellinger & Wahlster, 2013;Brizzi, Lotito, Ferrera, Conzon, Tomasi & Spirito, 2011) .

The main features (Kagermann, Helbig, Hellinger & Wahlster, 2013) of Industry 4.0 include the following: Horizontal Integration through value networks to facilitate inter-corporation collaboration, describing the cross-company and company-internal intelligent cross-linking and digitalization of value creation modules all over the value chain of the product life cycle and between value chains of adjacent product life cycles. Vertical Integration of hierarchical sub-systems inside a factory, describes the intelligent cross-linking and digitalization within the different aggregation and hierarchical levels of a value creation module from manufacturing stations via manufacturing cells, lines and factories, also integrating the associated value chain activities such as marketing and sales or technology development to create flexible and reconfigurable manufacturing system. And End-to-End Engineering Integration across the entire value-chain to support product customization, describes the intelligent cross-linking and digitalization completely all phases of a product life cycle – from the raw material acquisition to manufacturing system, product use, and the product end of life.

After all discussion about the definition, the Industry 4.0 is here and is becoming an increasing topic of interest among technology giants and business communities. The hype is not baseless as there are enough evidence to support the success of "Internet of Things" in the coming years. According the Bernard Marr (2017) the contributor of Forbes, today we have about 15 billion IoT devices. By 2020, it is predicted that the number will grow to between 50 and 70 billion devices connected to the Internet. This means the possibilities of how to use these devices and the data they generate are endless. And herein lies the problem: it is far too easy for companies to get overwhelmed and lost by all this. So, one might simply ask "What is Internet of Things" and how it is going to impact our lives and career opportunities. There is a lot of complicated technology and terminologies at work in IoT world.

Unfortunately, many companies are enthusiastic on the IoT trend without giving much thought to how this is aligned to their business strategy or how they could achieve the biggest business value. In the consequence, the following are the essential ways every company should use of the IoT in business:

- **Improve Decision-Making:** The huge volumes of new data from IoT sensors and devices simply adds to the massive pool of big data we now have in the world and intelligent companies use that data to inform strategic and operational decision making;
- **Understand Customers:** More and more companies make IoT-enabled products which connect them directly to their customers' behaviours and preferences;
- **Deliver New Customer** Value propositions: one step further from using IoT devices to better understand customers is to make them part of your offering;
- **Improve and Optimize Operations:** The data generated from all the IoT devices can also be used to improve the ways companies are run. It can help to automate factories and improve the efficiency of internal processes;
- **Generate an Income and Improve the Value of the Business:** The data and insights our IoT devices generate are often extremely valuable. The most direct way to appreciate this value is by selling the data or insights to others.

MAESTRI Total Efficiency Framework

In order to advance the sustainability of manufacturing and process industries, it is important to foster the introduction of a management system in the form of a flexible and scalable platform, aiming to guide and simplify the implementation of a MAESTRI Total Efficiency framework (MTEF) in organizations. The overall aim of the MTEF is to promote improvement culture within process industries by assisting decision-making processes, supporting the development of improvement strategies and helping on the definition of priorities to improve companies' performance from both an environmental and economic point of view.

The MTEF is based on four main pillars to overcome the current barriers and promote sustainable improvements, as illustrated by Figure 1:

- An *Effective Management System* targeted for process and continuous improvement;
- *Efficiency assessment* tools, helping to define improvement potentials and strategies in order to support decision making processes;

- Integration with an *Industrial Symbiosis* methodology, focusing on material and energy exchange;
- An *IoT Platform* to facilitate the concept implementation, ensuring an integrated control of the improvement process;

Based on a holistic approach, which combines different assessment methods and tools, the overall purpose of the MTEF is to support improvement on a continuous basis and increase eco-competitiveness by fostering sustainability in routine operations. Its conceptual approach will be based on a life cycle perspective, centred on models for dynamic simulation of both individual and complex systems, to better understand processes and the opportunities to add value. This life cycle approach is important to avoid problems (waste, environmental impacts, etc.) shifting from one life cycle stage to another.

The MTEF represents a flexible and scalable platform, which provides an effective management system that aims to advance the sustainability of manufacturing and process industries. It combines the four pillars in one holistic platform which enables an overall efficiency performance assessment from environmental (including resource and energy efficiency), value and cost perspectives. MTEF encompasses Environmental Performance Evaluation with Environmental Influence and Cost/Value assessment models through a life cycle perspective. The aim is to support the decision-making process, by clearly assessing resource and energy usage (valuable / wasteful) of all process elementary flows, and the eco-efficiency performance. Decision support via value-adding optimization is also foreseen among the integration of the modules.

Figure 1. MAESTRI total efficiency framework concept

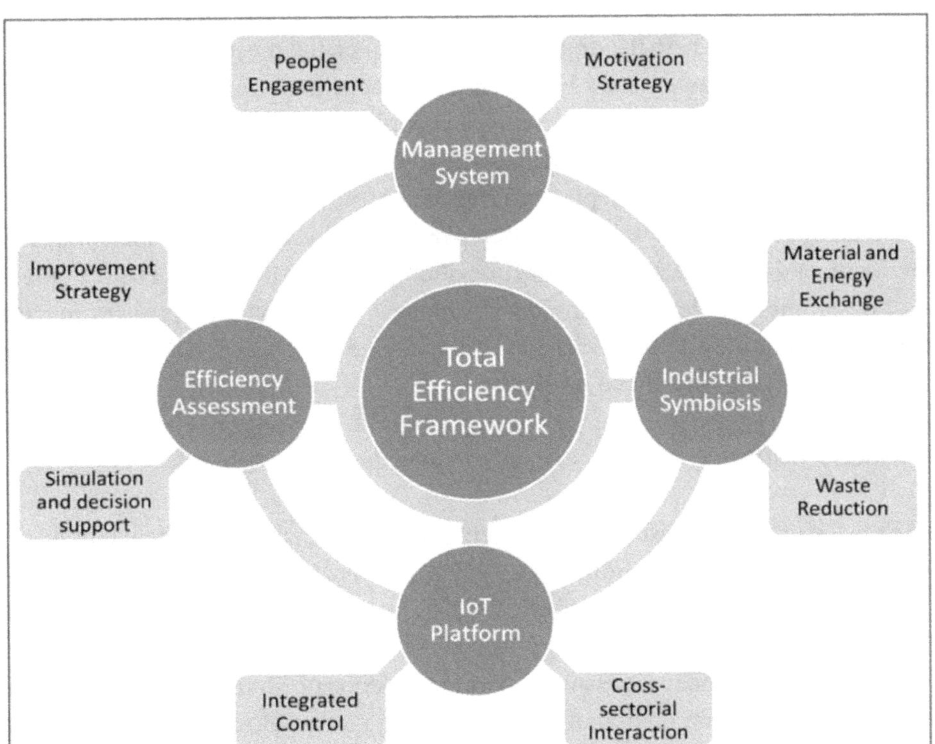

The Four Pillars of the MAESTRI Total Efficiency Framework

Effective Management System

Here the conventional Lean Management System is supplemented by some elements explicitly oriented on boosting eco-efficiency. These supporting methods and tools address the challenges and barriers for energy and resource management identified in the early stage of the MAESTRI project. The barriers are factors hindering the implementation of environmental strategies. Since the early 90s they have been identified and classified by many authors. One of the approaches includes classification of external and internal barriers. External barriers are associated with external factors like policy, market or economy which cannot be directly controlled by company; such as insufficient drivers, lack of support and guidance, cost of certification. On the other hand, internal barriers describe company specific factors (socio-technical aspects) like management system, organizational structure, culture which may be controlled by the organization; such as lack of sufficient human resources, lack of understanding or wrong perception of environmental behaviour, employee's resistance (Hillary, 2004; Shi, Peng, Liu, & Zhong, 2008). Although barriers have different degrees of difficulty, internal barriers are considered more problematic than external barriers (Dahlmann, Brammer, & Millington, 2008; Hillary, 2004; Murillo-Luna, Garcés-Ayerbe, & Rivera-Torres, 2007; Post & Altma, 1994). That is why several methods were proposed to face the internal barriers.

The Eco Orbit View (EOV) method links the company's business strategy with environmental perspective in order to achieve symbiotic gains in both aspects with the same improvement activities. Focused Gemba Walk and Retrospective Analysis aim to promote structured problem solving process and base process improvement activities in process industry on root cause analysis and the term "genchi gembutsu" (essentially "go and see", which stress the importance of empiricism) (Lean Enterprise Institute, 2014). Eco Lean Management Boards help the organization to focus on the most important goals and activities originating from company's strategy and EOV. They also contribute to cultural transformation of the company thanks to exercising management routines according to Lean Management principles, make the changes more sustainable and increase managers' motivation.

The MAESTRI management system plays an important role in the MTEF as it makes the company revise its goals, rethink the way the strategy is deployed to the lower levels of the company as well as brings the environmental perspective to the front line of improvement activities. The activities completed in the scope of designing or redesigning the management system should be primary in comparison to automation of data flow as they define which data are really needed, which KPIs should be built and for whom and in which form they should be presented.

Efficiency Assessment

Within the MTEF, the efficiency assessment, as mentioned before, comprises the integration of the eco-efficiency and efficiency – Efficiency framework, which encompasses several modules: i) eco-efficiency (environmental influence and economic performance); ii) resource and energy efficiency; and iii) overall efficiency assessment.

The main goal of this efficiency assessment, within the scope of MAESTRI, is to define and characterize a new approach for the integration of the two existing methodologies ecoPROSYS and MSM, in such a manner that enables the Efficiency Framework to be used to assess eco-efficiency and efficiency performance, as well as to assess the overall production system efficiency.

This integrated framework empowers a flexible and straightforward analysis facilitating implementation and use, and enable decision support analysis in different processes in industries of different types and sizes. Its originality is founded not only on the integration of results of the two original methodologies (ecoPROSYS and MSM), but also on the creation of a concurrent assessment of eco-efficiency (including LCA and LCC aspects) and operational efficiency (with foundations on Lean Manufacturing value/waste assessment) with focus on the creation of practical and user-friendly analysis and tools. This concurrent assessment converges into a compromise between cost effectiveness and cost efficiency, which can be used for informed decision-making.

The outline of the integration of ecoPROSYS and MSM concerns the exchange of information between efficiency and eco-efficiency assessments. In fact, such exchange of information is the focus of the integration of the two methods in order to support the decision-making process, considering efficiency, environmental and economic performance as whole and not as isolated domains.

As mentioned before, the outcomes of the Efficiency Framework will enable to simultaneously assess eco-efficiency and efficiency performance by using cross-parametrization of the individual tools and the direct results from ecoPROSYS and MSM, respectively. These results, in addition to support a more complete and informed decision-making process regarding sustainable development, also play an important role when it comes to reporting results through score-cards/dashboards within the Efficiency Framework.

The integration of different aspects related namely with, eco-efficiency, operational efficiency, value and waste identification, play an important role within the MTEF.

As a result of the combination of these two fundamental eco-efficiency and efficiency/performance methods (ecoPROSYS and MSM), the Efficiency Assessment intends to encourage businesses not only to search for environmental improvements that yield parallel economic benefits, but to put emphasis on value creation and capturing. In this sense, by increasing value of goods, business tend to maximize resource productivity, gain bottom-line benefits, and reward shareholders, rather than simply improve efficiency or eco-efficiency performance, or minimize waste or pollution.

Industrial Symbiosis

Being focused on increasing the efficiency of resources by performing multi-company exchanges, IS seems to be a highly adequate concept to implement in the process industry. This industry is often characterized as energy-intensive and resource-intensive, with a significant contribution to GHG emissions and a high dependence on resources availability (Garetti & Taisch, 2012). The study of complementary approaches, such as the MTEF, combining IS applications with other improvement actions based on eco-efficiency and lean techniques seem relevant for this industry. This brings additional chances to overcome barriers to IS implementation in industry. The development of IS can be seen as a dynamic process leading to the creation of industrial ecosystems, within which new opportunities are expected to emerge over time (Liu et al., 2011). It is mainly an incremental and iterative process in which independent resource exchanges are gradually established among the companies in the IS network. In MAESTRI we focus on fostering the creation of resource exchanges in a spontaneous manner between companies, i.e. without a third party intervention to either facilitate or drive the exchanges creation. Drivers of these spontaneous exchanges are revenue enhancement, cost reduction and business expansion (Chertow, 2007).

Three generic stages of IS development can be defined as: 1) the identification of opportunities for IS exchanges; 2) the design and planning of the exchanges and; 3) the actual implementation of the exchanges. MAESTRI has a strong focus on the first one as main barrier preventing the implementation of IS exchanges. Regarding this first step, MAESTRI aims to support companies to identify and characterize waste as well as obtaining ideas of how other companies have dealt with certain types of waste. IS refers most commonly to waste materials, wastewater, loss/dissipated heat or unused by-products of production processes. MSM and ecoPROSYS can support the analysis of those undesired outputs (what are they and how much are produced) that are potential subjects for IS. Additionally, MAESTRI has developed a library of case studies and an exchanges database that will provide companies looking at implementing IS in their processes insights and specific details regarding already analysed and planned exchanges by other companies (Benedetti et al., 2017). These mechanisms to identify IS opportunities are aligned with those identified by Grant et al. (2010), which are the discovery of a new process to transform a by-product into a usable resource, the identification of links between inputs and outputs needed / produced in different organizations and the replication of successful exchanges done previously by similar organizations.

Besides, the interactions among companies within IS networks are attracting more and more interest in the research community, with social factors such as culture change, inter-firm learning and social capital (Holgado et al, 2016). This relates strongly to leadership approaches and management systems, thus, linked to the implementation of a Lean Management approach for an effective management system in MAESTRI.

Technological Aspects in IS Implementation

Sakr et al. (2011) pointed out to technological and economic feasibility as necessary conditions for IS to happen. The technological factors, together with regulatory factors, have been identified as highly connected to the promotion of IS applications at its initial stages (Baas, 2008).

As mentioned above, there is a high degree of contextualization in the implementation of IS. There is a need to address this contextualization issue (e.g. by developing tools and methods) in order to promote IS design and planning (Holgado et al., 2016). Particularly, there are some attempts to develop tools for IS opportunity identification based on an input-output approach. These tools collect data regarding inputs and outputs of various entities systematically to make links across industries or providing profiles of typical inputs/outputs for a specific type of facility (Chertow, 2000). However, the lack of taxonomical classifications or common language for potentially exchangeable resources is the main challenge to develop ICT-based tools for input-output matching while there are more opportunities for the developed of ICT applications on the mimicking of existing cases in similar industries as the industries can be more easily codified / search within a database (Grant et al., 2010). Therefore, MAESTRI approach for IS, focusing on a mimicking approach for the development of the exchanges database, seems adequate to potentially extend in the future the results towards the MTEF.

Internet of Things

To provide the different end user software tools with data needed for energy and resource efficiency analysis and continuous KPI monitoring, the MAESTRI Total Efficiency Framework comprises a middleware platform based on IoT technology. On the one hand, this platform facilitates the integration of different sensors, measurement devices and existing enterprise applications in order to collect and pre-process the data from all relevant data sources, allowing a holistic analysis of the energy and resource efficiency of the company. On the other hand, such a platform will provide also a model-based approach and the event-based annotation of data in order to add meta-data to allow for the process-step based measurement and analysis. The IoT platform should be seen as complementary development, supporting the different tools from all the other pillars along the different stages of an improvement process.

The overall architecture of the MTEF is depicted in *Figure 2*. The central element is the IoT platform, which facilitates the data transfer from machines, systems, and sensors to end user software tools and applications at the industrial sites.

From the IoT Platform point of view, all systems to be connected can be seen either as data sources or as data consumers:

- Data sources will be hardware and software components in the shop floor as well as (pre-existing) enterprise applications of the industrial companies. They produce data to be analysed, processed and interpreted in order to calculate KPIs relevant for the evaluation and optimization of the industrial processes.
- Data consumers will be the MAESTRI Front-End Applications, but also other software tools in the companies could require access to data from the shop floor via the IoT Platform. Therefore, existing enterprise applications could possibly play both roles, i.e. they could be data sources and data consumers at the same time. Then, besides the MAESTRI Front-End Applications, the IoT Platform will expect also other (custom developed) applications to be integrated in the platform

Figure 2.MAESTRI total efficiency framework architecture

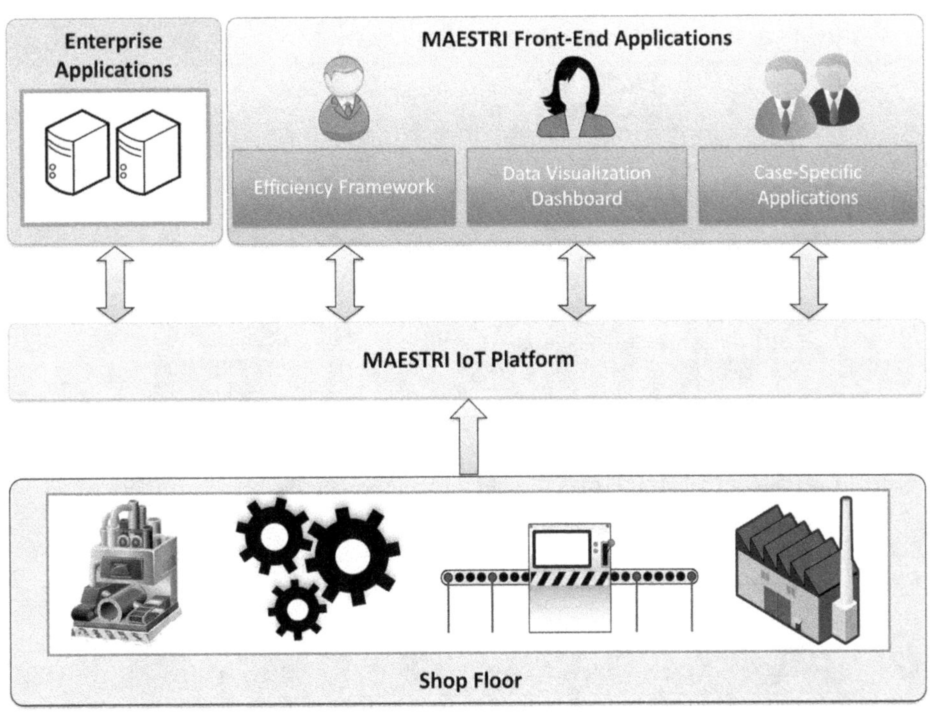

after or during the project's lifetime. In this way, one of the general requirements towards the IoT Platform is to be open for flexible integration of additional end-user applications.

Due to the need for companies to continually optimize their processes, it is not enough anymore just to look at e.g. energy consumption as an aggregated number, but it is necessary to have the possibility to split up the measurements according to the actual process steps (Fisseler & Reiners 2015). The approach used in the MTEF and its IoT Platform is based on the combination of a model-based approach and the event-based annotation of data in order to add meta-data to allow for the process-step based measurement and analysis. For the models we use a BPMN 2.0 based process model which will be executed during runtime. In addition to that a deployment model is used in order to connect to the data sources and define which events will be produced by which entity. The basic architecture building blocks (Fisseler, Reiners, & Kemény, 2016) can be seen in *Figure 3*. The Tools "Modeler" and "Mapper" on the right-hand side are only used during design time and are used to create/update the aforementioned models.

The Modeler tool allows for the creation of BPMN 2.0 models ("business process model") which are then exported into XMI format in order to be reused and extended by the Mapper Tool. In the business process model the process can be defined in as much detail as wanted. In the beginning one would start with a simpler model in order to identify the process steps for which an optimization would give the best cost/benefit ratio. This has a distinct connection to the MSM methodology and is described in detail in (Fisseler, Schneider, Lourenço, & Baptista, 2017). During further refinement of the process optimization the business process model can be specified in more detail allowing for even more fine-grained analysis and optimization.

Figure 3. Business process monitoring architecture

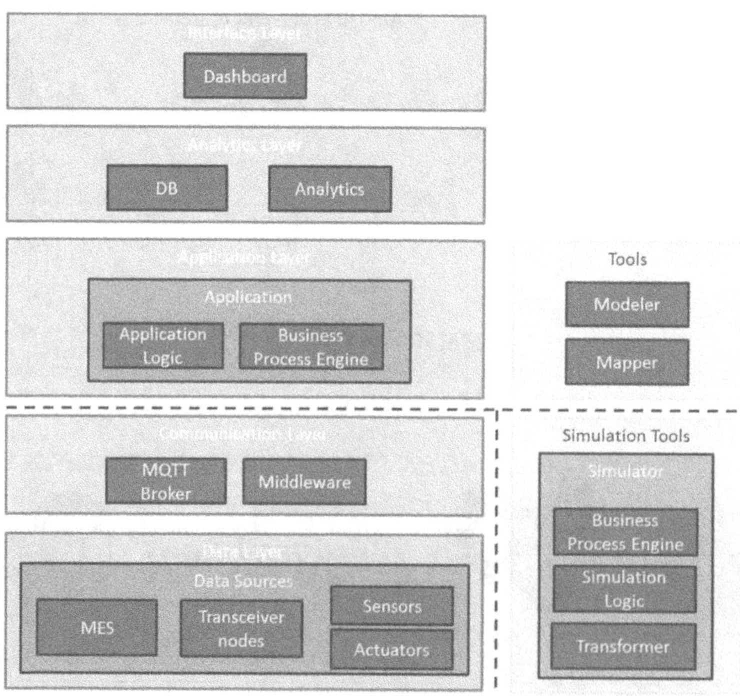

The Mapper tool is used to connect the business process model with the data measurement components as defined in the deployment model. This is done by putting the event topics containing the measurements from the sensors to the entities in the business process model such as start/stop events and process steps.

The Business Process Engine subscribes to all events defined in the business process model and which will be distributed through the MQTT broker in the Communication Layer. The event data will then be extended by the semantic information in which process step this event has occurred and a new event will be generated and transferred over the MQTT broker.

Finally, the MTEF, after the development and integration of the tools and methods, will be implemented and validated in four real industrial settings across a variety of different sectors. Those four industrial pilot cases (two of the chemical sector, one of injection moulding, and one from the metalworking sector) represent one of the main sources for identification of the requirements upon the methodological approach and the platform's functionalities as well as for their continuous refinement based on the testing results.

MAESTRI Total Efficiency Framework Implementation

The MASTRI Total Efficiency Framework aims to provide a modular and flexible toolbox, integrating the four pillars, as well as a methodology how to select and apply the different tools in industrial companies in order to support improvements of their energy and resource efficiency. Depending on the specific situation of a company and its existing processes and infrastructure, the most appropriate tools from the MTEF should be selected.

While some of the MAESTRI tools are to be applied in workshops to support an initial analysis of the company's situation and to detect improvement potentials, other tools are provided in the form of software, especially those that aim to support the efficiency and eco-efficiency assessment. The eventual goal of MAESTRI is to apply Internet-of-things technologies to connect all relevant systems and data sources, enabling an integrated assessment and continuous monitoring and improvement of the company's energy and resource efficiency.

Overview of the Implementation Process

Already in parallel to its development, the MTEF is initially implemented and validated in four real industrial settings, aiming to collect feedback for the further improvement of the toolbox and methodology. One of the first lessons learned from these four different cases was that there is no "one-size-fits-all" approach for the application of the MTEF. The selection of the appropriate tools and the approach for their application has to be tailored (combined) to the specific situation of each company, considering the types of production processes to look at (e.g. discrete manufacturing in contrast to batch production), as well as, previous activities regarding energy and resource efficiency. To facilitate this adjustment to the as-is situation, the MAESTRI Total Efficiency Framework will include a methodology and guidelines how to identify and apply the most appropriate tools.

When entering a company to apply the MTEF, the first step will be to assess in how far activities related to the topics of the MAESTRI pillars were already started (i.e. a kind of "maturity assessment" of the company regarding each of the four pillars), aiming to build upon the previous activities and to get synergies with existing procedures. This initial assessment will address, besides others, the following aspects:

- In how far KPIs regarding energy and resource efficiency are already defined in the company
- In how far the company's KPIs on operational level are based on and aligned with the overall strategic goals of the company
- In how far the company already defined eco-efficiency-related improvement actions that will allow to achieve goals resulting from the overall strategy deployment
- In how far waste and by-products reuse opportunities were already analysed
- In how far data about energy and resource consumption is already collected on a regular basis and how such data can be accessed and reused.

To support a systematic assessment of the as-is situation, the MTEF provides workshop tools like Eco Orbit View and the IS methodology. The aim of these tools is to support the initial identification of improvement potentials and of those areas in the company, where a more detailed analysis regarding improvement potentials would make sense. In addition, MAESTRI provides the library of case studies and the exchanges database, supporting companies to get initial ideas for IS opportunities, based on other companies' experiences and lessons learned.

As a follow-up to the initial workshops, the MAESTRI tools for efficiency and eco-efficiency assessment (MSM and ecoPROSYS) should be used to analyse the previously identified areas of the company in more detail, supporting the identification and assessment of improvement potentials, also facilitating a trade-off between expected gains and the cost of possible improvement measures. The same tools can also be used later to do an ex-post analysis of the improved situation after the implementation of specific improvement measures, comparing the actual improvements to the expectations based on the previous assessment.

After the first detailed analysis was finished and improvement potentials were identified, the application of MAESTRI tools should be introduced as a regular activity, facilitating a continuous stepwise improvement of the company's energy and resource efficiency. Such a regular application of MAESTRI tools needs to be supported by a continuous and consistent collection of all data that is relevant for monitoring the situation regarding energy and resource efficiency. This is facilitated by the implementation of the MAESTRI IoT platform, which supports the automatized collection and pre-processing of the relevant data from the shop floor and existing enterprise applications, storing and providing them in a format as needed by the MAESTRI front-end applications.

A full implementation of the MTEF in a company is expected to support the energy and resource efficiency in different ways and on different levels. It should provide support for strategic decisions about medium- to long-term improvement measures (e.g. investment in more efficient technology, implementation of IS practices), but it should also support the operational staff members to take energy and resource consumption into account when doing daily routine decisions (e.g. by making KPIs visible via displays in the shop floor, triggering alarms or sending notifications in case of KPIs exceeding pre-defined thresholds, providing suggestions for eco-efficient process control). The MTEF also aims to assure that all the eco-efficiency-related improvement actions are focused on and aligned with the strategic goals and challenges of the company, and to provide the proper motivation for managers to conduct daily improvements including eco-efficiency issues.

Application of MAESTRI Tools

Overall, it is worth mentioning that the main MAESTRI Front-End Applications to be connected to the IoT Platform are MSM and ecoPROSYS. These two applications will be integrated and together form the Efficiency Framework. The Efficiency Framework's connection to the IoT Platform is via MSM and EcoPROSYS, as it works on the integrated results of these two tools.

The Management System tools and the IS methodology will be primarily accessing data via the efficiency framework (ecoPROSYS, MSM), thus lean management and IS will make use of the IoT Platform in an indirect way. In addition, ultimately enabling an integral application of the MTEF.

The combination of the efficiency and eco-efficiency results will enable to assess the effectiveness of the eco-efficiency performance improvement. This will be done by monitoring, using the IoT platform, the deviations between the real eco-efficiency (real economic value over real environmental influence) and targeted eco-efficiency (target economic value over target environmental influence). Moreover, the goal is to evaluate if eco-efficiency performance variation is due to higher or lower environmental influence, or due to higher or lower economic value.

Targets for environmental influence and value should be defined for the novel indicators fulfilling. Eco-efficiency targets are the one defined for the VA efficiency assessment of the MSM approach. One major consideration for presenting the ratios and the targets is that these should always be self-contained. This means that the results for a certain processing unit are only dependent on the variances that occur in that processing unit. Therefore, major efficiency, environmental or value variations occurring in a certain processing unit, will not affect positively nor negatively the results of other processing units, thus ensuring a robust decision making process.

The Efficiency Framework results and scorecards should show different information according to the profile (e.g. manager, director, operator, team leader, etc.), and should enable the unfolding/breakdown of the scorecards through the organizational or functional hierarchy of the company. This means that logical links should be implemented from raw data (shop floor data, e.g. energy consumption from a specific machine), to unit process, sectors, all the way to plant, company or group level presenting the global aggregated data (overall business aspects).

Regarding the ultimate novel outcome from the Efficiency Framework, the MAESTRI metric called Total Efficiency Index (TEI), this index is calculated for each processing unit of the production system under analysis, but can be successively aggregated for the complete production process of a given product (through the organizational or functional hierarchy of the company). In quantitative terms, the TEI is obtained by multiplying the normalized eco-efficiency and the efficiency assessment results from MSM (see equation 1). The normalized eco-efficiency ratio represents, in percentage, the relation between real and targeted eco-efficiency performance (see equation 2) and is calculated dividing the real eco-efficiency ratio (see equation 3) by the targeted eco-efficiency ratio (see equation 4).

$$\text{Total Efficiency Index (\%)} = \text{Normalized eco-efficiency} \times \text{Process efficiency} \quad (1)$$

$$\text{Normalized eco-efficiency (\%)} = \text{Real eco-efficiency ratio} \,/\, \text{Target eco-efficiency ratio} \quad (2)$$

$$\text{Real eco-efficiency ratio (\%)} = \text{Real value (€)} \,/\, \text{Real environmental influence (Pt)} \quad (3)$$

Target eco-efficiency ratio (%) = Target value (€) / Target environmental influence (Pt) (4)

The logic behind this index is to combine two fundamental efficiency aspects, namely eco-efficiency, which considers the ecology and economy, with the ecoPROSYS methodology, and operational efficiency, which considers the NVA and VA activities aligned with the Lean Principles application by MSM methodology.

Figure 4 presents the calculus diagram for the Total Efficiency Index, integrating results from ecoPROSYS and MSM and also applying the algorithm of MSM for the integration and visual representation of the results scorecard.

Consequently, TEI main outcome is related to the ability of evaluating if eco-efficiency performance variation is due to higher or lower environmental influence, or due to higher or lower economic value. In practice, this results from the distribution variance of TEI results that occur on two major axes: the efficiency and eco-efficiency.

The Efficiency Framework supports the decision-making process and keeps track, simultaneously, of eco-efficiency and efficiency performance, enabling managers to take actions that will improve both, efficiency and eco-efficiency performance. For this reason, it is imperative to assure their correct implementation to provide an accurate determination of the above-mentioned index (Total Efficiency Index) considering the existing settings of different sectors of process industries.

Despite the fact, that the efficiency framework relies clearly on the IoT platform to provide the real time data, as the ecoPROSYS and MSM are the software components that aim to support the efficiency and eco-efficiency.

Nevertheless, "Doing more with less" is frequently used to describe the purpose and scope of the eco-efficiency. The same quote accompanies the Lean Management approach, which is commonly used to organize and manage the production and supporting processes in manufacturing industry. With this in mind the interconnection of the eco-efficiency principles, within ecoPROSYS, with Lean Principles, which foresees synergic aspects and the definition of potential improvement areas that suits simultaneously eco-efficiency (economic and environment) and Lean Principles (value and waste identification). Improvement of synergic aspects from lean principles implies using the IoT platform to feed the efficiency

Figure 4. Total Efficiency Index diagram calculus structure

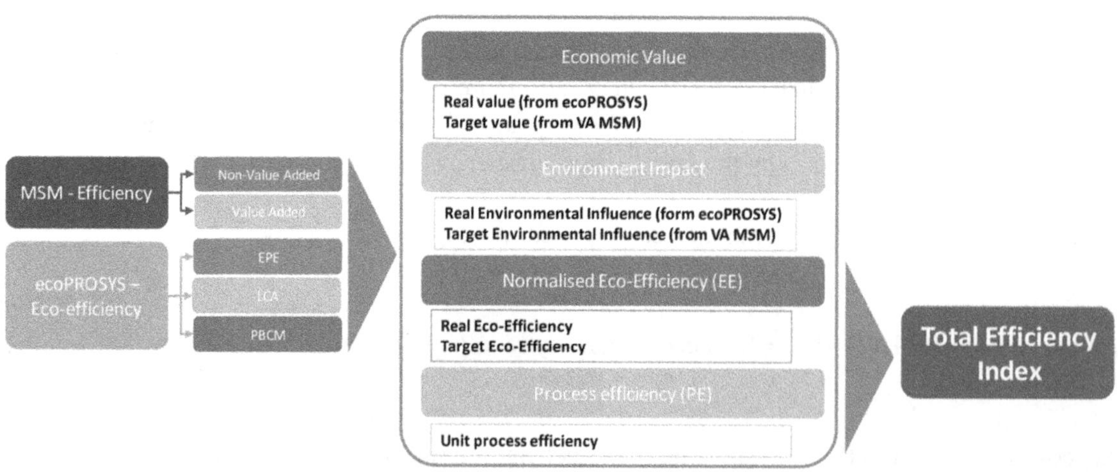

framework, and having results for eco-efficiency and operational efficiency (and the respectively cost and environmental influence reductions).

On the other hand, and although MSM and EOV demonstrate some similarities, for instance selecting proper aspects for analysis, they remain complementary approaches. EOV is designed as a quick, simple assessment method which can be performed during a workshop of a few days. It requires a basic understanding of the company's KPIs and KEPIs and interlinks between them in order to identify the potential improvement areas. The continuous monitoring using MSM may support the tracking and evaluation of the evolution of some of those KIPs.

The main connections between eco-efficiency and efficiency results have been identified so far with respect to the first two questions. In this sense, the identification of process waste (MSM) and of residues (ecoPROSYS) can contribute, together with a value analysis tool to the first question: "how to see waste?". The LCA analysis within the ecoPROSYS methodology can give support to the second question: "how to characterize waste?". MSM has the potential to provide a multi-dimensional analysis, including resource efficiency, operational efficiency and cost analysis, among other possibilities. Concretely, the resource efficiency analysis will provide useful information to identify process waste and it can contribute to the "How to see waste" within the IS methodology.

The Environmental Performance Evaluation from ecoPROSYS methodology is a process analysis of environmental aspects to characterize the intensity and significance of environmental aspects according the eco-efficiency principles. This evaluation will result in the identification of residues at factory level and thus, providing additional information for the IS methodology, particularly on "How to see waste".

Such application of the MTEF allows a clear understanding of the overall efficiency and eco-efficiency of the performance, in a real time scenario by retrieving the data via the IoT platform, and also enables companies to keep track of opportunities to reduce waste and find potential improvements considering Lean approaches.

CONCLUSION

The overall aim of the proposed MAESTRI Total Efficiency Framework is to support the sustainable improvement of industrial companies' environmental and economic performance, in particular focusing on the process industry. The MTEF will provide a modular and flexible toolbox, as well as guidelines how to identify and apply the most appropriate tools, allowing to adjust the MTEF to the specific situation of each company.

The MAESTRI approach to support IS applications in the process industry builds on widening the knowledge on IS industrial cases as main source of information and on a stepwise methodology to support the path from waste identification and valuing to exchange definition and exploitation phases. This will support companies willing to engage with the implementation of the IS concept in their business. The MTEF aims to allow an innovative and practical integration of eco-efficiency analysis, including LCA and LCC, with process efficiency assessment (including both resource efficiency and operational efficiency) analysis from the perspective of Lean Manufacturing, thus implicitly evaluating "value added" and non-value added" (waste) actions and re-source usage. The proposed solution leverages a middleware platform, which facilitates the automatization of the data transfer from machines, systems, and sensors at the industrial sites to end user software tools and applications. It allows to seamlessly interconnect heterogeneous devices, systems and subsystems in order to achieve a higher degree of interactions between the shop floor, the legacy management systems and the end users. This approach thus enables the integration of results of two original methodologies and related tools (ecoPROSYS and MSM) to produce a new metric defined as "Total Efficiency Index" (TEI). In order to support these tools to assess the overall performance and efficiency in order to reduce waste and costs, a process oriented monitoring and control framework is necessary to provide real real-time monitoring as well as an automatic annotation of the data required for the calculation of TEI. The MTEF acts as enabler of the hyper-connected factory, to support decision-making for the development and simulation of improvement strategies by facilitating the continuous analysis and monitoring of a companies' efficiency and eco-efficiency. Moreover, automated data collection for the energy and resource consumption, as well as other process related activities in near-real-time will enable to "trigger alarms", therefore managers and operators can act immediately. Thus, it helps to identify major inefficiencies and consequently underpins sustainability.

The further development and validation of the MTEF will be supported by its initial application in four real industrial settings across a variety of different sectors. Those four industrial pilot cases (two of the chemical sector, one of injection moulding, and one from the metalworking sector) represent one of the main sources for identification of the requirements upon the methodological approach and the platform's functionalities as well as for their continuous refinement based on the testing results.

ACKNOWLEDGMENT

This project has received funding from the European Union's Horizon 2020 research and innovation programme under grant agreement No 680570.

REFERENCES

Baas, L. (2008). Industrial symbiosis in the Rotterdam Harbour and Industry Complex: Reflections on the interconnection of the techno-sphere with the social system. *Business Strategy and the Environment*, *17*(5), 330–340. doi:10.1002/bse.624

Benedetti, M., Holgado, M., & Evans, S. (2017). MAESTRI Deliverable D4.2. Prototype library of case studies linked to a waste database. Retrieved from https://maestri-spire.eu/downloads/technical-materials/

Chertow, M. R. (2000). Industrial symbiosis: Literature and taxonomy. *Annual Review of Energy and the Environment*, *25*(1), 313–337.

Chertow, M. R. (2007). "Uncovering" industrial symbiosis. *Journal of Industrial Ecology*, *11*(1), 11–30. doi:10.1162/jiec.2007.1110

Dahlmann, F., Brammer, S., & Millington, A. (2008). Barriers to proactive environmental management in the United Kingdom: Implications for business and public policy. *Journal of General Management*, *33*(3), 1–20. doi:10.1177/030630700803300301

Fercoq, A., Lamouri, S., & Carbone, B. (2016). Lean/Green integration focused on waste reduction techniques. *Journal of Cleaner Production*, *137*, 567–578. doi:10.1016/j.jclepro.2016.07.107

Ferrera, E., Rossini, R., Baptista, A. J., Evans, S., Hovest, G. G., Holgado, M., ... Estrela, M. A. (2017). Toward Industry 4.0: *Efficient and Sustainable Manufacturing Leveraging MAESTRI Total Efficiency Framework. In International Conference on Sustainable Design and Manufacturing* (pp. 624-633). Cham: Springer.

Fisseler, D., & Reiners, R. (2015). Prozessorientierte Überwachung in der Produktion. In D. Cunningham, P. Hofstedt, K. Meer, & I. Schmitt (Eds.), *INFORMATIK2015* (pp. 917–927). Gesellschaft für Informatik.

Fisseler, D., Reiners, R., & Kemény, Z. (2016) Monitoring and control framework for business processes in ubiquitous environments. In *Proceedings of 2016 Intl IEEE Conferences on Ubiquitous Intelligence & Computing, Advanced and Trusted Computing, Scalable Computing and Communications, Cloud and Big Data Computing, Internet of People, and Smart World Congress* (pp. 863-868). 10.1109/UIC-ATC-ScalCom-CBDCom-IoP-SmartWorld.2016.0136

Fisseler, D., Schneider, A., Lourenço, E. J., & Baptista, A. J. (2017). Combining Process Based Monitoring with Multi-layer Stream Mapping. In G. Campana, R. Howlett, R. Setchi, & B. Cimatti (Eds.), *Sustainable Design and Manufacturing SDM 2017* (Vol. 68). Cham: Springer. doi:10.1007/978-3-319-57078-5_53

Franz, J. (2013). The Toyota Way to Continuous Improvement. In *Conference Proceedings, 13th International Lean Management Conference.* Wroclaw: Lean Enterprise Institute Poland.

Garetti, M., & Taisch, M. (2012). Sustainable manufacturing: Trends and research challenges. *Production Planning and Control, 23*(2-3), 83–104. doi:10.1080/09537287.2011.591619

Gibbs, D., & Deutz, P. (2007). Reflections on implementing industrial ecology through eco-industrial park development. *Journal of Cleaner Production, 15*(17), 1683–1695. doi:10.1016/j.jclepro.2007.02.003

Grant, G. B., Seager, T. P., Massard, G., & Nies, L. (2010). Information and communication technology for industrial symbiosis. *Journal of Industrial Ecology, 14*(5), 740–753. doi:10.1111/j.1530-9290.2010.00273.x

Growth Within: a circular economy vision for a competitive Europe. (2015, June 25). Retrieved from Ellen Macarthur Foundation. Retrieved from https://www.ellenmacarthurfoundation.org/publications/growth-within-a-circular-economy-vision-for-a-competitive-europe

Hauschild, M. (2015). Better – But is it Good Enough? On the Need to Consider Both Eco-efficiency and Eco-effectiveness to Gauge Industrial Sustainability. In *27The 22nd CIRP conference on Life Cycle Engineering* (pp. 1-7). Sydney: Procedia CIRP.

Hermann, C., Blume, S., Kurle, D., Schmidt, C., & Thiede, S. (2015). The Positive Impact Factory – Transition from Eco-efficiency to Eco–effectiveness Strategies in Manufacturing. In *The 22nd CIRP conference on Life Cycle Engineering* (pp. 19-27). Sydney: Procedia CIRP.

Hillary, R. (2004). Environmental management systems and the smaller enterprise. *Journal of Cleaner Production, 12*(6), 561–569. doi:10.1016/j.jclepro.2003.08.006

Holgado, M., Benedetti, M., Evans, S., Baptista, A. J., & Lourenço, E. J. (in press). Industrial symbiosis implementation by leveraging on process efficiency methodologies. Procedia CIRP, Life Cycle Engineering (LCE). *Conference, Copenhagen, Denmark.*

Holgado, M., Morgan, D., & Evans, S. (2016). Exploring the scope of industrial symbiosis: Implications for practitioners. *Smart Innovation. Systems and Technologies, 52*, 169–178.

Jekiel, C. M. (2011). *Lean Human Resources - Redesigning HR Processes for a Culture of Continuous Improvement*. New York: Productivity Press.

Kastner, C. A., Lau, R., & Kraft, M. (2015). Quantitative tools for cultivating symbiosis in industrial parks; a literature review. *Applied Energy, 155*, 599–612. doi:10.1016/j.apenergy.2015.05.037

Lean Enterprise Institute. (2014). *Lean Lexicon: a graphical glossary for Lean Thinkers (5th editio)*. Cambridge, USA: Lean Enterprise Institute.

Lehni, M., Schmidheiny, S., & Stigson, B. (2000). Eco-efficiency: creating more value with less impact. Geneva: World Business Council for Sustainable Development.

Liu, Q., Jiang, P., Zhao, J., Zhang, B., Bian, H., & Qian, G. (2011). Life cycle assessment of an industrial symbiosis based on energy recovery from dried sludge and used oil. *Journal of Cleaner Production, 19*(15), 1700–1708. doi:10.1016/j.jclepro.2011.06.013

Lourenco, E., Baptista, A., Pereira, J., & Dias-Ferreira, C. (2013). Multi-layer stream mapping as a combined approach for industrial processes eco-efficiency assessment. In Re-engineering Manufacturing for Sustainability (pp. 427-433).

Mann, D. (2015). *Creating a Lean Culture: Tools to Sustain Lean Conversions* (3rd ed.). Boca Raton, FL: CRC Press.

Michelsen, O., Fet, A. M., & Dahlsrud, A. (2006). Eco-efficiency in extended supply chains: A case study of furniture production. *Journal of Environmental Management, 79*(3), 290–297. doi:10.1016/j.jenvman.2005.07.007 PMID:16253417

Mirata, M. (2004). Experiences from early stages of a national industrial symbiosis programme in the UK: Determinants and coordination challenges. *Journal of Cleaner Production, 12*(8), 967–983. doi:10.1016/j.jclepro.2004.02.031

Murillo-Luna, J. L., Garcés-Ayerbe, C., & Rivera-Torres, P. (2007). What prevents firms from advancing in their environmental strategy? *International Advances in Economic Research, 13*(1), 35–46. doi:10.100711294-006-9059-6

Paquin, R. L., & Howard-Grenville, J. (2012). The evolution of facilitated industrial symbiosis. *Journal of Industrial Ecology, 16*(1), 83–93. doi:10.1111/j.1530-9290.2011.00437.x

Post, J. E., & Altma, B. W. (1994). Managing the environmental change process: Barriers and opportunities. *Journal of Organizational Change Management, 7*(4), 64–81. doi:10.1108/09534819410061388

Rother, M., & Shook, J. (1999). *Learning to See: Value Stream Mapping to Add Value and Eliminate MUDA*. Lean Enterprise Institute.

Sakr, D., Baas, L., El-Haggar, S., & Huisingh, D. (2011). Critical success and limiting factors for eco-industrial parks: Global trends and Egyptian context. *Journal of Cleaner Production, 19*(11), 1158–1169. doi:10.1016/j.jclepro.2011.01.001

Shi, H., Peng, S. Z., Liu, Y., & Zhong, P. (2008). Barriers to the implementation of cleaner production in Chinese SMEs: Government, industry and expert stakeholders' perspectives. *Journal of Cleaner Production*, *16*(7), 842–852. doi:10.1016/j.jclepro.2007.05.002

Shook, J., & Rother, M. (1999). Learning to See: Value Stream Mapping to Add Value and Eliminate MUDA.

United Nations. (1987). Report of the World Commission on Environment and Development.

U.S. Department of Commerce. (2014). *How does Commerce define Sustainable Manufacturing?* International Trade Administration.

Womack, J. P., & Jones, D. T. (1996). *Lean thinking: banish waste and create wealth in your corporation*. New York, NY: Simon & Schuster.

Womack, J. P., Jones, D. T., & Roos, D. (1990). *The Machine that Changed the World: The Story of Lean Production* (1st ed.). New York: Harper Perennial.

KEY TERMS AND DEFINITIONS

Business Process: A collection of related and structured activities that produce a specific service or product for a particular customer or customers.

Efficiency Assessment: Provides an objective analysis of the operating processes and technical capabilities of your current data.

Eco-Efficiency: Philosophy that aims to encourage entities achieve environmental improvements that yield simultaneous economic benefits.

Industrial Symbiosis: Fosters the sharing of resources (energy, water, residues and recycled materials etc.) between two different manufacturing processes.

Internet of Things: The extension of the network connectivity to physical devices, "smart devices", which enable these objects to collect and exchange data.

Lean Management: Management philosophy that aims to encourage companies to do more with less, focusing on reducing waste.

Process Industries: Industries concerned with processing bulk resources into other products.

Sustainable Manufacturing: The creation of manufactured products through processes that minimize negative environmental impacts.

Chapter 10
A Practical Approach to Manufacturing Execution Systems at Bosch AvP:
Scope, Structure, and Implementation

Maria João Lopes
Bosch Termotecnologia, Portugal & Universidade de Coimbra, Portugal

Duarte Almeida
Bosch Termotecnologia, Portugal

Francisco J. A. Cardoso
Universidade de Coimbra, Portugal

ABSTRACT

With Industry 4.0 related initiatives, a brand new array of opportunities has emerged for organizations to face the ordeals that come with managing ever-growing manufacturing needs. The exponential increase in the complexity of supply chain management has put a real strain on manufacturing operations. In order to succeed, organizations must turn to solutions such as manufacturing execution systems (MES) in order to stay competitive. In this research chapter, we discuss the impact of MES in organizations, whilst describing the process for going from a theoretical concept to a hands-on system which runs the shop floor operations.

DOI: 10.4018/978-1-5225-4936-9.ch010

ORGANIZATION BACKGROUND

Bosch Termotecnologia SA (referred to as AvP) is located in Aveiro, Portugal and is the leading competence center for water heating development in the Bosch Group. Part of its Thermotechnology division, it is also one of the most important manufacturers of gas water heaters, wall-hang boilers and heat pumps.

In Portugal, the Bosch Group owns three different plants which belong to different Business Units with a combined sales revenue of 1.1 Million Euros in 2016. The Bosch Group has a very firm belief on Research & Development (R&D). That is reflected in the investment in R&D for Bosch Portugal: 100 Million Euros in 2016 (Bosch, 2017).

Bosch is known as being a very strong advocate for Industry 4.0. In fact, the Bosch Group has become a leading example by showcasing the usefulness of Industry 4.0 technologies (GTAI, 2016). Actually, the Bosch Group has quite an interesting role: on one hand, it uses Industry 4.0 technology as a basis for managing its manufacturing needs, and, on the other hand, it also provides commercial solutions in this area through Bosch Rexroth and Bosch Software Innovations, just to name a couple of examples. Manufacturing Execution Systems (MES) constitute the backbone for most Industry 4.0 innovations, as we will see in the upcoming sections.

SETTING THE STAGE

Manufacturing today is very different from what it once was. Henry Ford once said that people could have their cars in any color of their choosing, as long as that color was black (Ford, H., & Crowther, S., 1922). – a quote which perfectly resumes the mindset of the 1920s. Then, Henry Ford was a pioneer of industrial management, having successfully implemented the assembly line concept, with a somewhat paradoxical aversion to change. After years of Model T production, he would eventually come around and release the Model A. However, his inertia would cost him the position of leading player of the industry: Ford would then be outrun by Chevrolet and Plymouth (Geldermann, 2017).

This small lesson from history only comes to show that there is only one thing that matters, when it comes to manufacturing: being competitive, i.e., staying ahead of competition – this was exactly what Henry Ford failed to do when it came to the Model A.

However, this is easier said than done, as managing a company is a tremendous challenge. For instance, one of the main problems which organizations struggle with is handling the complexity of their infrastructure. Henry Ford was well aware of this, since he knew that producing two car models would significantly increase the complexity of his organization. In fact, he was so concerned with this that he ended up owning the ore mines, railways and boats that supplied the Ford plant (Geldermann, 2017), all this aiming at keeping the supply chain under control without having to rely on external suppliers.

The ordeals that Henry Ford was facing in the first half of the 20th century to stay competitive are still valid today. Unfortunately, the magnitude of the challenge has increased exponentially: not only do we have to manufacture thousands of different references, but also having to keep track of ever more complex supply chains, in order to comply with the industry's norms and with the customer's demands, safety and satisfaction.

The need to handle this increase of complexity in the industrial scope was one of the main driving forces that led to the development and implementation of Industrial Information Systems such as MES: globalization and the general increase of competitiveness have made productivity (both monitoring and improvement) to became a major concern for industrial organizations. Regarding that point, MES have definitely something to add.

One of the gurus of industrial management was, undoubtedly, Frederick Taylor (1856-1915), this being the first to study manual work's productivity, and how it could be improved (Deci, & Ryan, 1991). Up until his studies, what we now regard as productivity as a whole, was thought of being a quality merely associated with the individual himself: whether he was either hard-working or lazy, or whether his physical shape was adequate or not.

Taylor's methods were, in fact, extremely simple: he would record the steps it would take to complete the task at hand, whose duration would be duly considered, and superfluous subtasks would be eventually removed. This approach was so effective that manual workers' productivity started to grow at a rate of 3% annually. This increase in productivity would have a significant impact in economic growth.

The revolution that Taylor's work sparkled is comparable to the revolution that Information Systems (IS) have brought about in industrial engineering (Short, 1990). Thus, IS can be defined as a set of interrelated components that collect, retrieve, process, store and distribute information in order to support decision making and control in an organization. Information Systems contain meaningful information about people, processes or products. Now, it is worth to remark the difference between Information and Data Systems, since the former deal with high-level data, whose semantics make it meaningful for humans, and the latter gather and validate low-level data representing both event collections and time series of process variables, thus before they are treated and made fit for human interpretation (Laudon & Laudon, 2010).

Thereby, everyone who works in an industrial setting is now facing a glorious opportunity to change the way in which information can be perceived, and the benefits one can extract from it. A brave new world of digitization is quickly unfolding before our eyes and it is our duty to use the tools which we are presented with, in order to achieve better results and work towards continuous improvement. In the present work, we start by understanding the functionalities of MES, their history and meaningfulness. Afterwards, we will be able to understand the pre-requisites in order to install a MES, and what needs to be done, engineering wise, to design the requirements for such a system. This research chapter will be wrapped-up with some considerations about the impact of a MES throughout an organization and some features which can be implemented later on.

AN INTRODUCTION TO MES

Historical Contextualization

Few innovations have revolutionized manufacturing the way Information Systems did. The idea of using an IT infrastructure to support manufacturing related operations is not a very recent one. In fact, as early as the 1940's, IS which performed Material Resource Planning (MRP) ran on mainframe computers (Inc., 2014), systematically calculating the amount of raw materials which are needed to manufacture a certain amount of finished products. Nowadays, MRP require not only knowing the amount of raw material already in stock but also the Bill of Materials (BOM) which encapsulate the various levels of the hierarchy of a product.

Thus, MRP constitutes one of the first stepping stones for digitizing operations. However, this sort of system formerly depended heavily on the manual inputs with which it was provided for. The same thing happens with a calculator: the user is solely responsible for the inputs there provided. This uncertainty was one of the reasons that led to the demise of this first generation of MRP systems. Alongside with this, since MRP systems belonged to the strict domain of manufacturing planning, they could not reach other aspects of a business which are just as important.

After these systems, other important IS started to rise. Computer Integrated Manufacturing (CIM) systems were one of them, an approach which became relatively popular in the late 1980s, though their underlying concept was significantly different from that of the MRP, the idea of CIM consisting in the integration of the various elements of the value chain in manufacturing operations. Engineers would use Computed Aided Design (CAD) tools for product development, a concept which is very much alive these days. The outputs of CAD would then be used to generate relevant product related information such as BOM, which would then be processed by the manufacturing planning department and purchasing (Salvendy, 2001). Besides CAD, the reader may also be familiar with the concept of Computer Aided Manufacturing (CAM) which also stemmed from the CIM methodology, acting as a counter part for CAD, but on the manufacturing side.

CIM presented an innovative holistic approach, but its concept was much too far ahead for the technology available at the time. Technology limitations and the unwillingness of managers to invest in such a system would lately undermine the ability for these systems to develop any further.

Following CIM, MRP II systems started gaining momentum. Despite the naming resemblance with MRP systems, MRP II systems had a very different approach where, instead of focusing merely on the MRP process, they had a more global scope, incorporating the activities of departments which were not directly related to manufacturing. These included finances and marketing, for instance (Salvendy, 2001).

In parallel with MRP II systems, the first Enterprise Resource Planning (ERP) systems started to make their way into the market. As the name implies, ERP systems focus on covering all core activities of an organization (SAP, 2016). Due to its uniquely broad range, ERP systems are usually split in a modular form, so that every module is ready to function in an independent manner. However, ERP systems are used for planning operations which use a timeframe of days, weeks and months. Hence, it soon became apparent that ERP systems were not fitted for the task of managing manufacturing operations.

This need would later be fulfilled by MES, which work with a timeframe of seconds, in order to actively control production and track results in order to assure traceability. The functionalities of MES are featured according to ANSI/ISA-95 – Enterprise-Control System Integration standard, which was firstly issued by the International Society of Automation (ISA), in 2000, as described in the next following section.

Functionalities (ANSI/ISA-95)

The ANSI/ISA-95 standard defines 12 fundamental functionalities for MES. Manufacturing Operations Management (MOM) is defined as the hierarchical level which is upper limited by the Business Planning and Logistics tier (where ERP systems can be found) and lower limited by the Automation tier, related to the shop floor machinery, as depicted in Figure 1. ERP systems are known to struggle with various data sources related to different suppliers, for instance. In this particular scenario, solutions like cloud-based MES's are known to unify these myriads of different supply chains (Helo, Soursa, Hao & Anussornnitisarn, 2014).

Figure 1. Levels of information systems in an industrial organization

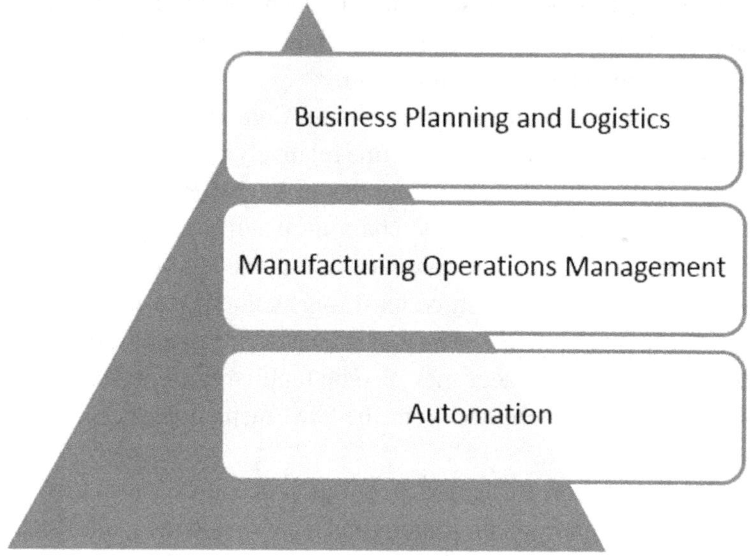

According to this standard, MES work at the MOM tier, supporting a very active interface, whose roles are highlighted below. One must bear in mind, however, that the classical MES structure struggles to meet changing demands and requirements. In this sense, a more dynamic architecture could be adopted according to recent research (Cupek, Ziebinski, Huczala, & Erdogan, 2016). While most of the MES's relied on the SOA (Service Oriented Architecture), RESTful (Representative State Transition) services are now more popular than SOA (Iarovyi, Mohammed, Lobov, Ferrer, & Lastra, 2016).

1. Resource Allocation and Control

MES are useful for the management of the resources needed for production, i.e., equipment, materials and people. They keep track of their statuses in real time, and ensure that they are ready to start producing.

2. Dispatching Production

This stage is concerned with how manufacturing related work is distributed amongst people and equipment. It can be done in form of orders, which are usually represented through Kanban cards, the work order being changed in real-time, depending on the available resources. The MES can use rework processes and buffers, which can assume the form of supermarkets, in order to even the production flows, and keep production at a desired level.

3. Data Collection and Acquisition

This task is intrinsically connected with the Automation level. Data gathered from the process is sent to the MES throughout the machines and networks. The MES is expected to save a history of the acquired data and to provide machine and process status in real time.

4. Quality Operations Management

Due to its data collection and acquisition requirements, a MES is expected to use real-time gathered information to generate several types of quality reports, such as Statistical Process Control (SPC) reports or Pareto analysis. In some cases, it can even analyse the outputs of such reports and recommend possible solutions to mitigate deviations in the process. It includes tracking, test monitoring and interaction with information from the Metrology department, for instance.

5. Process Management

This task is related to process deviations: in case a deviation is detected, the system either corrects it automatically or triggers an alarm, which will notify the personnel in charge that something is not within the planned acceptable deviation range. It can thus act either as a decision maker or it can help in the decision-making process of the machine operator.

6. Production Tracking

This function comprises two main tasks: (i) providing status for the overall process, thus encompassing number of parts produced in real time, real cycle times and resource allocation, and (ii) bringing about traceability of a product throughout the supply chain.

7. Performance Analysis

Performance analysis refers to using real-time generated information in the form of reports or Key Performance Indicators (KPI) to assess production performance. This function is intrinsically related to the "Production tracking" function as it analyses its outputs. Current statistics should be confronted with the goal for a certain time frame along with some past results, in order to allow a comparison to be made.

8. Operations and Detailed Scheduling

This function analyses the various tasks independently, and then arranges them in a sequence which considers not only the process requirements, but also some other restrictions, such as material, personnel and machine availability. This analysis can reveal parallel tasks and provide alternatives, whenever a constraint is disrupting the normal production flow.

9. Document Control

Document control relates to record keeping and storage of up-to-date information. The MES should be able to automatically fill-in forms that keep track of production, such as amounts produced every hour or the number of parts that failed determined quality tests. The MES should also constitute an information repository of valuable process related documentation, i.e., visual instructions, environmental standards, product recipes, best practices, etc.

10. Labour Management

At its core, this function is related to managing personnel's hours, tasks and training. Labour shifts and activity allocations are meant to be controlled here, as this task interacts closely with the module of "Resource allocation and control", since it contains the effective and planned working time for each worker.

11. Maintenance Operations Management

This function keeps track of machine availability, by sharing some of the responsibility for the equipment's maintenance with Level 4 systems. Scheduling for periodic, preventive and predictive maintenance can be carried out by the MES, thus issuing maintenance requests and providing some information which will be useful for machine diagnosis. The MES is also expected to keep a detailed log of all interventions previously done.

12. Movement, Storage and Tracking of Materials

The final function of a MES also overlaps with the top tier of Business Planning and Logistics. Some of the logistic functions enclosed here, such as receiving material, processing it and shipping the final product, can be performed by the MES.

As we were able to understand, MES provide an array of very different features due to its interfacial nature. This character is also what makes MES the standard basis for managing manufacturing operations and a strong argument for addressing it as an Industry 4.0 platform. It is, however, important to emphasize that not all MES offer all the functionalities listed in the ISA-95 standard.

Leverage and Competitive Advantages

During the 2011 Hannover Messe (Deutschland, 2014), the German government launched its *Industrie 4.0* initiative. This was one of the 10 initiatives which were featured on Germany's High-Tech Strategy 2020 Action Plan, a strategic approach to increase Germany's competitiveness over the upcoming years (GTAI, 2016). Industry 4.0 is commonly known as the 4th Industrial Revolution after Mechanization, Electrification and Automation. Industry 4.0 revolves around the smart factory concept, where production stations can rearrange themselves to produce different product references, Automated Guided Vehicles (AGV) automatically supply the various production lines while following optimized paths, and machines are constantly monitored through a wide range of sensors, in order to check on its need for a maintenance intervention. All the shop floor and its machines are trendily perceived as industrial cyber-physical systems able to communicate in a general network feeding us with all the information necessary to achieve complete shop floor transparency (Colombo, Karnouskos, Kaynak, Shi, & Yin, 2017).

As we know, knowledge is power, and information is definitely the business to address when it comes to Industry 4.0. Hence, MES are extremely well positioned to act as the backbone of such an ambitious revolution, given a number of invaluable capabilities: (i) controlling manufacturing in real-time, (ii) storing production data, (iii) performing data analysis, and (iv) managing production related resources. Moreover, there is now the possibility of performing some data crossover between plants, regions and even society itself, the so called social manufacturing (Tao, Cheng, Zhang & Nee, 2017). This enables information to surpass its initial plant scope.

However, it is important to understand that the hype surrounding Industry 4.0 is not merely related with the technological challenges and opportunities that arise from it. In fact, one of the main arguments for adopting Industry 4.0 related technologies is cost reduction. The PwC's 2016 Global Industry 4.0 Survey indicated that, on average, companies can expect an average cost reduction of 3.6% until 2020 (Pwc, 2016). This cost reduction stems not only from automating tasks that are traditionally manual, but, also, from increased operational efficiency, which enables plants to be more productive with the same resources. It is this ability to deal with extremely complex supply chains that proves to be the ultimate advantage: meeting the customer demands is paramount.

This competitive advantage is of such an order of magnitude that manufacturing giants such as those located in China, which have their biggest leverage on the low wages practiced there, may eventually feel threatened, as the cost of manufacturing starts to become similar in high wage countries, such as Germany (GTAI, 2016). This is truly the dawn of a new era: an era of increased productivity through technological innovation.

Despite this clear argument, the scope of Industry 4.0 reaches far beyond manufacturing operations, with a significant commercial impact arising from the support of a closer relationship between brands and their customers. For instance, if one is to buy a new car, the customer does not have to choose between the usual three equipment packages manufacturers used to offer; simply through a Web Browser, prospective buyers can customize their cars, thus opting between fabric and leather seats, choosing the desired infotainment system or the motorization. This ability to customise products empowers users and it is very well appreciated by its customer base.

However, this comes at a price: enabling personalised purchasing options takes its toll on manufacturing operations, and this has to be handled carefully dealing with thousands of different references, which is made feasible through a MES.

Besides consumer-centric experiences, and the growth in the market share that arises from there, Industry 4.0 provides a wide set of tools for continuous improvement in the form of methods, such as condition monitoring and data mining. Not only that, but a MES can identify the weaknesses in a production process, such as bottlenecks and efficiency losses, and well-defined actions can then be applied and their impact can be measured by the system itself, following a closed loop Plan-Do-Check-Act (PDCA) methodology.

MES IMPLEMENTATION: THE NECESSARY STEPS

Understanding MES

The first step to implement a MES should always be the full understanding of what "Manufacturing Execution Systems" really are. Here, the focus is on providing a clear overview of what a MES really can be, as based on the main components, rather than on providing a precise definition. Furthermore, a good description of its basic functions was already provided in the previous section.

Thus, the goal here is to produce an overview which can be used by different audiences: IT or manufacturing professionals, which are familiar with MES's and that have in this chapter the possibility of examining a use case of the implementation of such an Information System, while being used as a reference guide to help planning and structuring MES related tasks. The aim of this chapter is also to be understood and used by industrial managers, even with no technical background whatsoever. Managers will be able to gather a very wide array of knowledge on how MES's can bring extensive benefits for their companies and what resources are necessary to implement such a system, helping therefore in eventual decisions of acquiring, developing or extending these technologies.

And at last this chapter is also aimed at graduate students with no previous contact with industrial settings, thereby struggling to understand what 21st century industry is really about.

For someone which is only now starting to grasp the MES concept, one might say that a MES can be mainly thought of as an Information System. It encompasses the following major components:

1. A place to store data, typically as data servers, which one might picture as the hardware basis;
2. A way to organize this data, so as to upgrade from the data to the information level, with recourse to both data bases and specific application software running on the server machines;
3. Means to access this information, so that full advantage can be taken from it; this is typically done through user interfaces, also known as front ends, which consist of software components specifically designed for the end users, typically running on a browser, such as the Internet Explorer or the Mozilla Firefox. They can also exist in the form of Windows programs that can be accessed from a mere laptop.
4. At last, with a background role, the server will run other software packages, thus receiving and integrating data incoming from manufacturing monitoring and, in general, from both other computers and users.

It is quite important to realise that what really distinguishes a MES from other kinds of Information Systems lies in this ability to communicate with other incredibly specific systems: production machines. Therefore, it might be clear now that the above mentioned fourth component is truly important, as it acts as a gateway towards the production machines which are the main information source in a MES. Next, an allegory is provided in order to better illustrating this basic dialog mechanism between a MES and production machines on the shop-floor:

Thus, the MES might ask to a machine: "How are you?"

Then the machine would reply: "Pretty well, thank you. Producing as I should, at a rate of 20 parts per minute, no errors or problems during this shift, everything is perfect. Hope it stays that way."

However, the machine could have also replied: "Tough shift this one. I stopped one hour at 09:15 AM because of a broken pressure sensor. After its replacement and calibration, we kept on producing but at a slower pace than we should (average of 14 instead of 20 parts per minute). This happened because, often, we were receiving raw materials with quality problems, specifically the aluminium sheet, leading to leakage problems in the assembled parts".

After a couple of moments, the machine now triggers the interaction with the server: "Hey MES! I have now finished the 500 parts of the reference 111111111 (serial numbers 987000001 to 987000500) and I will start now with the production of 300 parts of the reference 222222222. Can I produce them?"

Where the MES replies: "Well, indeed you may start producing, since there are no blockage orders for that reference, from the Quality Department. Please assign these 300 parts with the serial numbers from 87500001 to 87500300. They need to pass through stations 1, 2, 3, 5, 6 and 7. The process parameters for production are as follows..."

The content of the dialog above contains examples of the kind of information that is exchanged between a MES and production machines. Obviously, this will not happen using words and sentences of plain English language; instead, it happens digitally, by using mere binary code. They have, however, a language of their own: not English, but their very specific communication protocols, which follow a set of rules and are conveyed through physical means such as cables or wireless radiofrequency.

Nonetheless, we should have in mind that we are now coming to an age where machine learning is not a mere product of our imagination. In fact, many companies use machine learning algorithms to solve problems such as filtering SPAM from e-mail inboxes (Blanzieri & Bryl, 2008). Thus, it should not come as a surprise for the reader that systems which are able to learn from their environment are now a reality. In this sense, the concept of digital twin seems to emerge in a rather natural way. A digital twin is a copy of a complex physical system which can be run digitally, thus being able to simulate beforehand certain usage conditions. Not only that but it can also learn from the physical system that it is emulating. NASA has demonstrated interest in these systems for testing the amount of strain that the components of an aircraft are subject to without flying its physical counterpart (Tuegel, Ingraffea, Eason & Spottswood, 2011). Like NASA's aircrafts, digital twins can also be used in the shop floor as they can learn from their surroundings: not only using sensor data but also from human input, being voice an example. All in all, we may not be very far from machines learning from the information that we give them through voice commands. In this sense, a more intelligible language such as English could be used to interact with such systems.

At this point, two main dimensions might be immediately identified regarding a MES implementation: (i) the IT Infrastructure (servers, databases and the MES software), and (ii) production machines capable of this sort of communication and programmed to communicate as pre-defined here. Therefore, we will first analyse a typical IT Infrastructure commissioning. Afterwards, we will go through the implementation of the needed functionalities on the machine side.

IT Infrastructure

In order to run a MES, an operational IT Infrastructure must be in place, basically consisting of a cabled network interlinking the different production machines; for the time being, wireless networks are strongly discouraged for that purpose due to safety concerns. In fact, at present it is extremely rare for a production machine to support wireless communications.

Generally, the production network interfaces with other network which contains the servers it needs to have access to. The interfacing is managed by a firewall, a program which manages communication between two machines in a network. Firewall management is quite critical and must be handled with the utmost care.

At the server level, there are four basic needs: storage, interfacing with production machines, running MES processes and providing a user interface.

Storage needs are fulfilled by a database which can rely on different technologies such as SQL or Oracle. This database is meant to store all the data regarding product, shop floor process and MES system itself. Database servers are usually implemented as physical servers, in the sense that they need specific hardware and require a high-performance CPU and RAM, as well as an enormous amount of space available, since they require a server room with special environmental conditions and room for the required level of redundancy, which encompasses both communication networks and power supply.

A database server can be thought of as the most critical aspect of a MES, since all production data is stored there, and losing it would be a tremendous loss. Due to this critical aspect, it is recommended to exchange the server hardware every couple of years in order to prevent hardware failures, so as to prevent data losses. In most industries, production might need to be recalled if this traceability data is lost.

Besides the physical installation, a database server needs to be fully configured before its use. This includes Operating System installation, antivirus installation and specific software for running database applications. The procedures of ordering, installing and configuring the database server can be quite complex and therefore may take a significant deal of time. Depending on organizational and bureaucratic aspects, as well on technical details, such as components availability, these procedures may take a few weeks at best, or 6 to 8 months at worst. When it is ready, the database is now available to hold the required schemas.

Machine-system interaction and running MES processes can be performed by the same server, which usually goes by the name of Application Server. Despite its many tasks, an Application Server is perfectly capable of running in a virtual environment, since it does not require so much computational power as a database server, for instance. Getting a virtual server ready is much faster than a physical one, since both the hardware and software infrastructure are already in place.

Thus, the Application Server runs the core processes needed for a MES server and, additionally, has also the ability to communicate with production machines, thereby answering to their requests. The Application Server should be able to communicate with devices over the production network. This is yet another requirement to be fulfilled in order to complete the IT infrastructure.

Besides this, web hosting is to be provided as well, which is carried out through a Web Server responsible for querying the database for data which will be displayed in the form of a report or other visualization. The Web Server can also run as a virtual machine, being independent from the Application Server.

From the above, one can understand that getting the necessary IT infrastructure in place is a complex process, which requires a lot of work and a considerable deal of time to be completed. As an end note, it is crucial to emphasize that there is no standard server solution and, therefore, it has got to be adapted to the specific needs of each plant, and the current server infrastructure must also be checked in order to find the best possible solution.

Production Machine-MES Integration

In parallel with the IT Infrastructure implementation, integration of production machines must be taken into consideration, thus encompassing the specification of both functionalities and communication protocols. This must be clearly defined in advance, so that the two entities can work according to the previously defined, in an efficient manner, thus involving both the machine programmer and the MES applications engineer.

Figure 2. Interaction scheme between the servers and user interfaces for a MES

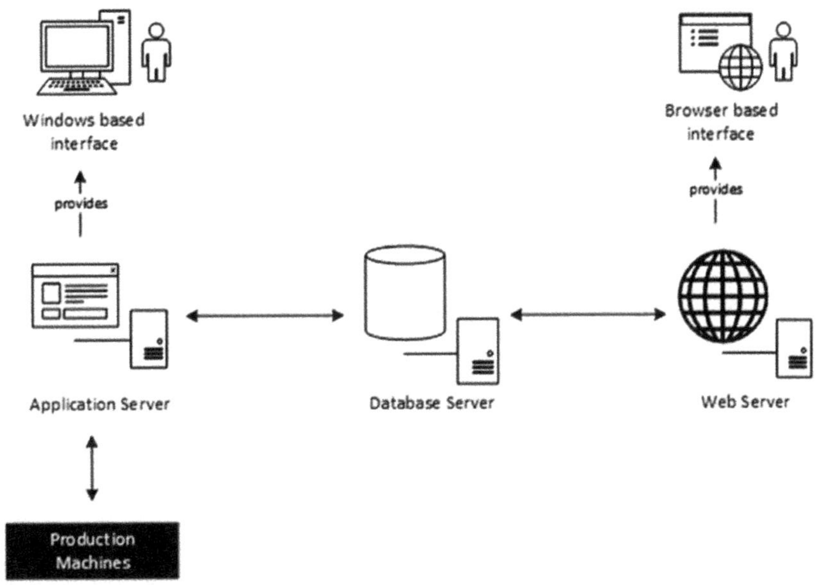

The machine programmer implements the defined functionalities on the machine side, which typically implies programming a Programmable Logic Controller (PLC) so that it becomes capable of sending 'telegrams' to a remote network entity (in this case, the MES), and, in the opposite direction, receive, process and understand the incoming 'telegrams' from the MES.

The MES applications engineer has the mission of implementing the same communication primitives and functionalities, but on the MES side, therefore working on the MES specific software running on the system servers.

Different kinds of MES support different kinds of communications. A broad range of different functionalities can be found in the market, which may be classified in two broad groups according to the respective scope:

1. Series Production Communication
2. Machine State Information

The engineering activities at both the Machine and MES sides consists on the implementation of the functionalities referred to above.

The engineering work may typically take a few days or a few weeks, depending on the technologies being used, as well as the complexity of communications, the experience of the engineers and the number of machines to be connected. In this use case, two PLC engineers were assigned and used 3 weeks to understand the machine source codes, prepare all the needed adaptations and two weeks to deploy this features in all the 15 stations of the connecting assembly line. A process engineer is also required to support in the specifics of the machines, guaranteed the correctness of their function and to speed up the test and debug activities. After this commissioning a follow up stage comes in place, to check all the functionalities for all the stations and to identify and implement possible improvements. This last stage took 3 weeks during regular series production period.

Regarding the investment, typically the MES communication integration is quoted together with all the remaining aspects of the MES itself (the IT Infrastructure and the software running on the servers).

However, the machine side adaptation might not be included in the MES commercial package, since a MES is typically completely independent from the machines and the kind of automation they carry out. Due to this constraint, it must be taken into consideration that engineering work is necessary to implement such communication capabilities at the production machines, which requires specific knowledge in the field of automation engineering. Hence, the investment may vary depending on the market or country where the engineering services are hired from — these may be ascribed to either internal human resources or outsourced. Apart from the investment for the MES system itself, typically this machine engineering work might have a cost ranging from 15.000€ to 50.000€, depending on the market, the complexity and the number of machines to be integrated.

Series Production Communication

The MES-machine interaction is mainly about keeping track and control of production. Therefore, it is necessary to connect all the production equipment, or stations, to the MES so that, during production, all the different stations for each one of the assembly lines are exchanging information with the MES on a regular basis.

This information may be concerned with both the regular production follow-up and the machine status (machine specific "health" parameters, such as the operation mode, error state and condition monitoring measurements, such as engine temperature, rotations, oil temperature, viscosity, etc).

During a regular series production, the interaction between each of the production machines (Stations) may be described according to the diagram below (Figure 3), where:

Figure 3. Interaction between a production machine and the MES server

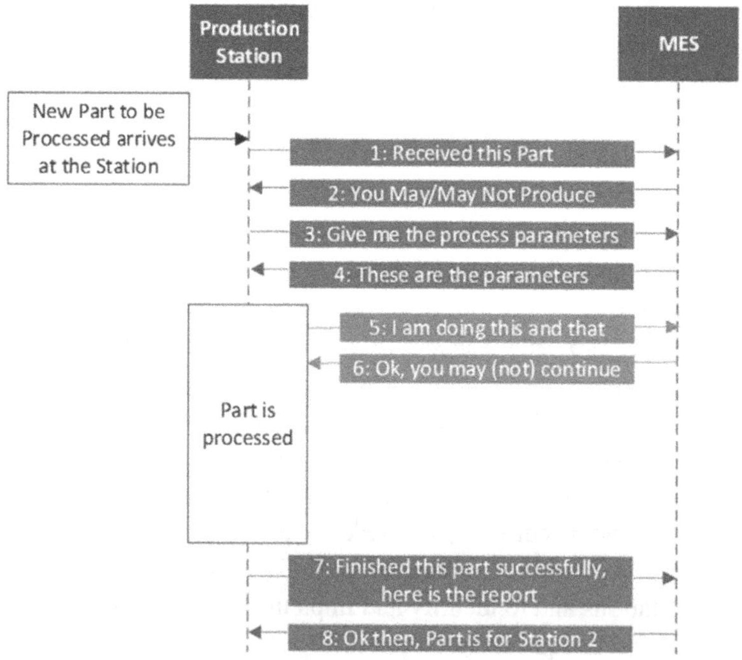

1. Whenever a new part arrives to a station in order to be processed (assembled, manufactured or tested), a notification is sent to the MES, so that the system knows precisely at what time that precise part (identified with a complex number containing, at least, the type reference code, batch number and serial number) arrived to that specific station.
2. With this notification, the server may trigger a set of system verifications, thus checking whether or not that part may be produced at this station. This set of verifications typically includes information such as:
   - The production of this reference type is not blocked by either the Quality or the Production departments;
   - This particular part passed successfully through all the process steps beforehand;
   - This part is supposed to pass through this station;
   - etc.

Only after all these verifications are concluded, is a 'telegram' sent to this station, thus allowing production to proceed. If any of these verifications fail, then a 'telegram' is prompted at the machine with the relevant details, to the benefit of the machine operator and/or the line manager.

3. Typically, different kinds of products may be processed at the same line and on the same stations. Naturally, different products will cause different process parameters to process each part, often referred to as the product recipes for each station. It is possible to manage over them locally, at each individual station, by having a set of parameters for each product type. Moreover, the MES allows the management of all these parameters for all the different stations and products to be centrally handled, by taking advantage of stored recipe libraries.
4. Therefore, for each individual part, information can be exchanged so that all the recipe values/parameters are sent from the MES to the specific machine, so as to determine its operation.
5. During the process of assembling, manufacturing or testing, communication can also take place.

In an assembling station, for example, if a subcomponent is being assembled, the machine may inform the MES about what subcomponent (reference and serial number) is being assembled, so that the MES can reply (6.) with the confirmation that this is the correct component (checking for example at the ERP SAP at the updated Bill Of Material of the main component), or even if this subcomponent is blocked for any reason, for example, as determined at the Quality Department.

During complex processes with several different steps, the machine can inform about the ongoing process steps, sending reports of the steps already completed so far, or even asking about details regarding the future steps.

6. Based on the information sent at the previous step, the MES can evaluate the correctness of the results. If, for example, there is a problem in a specific step of the process, then the MES can detect it, and trigger the immediate cancelation of the current process running at the machine. This may allow some time and materials savings as, otherwise, it could have been necessary to wait until the end of the whole process to conclude about the non-correctness of a particular step and, therefore, the whole process.

7. Once the processing of the part is finished, the MES is informed with a precise time-stamp. Therefore, now it is known exactly the time it took to process this part at this station, allowing it then to compare it with what was previously and theoretically defined / planned.
8. The MES may acknowledge this process conclusion, after verification. The MES will also inform the station about the next station for the finished product, so that this piece of information may be automatically presented to the respective machine operator. This is a way of avoiding a source of potential error, which results from taking the part to a wrong station.

This communication mechanism exists permanently with all the stations in a plant. For someone interested or in need of these data, all the production details can be checked in real time, anywhere in the world, just by opening a browser page.

Machine State Information

Besides production related communication, information on working condition of machines and other equipment may also be exchanged in this Machine–MES interaction.

Thus, in case of a downtime situation in a production line, assistance from Maintenance is required and a technician logs onto the equipment and operates it in manual mode (this will trigger 'telegrams' from the station to the MES, with the log on information which includes both user ID and the exact time, as well as information on the current operation mode). Once the problem is solved, the technician logs out from the machine, which is then set back to automatic mode, and production starts running again. For each of these steps, the MES receives a notification.

In the scope of this example, some other useful information may be deduced: (i) the time elapsed between the instant the downtime situation was firstly reported and the login time of the Maintenance technician represents one of the main KPI related with a maintenance request, and it is tracked in most of the main industrial organizations, and (ii) the time it took to diagnose and solve the problem (time between the login on the machine and the switch of the machine's operation mode to automatic).

Another typical interaction example related to machine state is aligned with one of the main industrial trends: condition monitoring and predictive maintenance. Some critical machines have a major impact in case of stoppages, and, therefore, the maintenance of such machines and respective parts represents a big concern for the whole organization, as stoppages of either category (non-planned and planned) for equipment of this kind will always represent an enormous amount of money.

This led to the development of the concept of machine condition monitoring, where sensors are installed on the equipment and are constantly measuring and evaluating their conditions. A lot of operational variables may be measured, such as temperature (of parts, fluids, engines, cooling systems, etc.), pressure, vibrations, noise, visual signs (using artificial vision), etc. Provided that regular working conditions are well known and duly programmed, in what concerns to both current monitoring results and trend analysis, then it is possible to react very quickly and in proportion to the respective grade of risk in the industrial line.

In cases where this knowledge is not available, Industry 4.0 has a proposal to solve this problem: by acquiring relevant data from sensors and keeping a record of it (Historian), then cross-correlating it with failure data, and using data analytics and some artificial intelligence/machine learning algorithms over it, it might be possible to assess this cause-effect correlations, and even to achieve some predictive models to foretell machine's problems, thus preventing failures or stoppage.

With such predictive models, the paradigm of the ubiquitous preventive maintenance (time or utilization based) may dramatically change: a fixed and pre-defined lifespan of equipment would be abandoned in favour of an asset management policy, based on the real time machine data and respective condition diagnostic.

Machine-MES Communication Protocols

As referred to above, during the normal functioning of a MES during production, each machine is continuously exchanging digital data with the MES on the server side. Therefore, this digital communication has a major role in the system, being very important to clearly define how this communication will take place when planning the commissioning of a MES.

Network communications may be supported by different technologies, where several layers may be defined according to the ISO-OSI model (Open System Interconnection) developed by ISO (International Organization for Standardization) (ISO/IEC, 1994).

Regarding this complexity and the wide range of digital communication solutions, at first it is really important and recommended to define which technology is to be used to support the communication. Some of the most used are simple TCP/IP messages or XML 'telegrams', where it is necessary to define the data structures that are to be exchanged. This is quite a novelty in industrial systems due to the fact that this kind of communication is widely used in the scope of the Internet. For example, XML was initially created to establish a common ground for publishing on electronic formats typically related with the World Wide Web. Currently, its use is very widespread, and it has been gradually helping to shape the way we look at data (W3C, 2016), its structured syntax being a major advantage as it greatly increases readability as will be seen below.

On the other hand, standard and widely spread industrial communication protocols may continue to be used, such as OPC (Open Platform Communication), developed by the OPC Foundation (OPC-UA, 2017). Nowadays, the main trend consists in using the latest development of the foundation: the OPC Unified Architecture (OPC-UA), and most of MES packages available on the market are now OPC-UA compliant. Machine manufacturers are as well increasingly adding OPC-UA communication capabilities to their machinery portfolio.

Once the underlying communication technology is defined, it is of major importance to go into the very details of communication, thus defining the contents of every message, regarding its structure and the meaning of each data set. The type of the messages must also be defined, and, for each message type, all of its message fields. The level of detail must allow that, for each and every sensor measurement, a clear specification of how the value is to be presented (its data type format: as a string, integer or floating point and how many bits are to be considered in the data field), the scientific units (for example, for a pressure sensor, it must be defined whether the value is in Bars, mBars, Pascal or Pounds per Square Inch), and so on.

This is a massive job, but definitely one worth doing in advance, in a concerted manner, so that the commissioning of the two-sided communications occurs in a much smoother and more efficient way.

AFTERMATH

Impact in the Organization

At the starting point of this project, it was clear that this MES implementation would have a very significant impact in various aspects of how to run our industrial organization. Most importantly, being able to show professionals throughout the different departments how the MES could impact their daily working life: if there was one thing that was made clear in the course of this project is that, in order to teach people how to use the MES, it is mandatory to understand their needs and prove the system's usefulness. It is usually a bad omen to simply chose and use a system just because it is a novelty: users must understand its benefits very clearly, in order to take the most from it.

Having a MES in the organization had a very profound impact in the way things are run. Very specific situations for each department are to be address below, which mirror the benefits of a MES. Thus, starting with the most obvious one — the Production department — each organizational unit will have a maximum of two examples each.

The Production department is responsible for most of the aspects related to manufacturing. All KPI calculations were performed manually, with the available data. This has two disadvantages: (i) one must wait for the calculation to be done in order to get the KPI, with the added difficulty of obtaining KPI data for the previous shift, and (ii) the KPI can be miscalculated, since it is done manually or there may be something wrong with the inputted information.

Besides KPI, the Production department is also responsible for keeping up with production plans and inputting information in different places. With a MES, all services become more integrated and some of the information is even filled in automatically. For instance, part count used to be done manually or by other counting system but, with a MES, the machine performs the calculation and the part count number is used by other MES processes to calculate the Overall Equipment Efficiency (OEE) for that line.

As for the Maintenance department, a MES can act as an order management system, where Maintenance technicians can consult the available spare parts to perform a certain repair. This avoids unnecessary dislocations to the storage area only to receive this input. Moreover, with condition monitoring, the status of a machine can be assessed and its scheduled maintenance intervention can be adjusted to the machine's real needs, bearing in mind its utilization rate.

Also, the Quality department will no longer have to spend very long hours crafting Quality reports as the MES is able to output meaningful ones in a matter of seconds. This allows Quality personnel to focus on the causes and finding ways to mitigate them. Moreover, these corrective measures can be traced through time in order to check their impact. MES are also able to perform Statistical Process Control which is an important tool in Quality Control, as data are directly sent from machines to the server, with benefits in terms of data integrity as paperwork is left behind.

The Logistics department could also benefit from more realistic data, when it comes to the cycle time of each line, since one of their activities is to create the production schedule: if they know that 95 parts can be produced per hour rather than the 90 which were planned, then there is an output gain.

The Process and Product Development departments can have instant access to meaningful data in plot form, which allows them to better understand the relationship between the product and the process itself. Both areas are known for being iterative, so having all of this data to support their work is definitely important.

The Industrialization department can closely monitor the start of production and the ramp-up stage of a new line by following important KPI on a daily basis. Moreover, current knowledge on machine condition status can tell the stations which are the most problematic, and which need to be intervened first.

The Infrastructures department will be able to track energy consumption from the various machines, and warnings and alarms concerning deviations will be issued, and their causes investigated.

Even at the Top Management level, plant managers can get an overview of various plants of the same division, which are scattered all over the world, and compare their efficiency and outputs.

As we were able to see, a MES is the sort of system which creates a plant-wide impact which must be carefully introduced, with proper training which must be regarded as continuous and not limited to a couple of sessions.

CONCLUSION

In the last few decades, MES have grown in terms of importance and propagation in Industry. They represented a way to achieve competitive advantages by bringing transparency to the production, and therefore efficiency improvements. Hence, following the *Industrie 4.0* initiative by the German government, in 2011, the MES dissemination rhythm suffered a significant increase, as MES are considered the main tool towards the digitalization and interconnectivity. New concepts are emerging such as the digital twin of an assembly line or even an entire value stream, but MES are still considered as a major framework to encompass some recent industrial trendy fields such as Big Data, Data Mining or Business Intelligence. Predictive Maintenance, planning systems based on market prediction, or strategic decision support systems based on trend prediction on global markets, are some of the examples where all these new technologies might be applied on. All these facts help to explain the rush to this kind of systems felt in the last few years.

In the present work, it was explained what a MES is about, the functionalities and the technological components. A case study and implementation were presented, along with what is required for a MES implementation, since the introduction of such a system in one organization has a massive impact, for it will influence significantly the way people work, throughout the different departments of the organization. People will need to be trained to use and maintain the system. The way industrialization takes place will also suffer a significant change, with major advantages only in case the system is well structured and documented, and the connection of new equipment can be smoothly and efficiently done. Nevertheless, it may also represent some loss of flexibility, given that the increase of complexity may bring about longer industrialization processes.

The production itself relying on a MES, one must be aware that a technical problem in the system may represent production stoppages, or, even at the worst scenario, may represent products recall in case traceability data are lost.

Furthermore, a recent trending topic regarding MES and digitalisation of industry is data security. All this technological innovation may represent a threat to industrial organizations, since they may represent a back door to a company's sensitive data: methods and technologies addressing both data security and protection lay in the broad scope of the IT infrastructure.

Regarding the MES implementation, it was presented as a complex procedure, involving a wide range of technologies, processes and implications. The time and investment necessary to implement it may significantly vary in duration from 3 to 12 months, and in cost from 50.000€ to 500.000€, depending on several factors, most of them referred to along the text.

REFERENCES

Bosch (2017). *Bosch in Portugal*. Retrieved June 25, 2017, from http://www.bosch.pt/en/pt/our_company_10/our-company-lp.html

GTAI. (2016). *Industrie 4.0 – Smart Manufacturing for the Future*. Retrieved June 15, 2017 from https://www.gtai.de/GTAI/Content/EN/Invest/_SharedDocs/Downloads/GTAI/Brochures/Industries/industrie4.0-smart-manufacturing-for-the-future-en.pdf

Ford, H., & Crowther, S. (1922). My Life and Work. Garden City, NY: Doubleday, Page & Co.

Geldermann, C. (2017). Henry Ford. Britannica. Retrieved May 10, 2017, from https://www.britannica.com/biography/Henry-Ford

Drucker, P. F. (1991). Knowledge-Worker Productivity: The Biggest Challenge. *California Review Management*, *41*(2), 79–80.

Short, T. (1990). The New Industrial Engineering: Information Technology and Business Process Redesign. *MIT Sloan Management Review*.

Laudon, K. C., & Laudon, J. P. (2010). *Management Information Systems: Managing the Digital Firm, Global Edition*. Pearson Higher Education.

Inc. (2014). *Material Requirements Planning (MRP)*. Retrieved May 10, 2017, from https://www.inc.com/encyclopedia/material-requirements-planning-mrp.html

Salvendy, G. (2001). *Handbook of Industrial Engineering: Technology and operations management*. New York: Wiley. doi:10.1002/9780470172339

Cupek, R., Ziebinski, A., Huczala, L., & Erdogan, H. (2016). Agent-based manufacturing execution systems for short-series production scheduling. *Computers in Industry*, *82*(October), 245–258. doi:10.1016/j.compind.2016.07.009

Iarovyi, S., Mohammed, W., Lobov, A., Ferrer, B., & Lastra, J. (2016). Cyber-physical systems for open-knowledge-driven manufacturing execution systems. *Proceedings of the IEEE*, *104*(5). 10.1109/JPROC.2015.2509498

SAP. (2016). *What is ERP?* Retrieved May 15, 2017, from https://www.sap.com/products/what-is-erp.html

Deutschland. (2014). *Industry 4.0 at Hannover Messe*. Retrieved May 18, 2017, from https://www.deutschland.de/en/topic/business/globalization-world-trade/industry-40-at-hannover-messe

Helo, P., Soursa, M., Hao, Y., & Anussornnitisarn, P. (2014, May). Toward a cloud-based manufacturing execution system for distributed manufacturing. *Computers in Industry*, *65*(4), 646–656. doi:10.1016/j.compind.2014.01.015

Colombo, A., Karnouskos, S., Kaynak, O., Shi, Y., & Yin, S. (2017). Industrial cyberphysical systems: A backbone of the fourth industrial revolution. *Industrial Electronics Magazine IEEE*, *11*(1), 6–16. doi:10.1109/MIE.2017.2648857

Tao, F., Cheng, Y., Zhang, L., & Nee, A. (2017, June). Advanced manufacturing systems: Socialization characteristics and trends. *Journal of Intelligent Manufacturing*, *28*(5), 1079–1094. doi:10.100710845-015-1042-8

PWC. (2016). *Global Industry 4.0 Survey: Building the digital enterprise*. Retrieved May 18, 2017, from www.pwc.com/gx/en/industries/industries-4.0/landing-page/industry-4.0-building-your-digital-enterprise-april-2016.pdf

ISO/IEC. (1994). *Standard 7498-1:1994*. Retrieved June 5, 2017, from http://standards.iso.org/ittf/PubliclyAvailableStandards/s020269_ISO_IEC_7498-1_1994(E).zip

Blanzieri, E., & Bryl, A. (2008). A survey of learning-based techniques of email spam filtering. *Artificial Intelligence Review*, *29*(1), 63–92.

Blanzieri, E., & Bryl, A. (2008, March). of email spam filtering. *Artificial Intelligence Review*, *29*(1), 63–92. doi:10.100710462-009-9109-6

Tuegel, E., Ingraffea, A., Eason, T., & Spottswood, S. (2011). Reengineering Aircraft Structural Life Prediction Using a Digital Twin. International Journal of Aerospace Engineering.

W3C. (2016). *Extensible Markup Language (XML)*. Retrieved June 11, 2017, from https://www.w3.org/XML/

OPC-UA. (2017). *Unified Architecture*. Retrieved June 13, 2017, from https://opcfoundation.org/about/opc-technologies/opc-ua/

KEY TERMS AND DEFINITIONS

Enterprise Resource Planning: Normally associated with the namesake system, it provides the backbone for running most activities which are not directly related to manufacturing such as finances or human resources.

Industry 4.0: A strategic initiative started by the German government in order to increase German's competitiveness in manufacturing over the upcoming years through increased shop floor digitalization.

Key Performance Indicators: A set of numerical values which reflect the performance of a certain aspect of manufacturing operations.

Manufacturing Execution System: An information system supporting real-time production control, thereby running most manufacturing operations, and storing traceability data.

Traceability: The act of storing process related data for every single product in order to assure legal and safety compliance.

Chapter 11
The Logistic Model Based on Positions (LoMoBaP [MoLoBaC]) and Industry 4.0

José G. G. Hernández R.
Minimax Consultores, Venezuela

Gilberto J. Hernández G.
Minimax Consultores, C.A., Venezuela

María J. García G.
Minimax Consultores, C.A., Venezuela

ABSTRACT

This chapter relates business logistics, Industry 4.0 and multicriteria decision making. It is analyzed how the Logistic Model Based on Positions, it is adapted to the approach of Industry 4.0. Business logistics is related practically with every area of an organization, which making it appropriate to study organizations. But at the same time these multiple relationships make its study complex.To minimize the complexity of business logistics learning, four models were created to facilitate their teaching. One of them, MoLoBaC, which studies logistics based on the functions performed by those who have positions associated with it, has repercussions in all areas of the organization. This impact makes the MoLoBaC an ideal candidate to study the new business approaches. At the same time it generates interest to know as the MoLoBaC, can adapt to the approach of Industry 4.0. From everything previous the general objective of this work arises: Perform a qualitative-quantitative analysis of the adaptation of the Logistic Model Based on Positions to the new business approach, Industry 4.0.

DOI: 10.4018/978-1-5225-4936-9.ch011

INTRODUCTION

To understand the concept of Industry 4.0, it must be taken into account that it is considered the fourth revolution of the industrial age. The first three industrial revolutions were: the introduction of water and steam as mechanical energy at the end of the XVIII century (Brettel et al., 2014); the industrial use of electricity, which allowed and facilitated mass production as early as the end of the XIX century (Rüßmann et al., 2015; Schmidt et al., 2015) and the introduction of automation of the manufacturing processes through the programmable logic controllers (PLC) almost at the end of century XX (Borisov et al., 2016; Brettel et al., 2014; Flores et al., 2015; Hernández, Hernández & García, 2017; Rüßmann et al., 2015; Schmidt et al., 2015). And the fourth revolution (Industry 4.0) arises when, through technology, the real world is integrated with the era of computer development (Zhou, Liu & Zhou, 2015).

The concept of Industry 4.0 is based and involves many fields of knowledge and industrial activity, one of them being the logistics (Posada et al., 2015; Seitz & Nyhuis, 2015; Valdez et al., 2015). And, in particular, it is often referred as smart logistics, either in the context of Industry 4.0 (Menicocci, 2015; Stock et al., 2015; Stock & Seliger, 2016) or in related contexts (Kodym & Unucka, 2016; Shafiq et al., 2015), and even usually uses the term Logistics 4.0 (Domingo, 2016). In any case it is considered the logistics as a differentiating element (Seitz & Nyhuis, 2015).

With regard to business logistics, as discussed in different works (Barreto et al., 2016; De Burgos et al., 2016; García et al., 2017; Hernández, Hernández & García, 2017; Schwarz et al., 2016; Tarazón et al., 2016), it is possible to see that it is related to practically all the areas of an organization, this is an advantage, since the same can study the organization as a whole. But at the same time is a disadvantage, because it makes its complex study.

To facilitate the teaching of logistics, in academia, four qualitative-quantitative models were created: Logistic model Supply, Production, Distribution and Inverse (LSPDI, in Spanish el modelo Logística de Abastecimiento, Producción, Distribución e Inversa [LAPDI]) which focuses on logistical flows to analyze business logistics; the Logistic Model Based on Positions (LoMoBaP, in Spanish Modelo Logístico Basado en Cargos [MoLoBaC]), which takes into account all the functions performed by those who hold positions related to business logistics; the Logistic Model Based on Indicators for Positions (LoMoBaIPo, in Spanish Modelo Logístico Basado en Indicadores de Cargos [MoLoBaICa]), which in a way overlaps the MoLoBaC and analyzes the logistics through management indicators and the Logistic, Strategic, Tactical, Operational with Inverse Logistics Model (STOILMo, in Spanish Modelo Logístico, Estratégico, Táctico, Operativo con logística Inversa [MoLETOI]), which is based on the three components of the administrative pyramid the: Strategic, Tactical and Operational and including reverse logistics, analyzes the business logistics. This work will focus on the second of these models the MoLoBaC.

The relevance that Industry 4.0 takes every day and the presence of these models to facilitate the understanding of logistics motivates this research. It is desired to study the application possibilities that one of these models, the MoLoBaC, could continue to have in the field of Industry 4.0 and it is important that this analysis is done both qualitatively and quantitatively.

To perform this quantitative analysis in addition to Industry 4.0 and the MoLoBaC, another aspect that is relevant for this work are the multicriteria models. These multicriteria models should be used when there are a set of alternatives to evaluate, which must meet several criteria simultaneously, regardless of whether these criteria can be counterpoised.

Of the many existing multicriteria models (Ho, Xu & Dey, 2010; Sanayei, Mousavi & Yazdankhah, 2010; Tarazón et al., 2016) in this research only two of the simplest to implement and use will be worked upon, the Matrixes Of Weighing with multiplicative factors (MOWwMf [MDPcFm]) (García et al., 2017; Hernández & García, 2010; Jeney, 2014) and the Multiattribute Models with multiplicative factors (MMwMf [MMcFm]) (García, Hernández & Hernández, 2015; Hernández & García, 2010; Tarazón et al., 2016).

The Matrixes Of Weighing (MOW [MDP]), as its name indicates and can be seen in Table 1, are simply an array of rows and columns, where through a set of criteria the alternatives are weighted according to the level of compliance that they have of each of the respective criteria (García et al., 2017; Hernández & García, 2010). These MOWs are transformed into MOWwMf, when the results obtained for each alternative are corrected through the multiplicative factors, which is a concept taken from the MMwMf.

Because of their structure, the MOWwMf are perhaps the easiest multicriteria model to implement, use and interpret. By which lends it to facilitate decision-making in various fields of human performance.

Supported in its simplicity, the MOWwMf can be used in group decision-making, and from them can even emerge the MOWwMf of multiple layers (García et al., 2017), that allow a simultaneous assessment of the alternatives and parameters that are being used as criteria.

In Table 1 it can see that, in a MOWwMf, for each of the n criteria (C_j), there is a weight range that goes from PC_{ji} to PC_{jf}, where PC_j mean weight of the criterion j and i represents the initial and f final. Also, in Table 1, it can be seen each one of the payments ($P_{k,j}$), which are the values obtained for each alternative k, according to each criterion j, the multiplicative factors (fg_h), the alternatives (A_k) and the column Total where are reflected the values obtained for each alternative (Total A_k). This value, Total A_k, is evaluated by using the equation 1.

$$\text{Total } A_k = \prod\nolimits_{h=1,H} fg_h * \sum\nolimits_{j=1,n} v_j * P_{k,j} \tag{1}$$

where, in addition, the symbology, previously mentioned, are the products of the fg_h, that go from 1 to the total number of factors H and are presented the summations of the products of the multiplicative values (v_j) for each one of the $P_{k,j}$.

For more details of the MOWwMf, it is advisable to review the work of García et al. (2017), Hernández and García (2010) and those cited by them, related to this topic.

Table 1. General characteristics of a MOW

		Criteria (C_j) and Its Weights				Multiplicative Factors (fg_h)	Total
		Criterion 1	**...**	**Criterion n-1**	**Criterion n**		
	Weight	PC_{1i} to PC_{1f}		PC_{n-1i} to PC_{n-1f}	PC_{ni} to PC_{nf}	---	---
Alternatives (A_k)	A_1	$P_{1,1}$	...	$P_{1,n-1}$	$P_{1,n}$	fm_1	Total A_1
	A_2	$P_{2,1}$	...	$P_{2,n-1}$	$P_{2,n}$	fm_2	Total A_2
	..	...	$P_{k,j}$	...	...	---	...
	A_{m-1}	$P_{m-1,1}$	...	$P_{m-1,n-1}$	$P_{m-1,n}$	fm_{m-1}	Total A_{m-1}
	A_m	$P_{m,1}$	...	$P_{m,n-1}$	$P_{m,n}$	fm_m	Total A_m

Source: García et al. (2017).

The other multicriteria model that will be used in this work, the MMwMf, are just as simple to work as the MOWwMf and as we could see in García, Hernández and Hernández (2015) and Hernández and García (2010), they evaluate the alternatives through the valuable attributes, which must be evaluated as components of the criteria.

This evaluation of the alternatives is done through equation 2, where, for k general multiplicative factors and i criteria, each of them with j attributes, Pts represents the value reached by the alternative, $\prod_k fg_k$ represents the product of the general multiplicative factors (fg_k), which are multiplied by the sum of the $\prod_h f_{ih}$ which represents the product of the multiplicative factors of the criteria, which in turn are multiplied by the weights of each one of the criteria (pc_i) and by the sum of the products of the weights of the attributes j of criterion i (pa_jc_i) by the value of each one of the attribute j of criterion i (va_jc_i).

$$Pts = \prod_k fg_k * (\sum_i \prod_h f_{ih} * pc_i * (\sum_j pa_jc_i * va_jc_i)) \qquad (2)$$

Like the MOWwMf, the MMwMf, although starting as additive models, with the correction of the multiplicative factors, reach a great field of application and a very high flexibility, which makes them candidates to be used in many multicriteria problems, with the great advantage that both, MOWwMf, and MMwMf, allow to recognize, through the percentage of the total value reached, whether or not the alternative evaluated is a good alternative.

Another great advantage of both models is that they evaluate each of the alternatives without depending on the remaining alternatives, which gives them great independence of use always and especially when the alternatives are not all defined from the beginning, or when there are alternatives that are discarded a posteriori.

As discussed in introducing the MOWwMf, if someone wants to have more details about the MMwMf, it is recommended to read the works of García, Hernández and Hernández (2015), Hernández and García (2010) and those cited by them, which are related to this subject.

With the brief presentation of the MOWwMf and the MMwMf and the previous introduction that has been made of some models for the study of business logistics and the concept of Industry 4.0, the elements on which this research is centered are complemented by which it is able to express its general objective: Perform a qualitative-quantitative analysis of the adaptation of the Logistic Model Based on Positions to the new business approach, Industry 4.0.

From this general objective three specific objectives arise:

- To show the relevancy of the concept Industry 4.0.
- To present the fundamental aspects of the MoLoBaC.
- Conduct a qualitative-quantitative analysis of how well MoLoBaC is adapted to the Industry 4.0 environment.

METHODOLOGY

To achieve the general objective and its three specific objectives, it will be used the Integrated-Adaptable Methodology for the development of Decision Support System (IAMDSS, in Spanish, Metodología Integradora-Adaptable para desarrollar Sistemas de Apoyo a las Decisiones [MIASAD]), which has proved to be very useful, to complete research in different fields of knowledge (De Burgos et al., 2016;

García et al., 2017; Hernández, Hernández & García, 2017; Schwarz et al., 2016; Tarazón et al., 2016). Because of its flexibility, MIASAD can adapt to different types of research, where it can select from its twenty basic steps, only those that are considered indispensable at the time of application.

Analogously to what was done in other research (De Burgos et al., 2016; García, Hernández & Hernández, 2015; Jeney, 2014; Schwarz et al., 2016; Tarazón et al., 2016) in this work the following steps will be followed:

1. Define the problem, which as indicated in the objectives is to perform a qualitative-quantitative analysis of the adaptation of the Logistic Model Based on Positions (MoLoBaC) to the new business approach, Industry 4.0;
2. Elaborate a first prototype, which includes among its main functions to offer a clear vision of what will be the result of the research, so that among the first aspects to be carried out is to identify the users of the final product that, in this case, because it is a scientific article, will be the its readers. Among the main readers of the article, should be all interested in the new approach of Industry 4.0 and all the aspects that this concept involves, all those interested with the study of logistics are added, especially those who in one way or another are associated with teaching logistics and to the models that exist to explain it. Finally, to them will be added all the scholars of the mathematical models, especially those interested in multicriteria models. The first prototype, on the other hand, serves as a starting point for establishing the structure of the article, which in addition to this introduction and the background consists of three central sections. Emphasizing that already in the Introduction there was covered the relative thing to the models multicriteria, in particular the two models to use in this investigation, the Matrixes Of Weighing with multiplicative factors (MOWwMf) and the Multiattribute Models with multiplicative factors (MMwMf). Therefore, in the first of the sections will show the relevance of the concept Industry 4.0 and its surrounding environment. The second section will be dedicated to presenting the MoLoBaC, with special emphasis in each of their areas. And the third of the sections, which will be the focus of the work, will be devoted to presenting an analysis, first qualitative, which will highlight the common aspects between MoLoBaC and Industry 4.0 and then a quantitative analysis, where through the use of the MOWwMf and the MMwMf, will quantify the impact that can MoLoBaC in the teaching of logistics and industrial applications of it, in an environment of Industry 4.0. To close the chapter some conclusions and recommendations for future research will be presented and the references of the mentioned works will be listed;
3. Get Data, in this case about, Industry 4.0, the MoLoBaC, the MOWwMf and the MMwMf;
4. Establish alternatives, which would be the different ways to analyze whether MoLoBaC is compatible with Industry 4.0. This analysis should be both qualitative and quantitative;
5. Evaluate the alternatives, taking into consideration that the qualitative component can be simply a comparison, but that the quantitative analysis must yield some number that allows to obtain some clear conclusion as for the possibilities of use of the MoLoBaC in Industry 4.0;
6. To select the best alternative, according to the previous evaluation and considering the secondary objectives, whether tacit or explicit. It is important to clarify that the selected alternative may be several models;
7. To implement the selected alternative. Although it is not an implementation as such, it must be possible to arrive at a clear conclusion of the potential of use of the MoLoBaC in the environment of Industry 4.0 and

8. Establish controls or mechanisms which allow to recognize if the conclusion obtained can remain valid during the time.

As a limitation of the scope it has that it will not realize field works, but the analysis of the MoLoBaC will be done through its features, as broadly as possible, in a certain way, in a hypothetical way, to give the study a greater generality.

BACKGROUND

Three main aspects are intertwined in this chapter: Industry 4.0 and its environment, logistic models, in particular the Logistic Model Based on Positions (MoLoBaC) and a couple of multicriteria models, the Matrixes Of Weighing with multiplicative factors (MOWwMf) and the Multiattribute Models with multiplicative factors (MMwMf).

Of these two multicriteria models already in the introduction, their antecedents were presented and it can be added that they have been used to support research on some aspects of logistics, as seen in García et al. (2017), Jeney (2014) and Tarazón et al. (2016) although these are not the only multicriteria models that have been used to study aspects of logistics, to mention only a few it can be considered the works of De Felice & Petrillo (2012), Ho, Xu and Dey (2010) and Sanayei, Mousavi and Yazdankhah (2010).

And with direct reference to MoLoBaC, there are also many backgrounds that can be mentioned, since this model has been used in a lot of occasions to study different aspects of and from the business logistics, as can be seen in the works of Barreto et al. (2016), whom, starting from their different areas, support logistics as a whole to study the safety and security (S&S) business, García et al. (2017) and Schwarz et al. (2016) whom support the study of inventory control, Jeney (2014), who, through one of its positions, studies the generation and management of knowledge, making use of a MOWwMf, Hernández, García and Hernández (2016) whom, through an area of MoLoBaC, study the generation and management of knowledge and Tarazón et al. (2016) whom use it to study the relationship between logistics, generation and knowledge management and business innovation.

With respect to Industry 4.0, if the study goes back to the beginning of the industrial era, the antecedents would be countless, but limited to the works related to the concept of Industry 4.0 itself there are also many works to which reference can be made, in particular, in the introduction of this research, reference has already been made to the works of Brettel et al. (2014), Hernández, Hernández and García (2017), Rüßmann et al. (2015), Schmidt et al. (2015), which make use of the concept of Industry 4.0 from different perspectives.

In any case, the next two sections will be dedicated to comment, with a little more depth, first the concept of Industry 4.0 and then the MoLoBaC.

RELEVANCE OF THE INDUSTRY 4.0 CONCEPT

The term Industry 4.0, could be said that at present, is in permanent construction (Roblek, Meško & Krapež, 2016), where the real world and the virtual world are growing together (Wang, 2016). Industry 4.0 is seen as the new paradigm of autonomous and decentralized control in production (Gölzer, Cato &

Amberg, 2015), making use of self-optimization. These new paradigms of production it allows to reach higher levels of efficiency in the production processes and in the production itself.

The work of Shafiq et al. (2015) shows that Industry 4.0 has as its objective the smart factory, which is characterized by adaptability capacity, resource efficiency and ergonomics, as well as the integration of customers and business partners.

According to Kolberg and Zühlke (2015), the term Industry 4.0, was coined in Germany and is related to a greater integration of the Information and Communication Technologies (ICT) with the productive processes. In a way, this is confirmed for Lee, Kao and Yang (2014), whom say that Germany leads the transformation toward the fourth industrial revolution (Industry 4.0), which is based on the availability, for the manufacture of the Cyber-Physical Systems (CPS), which means a higher availability and possibilities of integration of new software, technologies, wireless devices and algorithms that allow it has a smarter production, in the sense of efficient.

Then these authors (Lee, Kao & Yang, 2014) refer to other aspects that are involved with Industry 4.0 and point to five categories: Operator manager integration; Fleet of machinery; Quality of products and processes; Big data and the cloud and Network of sensors and controllers (S&C). The latter is reinforced by Seitz and Nyhuis (2015), whom emphasize that the development of the CPS and Cyber Physical Production Systems (CPPS), as well as the applications of the analysis of Big data has great potential in the industry today and in the future.

For its part Lom, Pribyl and Svitek (2016), also present an approximation of what is Industry 4.0 and say that can be defined as a collective term for technologies and concepts of a value chain that is created in conjunction with the CPS.

The frequent mention to the CPS, makes it necessary to generate some general comments on them, since they are a fundamental pillar of Industry 4.0 (Almada-Lobo, 2016; Brettel et al., 2014; Grangel-González et al., 2016; Hermann, Pentek & Otto, 2016; Kolberg & Zühlke, 2015; Posada et al., 2015; Shafiq et al., 2015; Waibel et al., 2017; Zhou, Liu & Zhou, 2015), as has already been pointed out and can be seen below.

It begins with Monostori (2014), who comments that the CPS are systems of collaborating computational entities that are in intense connection with the surrounding physical world and its ongoing processes, providing and using, at the same time, data access and data processing services available on the Internet. And he adds those CPS have extraordinary potential to change many aspects of life today; Understand the beginning of the century XXI.

Another relevant comment is of Hermann, Pentek and Otto (2016) is to point out that in the manufacturing environment, the CPS, include intelligent machines, storage systems and facilities of production capable of exchanging information of autonomous form, unleashing actions and being controlled of independent form.

To try to present a better characterization of the CPS, it is used Zhou, Liu and Zhou, (2015), whom clarify that they connect all the physical devices to Internet and incorporate five functions: Computing, Communications, Precision control, Coordination and Autonomy. And they add that the CPS trough the integration of the virtual world and the physical world, do that the intelligent products and the intelligent production are possible.

Similarly, Borisov et al. (2016), talk about an architecture of 5C: Connection, referring to the data acquisition from sensors and machines; Conversion, transformation of data into information; Cyber, which manages and analyzes information; Cognition, presents the knowledge acquired and Configuration, self-configuration and self-adaptation to changes.

Looking to clarify a little more, it will follow to Lee, Bagheri and Kao (2015), who point out that in general a CPS consists of two major components: first; advanced connectivity, that ensures the acquisition of real-time data from the physical world and the feedback of information from cyberspace and second; the intelligent management and analysis of data, which is the computational capacity and that builds the cybernetic space.

More directly, it can be said that CPSs include sensors and active elements ("actuators") through which they can collect and send data (Domingo, 2016). At the same time, they allow the convergence of the physical world with the virtual world, and by enabling virtual simulations of products and processes, before and during the operation, they become a key element in achieving critical goals and offering flexible production (Posada et al., 2015).

With respect to intelligent logistics, Stock and Seliger (2016) comment that the CPS can be used to support the flow of materials within the factory and between factories and with customers and other stakeholders.

In other side, Lom, Pribyl and Svitek (2016), in addition to emphasizing that the CPS are an important part of Industry 4.0, they give them a leading role in Interoperability, Virtualization, Decentralization and Service Orientation and they add that they can facilitate the self control through the exchange of autonomous information, allowing, through an IP address (Identifier assigned to each computer), the monitoring and control on line.

Following Gölzer, Cato and Amberg (2015) this information can be complemented, since they point out that the CPS can communicate with each other, at the same time, by detecting the environment that surrounds them, interpret the availability of data and act on the real world. In this way the products and the entire production system, including machines and warehouses, are integrated into global production networks, thus becoming intelligent entities that can self-monitor and self-optimize. In addition, with CPS it can integrate and manage the Big data through the cloud.

Already to close this brief discussion on the CPS Monostori (20140) commented that the CPPS, based on the latest developments in computers science and information technology, contribute greatly to the sciences and technologies of the manufacturer, that is to say to Industry 4.0 and although it makes an exception, that some cases may be simple speculations, it can be had of the CPS and the CPPS, expectations with regard to: Robustness in all levels (of the industry, for the interest of this work); Self-organization-maintenance-repair and other self, where it can mention optimization and navigation; Safety (it can be said Security & Safety [S&S]); Remote diagnosis; Real-time control; Transparency; Predictability; Efficiency; Global Tracking and Tracing; Interoperability and correction of models, as well as can be expected new business models and services among other expectations.

But, although the CPS and CPPS are fundamental elements in Industry 4.0, are not the only ones, as noted at the beginning of this section and in the introduction.

So, trying to follow this line to detect elements on which is based Industry 4.0, it can be noted that Zhou, Liu and Zhou, (2015), in addition to the CPS and CPPS, include three other key technologies in the development of Industry 4.0: Mobile Internet and Internet of Things (IoT); Cloud Computing technology and Big data and advanced techniques for the analysis. To these three technologies, Hermann, Pentek & Otto (2016) added: Internet of Services (IoS); Intelligent Manufacturing; Intelligent production; Intelligent products and communication Machine-to-Machine (M2M).

For then part, when they try to submit their expressions relating to the fundamentals of Industry 4.0, Roblek, Meško and Krapež (2016), in addition to including the smart city, discuss a series of fundamental concepts among which stand out: Smart factory, Smart manufacturing, Intelligent factory, Factory of the future; New systems in the development of products and services; Self-organization; Intelligent products; New systems of procure and distribution; Adaptation to the human needs and Digital sustainability.

Other elements are highlighted by Lee, Bagheri and Kao (2015), whom point to five levels:

I Connection intelligent, where the condition is monitoring, by which stand out the sensors; II Conversion of data to information, is the level of self-consciousness, where are the machines and their components; III Cyber, here is the auto comparison and emphasize the fleet of machines and the comparison of the pairs; IV Cognition, where the decisions are optimized and have special leading role the Decision Support Systems (DSS) and V Configuration, in which appear the flexible control systems.

For its part, in addition to those already mentioned Big Data, Gölzer, Cato and Amberg (2015), present other terms related to Industry 4.0, such as: Sensors, Self-control, Self-optimization, Support and Requirement of information technologies (IT), IoT, Cloud solutions, Intelligent machines, intelligent Products, Systems of intelligent production and Extraction and acquisition of date.

While Lom, Pribyl and Svitek, (2016), in addition to many of the already presented terms they emphasize those related to Internet and talk of IoT, IoS, Internet of Energy (IoE), Internet of the Persons (IoP). In addition, they explain that Industry 4.0 is based on six principles: Interoperability, Virtualization, Decentralization, Real-Time Capability, Service Orientation and Modularity.

With a similar vision Wang (2016), lists ten technological concepts, associated with Industry 4.0, where, in addition to the CPS and others have already mentioned, makes mention of: the radio frequency identification (RFID) technology; Computational Intelligence (CI) and the Data mining (DM). But more than these concepts, it is important to note that his work focuses on the Intelligent predictive maintenance (IPdM). Another additional concept, Artificial Intelligence (AI), is added by Wang et al. (2016).

Although it can be mentioned more authors that add concepts or elements which integrate or that are an important part of Industry 4.0, but with those listed here, a balance can be presented of the main features of Industry 4.0, which will be summarized in Table 2. To understand a little better the reason for including the elements listed in Table 2, it is appreciated to read the explanatory notes at the end of it.

The information of the table 2 has been extracted from all the authors whom cover the topic of Industry 4.0 and the CPS and whom till now have been quoted, nevertheless, it is necessary to emphasize that the biggest volume of information has taken of Brettel et al. (2014); Hernández, Hernández and García (2017); Posada et al. (2015); Rüßmann et al. (2015); Shafiq et al. (2015) and Zhou, Liu and Zhou (2015) and in addition of Cheng et al. (2016); Sommer (2015) and Speicher et al. (2015), that still had not been quoted. To facilitate reading the table 2 has been divided into 2 A, where the components are highlighted and 2 B where the objectives and strategic aspects are mentioned.

It should also be noted that some researchers may consider that some concepts are ignored, others repeated and some that are already expressed by another concept that wraps them. The intention is to emphasize all those concepts that, without any doubt characterize Industry 4.0 and may be useful for the analysis that will be done in the final section.

One aspect that was mentioned in Table 2 and which has now gained more relevance is Safety & Security (S&S), as it has been presented in the work of Barreto et al. (2016). In particular, there must be rapid responses to cyberattacks, such as those that have occurred worldwide in the middle of May 2017 and which have affected thousands of organizations from different countries (Wong & Solon, 2017).

With the information included in Table 2, and these later statements close the comments, about the relevance of Industry 4.0.

Table 2. A) Some components of Industry 4.0; B) Some objectives and strategic aspects of Industry 4.0

Components	Commentary
A01. Cyber-Physical Systems (CPS) and Cyber Physical Production Systems (CPPS)	The CPS are responsible for the communication between the physical world and the virtual world.
A02. Internet of things (IoT)	That facilitates smarter communication.
A03. Internet of Services (IoS)	Contributes to integration with the physical world.
A04. Internet of objects (IoO)	Facilitates integration with the physical world.
A05. Internet of Energy (IoE)	Helps determine the use of energy resources.
A06. Internet of the Persons (IoP)	Key to recognizing human needs.
A07. Self-organization and self-optimization	The system is organized and optimized autonomously.
A08. Intelligent factory	That itself works with the staff and the machines to achieve their goals.
A09. Products and machines intelligent	Which are coordinated in an autonomous way.
A10. Intelligent production scenarios	Associated with products and machines intelligent.
A11. Computational Intelligence and decentralized computing	Achieved through the Internet and the CPS. And based on the self-adaptation.
A12. Cloud Computing Technology (CCT)	It is necessary to take advantage of the ubiquity offers.
A13. Cloud-based manufacturing	Supported by the CCT, it allows, among other aspects: the self-service, increased agility, resource sharing, scalability, data storage.
A14. Big data and advanced analysis techniques	High volumes of data are a resource of great interest. It is necessary to be prepared to handle them.
A15. Availability and possibilities of integration of new software	Growth in computer tools is an aspect to be exploited.
A16. High performance computers	This includes the integration of the equipment.
A17. The robotics and the presence of autonomous robots	The integration of the humans with robots allows for a more intelligent production.
A18. Information and Communication Technologies (ICT)	Working in conjunction with the CPS, to provide mutual support.
A19. High-precision controllers and sensors in general	Detect and make changes in the environment, with precision.
A20. Machinery and whole plant collectors of data	The organization as a whole is a receiver of data, which must be transformed in information.
A21. Virtual engineering object (VEO)	A VEO is a knowledge representation of an artifact of engineering.
A22. Virtual engineering process (VEP)	A VEP is the representation of knowledge of the manufacturing process.
A23. Machine to machine (M2M) communication	With the support of the ICT.
A24. Synchronous and Asynchronous communication	That, sometimes, allows to continue actions without waiting for response.
A25. Individualized data tracking	Taking advantage of the ICT, Internet and the CPS.
A26. Miniaturization	The reduction of the size of many components makes it easier to perform many tasks.
A27. Augmented Reality	In the beginning will help operators to better understand the processes and facilitate repairs.
A28. Data mining (DM)	To find patterns in data.
A29. End-to-end Engineering	Required through the whole value chain.
A30. Collaborative Networks	Integrating the different actors through the Internet.
A31. Interoperability and Integration person-machine- products	Machines and products with high communication capacity.
A32. Virtualization	Thanks to the appropriate interfaces to communicate the real world with the virtual.
A33. Decentralization	Facilitated by the autonomous machines.
A34. Capacity in real-time	Thanks to the multi-sensor data is captured and analyzed in real time.

continued on following page

Table 2. Continued

A35. Service orientation	Both internally and externally.
A36. Modularity	This provides greater flexibility.
A37. Diversity of resources	Those allow to produce multiple small lots.
A38. Radio frequency identification (RFID)	Contain relevant information that facilitates communications in general.
A39. Scanners	Help to facilitate the continuity of the process.
A40. Decision Support Systems (DSS)	Integrate into productive systems to facilitate decisions.
A41. Simulation	It even will be used to analyze the operability and the production in plant.
A42. Printers 3D	New technology that develops in parallel with Industry 4.0. And that will facilitate the simulation of products.
Objectives and Strategic Aspects	**Commentary**
B01. Factory intelligent, adaptable, flexible, efficient, green, scalable, consistent, with high quality at low cost, efficient management of resources and ergonomics, as well as the integration of customers and business partners in business processes and value.	It is the objective macro that is pursued.
B02. Flexible Manufacturing Systems (FMS) & Reconfigurable Manufacturing System (RMS) (FMS & RMS)	To achieve the macro objectives, production systems of great flexibility and adaptability are needed.
B03. Optimization of production	Supported by CPS and IoT.
B04. Rapid, autonomous, individualized and decentralized production	To serve ever more complex markets and more demanding customers.
B05. Intelligent products or with digital memory	They help to drive the productive process.
B06. Intelligent production systems	Looking for greater autonomy.
B07. Utilize and integrate the quality of products and processes	The quality of one must serve to learn and improve the quality of the other.
B08. Ensure self-evaluation, self-prediction, self-comparison, self-re-configuration and self-maintenance. Reaching even the IPdM	Ensure the autonomy of men and especially of machines, throughout the organization. It will be the machines themselves that report the need for maintenance.
B09. Broad and reliable communication network	It depends on the communication network, so it must be reliable and of high capacity.
B10. Mass customization and customization of manufactured products	Production directed to the masses, but with a greater individualization, working with small lots.
B11. Simultaneous planning of products and processes	Obtaining, major efficiency and minor costs.
B12. Adaptive production chain	Automatic adaptation and flexible to new changes.
B13. Additive Manufacturing	Directed toward the production of prototypes and individualized components.
B14. Standardization	A set of standards should be developed to facilitate communication and integration.
B15. Increase interest in digitization	It could be said in everything cybernetic in general. And this can be a problem for some companies.
B16. Cooperative production systems	That they can collaborate with each other, even at a distance.
B17. Increase production flexibility	That allows a more personalized attention, adaptation to new business models and to demographic changes.
B18. Have an adequate network of sensors and controllers (S&C)	Also, it is necessary to guarantee that the S&C work perfectly.
B19. Remote diagnosis	Not only is diagnosis made remotely, but corrective measures are implemented.
B20. Tracking of parts and products	Taking advantage of the communication between machines and products.
B21. Improvements in logistics and supply chain (SC)	Thanks to better information through the IoT.
B22. New systems in distribution and procure	Supported in the integration with stakeholders.

continued on following page

Table 2. Continued

B23. SC flexible and visible	Adaptable to the demands of the customers.
B24. Efficient management	To the being the most complex productive systems need a more efficient management.
B25. Organization and design of the work	The production system must be adapted to the needs of today and of the future.
B26. Provide value-added services	For all the stakeholders.
B27. Safety & Security (S&S)	S&S must be provided to people and the environment and the S&S of the facilities must be guaranteed.
B28. Continuous staff training	Who should be trained to face major challenges.
B29. Improve the efficiency of resource management	In addition to the human resource, one must be very efficient in handling all new resources.
B30. Adapting to human needs	Smarter tracking of consumer needs.
B31. Positive impact on the economy	Both the company, the nation, and the region.
B32. Decentralization of productive processes	It is facilitated by the autonomy of the parties. And it allows a greater approach to the customer.
B33. Share data and information	Facilities are created for this to happen.
B34. Verification and self-checking and correction	To ensure that everything works properly.
B35. Radical changes in the types of services and business models	Changing the way of interacting in SC.
B36. Improve man machine interaction and integration operator manager	In flatter organizations. It includes coexistence with robots and greater attention to the health of machines.
B37. To take care and to learn of the fleets of machines	To classify the machines, for better learning from his common conditions.
B38. Reduce waiting times	The easy use of relevant information improves the decision making.
B39. Generalized reduction of costs	To optimize production processes.
B40. Horizontal Integration	Through value-added networks that facilitate collaboration.
B41. Vertical integration	To create flexibility and a manufacturing system that can be reconfigured.
B42. Digital integration End to end	Integration through the entire value chain. To facilitate the customerization.
B43. Interoperability	Machinery, data, products and components must operate together during the production process.
B44. Waste reduction	Owing to better planning and control.
B45. Friendly with employees	Greater automation facilitates employee work.
B46. The development of intelligent devices	Developing smart devices will become more and more necessary. This is a great challenge.
B47. Competitive prices	Through better production systems.
B48. A closer approach of suppliers and customers to the production process	What allows fitting more rapid to changes on the market.
B49. To increase the international competitiveness	The flexibility of the production and better adapted to the needs of consumers.
B50. Service orientation	For both internal and external customers.
B51. Reduce costs in production processes	Mainly because the production is only what it wants.

Note, in Table 2, that there are aspects such as intelligent products and machines and end-to-end integration that are both a component (Table 2A) and an objective (Table 2B) of Industry 4.0. Hence it has been commented, that it is a concept in formation and in permanent evolution.

Although there has been no further discussion on this point, it should be noted that Industry 4.0 is changing the profile of future employees. This requires that it is necessary to train, with a new approach, the personnel that will work in an Industry 4.0 environment. In this sense, it can see the relevance of the aspect to be covered in the next section the Logistic Model Based on Positions (MoLoBaC).

BRIEF COMMENTS OF THE LOGISTIC MODEL BASED ON POSITIONS

Already in the introduction were presented four qualitative-quantitative models created to facilitate the teaching of logistics: the Logistic model Supply, Production, Distribution and Inverse (LAPDI); the Logistic Model Based on Positions (MoLoBaC); the Logistic Model Based on Indicators for Positions (MoLoBaICa) and the Logistic, Strategic, Tactical, Operational with Inverse Logistics Model (MoLE-TOI). This work, as already mentioned, will focus on the MoLoBaC.

As they point out Barreto et al. (2016), De Burgos et al. (2016), Hernández, García and Hernández (2016) and Tarazón et al. (2016) and it can be seen in Figure 1, the MoLoBaC is a very simple model, made up of forty-four positions (each one identified by a number), that are grouped in twelve areas and these at once in six stages.

The areas can be mixed if they include positions of different stages or pure if all their positions belong to the same stage.

The six stages, which make up the MoLoBaC, are:

1. Supplying, with a single area, Procure, which is pure;
2. Production, formed by two pure areas, Maintenance and Inventories;
3. Distribution, the largest with respect to the number of areas, four: Order processing, first of the mixed areas, integrated by positions of the stages Production and Distribution, Physical distribution, another pure area, Transportation, the second mixed area shaped by positions of the stages Distribution and of General of the company and the area that comes to be the reason of the being of the model, Costumer service, that also is pure;
4. Inverse, shaped by only one pure area, Inverse logistics;
5. General of the company, conformed by three areas, Intrinsic to logistics, Supported by logistics and Supporting logistics, all them pure and
6. General of information (Information), integrated by the last one of the areas, Information, which is mixed to hold positions of the stages General of the company and the own of General of information.

In general, when working with the MoLoBaC, the study focuses on identifying and studying the functions of one of its positions or the positions of an area. This is usually done by building tables for each position, summarizing their functions.

It is the case of the works of De Burgos et al. (2016), with the Packing manager, García et al. (2017), with the Inventory manager, Jeney (2014), with the System Information and Networks manager, or Schwarz et al. (2016) whom analyze the Inventories models manager and Hernández, García and Hernández (2016), whom study the three positions of the Inverse logistics area.

But in this research, as the situation that was presented in the work of Barreto et al. (2016), it is desired to identify functions, which in one way or another, help relate the MoLoBaC with specific aspect, in this case with Industry 4.0, by which the functions that are reflected in Table 3 shall cover one or several aspects relating to Industry 4.0 rather than specific aspects of a position.

How did Barreto et al. (2016), since it is a question of forty-four positions and that some functions that could be shared by several of them, the Table 3, is organized by areas of the MoLoBaC, however, at the end of each function will include a number or several numbers that represent the position or positions that with greater depth performs this function. At all times, some of the numbers may correspond to positions of other areas. In this case they will appear with high frequency the positions of the Information area, given that in Industry 4.0, some of its greatest strengths are in research and development (R&D) and communication.

It is necessary to clarify now that in the citations related to the MoLoBaC and the other logistic models, as well as in the MIASAD and with the two multicriteria models discussed, it has been tried only to present the last works that are related to the subject and that preferably they are written in English language, for which it is recommended to those who wish to have more details on these topics to review the citations of these works and in turn the citations of these and so on until finding the original sources of the same, according to their respective interest.

In this Table 3, some abbreviations will be used, many of which will continue to be used in the rest of the work: CPPS by Cyber Physical Production Systems, C&R by Collection and Reception, E&M by Equipment and Machines, ICT by Information and Communication Technologies, IL by Inverse logistics, IPdM by Intelligent Predictive Maintenance, M2M by Machine-to-Machine, MH by Material Handling, NT by New Technologies, P-S, by Plant Shutdown, R&D by Research and Development,

Figure 1. General outline of the Logistic Model Based on Positions (MoLoBaC)

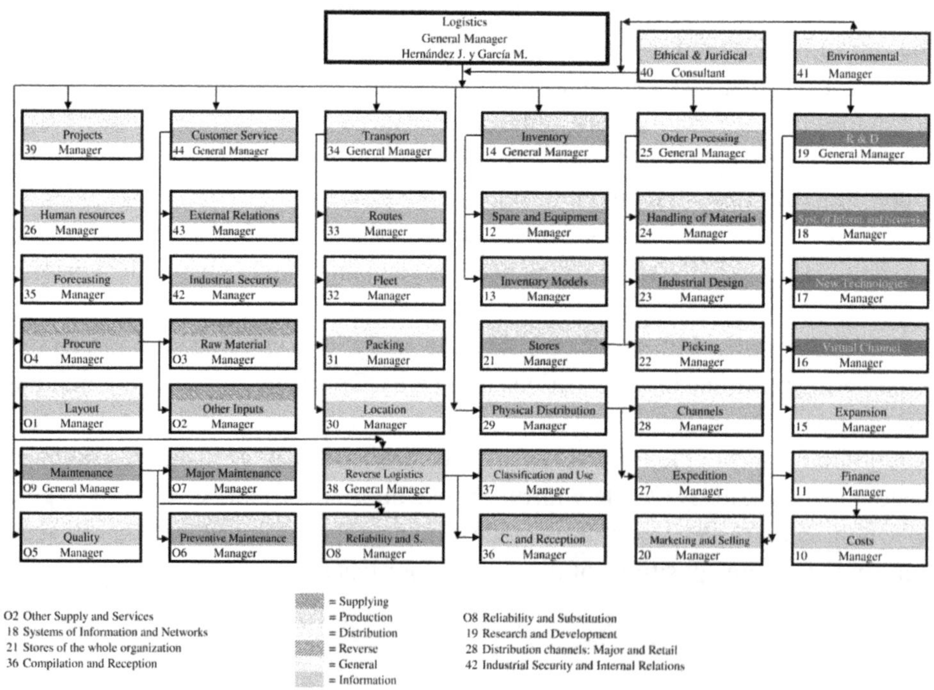

S&C by Sensors and Controllers, S&S by Safety and Security. Some of these abbreviations were used previously in the work.

The functions shown in Table 3, have been taken, modified or inspired by works that have been analyzed functions of positions of the MoLoBaC (De Burgos et al., 2016; García et al., 2017; Hernández, García & Hernández, 2016; Jeney, 2014; Schwarz et al., 2016) or works that relate the logistics or some of its aspects to Industry 4.0 (Buhr, 2015; Domingo, 2016; Gath, Herzog & Edelkamp, 2014; Gilchrist, 2016; Mucha, Büttner & Röcker, 2016; Zawadzki & Żywicki, 2016), but particularly of Tarazón et al. (2016) and fundamentally of Barreto et al. (2016).

With the functions presented in Table 3, it can have a fairly clear idea, how it can be MoLoBaC interaction with Industry 4.0, which will be developed in the next section.

QUALITATIVE-QUANTITATIVE ANALYSIS OF THE ADAPTATION OF MOLOBAC TO THE ENVIRONMENT OF INDUSTRY 4.0.

The qualitative analysis will emphasize the coexistence of the Logistic Model Based on Positions (MoLoBaC) with the Industry 4.0 environment, while the quantitative analysis will try to measure two possibilities of application for MoLoBaC, in training the new staff that needs the use Industry 4.0 and also in the industries themselves adaptable to the concept of Industry 4.0.

In order to carry out these analysis, seven basic aspects have been selected: Autonomy (Aut), Communication (Com), Flexibility (Fle), Innovation (Inn), Productivity (Pro), Safety & Security (S&S) and Technology (Tec), since they constitute the main bridges between the two aspects to be compared.

In all cases it is a question of presenting arguments that reinforce a certain position. To do these Tables 2 and 3 will be used, from which the different elements are taken. In order not to repeat information, three characters will be used to identify the subject to which reference is made, as was done in Tables 2 and 3. The letter A followed by two numbers corresponds to Industry 4.0 components (Table 2A), the letter B and two numbers will identify Industry 4.0 objectives (Table 2 B) and three numbers will refer to the MoLoBaC functions (Table 3).

First will be the qualitative analysis and subsequently the two quantitative.

Qualitative Analysis to Compare the Industry 4.0 Environment With the MoLoBaC

To facilitate the visualization of the qualitative analysis, Table 4 which presents the most relevant aspects between Industry 4.0 and the Logistic Model Based on Positions (MoLoBaC) will be used. This qualitative analysis is presented in a very simple way, following what its name indicates, an analysis based on qualities, since we only want to make a comparison that allows us to perceive the common areas between these two fields of knowledge: The MoLoBaC and Industry 4.0.

The first part of Table 4 will highlight the aspects where there is greater distance between the MoLoBaC and Industry 4.0 and in the second part will show the aspects of greater coincidence.

Table 4, the same as Tables 5 and 7 that will be used later, to perform the quantitative analysis, when using elements taken from Tables 2 and 3, should be understood as an argument to reinforce any affirmation or denial that is made from these variables. These arguments are placed in parentheses and always refer to the preceding tables. Therefore, the following identifications will be used: (1): See the

Table 3. Functions of the positions of the MoLoBaC that help it to interact with Industry 4.0

Area: Procure	
001	To help facilitate the integration of suppliers, with all levels of the organization. 02, 03, 04, 19.
002	Make use of NT, which allow them to be innovative in terms of supply. 04, 03, 02, 17, 19.
003	Ensure that all providers are managed under platforms compatible with the organization. 04, 19, 18.
004	Collaborate with suppliers to maintain up-to-date their communication platforms. 19, 04.
005	Participate in the evaluation of NT for the management of supply. 19, 17, 04, 03, 02.
006	Acquire S&C to facilitate handling of hazardous inputs. 19, 18, 17, 04, 02, 42.
007	Ensure that conditions and E&M are available for agile and S&S reception of all the inputs that come to the organization. 19, 17, 04, 02, 01, 42.
008	Prepare the organization to have a strict control of all the inputs that are acquired. 02, 04, 19.
009	Ensure, with suppliers, must be an exchange of information S&S. 19, 18, 17, 04, 03, 02.
010	Take care that providers have sufficient autonomy to supply the Organization. 04, 03, 02, 19.
Area: Maintenance	
011	Establish Maintenance policies that lead to an IPdM. 06, 09, 19.
012	Automate the performance binnacle of all E&M of the organization. 19, 18, 09, 06.
013	Implement S&C to avoid any possibility of sabotage to E&M. 19, 17, 09, 06, 42.
014	Supervise that dangerous machinery, has the necessary controls and protective measures. 19, 09, 42.
015	Ensure that preventive maintenance is carried out autonomously. 06, 09, 19.
016	Be alert to introduce innovations in all phases of maintenance. 09, 08, 07, 06, 19.
017	Ensure that E&M throughout the organization is flexible and adequately performing. 09, 06, 08.
018	Help in the self-programming of the replacement of E&M. 08, 19, 17, 09.
019	Provide the necessary elements for a good communication man-machine and M2M. 19, 17, 09, 08, 07.
020	Care for the good performance and integration of the staff recruited for the P-S. 07, 09.
021	To encourage the installation of S&C to minimize risks in maintenance tasks. 09, 06, 07, 19, 17, 42.
Area: Inventories	
022	Establish policies for the management of inventories, the most autonomous possible. 14, 19.
023	Seek to implement innovations in the management of inventories, through ICT. 14, 13, 17, 19, 18.
024	Implement models of inventories to facilitate control and intelligent production. 13, 14.
025	Use the inventory models that allow the organization to have better controls. 13, 14, 19, 18.
026	To seek a self-handling equipment, parts and spare parts. 12, 14, 19.
027	Include S&C that will help to maintain the availability of inputs that the organization required. 12, 19, 14.
028	Make sure that what is contained in inventories to match what is stored in warehouses. 14, 12, 21, 19.
029	Manage inventory systems autonomous, flexible and S&S. 14, 18, 19, 13.
030	Providing S&C and the necessary equipment to avoid fire and its spread. 12, 02, 04, 42.
Area: Order Processing	
031	Establish policies to facilitate good communication with stakeholders, in particular customers. 25, 18.
032	Maintain a permanent follow-up of clients. 25, 18.
033	Seek the support of the NT, to achieve greater innovations in design, MH and picking. 24, 23, 22, 19, 17.
034	Establishing S&C to facilitate the MH. 24, 19, 17, 18.
035	Have a good network of S&C to facilitate the movement of products and E&M. 24, 19, 17, 25.
036	Promote an autonomous picking. 22, 17, 19, 25, 21.
037	Encourage the design of products, equipment and processes, flexible, S&S and intelligent. 23, 19, 17, 25, 42.
038	Design and work with products and processes, which in addition to intelligent, be ergonomic. 23, 19, 17, 18.

continued on following page

Table 3. Continued

039	It has a policy of industrial design that promotes the autonomy and flexibility. 23, 19, 18, 25.
040	Ensure a storage autonomous and integrated with MH and picking. 22, 21, 19, 17, 18.
041	Establish special S&C for hazardous MH or of particular conditions. 22, 24, 42.
	Area: Physical Distribution
042	Establish policies that guarantee a dispatch and distribution as autonomous as possible. 27, 19, 18, 17, 29.
043	To integrate to the physical distribution of the organization, to all the members of the channel. 28, 29, 19.
044	Try to be sustained in NT and the S&C, to be innovative in the physical distribution. 29, 28, 27, 19, 17, 18.
045	Try to work with the channels that provide better communication and greater integration. 28, 29, 16.
046	Maintain a strict control of the distribution channels in regard to S&S it is refers. 28, 29, 16.
047	Take care of maintaining open communication channels, which reach the final customer. 29, 19, 44.
048	Ensure an autonomous and flexible dispatch. 27, 29.
049	Integrate and provide autonomy to Cross-docking points, through S&C. 19, 17, 29, 28, 27.
	Area: Transportation
050	Establish transport policies that encourage autonomy, flexibility and S&S. 34, 19, 18, 17, 33, 42.
051	Look for a permanent innovation in everything related to transport. 34, 33, 32, 31, 30, 19, 17.
052	Create S&C to help transport S&S of hazardous materials. 34, 19, 33, 18, 32, 17, 42.
053	Consider aspects of flexibility, autonomy and S&S to establish the organization fleet. 32, 19, 34, 42.
054	Ensure that all the routes used are flexible, S&S and can be self-optimize. 33, 34, 19, 18, 17, 42.
055	Make efforts to use intelligent process of packing. 31, 19, 17.
056	Use flexible packaging, intelligent and benign with the staff and the environment. 31, 41, 19, 17, 42.
057	Locate all the facilities of the organization in places that facilitate the integration and the S&S. 30, 19, 42.
058	Establish measures to ensure good communications with the staff of transport. 34, 33, 32, 19, 18.
	Area: Inverse Logistics
059	Establish S&S, flexible and autonomous policies for the management of IL. 38, 36, 37, 19, 42.
060	Ensure that C&R is carried out in a flexible and autonomous manner and S&S for staff. 36, 38, 19, 42.
061	Have S&C to assist in the classification and future use of IL products. 37, 38, 19, 17.
062	Ensure that the best possible use of IL products is achieved in an autonomous and flexible manner. 37, 19.
063	Innovate permanently in all aspects of IL. 38, 37, 36, 19.
064	Placing S&C to assist in the minimization of by-products, waste and scrap. 37, 10, 19, 17, 42.
065	Ensure the elimination of waste the most possible Safety and in a benign way with the environment. 37, 38.
066	Use S&C for an adequate management of wastes and effluents of the organization. 37, 38, 19, 42.
	Area: Intrinsic to Logistics
067	Establish policies for distribution and use of the spaces that facilitate the flexibility and autonomy. 01, 19, 42.
068	Create quality policy, through S&C, with autonomy of action on the processes. 05, 19, 18, 17.
069	To ensure that the different areas have enough S&C to facilitate operations. 01, 19, 17.
070	To care that the E&M have the space necessary to facilitate their autonomy and the IPdM. 01, 19, 09.
071	To intensify the use of ICT to impulse innovation in this area. 01, 05, 30, 19, 17, 18.
072	Do have immediate shelters in good condition, for the case of disasters. 01, 42.
073	Perform a constant control of the processes and products to ensure their quality. 05, 19, 42.
074	Choose local to the different facilities that simplify the autonomy and flexibility. 30, 19, 42.
	Area: Supported by Logistics
075	Take into consideration the autonomy, flexibility and S&S at any expansion project. 15, 39, 19, 42.

continued on following page

Table 3. Continued

076	Make forecasts that it allows to anticipate the needs of E&M and S&C. 35, 19, 42.
077	Implement projects that contribute to autonomy, integration and flexibility. 39, 19, 18, 17.
078	Maintain and encourage innovation in the whole organization through new projects. 39, 19.
079	Use S&C that through forecasts detect situations that can compromise the S&S. 35, 19, 42.
080	Promote projects that improve the capacity of ICT and technology in general. 39, 19, 18, 17.
081	Make intensive use of S&C to improve forecasts. 35, 19, 17.
Area: Supporting Logistics	
082	Guarantee the minimization of costs, through an intelligent production system. 10, 19, 18, 17.
083	Ensure that budgets are generated autonomously. 11, 19, 18.
084	Use the analysis of the markets and consumers to improve productivity. 20, 19, 18, 17.
085	Provide the necessary training to the personnel, for the proper handling of the NT. 26, 19, 17.
086	Create workshops and courses for staff to improve the management of their communications. 26, 19.
087	Reviewing the legal framework that may affect the organization, especially in the management of the ICT. 40.
088	Set all policies of legal protection against situations of risk and S&S. 40.
089	Create S&C to assist in the defense of the environment. 41, 19, 18, 17.
Area: Information	
090	Set the policies for the management of information and of the NT for the entire organization. 18, 19.
091	Encourage innovation in the whole organization, especially through the NT. 19, 18, 17, 16, 15.
092	Create systems and mechanisms that make it possible to react to any failure in communications. 18, 19, 17.
093	To provide the organization with a platform that facilitates the smart production. 17, 18, 19.
094	Perform a constant R&D, to provide improvements in the field of NT. 19.
095	Analyze all accidents and failures of systems that have occurred, however trivial they may seem. 18, 42.
096	Ideate S&C and systems to detect any failure as early as possible. 18, 17, 19.
097	Supporting the virtual channels to be a way of integration with society. 16, 18, 42.
098	Constantly analyze of all progress in CPPS and ICT. 19, 18, 17.
099	Maintain a constant monitoring of the information circulating in and from the organization. 18.
100	Work on the evaluation and acquisition of E&M and S&C, that offering higher yields. 17, 18, 19.
101	Create models to detect any violation of preventive measures. 18, 42.
Area: Costumer Service	
102	Set all policies concerning relations with stakeholders and between internal clients, as well as the S&S of the organization. 44, 43, 42.
103	Work with State institutions to improve the S&S and the flexibility of the organization. 43, 44, 19, 18.
104	Seek a better relationship with all stakeholders through innovation. 44, 43, 42, 19, 18.
105	Get the greatest possible integration, of the consumer, with the organization. 44, 43, 19, 18.
106	Receiving support from the stakeholders, to improve the integration, flexibility, and the S&S of the organization. 43.
107	View contingency plans and set S&C that minimize negative impacts on society. 43, 19.
108	Permanently make analysis of risks that the organization may be subject. 42, 18.
109	Develop methods and S&C for the prevention of accidents and unsafe situations. 42, 19, 18, 17.
110	Establish management indicators to measure the performance of the organization, particularly in S&S. 42, 19.
111	Ensure an autonomous and S&S management of all the flows of the organization. 42, 19, 18, 17.

works of García et al. (2017), Jeney (2014) and Schwarz et al. (2016); (2): See the work of Tarazón et al. (2016); (3): See the work of Barreto et al. (2016).

These identifications shall have the same meanings in Table 5, where in addition it will be used (4) to indicate that it can reach some arguments in Table 4. In Table 7, which will be incorporated (5) to indicate that there is a position of MoLoBaC dedicated to this attribute, (6) to point out that all the positions of the MoLoBaC cooperate in this and (7) to indicate that there are arguments presented in Table 5.

It is enough to look with some attention at Table 4, to understand that Industry 4.0 and the MoLoBaC have many aspects in common, perhaps, many more, than one could think at first. This allows to affirm, without any doubt, that a total coexistence of MoLoBaC with the environment of Industry 4.0 is possible.

Perhaps the aspect in which they find less coincidence is the handling of technology, since Industry 4.0 does it with greater intensity than the MoLoBaC, however, it should be noted the statement of Buhr (2015), which says that Industry 4.0 is more than pure technology. On the other hand, it can be affirmed that MoLoBaC is more than logistics, since the model tries to cover functions that relate to practically all the positions of an organization. That is to say, both Industry 4.0 and MoLoBaC are aimed at improving industrial performance, hence there are so many aspects in which they coincide or complement each other.

With this brief comment, the qualitative analysis is completed and then, the two quantitative analyzes will be presented.

Comparing the Industry 4.0 Environment With the MoLoBaC, Through an MOWwMf and a MMwMf

First, it will analyze the possibilities of using MoLoBaC directly in industries that follow an approach of Industry 4.0. That is, through Matrixes Of Weighing with multiplicative factors (MOWwMf), it will evaluate the possibility of applying the MoLoBaC directly in industries that operate in an environment of Industry 4.0. Second, the possibilities of using the MoLoBaC in teaching will be measure through Multiattribute Models with multiplicative factors (MMwMf) that will determine the strength of the MoLoBaC, to participate in the training of new staff that requires Industry 4.0.

A MOWwMf to Evaluate the Application of MoLoBaC in Industry 4.0

It will use a Matrixes Of Weighing with multiplicative factors (MOWwMf) type four, that is to say with different evaluation ranges and with multiplicative values. The eight aspects presented in Table 4 are used as criteria and the two alternatives to be evaluated will be: MoLoBaC must be used in the Industry 4.0 environment (Yes used) and MoLoBaC should not be used in the Industry 4.0 environment (Do Not Use). These two alternatives will not be seen as complementary, one would obtain the points that are missing from the other, but each one will be measured, according to the disadvantages and advantages they represent for Industry 4.0.

Criteria, by ordering them from major to minor, with their ranks and its multiplicative values are: Technology (Tec, 9) (0 to 100), Communication (Com, 7) (0 to 70), Productivity (Pro, 5) (0 to 60), Innovation (Inn, 5) (0 to 50), Safety & Security (S&S, 4) (0 to 50), Autonomy (Aut, 4) (0 to 30), Flexibility (Fle, 3) (0 to 30) and Other aspects (OtA, 2) (0 to 20), where the numbers next to the three letters representing each criterion, correspond to the respective multiplicative value.

Table 4. Qualitative analysis between Industry 4.0 and MoLoBaC

Less Common or Least Coincident Elements	
Aspect	**Comment (Reinforcement Arguments)**
Autonomy (Aut)	Essential in Industry 4.0, while the MoLoBaC analyzes the logistics through the functions of the positions. These positions interact with each other, which in a way makes them less autonomous. (A07, A17, A36, B04, B06, B08) (1).
Communication (Com)	Mainly between humans in MoLoBaC, in Industry 4.0, also gives importance to the communication between machines. (A23, A24, A31, B36, B43).
Flexibility (Fle)	Industry 4.0 looks for flexibility. The MoLoBaC is essentially flexible. (B01, B02, B12, B17, B23) (1).
Innovation (Inn)	The MoLoBaC creates the ambit to generate it in the whole organization through logistics, while Industry 4.0; it does, mainly, through technology. (A15, A18, B21, B25) (2).
Productivity (Pro)	For Industry 4.0 is a goal in itself. MoLoBaC seeks to improve logistics performance. And through its relationship with the whole organization, it is possible to improve productivity in general. (A29, B01, B03, B04, B12, B16) (1).
Safety & Security (S&S)	For MoLoBaC is one of the main objectives to pursue. Industry 4.0 is one more objective. (B27) (3).
Technology (Tec)	It could say that it is the main objective of Industry 4.0. In the MoLoBaC the reason being is the customer attention. (Almost all the A and many of the B) (1).
Other aspects (OtA)	Industry 4.0 and the MoLoBaC are in permanent evolution and both have as their ultimate aim improve of industrial performance, but the MoLoBaC searches through the logistics and Industry 4.0, makes it through technology. (1) (2) (3).
Most Common or Most Coincident Elements	
Aspect	**Comment (Reinforcement Arguments)**
Autonomy (Aut)	For Industry 4.0 is a goal in itself, is not so for the MoLoBaC, however, through a large number of its functions, is an objective to pursue. (A07 to A09, A11, A13, A17, A33, B04, B06, B08, B32, B45, 010, 012, 015, 018, 022, 026, 029, 036, 039, 040, 042, 048 to 050, 053, 054, 059, 060, 062, 067, 068, 070, 075, 077, 083, 111).
Communication (Com)	Industry 4.0 achieves this through intensive use of the Internet and CPS. The MoLoBaC has an entire area, Information, dedicated to improving it. (A01 to A06, A11, A18, A23 to A25, A30 to A32, A38, B03, B09, B21, B33, 001 to 005, 009, 019, 023, 031, 045, 047, 058, 071, 080, 086, 087, 090 to 102).
Flexibility (Fle)	It is one of the objectives of Industry 4.0. The MoLoBaC is flexible by definition and although the positions are interrelated, each one it has complete independence. (A13, A36, A37, B01, B02, B04, B10, B12, B13, B17, B48, B49, 011, 017, 029, 037, 039, 048, 050, 053, 054, 056, 059, 060, 062, 067, 070, 074, 075, 077, 103, 106) (1).
Innovation (Inn)	In the MoLoBaC is vital to achieve better customer service. In Industry 4.0 is a necessity, especially from the technical point of view. (A01, A11, A15, A21, A22, A41, B04, B12, B22, B30, B35, B37, B48, B49, 002, 016, 023, 033, 044, 051, 063, 071, 078, 091, 100, 104) (2).
Productivity (Pro)	For Industry 4.0, it is a goal to pursue. The MoLoBaC helps to improve it to follow its main goal of providing better service to the customer. (A08, A13, A24, A29, A37, A40, A41, B01 to B07, B10 to B13, B16, B17, B20, B24, B25, B32, B39, B41 to B44, B47, B49, 001, 007, 010, 012, 024, 037 to 039, 055, 061, 062, 064 to 066, 068, 069, 073, 082, 084, 093, 096, 100) (1).
Safety & Security (S&S)	It is a fundamental need for both, the MoLoBaC and Industry 4.0. (B27, 006, 007, 009, 013, 014, 021, 029, 030, 037, 041, 046, 052 to 054, 057, 059, 060, 065, 072, 075, 079, 088, 095, 099, 101 to 103, 106 to 111) (3).
Technology (Tec)	The MoLoBaC has all a stage, General of information, for technological improvements and for Industry 4.0, through the CPS and CPPS is its main engine. (Almost all the A, in particular: A01, A07 to A27, some B and many functions of table 3, in particular: 090 to 101).
Other aspects (OtA)	Both Industry 4.0, as the MoLoBaC, supported especially in the quality and respect for the environment, pursue adapt to new markets and provide better products and services to the client. (A05, A09, A35, B01, B04, B07, B10, B11, B21 to B24, B26, B30, B32, B35, B38, B44, B48 to B51, 031, 032, 056, 065, 068, 073, 087 to 089, 102) (1) (2) (3).

The order of the criteria, as well as their valuation ranges and multiplicative values, are given by the relevance to Industry 4.0. Multiplicative values, in this case, have the function of reward alternatives that comply with the respective criteria. Therefore, have higher multiplicative values the most important criteria.

The multiplicative factor Release of Resources (RoR) will be used, which will be linear from zero to one, where zero is whether the consumption of new resources is high and one if the release of resources is high. In this case, the high or not, can be quantified by the amount of money saved or invested.

Before constructing the MOWwMf, in Table 5, will present the arguments that justify the values with which it shall be weighted each one of the two alternatives.

As mentioned earlier these arguments will be reinforced by aspects of Table 2 and functions of the Table 3, which are showed in Table 4, for which, as already indicated, it will be used (4) to point out that these arguments are shown in Table 4. Table 5 presents a qualitative assessment, but it will be placed in brackets arguments and parentheses a numerical rank equivalent, in a 0 to 100 scale, which must be reinterpreted, at the time of building the MOWwMf.

Table 5. Values to build the MOWwMf that compares Industry 4.0 with MoLoBaC

Criteria and Multiplicative Factor	Alternatives	
	Do Not Use	Yes Used
	Impact on Industry 4.0 [Arguments] (Value Range)	
Technology	High [Although MoLoBaC does not make very important contributions, it not to use it neither improvement the situation (4)] (71 - 90)	Medium [Perhaps the weakest of the MoLoBaC aspect, but use the same help to incorporate them (4)] (45 - 60)
Communication	Low [Opportunities for improving it are lost by not to incorporate to the MoLoBaC (4)] (30 - 44)	Very High [The MoLoBaC can be a great contribution for them (4)] (92 - 100)
Productivity	Low [Not using MoLoBaC, does not benefits it (4)] (30 - 40)	High [The use of MoLoBaC would improve it, especially in logistics (1) (4)] (72 - 85)
Innovation	Very Low [Not using MoLoBaC would lose many opportunities (2) (4)] (10 - 20)	Very High [MoLoBaC offers great opportunities (2) (4)] (95 - 100)
S&S	Very Low [Not using MoLoBaC miss opportunities to improve (3) (4)] (10 - 20)	Very High [The MoLoBaC offers great contributions (3) (4)] (97 - 100)
Autonomy	Low [Some opportunities for improvement would be lost by not using the MoLoBaC (4)] (30 - 40)	High [The MoLoBaC can offer opportunities for improvement (4)] (75 - 90)
Flexibility	Low [Missed opportunities to improve it by not using the MoLoBaC (4)] (35 - 44)	High [There are good opportunities for improvement when using the MoLoBaC (4)] (73 - 90)
Other aspects	Medium [By not using MoLoBaC, opportunities would be lost in relation to the quality, the ethical, the environmental (4)] (50 - 70)	High [The use of MoLoBaC would guarantee improvements, in the environmental, in quality, in the ethics, among other aspects (4)] (71 - 85)
RoR	Does not generate higher costs, but does not produce savings. 0.90	Does not generate higher costs, but can produce small savings. 0.95

Regarding this scale will be based on: Very Low (0 - 20), Low (21 - 44), Medium (45 - 70), High (71 - 90) and Very High (91 - 100), which is often used with this type of models.

With the values in Table 5 is passed to build the MOWwMf (Table 6), which is very easy because it is only two alternatives.

For each of the payments two values will be reflected, one direct and the other, in parentheses, corresponding to the product of multiplying by the respective multiplicative value, to the direct value obtained. In the first row and below Total two values have been included, which correspond to the maximum achievable values, either directly and after multiplying by each of the multiplicative values. This second value appears in parentheses.

In Table 6, under the column Total it can see the value obtained by each alternative, before applying the multiplicative factor, in this case for Do Not Use 176.8 without multiplicative values and 1180.0 with the multiplicative values that represent percentages of 43.12% and 49.37% of the total achievable.

While for Yes used were obtained 329.8 without multiplicative values and 1823.1 with multiplicative values, which represent the 80.44% and the 76.28% respectively.

In a normal analysis, for a MOWwMf, alternatives are often rejected with percentages of compliance less than 40% and more than 70% of compliance is required to consider that an alternative is good. Therefore, with the totals obtained, the alternative Yes used, should be considered a good alternative, while Do Not Use, should be checked more carefully.

However, when applying multiplicative factors, Do Not Use would be reduced to 159.1 (38.81%) and 1062.0 (44.44%), what would approach it to become an alternative to be discarded. Meanwhile Yes used would fall to 313.31 (76.42%) and 1732.0 (72.47%), although this last value is slightly low, would still remain a good alternative. Thus, it can be said that the MoLoBaC will have a good field of application within industries, the ambit of the Industry 4.0.

To finalize next is realized the evaluation of the possibilities of use in education that can have the MoLoBaC.

Evaluation of the Use of MoLoBaC in the Training of Personnel for Industry 4.0.

To measure the possibility of using the MoLoBaC, as an additional tool to train the personnel of the industries in the environment of Industry 4.0, Multiattribute Models with multiplicative factors (MMwMf) will be used.

Table 6. MOWwMf to evaluate the possibility of applying MoLoBaC in Industry 4.0

Alternatives	Criteria								Total 410 (2390)
	Tec (9) (0 - 100)	Com (7) (0 – 70)	Pro (5) (0 – 60)	Inn (5) (0 – 50)	S&S (4) (0 – 50)	Aut (4) (0 – 30)	Fle (3) (0 – 30)	OtA (2) (0 – 20)	
Do Not Use	80.5 (724.5)	25.9 (181.3)	21.0 (105.0)	7.5 (37.5)	7.5 (30.0)	10.5 (42.0)	11.9 (35.7)	12.0 (24.0)	176.8 (1180.0)
Yes used	52.5 (472.5)	67.2 (470.4)	47.1 (235.5)	48.8 (244.0)	49.3 (197.2)	24.8 (99.2)	24.5 (73.5)	15.6 (31.2)	329.8 (1823.1)

The used criteria will be the eight previously presented in Table 5 and Table 6 and next to the attributes are presented in Table 7, where, as already mentioned, it will be used (1), (2), (3) and (4), with the same meanings as in Table 5 and in addition they are incorporated (5) to indicate that there is a position of the MoLoBaC dedicated to this attribute, (6) to point out that all positions of MoLoBaC cooperate in this and (7) to indicate that they were arguments shown in Table 5.

As done in Table 5, will present a qualitative assessment (with the representation of a numerical range in parentheses) of each one of the attributes. As in the previous case a unique multiplicative factor will be used, in this case, Integrate, which will be interpreted as the ability or not has the MoLoBaC to integrate different knowledge of the industry.

With regard to the weights of the criteria and the attributes, was used as a starting point the Decreasing primes (Dp) and were adjusted according to the judgment of the experts. To get a better idea of the use of DP and its adjustments it is recommended to review the work of García, Hernández and Hernández (2015).

These values in the Table 7 help to complete the evaluation of the MMwMf shown in Table 8, while only considering an alternative, use of the MoLoBaC in teaching in the environment of Industry 4.0 (Use MoLoBaC).

To obtain the total value is use Excel™, but can be done by hand, as they are very simple calculations.

Table 7. Attribute values and multiplicative factor of the MMwMf used to compare Industry 4.0 with the MoLoBaC

Criteria	Weight Criteria	Attributes	Weight Attributes	Qualitative Evaluation (Arguments) (Value Range)
Technology	0.25	New technologies	0.50	High (4) (5) (75 - 90).
		Information systems	0.30	Very High (5) (95 - 100).
		Virtual channels	0.20	High (5) (85 - 90).
Communication	0.19	Internal	0.55	Very High (4) (5) (95 - 100).
		External	0.45	Very High (4) (5) (92 - 100).
Productivity	0.16	Through the logistics	0.80	Very High (4) (6) (7) (92 - 100).
		Through other activities	0.20	High (4) (6) (7) (75 - 90).
Innovation	0.14	In logistics	0.70	Very High (2) (6) (7) (97 - 100).
		Throughout the organization	0.30	Very High (2) (6) (7) (95 - 100).
S&S	0.13	Safety	0.60	Very High (3) (6) (7) (97 - 100).
		Security	0.40	Very High (3) (6) (7) (97 - 100).
Autonomy	0.06	In logistics activities	0.75	Very High (4) (6) (7) (95 - 100).
		In other activities	0.25	Very High (4) (6) (7) (92 - 100).
Flexibility	0.05	In the strategic	0.40	Very High (4) (6) (92 - 100).
		In the tactical-operative	0.60	High (4) (6) (80 - 90).
Other aspects	0.02	Environmental	0.50	Very High (5) (7) (97 - 100).
		Juridical	0.30	Very High (5) (7) (97 - 100).
		Ethical	0.20	Very High (5) (7) (95 - 100).
Integrate				Very High (1) (2) (3) (4) (7) (0.97)

In Table 8 the value obtained by the alternative evaluated, Use MoLoBaC, is 91.32, after applying the multiplicative factor.

This value is extremely high, which indicates that there are very good expectations for the use of the MoLoBaC, to assist in the training of the staff of the environment of Industry 4.0.

However, this very high valuation is not a surprise, not only by the above results, the qualitative analysis and the first quantitative analysis, but by the same characteristics of the MoLoBaC.

Having completed this second part of the quantitative analysis, one can go on to present some lines of future research and to have some conclusions.

FUTURE RESEARCH DIRECTIONS

This chapter covers mainly two major themes, business logistics and Industry 4.0, but in addition a couple of multicriteria models have been used. That is why three major areas of research are opened for the immediate future.

With respect to Industry 4.0, it is a field of industrial activity, which is in full development, so that many aspects can be and should be deepened. In relation to what is presented in this chapter should

Table 8. MMwMf to assess the possibility of applying MoLoBaC in Industry 4.0

Criteria	Weight Criteria	Attributes	Weight Attributes	Valuation of Use MoLoBaC
Technology	0.25	New technologies	0.50	82.5
		Information systems	0.30	97.5
		Virtual channels	0.20	87.5
Communication	0.19	Internal	0.55	97.5
		External	0.45	96.0
Productivity	0.16	Through the logistics	0.80	96.0
		Through other activities	0.20	82.5
Innovation	0.14	In logistics	0.70	98.5
		Throughout the organization	0.30	97.5
Safety & Security (S&S)	0.13	Safety	0.60	98.5
		Security	0.40	98.5
Autonomy	0.06	In logistics activities	0.75	97.5
		In other activities	0.25	96.0
Flexibility	0.05	In the strategic	0.40	96.0
		In the tactical-operative	0.60	85.0
Other aspects	0.02	Environmental	0.50	98.5
		Juridical	0.30	98.5
		Ethical	0.20	97.5
Factor: Integrate				0.97
Total				**91.32**

continue to deepen the study of Industry 4.0 as a whole. In particular, it must investigate the potential of Industry 4.0 based on logistics, but also from other fields of industrial concern, as could be the markets, communications in general, innovation and the generation and knowledge management. Three fields are mentioned which were discussed in this paper, but which require many more deeper studies, the first is the technology, in particular the Cyber-Physical Systems (CPS) and the Cyber Physical Production Systems (CPPS) and their integration through the Internet, the second is the Safety & Security (S&S) and the third is the training of staff who must perform in this new industry.

In terms of logistics, mechanisms should be implemented to facilitate the incorporation of the Logistic Model Based on Positions (MoLoBaC) into the Industry 4.0 environment, both for the training of the staff, as well as a tool to be used in the industries. Additionally, it can explore the application in Industry 4.0, the other three logistic models mentioned in this work: the Logistic model Supply, Production, Distribution and Inverse (LAPDI); the Logistic Model Based on Indicators for Positions (MoLoBaICa) and the Logistic, Strategic, Tactical, Operational with Inverse Logistics Model (MoLETOI).

Finally there are the multicriteria models, in which case two possibilities are presented, one is to use the Matrixes Of Weighing with multiplicative factors (MOWwMf) and the Multiattribute Models with multiplicative factors (MMwMf) and other multicriteria models to further analyze the Industry 4.0 environment, in greater depth and the other is to make a deeper use, even outside of Industry 4.0, of the two models presented here, the MOWwMf and the MMwMf, view their robustness, simplicity and ease of implementation.

In addition to each of these three fields separate, Industry 4.0, logistics and Multicriteria models, is the integration of the same. Already mentioned, it is advisable to deepen the study of Industry 4.0, from different points of view of the logistics, these studies could be complemented with quantitative assessments, carried out through multicriteria models or other mathematical models. It would also be interesting to the valuation of different aspects of Industry 4.0, through multicriteria models whether, or not, with the presence of the Logistics.

Finally, it is advisable to open lines of research which make use of the material presented in this chapter, together with the presented in other chapters of the book and in this way to achieve greater integration of knowledge of Industry 4.0.

CONCLUSION

After having presented the Industry 4.0 environment, the Logistic Model Based on Positions (MoLoBaC) and a couple of multicriteria models, very easy to implement, the Matrixes Of Weighing with multiplicative factors (MOWwMf) and the Multiattribute Models with multiplicative factors (MMwMf), a qualitative analysis and a pair of quantitative analyzes were made to measure the relationship between MoLoBaC and Industry 4.0. It is fulfilling, with all this, the objectives of this research.

From all this study, there are several conclusions that can be obtained; the first is the challenge which means the environment of Industry 4.0, which receives this name to be considered the fourth industrial revolution.

But, as more closely related to the objectives of this work, it can be stated, through the qualitative analysis, that the MoLoBaC and Industry 4.0, share a lot of common aspects, which would certainly facilitate their coexistence.

This coexistence was ratified by the two quantitative analyzes, the first of them, carried out through a MOWwMf, confirmed that if there is a good field of application for the MoLoBaC in the industries that perform in an environment of Industry 4.0.

And the second quantitative analysis, carried out through a MMwMf, left in clear evidence, that MoLoBaC is an excellent tool, to be applied in the process of training the personnel that will be part of Industry 4.0.

And both in the training of the personnel needed for Industry 4.0, and for the industry itself the MoLoBaC offers a series of fields where its contribution can be used directly, with a great impact, just to mention the most outstanding ones, it can be said that with the help of the MoLoBaC, better performances can be achieved in: Autonomy, Communication, Environment, Flexibility, Innovation, Productivity, Quality, Safety & Security (S&S) and Technology.

REFERENCES

Almada-Lobo, F. (2016). The Industry 4.0 revolution and the future of manufacturing execution systems (MES). *Journal of Innovation Management, 3*(4), 16–21.

Barreto, O. E. A., Hernández, G. G. J., García, G. M. J., & Hernández, R. J. G. (2016). Security & Safety (S&S) and the Logistics Business. In *Proceedings ICIL2016*, AGH University of Science and Technology in Krakow, Krakow, Poland (pp. 12-23).

Borisov, O. I., Gromov, V. S., Kolyubin, S. A., Pyrkin, A. A., Bobtsov, A. A., Salikhov, V. I., … Petranevsky, I. V. (2016). Human-free robotic automation of industrial operations. In *Industrial Electronics Society, IECON 2016, 42nd Annual Conference of the IEEE* (pp. 6867-6872). 10.1109/IECON.2016.7793922

Brettel, M., Friederichsen, N., Keller, M., & Rosenberg, M. (2014). How virtualization, decentralization and network building change the manufacturing landscape: An Industry 4.0 Perspective. *International Journal of Mechanical, Industrial Science and Engineering, 8*(1), 37–44.

Buhr, D. (2015). *Social innovation policy for Industry 4.0. Friedrich-Ebert-Stiftung*. Germany: Division for Social and Economic Policies.

Cheng, F. T., Tieng, H., Yang, H. C., Hung, M. H., Lin, Y. C., Wei, C. F., & Shieh, Z. Y. (2016). Industry 4.1 for wheel machining automation. *IEEE Robotics and Automation Letters, 1*(1), 332–339. doi:10.1109/LRA.2016.2517208

De Burgos, J. J., García, G. M. J., Hernández, G. G. J., & Hernández, R. J. G. (2016). Generation and management knowledge. A view from the manager of packing. In *Proceedings ICIL2016*, AGH University of Science and Technology in Krakow, Krakow, Poland (pp. 32-44).

De Felice, F., & Petrillo, A. (2012). Hierarchical model to optimize performance in logistics policies: Multi attribute analysis. *Procedia: Social and Behavioral Sciences, 58*, 1555–1564. doi:10.1016/j.sbspro.2012.09.1142

Domingo, G. L. (2016). *The Challenges of Logistics 4.0 for the Supply Chain Management and the Information Technology* [Master dissertation]. Norwegian University of Science and Technology (NTNU), Trondheim, Norwegian.

Flores, S. A. A., Li, Y., Chen, W. N., Zhan, Z. H., Zhang, J., & Chen, L. Y. (2015). Industry 4.0 with cyber-physical integration: A design and manufacture perspective. In *2015 21st international conference on automation and computing (ICAC)* (pp. 1-6). IEEE.

García, G. M. J., Hernández, G. G. J., & Hernández, R. J. G. (2015). An approach to the ranking of projects to undertake. In A. Cobos & A. Vanti (Eds.), *Gobernanza empresarial de Tecnología de la Información* (pp. 427–441). Santander, España: Universidad de Cantabria.

García, G. M. J., Schwarz, I. L. M., Schwarz, I. T. M., Hernández, G. G. J., & Hernández, R. J. G. (2017). Inventories control, the Inventory manager and Matrixes Of Weighing with multiplicative factors (MOWwMf). *Journal of Global Business and Technology, 13*(1), 40–56.

Gath, M., Herzog, O., & Edelkamp, S. (2014). Autonomous and flexible multiagent systems enhance transport logistics. In *11th International Conference & Expo Emerging Technologies for a Smarter World (CEWIT) 2014* (pp. 1-6). IEEE. 10.1109/CEWIT.2014.7021143

Gilchrist, A. (2016). *Industry 4.0. The industrial internet of things*. New York: Apress.

Gölzer, P., Cato, P., & Amberg, M. (2015). Data Processing Requirements of Industry 4.0-Use Cases for Big Data Applications. In *Twenty-Third European Conference on Information Systems (ECIS)*, Münster, Germany (pp. 1-13).

Grangel-González, I., Halilaj, L., Coskun, G., Auer, S., Collarana, D., & Hoffmeister, M. (2016). Towards a semantic administrative shell for industry 4.0 components. In *IEEE Tenth International Conference on* Semantic Computing (ICSC) 2016 (pp. 230-237).

Hermann, M., Pentek, T., & Otto, B. (2016). Design principles for industrie 4.0 scenarios. A literature review. In *49th Hawaii International Conference on* System Sciences (HICSS*)* (pp. 3928-3937). IEEE.

Hernández, J. G., & García, M. J. (2010). Mathematical models generators of Decision Support Systems for help in case of catastrophes. An experience from Venezuela. In E. Asimakopoulou & N. Bessis (Eds.), *Advanced ICTs for Disaster Management and Threat Detection: Collaborative and Distributed Frameworks* (pp. 201–220). Hershey, PA: IGI Global. doi:10.4018/978-1-61520-987-3.ch013

Hernández, J. G., García, M. J., & Hernández, G. J. (2016). Social aspects of reverse logistics and Knowledge Management. In A. K. Goel & P. Singhal (Eds.), *Product Innovation through Knowledge Management and Social Media Strategies* (pp. 65–93). Hershey, PA: IGI Global. doi:10.4018/978-1-4666-9607-5.ch004

Hernández, R. J. G., Hernández, G. G. J., & García, G. M. J. (2017). Some sins in the negotiations in times of Industry 4.0: Soccer and BREXIT. In Delener et al. (Eds.), *Nineteenth annual International Conference Global Business and Technology Association* (pp. 224-232). Reading Book.

Ho, W., Xu, X., & Dey, P. K. (2010). Multi-criteria decision making approaches for supplier evaluation and selection: A literature review. *European Journal of Operational Research, 202*(1), 16–24. doi:10.1016/j.ejor.2009.05.009

Jeney, A. (2014). *Impacto del Gerente de Sistemas de Información y Redes del Modelo Logístico Basado en Cargos en la gestión del conocimiento de una organización, medido a través de una Matriz De Ponderación* [Master dissertation]. Universidad Metropolitana, Caracas, Venezuela.

Kodym, O., & Unucka, J. (2016). Gathering Information from Transport Systems for Processing in Supply Chains. *Ocean Engineering*, *6*, 676–680.

Kolberg, D., & Zühlke, D. (2015). Lean automation enabled by industry 4.0 technologies. *IFAC-PapersOnLine*, *48*(3), 1870–1875. doi:10.1016/j.ifacol.2015.06.359

Lee, J., Bagheri, B., & Kao, H. A. (2015). A cyber-physical systems architecture for industry 4.0-based manufacturing systems. *Manufacturing Letters*, *3*, 18–23. doi:10.1016/j.mfglet.2014.12.001

Lee, J., Kao, H. A., & Yang, S. (2014). Service innovation and smart analytics for industry 4.0 and big data environment. *Procedia CIRP*, *16*, 3–8. doi:10.1016/j.procir.2014.02.001

Lom, M., Pribyl, O., & Svitek, M. (2016). Industry 4.0 as a part of smart cities. In *Smart Cities Symposium Prague (SCSP)* (pp. 1-6). IEEE. 10.1109/SCSP.2016.7501015

Menicocci, G. (2015). *Disruptive technologies and their impact on the labor market* [Master dissertation]. Libera Università Internazionale degli Studi Sociali Guido Carli, Rome, Italy.

Monostori, L. (2014). Cyber-physical production systems: Roots, expectations and R&D challenges. *Procedia CIRP*, *17*, 9–13. doi:10.1016/j.procir.2014.03.115

Mucha, H., Büttner, S., & Röcker, C. (2016). Application Areas for Human-Centered Assistive Systems. In Human-Computer Interaction–Perspectives on Industry 4.0. Workshop at i-KNOW 2016, Graz, Austria (pp. 1-4).

Posada, J., Toro, C., Barandiaran, I., Oyarzun, D., Stricker, D., de Amicis, R., ... Vallarino, I. (2015). Visual computing as a key enabling technology for industrie 4.0 and industrial internet. *IEEE Computer Graphics and Applications*, *35*(2), 26–40. doi:10.1109/MCG.2015.45 PMID:25807506

Roblek, V., Meško, M., & Krapež, A. (2016). A Complex View of Industry 4.0. *SAGE Open*, *6*(2), 1–11. doi:10.1177/2158244016653987

Rüßmann, M., Lorenz, M., Gerbert, P., Waldner, M., Justus, J., Engel, P., & Harnisch, M. (2015). Industry 4.0: The future of productivity and growth in manufacturing industries. Boston Consulting Group.

Sanayei, A., Mousavi, S. F., & Yazdankhah, A. (2010). Group decision making process for supplier selection with VIKOR under fuzzy environment. *Expert Systems with Applications*, *37*(1), 24–30. doi:10.1016/j.eswa.2009.04.063

Schmidt, R., Möhring, M., Härting, R. C., Reichstein, C., Neumaier, P., & Jozinović, P. (2015). Industry 4.0-potentials for creating smart products: empirical research results. In *International Conference on Business Information Systems* (pp. 16-27). Springer International Publishing. 10.1007/978-3-319-19027-3_2

Schwarz, I. L. M., Schwarz, I. T. M., García, G. M. J., Hernández, G. G. J., & Hernández, R. J. G. (2016). Social Impact of Restrictions on Inventory Management. In *Proceedings ICIL2016*, AGH University of Science and Technology, Krakow, Poland (pp. 272-282).

Seitz, K. F., & Nyhuis, P. (2015). Cyber-physical production systems combined with logistic models–a learning factory concept for an improved production planning and control. *Procedia CIRP*, *32*, 92–97. doi:10.1016/j.procir.2015.02.220

Shafiq, S. I., Sanin, C., Szczerbicki, E., & Toro, C. (2015). Virtual engineering object/virtual engineering process: A specialized form of cyber physical system for Industrie 4.0. *Procedia Computer Science*, *60*, 1146–1155. doi:10.1016/j.procs.2015.08.166

Sommer, L. (2015). Industrial revolution-industry 4.0: Are German manufacturing SMEs the first victims of this revolution? *Journal of Industrial Engineering and Management*, *8*(5), 1512–1532. doi:10.3926/jiem.1470

Speicher, M., Tenhaft, K., Heinen, S., & Handorf, H. (2015). Enabling Industry 4.0 with holobuilder. In Informatik 2015 (pp. 1561-1575).

Stock, T., Bietz, T., Alva, D., Swat, M., Seliger, G., & Bähre, D. (2015). Design and Control of Manufacturing Systems for Enabling Energy-Efficiency and-Flexibility. *KES Transactions on Sustainable Design and Manufacturing*, *II*, 438–449.

Stock, T., & Seliger, G. (2016). Opportunities of sustainable manufacturing in Industry 4.0. *Procedia CIRP*, *40*, 536–541. doi:10.1016/j.procir.2016.01.129

Tarazón, C. J. C., Hernández, G. G. J., Lee, B. C., García, G. M. J., & Hernández, R. J. G. (2016). *Innovation Manager, Management Knowledge Generation, Logistics and some Multicriteria models. In Eureka 2016* (pp. 1–13). México: Torreón.

Valdez, A. C., Brauner, P., Schaar, A. K., Holzinger, A., & Ziefle, M. (2015). Reducing complexity with simplicity-usability methods for industry 4.0. In *Proceedings 19th Triennial Congress of the IEA* (pp. 14-21).

Waibel, M. W., Steenkamp, L. P., Moloko, N., & Oosthuizen, G. A. (2017). Investigating the Effects of Smart Production Systems on Sustainability Elements. *Procedia Manufacturing*, *8*, 731–737. doi:10.1016/j.promfg.2017.02.094

Wang, K. (2016). Intelligent Predictive Maintenance (IPdM) System–Industry 4.0 Scenario. *WIT Transactions on Engineering Sciences*, *113*(1), 259–268.

Wang, S., Wan, J., Li, D., & Zhang, C. (2016). Implementing smart factory of industrie 4.0: An outlook. *International Journal of Distributed Sensor Networks*, *12*(1).

Wong, J. C., & Solon, O. (2017, May 12). Massive ransomware cyber-attack hits nearly 100 countries around the world. *The Guardian*.

Zawadzki, P., & Żywicki, K. (2016). Smart product design and production control for effective mass customization in the Industry 4.0 concept. *Management and Production Engineering Review*, *7*(3), 105–112. doi:10.1515/mper-2016-0030

Zhou, K., Liu, T., & Zhou, L. (2015). Industry 4.0: Towards future industrial opportunities and challenges. In *12th International Conference on* Fuzzy Systems and Knowledge Discovery (FSKD) (pp. 2147-2152). IEEE.

ADDITIONAL READING

Aisyati, A., Jauhari, W. A., & Rosyidi, C. N. (2014). Periodic Review Model for Determining Inventory Policy for Aircraft Consumable Spare Parts. *International Journal of Business Research & Management*, *5*(3), 41–51.

Aljazzar, S. M., Jaber, M. Y., & Goyal, S. K. (2016). Coordination of a two-level supply chain (manufacturer–retailer) with permissible delay in payments. *International Journal of Systems Science: Operations & Logistics*, *3*(3), 176–188.

De, S. K., & Sana, S. S. (2015). Multi-criterion multi-attribute decision-making for an EOQ model in a hesitant fuzzy environment. *Pacific Science Review A. Natural Science and Engineering*, *17*(2), 61–68.

Dobrin, C. (2013). Industry 4.0, M2M, IOT&S - All equals? In IPC'13 (pp. 1-6).

Dye, C. Y., & Yang, C. T. (2016). Optimal dynamic pricing and preservation technology investment for deteriorating products with reference price effects. *Omega*, *62*, 52–67. doi:10.1016/j.omega.2015.08.009

García, G. M. J., Hernández, G. G. J., & Hernández, R. J. G. (2014). A Methodology of The Decision Support Systems applied to other projects of Investigation. In Encyclopedia of Information Science and Technology (3rd ed., Vol. 3, pp. 1978-1990). Hershey, PA: IGI Global.

García, G. M. J., Hernández, G. G. J., & Hernández, R. J. G. (2014). Knowledge management through the Material Handling Manager. In G. Dukic (Ed.), *Proceedings ICIL'*2014, University of Zagreb, Croatia (pp. 264-272).

García, M. J., Hernández, G. J., & Hernández, J. G. (2013). Enterprise diagnosis and the STOILMo. In Delener et al. (Eds.), *Fifteenth annual International Conference Global Business And Technology Association* (pp. 330-338). Reading Book: GBATA.

García, M. J., Hernández, G. J., & Hernández, J. G. (2014). Logistics flows through the functions of the Cost Manager. In Delener et al. (Eds.), *Sixteenth annual International Conference Global Business And Technology Association* (pp. 185-192). Reading Book: GBATA.

Guajardo, M., Rönnqvist, M., Halvorsen, A. M., & Kallevik, S. I. (2015). Inventory management of spare parts in an energy company. *The Journal of the Operational Research Society*, *66*(2), 331–341. doi:10.1057/jors.2014.8

Guerrero, M. L. E., Hernández, G. G. J., García, G. M. J., & Hernández, R. J. G. (2014). Indicators and the Picking manager of the Logistic Model Based on positions. In G. Dukic (Ed.), *Proceedings ICIL'2014*, University of Zagreb, Croatia (pp. 273-282).

Hernández, J. G., García, M. J., & Hernández, J. G. (2012). Dynamic knowledge: Diagnosis and Customer Service. In N. Delener (Ed.), *Service Science Research, Strategy, and Innovation: Dynamic Knowledge Management Methods* (pp. 558–584). Hershey, PA: IGI Global. doi:10.4018/978-1-4666-0077-5.ch030

Hernández, R. J., García, G. M., & Hernández, G. G. (2013). Enterprise logistics, indicators and Physical distribution manager. *Research in Logistics & Production*, *3*(1), 5–20.

Hernández, R. J. G., García, G. M. J., & Hernández, G. G. J. (2012). The Utilization manager and logistic flow. In G. Dukic (Ed.), *Proceedings ICIL'*2012, University of Zagreb, Croatia (pp. 456-462).

Hernández, R. J. G., García, G. M. J., & Hernández, G. G. J. (2016). Logistics, Marketing and Knowledge Management in the Community of Consumers. In S. Buckley, G. Majewski, & A. Giannakopolus (Eds.), *Organizational Knowledge Facilitation through Communities of Practice in Emerging Markets* (pp. 242–268). Hershey, PA: IGI Global.

Hernández, R. J. G., García, G. M. J., & Hernández, G. G. J. (2016). Enterprise diagnosis and the Environmental manager of LoMoBaP. *Journal of Global Business and Technology*, *12*(1), 60–76.

Hudnurkar, M., Jakhar, S., & Rathod, U. (2014). Factors affecting collaboration in supply chain: A literature review. *Procedia: Social and Behavioral Sciences*, *133*, 189–202. doi:10.1016/j.sbspro.2014.04.184

Khedlekar, U. K., Shukla, D. & Namdeo, A. (2016). Pricing policy for declining demand using item preservation technology. *SpringerPlus*, *5*(1957), 1-11.

Mousa, S. A. (2016). Macro Perspective of Industry 4.0: Factors Affecting Value Creation Modules. *The China Business Review*, *15*(10), 457–465.

Nagashima, M., Wehrle, F. T., Kerbache, L., & Lassagne, M. (2015). Impacts of adaptive collaboration on demand forecasting accuracy of different product categories throughout the product life cycle. *Supply Chain Management*, *20*(4), 415–433. doi:10.1108/SCM-03-2014-0088

Saldivar, A. A. F., Goh, C., Chen, W. N., & Li, Y. (2016). Self-organizing tool for smart design with predictive customer needs and wants to realize Industry 4.0. In *2016 IEEE Congress on Evolutionary Computation (CEC)* (pp. 5317-5324). IEEE.

Sarkar, B., & Moon, I. (2014). Improved quality, setup cost reduction, and variable backorder costs in an imperfect production process. *International Journal of Production Economics*, *155*, 204–213. doi:10.1016/j.ijpe.2013.11.014

Schlechtendahl, J., Keinert, M., Kretschmer, F., Lechler, A., & Verl, A. (2015). Making existing production systems Industry 4.0-ready. *Production Engineering*, *9*(1), 143–148. doi:10.100711740-014-0586-3

Shekarian, E., Olugu, E. U., Abdul-Rashid, S. H., & Kazemi, N. (2016). Analyzing optimization techniques in inventory models: the case of fuzzy economic order quantity problems. In *Proceedings of the 2016 International Conference on Industrial Engineering and Operations Management* (pp. 1229-1240).

Sibatrova, S. V., & Vishnevskiy, K. (2016). Present and Future of the Production: Integrating Lean Management into Corporate Foresight (Working paper BRP 66/STI/2016). National Research University Higher School of Economics, Moscow, Russia.

Veža, I., Gjeldum, N., & Mladineo, M. (2017). Integrating Digital Factory, Lean Management and Industry 4.0 into the Learning Factory–Case Study at University of Split. In *The 7th International Congress on Engineering and Information* (pp. 1-12).

Wang, K., Dai, G., & Guo, L. (2016). Intelligent Predictive Maintenance (IPdM) for Elevator Service. In *International Workshop of Advanced Manufacturing and Automation (IWAMA 2016)* (pp. 1-6).

Waschneck, B., Altenmüller, T., Bauernhansl, T., & Kyek, A. (2016). Production Scheduling in Complex Job Shops from an Industry 4.0 Perspective: A Review and Challenges in the Semiconductor Industry. In SAMI@ iKNOW (pp. 1-12).

Wu, L., Garg, S. K., Buyya, R., Chen, C., & Versteeg, S. (2013). Automated SLA negotiation framework for cloud computing. In *2013 13th IEEE/ACM International Symposium on Cluster, Cloud and Grid Computing (CCGrid)* (pp. 235-244).

KEY TERMS AND DEFINITIONS

Enterprise Logistics: Is centered in searching and achieving a greater satisfaction, present and future of the final costumer, and includes socio-environmental and ethical-legal aspects, organization planning, execution and control of all related activities related to the attainment, flow, gathering and maintenance of materials, products and services, since the raw material source, including there the costumers through inverse logistics, to the sale point of the finished product local or international, massive or enterprise, in a more effective and efficient, minimizing the times and costs and using modern information technologies.

Industry 4.0: It is a concept in formation, therefore in evolution. It identifies the fourth industrial revolution, which is characterized by the integration of the virtual world with the real world, through the Internet and the Cyber-Physical Systems (CPS). Among other objectives Industry 4.0 pursues productive systems characterized by intelligent, adaptable, flexible, efficient, green, scalable, coherent manufacturing with high quality at low costs, with efficient resource management and ergonomics, as well as integration of customers and business partners in business and value processes.

Logistics Model: Are all those created, explicitly or accidentally, to justify and understand the business logistics. In particular, four qualitative-quantitative models created in the academy, to explain the logistics, should be highlighted: Logistic model Supply, Production, Distribution and Inverse (LSPDI [LAPDI]); Logistic Model Based on Positions (LoMoBaP [MoLoBaC]); Logistic, Strategic, Tactical, Operational with Inverse Logistics Model (STOILMo [MoLETOI]) & Logistic Model Based on Indicators for Positions (LoMoBaIPo [MoLoBaICa]).

Logistic Model Based on Positions (MoLoBaC): Studies enterprise logistic through the functions relative to the positions. It consists of forty-four positions which are grouped into twelve areas that meet in six stages.

Matrixes Of Weighing With Multiplicative Factors (MOWwMf [MDPcFm]): A Matrixes Of Weighing (MOW) is one of the multicriteria models simplest to build and easy to implement and are nothing else, as its name implies, then an array of rows and columns, through which and following a set of weights on the appropriate criteria, different alternatives can be weighted and hierarchized, where both these alternatives and the criteria can have a quantitative or qualitative origin. A MOW becomes a MOWwMf when multiplying factors are applied to the results obtained by the same.

Multiattribute Models With Multiplicative Factors (MMwMf [MMcFm]): It is another of the multicriteria models very simple to build and implement. With them the alternatives are evaluated through the attributes, which correspond to the different criteria. In the evaluation process, each alternative obtains a payment for each of the attributes. From there the evaluation of each criterion is carried out, making the weighted sum of the products of the values of the alternative by the weight of the respective attribute. When obtaining the evaluation of each one of the criteria, which can be corrected by the multiplicative factors of the criterion, the total valuation is made. The values of each of the criteria produce the general payment through a weighted sum, where the result of each criterion is multiplied by its respective weight. The final result is corrected through the general multiplicative factors.

Safety & Security (S&S): Security is referred to situations those threatens the organization, as well as the lives of its members and is associated with intentional facts (malicious facts), while Safety is also associated with detriment to organizations and that especially affects humans, but are accidents that are not caused intentionally. The Safety of the plant is one of the biggest concerns for industrial activities, because they can involve major economic losses and what is more important, human lives. And the security takes as a principal intention to protect the company and his assets of threats. Both Security and Safety are focused on protecting the assets of an organization, the most important being the human resource, which is why S&S must cover all areas of the organization.

Chapter 12
Ceramic Industry 4.0:
Paths of Revolution in Traditional Products

João Barata
Technological Centre for Ceramics and Glass (CTCV), Portugal & University of Coimbra, Portugal & Polytechnic Institute of Coimbra, Portugal & Miguel Torga Institute (SMT), Portugal

Francisco Silva
Technological Centre for Ceramics and Glass (CTCV), Portugal & University of Minho, Portugal

Marisa Almeida
Technological Centre for Ceramics and Glass (CTCV), Portugal

ABSTRACT

Industry 4.0 presents new challenges for traditional sectors of the economy, for example, the production of ceramic products. This chapter reveals how traditional ceramic industries can (1) assess, (2) plan, and (3) execute Industry 4.0 adoption. The findings are based on the Portuguese ceramic sector. Three interrelated dimensions of the fourth industrial revolution are studied, namely, (1) digital ecosystems, (2) security and safety, and (3) digital sustainability. Industry 4.0 is not restricted to high-tech products and cannot be addressed by one-size-fits-all solutions. Moreover, it requires cooperation within business ecosystems. The authors propose a model for Ceramic Industry 4.0 and accessible guidelines for managers involved in global supply chains. This chapter suggests emergent research opportunities for (1) sectorial maturity models, (2) data quality and regulatory compliance, (3) cyber-security and risk management, and (4) an integrated vision of sustainability in the digital era.

INTRODUCTION

Industry 4.0 is changing traditional sectors of the economy (Brettel & Friederichsen, 2014). The impact of the forth industrial revolution is particularly relevant in small and medium sized enterprises (SMEs) with high levels of manual work. This is the case of ceramic companies that export the majority of their production and must be prepared to compete at a global scale. The ceramic industry from the European Union (EU)-27 accounts for 23% of global ceramics production. According to the Eurostat, it represented a production value of 28 billion Euros in Europe and over 200.000 direct jobs in 2015.

DOI: 10.4018/978-1-5225-4936-9.ch012

Ceramic industry could be divided in ten major sub-groups: bricks and tiles, floor and wall tiles, sanitaryware, pottery & tableware, refractories, abrasives, clay pipes, expanded clay, porcelain enamel, and technical ceramics. All these ceramic industry subsectors are energy intensive, namely due to the drying and firing processes, which involve firing temperatures between 800 and 2000 ºC. The manufacture of ceramic products is a complex interaction of raw-materials, technological processes, people, and economic investments. It includes the transport and storage of raw materials, ancillary materials and additives (e.g. deflocculating agent – sodium silicate for preparation of raw materials), preparation of raw materials, shaping, drying, surface treatment, firing, and subsequent treatment (Quinteiro, Almeida, Dias, Araújo, & Arroja, 2014). Complexity of the production process is diverse and also the market requirements are different for each ceramic industry sub-group. Yet, the entire sector is affected by the fourth industrial revolution.

There are new technological opportunities for ceramic production. Recent examples include the use of mobile technologies in maintenance and product traceability (Barata, Cunha, Gonnagar, & Mendes, 2017), additive manufacturing, 3D printing, and simulation platforms (Smit, Kreutzer, Moeller, & Carlberg, 2016). However, Industry 4.0 in mineral non-metal manufacture raises many challenges for managers. We subscribe to the view of Oesterreich and Teuteberg (2016, p. 136) about the "urgent need for the development, understanding and assessment of frameworks, business models, reference models and maturity models for Industry 4.0 implementation with focus on technology, people and processes". Industry 4.0 assessment models tailored for specific sectors of the economy will be essential. Other challenges include the creation of digital competencies (Prifti, Knigge, Kienegger, & Krcmar, 2017), the development of digital ecosystems (Andersen & Ross, 2016), improvement of work practices, and sustainable development (Chen et al., 2015). Moreover, there is an urgent need to identify and deploy pilot cases to guide the major changes towards industry of the future.

This chapter addresses Industry 4.0 in traditional sectors and specificities of mineral non-metal production in Portugal. The next section presents the background of our research. Afterwards, we identify challenges and opportunities in three key dimensions for the ongoing industrial revolution in ceramic, namely, digital ecosystems, safety and security, and digital sustainability. Next, we present the results of a field study and propose strategic recommendations. These developments emerged from a 120 participants' workshop that mobilized the entire industry. The chapter concludes revealing future research directions in the scope of digital transformation of ceramic production.

Background

Industry 4.0 is gaining increasing attention by researchers worldwide. A keyword search made in Google Scholar using a combination of the terms "Industry 4.0", "Industrie 4.0", and "Fourth Industrial Revolution" reveals a constant growth in the recent years, especially since 2013. There are several databases available for scientific research (e.g. Scopus, EBSCO, B-on …), but we decided to start with Google Scholar because it presents a broad search result of both academic and practitioners contributions. Figure 1 illustrates this trend.

Industry 4.0, *Industrie 4.0*, or *Usine du Futur* are examples of the terms used to identify a priority for Europe: Industry digitalization. The examples include the digital single market (European Commission, 2016c), the mobility of business processes within the entire supply chain and the upgrade for an integrated digital world with profound socio-technical implications. The term Industry 4.0 was initially coined as a reference for high tech policies of the German government. However, digitalization and

Figure 1. Evolution of Google scholar results about the topic of Industry 4.0

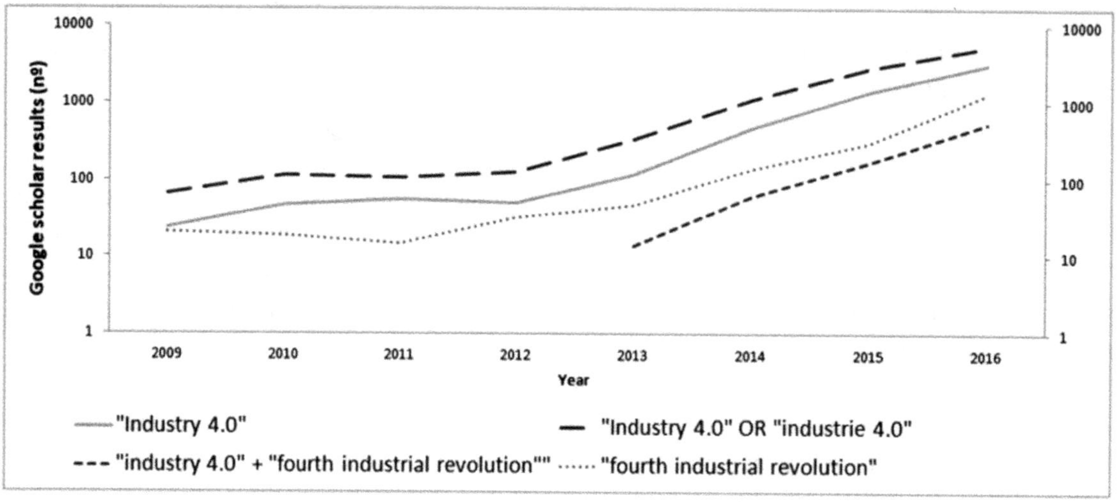

cyber-physical systems are still in an immature state, especially in traditional sectors of the economy, for example, ceramic and glass manufacturing. Later, the German government include other policies in parallel to Industry 4.0, namely, sustainability, nanotechnology, and internet-based services, requiring an integrated approach by managers. As stated by Prof. Klaus Schwab, "…we are in the midst of the Fourth Industrial Revolution, which will affect governments, businesses and economies in very substantial ways. We should not underestimate the change ahead of us…" (Schwab, 2015).

In November 2016, Jean-Claude Juncker reinforces the idea that "…digital technologies are going into every aspect of life. All they require is access to high speed internet. We need to be connected, our economy needs it, people need it…" (European Commission, 2016a). In his speech we can identify several important figures, for example, 90% of the professions will soon require digital qualifications, online commerce represents a saving of eleven billion Euros, and there is a priority to support cloud initiatives and the Internet of Things (IoT). Three main goals are identified: (1) improve connectivity, (2) create a better context for business, and (3) promote growth and employment. The European Commission mobilizes fifty thousand million Euros of public and private funding for industry digitalization. Moreover, the agenda aims at the creation of new competencies for the digital era (European Commission, 2016b).

According to Smit et al. (2016), the scenario in each country and each economic sector differs and it is possible to identify specific requirements for Industry 4.0 implementation, as presented in Figure 2.

The standardization is a top priority, referring to the open systems and platforms needed to connect the different elements of supply chains. Next, it is necessary to redesign business processes (Vidgen & Wang, 2006). It is also necessary to create new business models supported by information systems. Mobile and cloud also poses new challenges for ciber-security. Finally, the study also points to the importance of research investments, social aspects, and a legal framework.

And what happens in the case of traditional products that represent a significant part of the world economy? For example, in construction "only 19% of engineering and construction companies have advanced data analytics capabilities" and "it simply won't be possible for companies to achieve advanced digitization without making a step change in investment, given the continued rapid progress anticipated by companies who are already leading" (PwC, 2016). There are several barriers to consider, as presented in Figure 3.

Figure 2. Requirements for Industry 4.0 – Results presented by BITKOM, VDMA and ZVEI in 2013, considering 278 companies (Acatech, 2013; Smit et al., 2016)

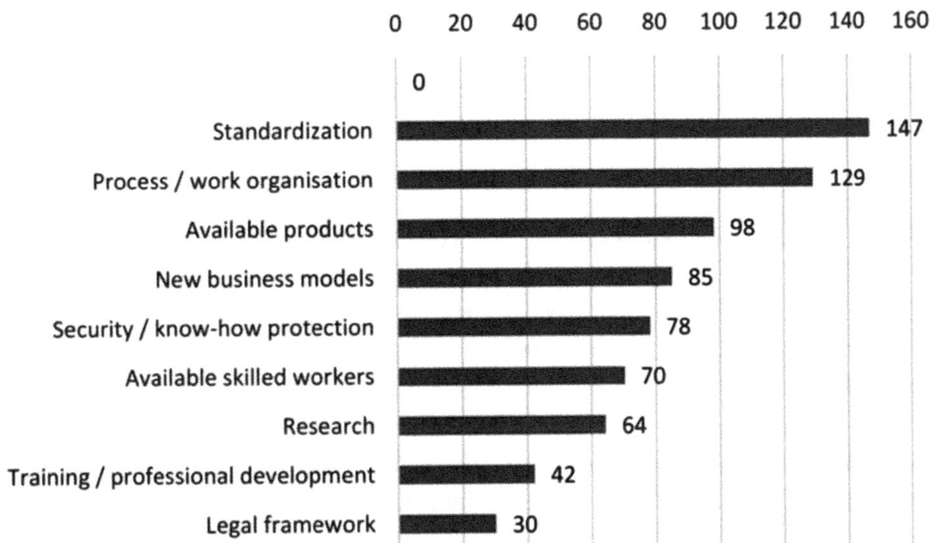

Figure 3. Barriers for industry digitalization: Global Expert Survey 2016 by McKinsey (Bauer et al., 2016)

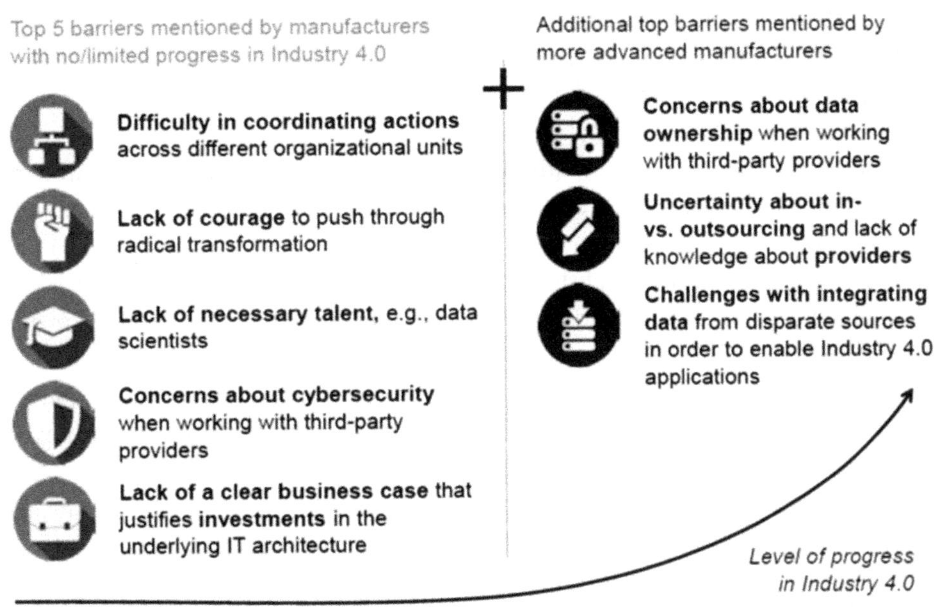

There are several barriers that managers face to implement Industry 4.0, including organizational aspects, confidentiality, and integration in the supply chain. These barriers are particularly relevant in traditional sector of the economy, mostly supported by small and medium sized companies; more difficulties to cooperate with universities in advanced research; and fewer resources to invest in technological innovation. As a consequence, traditional sectors must joint efforts with industry associations to

(1) diagnose their Industry 4.0 maturity level, (2) create a tailored roadmap according to the needs and opportunities of the supply chain, and (3) implement pilot projects to guide their Industry 4.0 efforts. The importance of pilot projects is also highlighted by Bauer et al. (2016) "providing the right implementation support both for an initial pilot, and for scaling the efforts across different sites is crucial to succeed".

The next section presents specific challenges for traditional products that we identified in our contacts with Portuguese ceramic industries.

INDUSTRY 4.0: THE CHALLENGE FOR TRADITIONAL PRODUCTS

This section addresses three key dimensions for Industry 4.0 research in ceramic industry: (1) digital ecosystems, (2) safety and security, and (3) digital sustainability. The first dimension includes technological aspects of Industry 4.0 and specific guidelines for industry digitalization. Next, we approach the social perspective of Industry 4.0, considering the new risks for people but also opportunities to improve work conditions. Lastly, we address the sustainability element that is so crucial to ceramic industries, namely, the environmental and energy elements of Industry 4.0.

Digital Ecosystems

Digitalization is determinant to Industry 4.0 (Smit et al., 2016; Zhou, 2013) and involves social (Degryse, 2016; Prifti et al., 2017), technical (Leyh, Schäffer, Bley, & Forstenhäusler, 2017), and organizational (Weill & Woerner, 2015) challenges. According to Weill and Woerner (2015), companies should use digital technologies to increase their knowledge about the consumers and lead the creation of digital ecosystems involving multiple business partners.

A vision of business ecosystem was initially suggested by Moore (1996) and is inspired in biological ecosystems. In this context, businesses cannot be planned and managed apart from the environment. Organizations evolve trough symbiotic relationships between the business and other elements of the ecosystem, for example, their partners, customers, and suppliers. As a consequence, business innovation requires moving beyond the organizational borders and managers are advised to plan an ecosystem approach to digital success (PwC, 2016). As stated by Bharadwaj, El Sawy, Pavlou, and Venkatraman (2013) in "a digitally intensive world, firms operate in business ecosystems that are intricately intertwined such that digital business strategy cannot be conceived independently of the business ecosystem, alliances, partnerships, and competitors".

There are important pillars of industry 4.0, namely, cloud computing, mobile connectivity, social aspects, big data and associated analytics, and innovation accelerators such as robotics, additive manufacturing, or the internet of things (Brettel & Friederichsen, 2014; Lasi, Fettke, Kemper, Feld, & Hoffmann, 2014; Smit et al., 2016). Big data means that large amounts of digital data can be analysed to assist the business strategies. Cloud computing and mobile devices are essential to create digital platforms that connect people and businesses around the globe. Smart sensors and location-based technologies are generating new digital data in real time that can be used to assist robot operation and support human decisions. 3D printing is now changing prototyping of new products and also the production process in many industries.

There are several contributions that explain how each pillar of industry 4.0 can contribute to digital ecosystems. Big data will be essential for additive manufacturing with ceramics, one of the most common materials for this form of manufacturing (Wang & Alexander, 2016). Cloud and mobile can support the development of new MES (manufacturing execution systems) platforms, more accessible and tailored for small and medium sized ceramic companies (Barata, Cunha, Gonnagar, & Mendes, 2017). Internet of things can assist energy management in ceramics, for example, building ceramics (Wang, Huang, Chen, Liu, & Xu, 2016) where the energy contributes for an important part of the final product cost, namely it can represent till 30% of the total cost. Huson and Hoskins (2014) studied 3D printing for concept models and ceramic artworks. Each pillar can contribute for the creation of digital ecosystems (Andersen & Ross, 2016; Weill & Woerner, 2015) in the perspective of (1) the digital infrastructure (e.g. Wang & Alexander, 2016) or (2) digital services. Moreover, it is possible to find studies that use a combination of industry 4.0 enablers to propose new solutions for ceramics (e.g. Wang, Huang, Chen, Liu, & Xu, 2016).

The list is vast, but technologies are only one part of the equation. According to several authors (Acatech, 2013; Brettel & Friederichsen, 2014; Lasi et al., 2014; Smit et al., 2016), industries must evaluate its maturity, create a comprehensive digital strategy and implement pilot projects to evolve in their Industry 4.0 efforts. The next section addresses possible tools and approaches to evaluate industry 4.0 readiness and prepare a roadmap.

Design-Time of Industry 4.0: Assessing Maturity Levels and Establishing the Strategy

There are several maturity models for Industry 4.0, however, the existing proposals are recent and there is a lack of proposals that completes the procedure model of development for transfer, evaluation, test, and maintenance (Becker, Knackstedt, & Pöppelbuß, 2009). In fact, some models are still under development, for example, FIR (2017), INTRO4.0 (WP5) Eureka project "Introduction strategies of Industry 4.0 methodology and technology for SMEs" to end in 2018 (KIT, 2016), COTEC maturity model for Portuguese industry, or IPH Hannover (IPH, 2017). Table 1 presents a list of academic and practitioners' models that are available to assess Industry 4.0 and establish a digital strategy.

The studies presented in Table 1 highlight the importance of assessment for Industry 4.0. There are also other studies focusing specific technologies, for example, a reference model and roadmap for Internet-of-Things in manufacturing (Soldatos, Gusmeroli, Malo, & Di Orio, 2016), industrial internet (Menon, Kärkkäinen, & Lasrado, 2016), and cyber-physical systems (Westermann, Anacker, Dumitrescu, & Czaja, 2016). Assessing is the first step, then, it is necessary to take actions to go digital, as presented in the next section.

Run-Time of Industry 4.0: Pilot Projects in the Ceramic Industry

After evaluating Industry 4.0 readiness, organizations must implement their digital strategy. Nevertheless, there are challenges for SMEs. A study promoted by the European Parliament recognizes that one obstacle to the participation of SMEs in the supply chain of Industry 4.0 is the "...capacity to run pilot projects to test out Industry 4.0 mechanisms and potentially limited access to facilities to test advanced solutions..." (Smit et al., 2016). Other barriers to develop Industry 4.0 projects in SMEs include the lack of awareness about technologies, high investments required, the need for specialized IT staff, and the dependency from big companies (Smit et al., 2016). We also found these evidences in the ceramic

Table 1. Models to assess Industry 4.0 maturity

Industry 4.0 Assessment Model	Model Description	Model Stages
(PwC, 2017)	An online self-assessment model for industry 4.0 created by a global consulting company. It suggests to (1) conduct the online self-assessment, (2) identify needs for action, and then (3) benchmark against other companies	Includes four stages (I Digital Novice, II Vertical Integrator, III Horizontal Collaborator, and IV Digital Champion) and six dimensions, namely, (1) Business Models, Product & Service Portfolio; (2) Market & Customer Access; (3) Value Chains & Processes; (4) Information Technology (IT) Architecture; (5) Compliance, Legal, Risk, Security & Tax; and (6) Organization & Culture
(Rockwell Automation, 2014)	The Connected Enterprise Maturity Model is a practitioner's model proposed by Rockwell Automation, a leading company in industrial automation. The model offers guidelines to implement advanced networks of operations technology (OT) and information technology (IT)	Five stages (from 1-assessment to 6-collaboration)
(Isaka, Nagayoshi, Yoshikawa, Yamada, & Kakeno, 2016)	A maturity model for production systems developed by Hitachi. It suggests the use of image analysis as a sensing technique	A plant at level 1 uses data for visualization of their site. Level 2 connection, allows product traceability and level 3 analysis, work automation and process optimization. The following stages are 4 measurement, to identify and solve production bottlenecks, 5 prediction, and the most advanced 6 symbiosis, where resources are optimized and production plans coordinated with company stakeholders
(IMPULS, 2017)	Industry 4.0 readiness self-assessment commissioned by the IMPULS Foundation of the German Engineering Federation (VDMA). It is a comprehensive model that addresses social, organizational, and technological aspects	There are six readiness levels ranging from 0 – insignificant Industry 4.0 activities to 5 - top performers
(Leyh et al., 2017)	The System Integration Maturity Model Industry 4.0 (SIMMI 4.0) considers four dimensions to assess the IT system landscape – vertical integration, horizontal integration, digital product development, and cross-sectional technology criteria	Five stages that start at Stage 1 – Basic digitization level and reaches a maximum of 5 - Optimized full digitization.
(Knoke, Missikoff, & Thoben, 2017)	Collaborative Innovation Capability Maturity for virtual manufacturing enterprises	Five stages aligned with the Capability Maturity Model (Paulk, Curtis, Chrissis, & Weber, 1993)
(Schumacher, Erol, & Sihn, 2016)	Model that includes social, technical, and organizational dimensions to assess Industry 4.0 readiness in manufacturing. These authors considered a total of nine dimensions, each one calculated as a weighted average of different items (62 in total). It is a comprehensive model with radar charts to visualize data and identify improvement priorities	The stage is assessed as a continuous result from 1 to 5.
(Ganzarain & Errasti, 2016)	Focuses on the process of change in diversification strategies. It is necessary to (1) define a vision, (2) establish a roadmap, and (3) implement Industry 4.0 projects ensuring training and risk management	The first stage is 1 – Initial (inexistent industry 4.0 vision) and can reach a maximum of 5 – detailed transformation of business model.

industry, requiring significant efforts from governments and associations to put Industry 4.0 in the managers' agenda.

A tailored roadmap for Industry 4.0 implementation must involve the creation of a digital infrastructure, digital business services (Immonen, Ovaska, Kalaoja, & Pakkala, 2016) and digital business processes (Vidgen & Wang, 2006). The need to complement infrastructure investment (e.g. cloud platform, IoT) with new services and processes becomes clear with the LEGO case study presented by Andersen and Ross (2016). While the initial focus was in the creation of new software platforms and infrastructure acquisition, recent advances include the creation of new digital services accessible to customers and business partners. Yet, a roadmap that expands company borders also requires implementation mechanisms of trust, a key element to ensure cooperation between elements of the supply chain (Grzybowska, Kovács, & Lénárt, 2014) and integration, which is an essential aspect for Industry 4.0.

Portuguese ceramic industry already started to implement Industry 4.0 but the context of ceramic production poses particular problems, as presented in Figure 4.

Figure 4 present the moulding phase of the ceramic production (on the left) and a plant area that mixes moulds used in the process (bellow, on the right) and products that are still under development process. This case is specific to the table and ornamental ware ceramic subsector (other production processes such as tiles of technical ceramic have different characteristics) but it can be used to illustrate the (1) multiple product references simultaneously under production, (2) the fragile characteristics of the product (consistency of the ceramic material) in all the stages (e.g. presenting difficulties to use robots), (3) the low cost of each unit, posing difficulties, for example, in the use of traceability devices such as RFID tags, and (4) the highly manual process that is mostly supported by paper records in SMEs.

In spite of the difficulties that are evident in traditional products such as ceramic, Industry 4.0 is not restricted to high tech industries. In fact, ceramic industries are emerged in global supply chains and must implement systems that adhere to the digital ecosystem needs. Moreover, the complexity to directly implement Industry 4.0 technologies (e.g. robots, simulation, mobile technologies) demands for new cooperation efforts between the industry, the university, ceramic associations and technological centres.

Figure 4. Going digital in traditional manufacturing: The challenge of numerous products

The result of this cooperation is already visible and new pilot projects are under development. Table 2 presents examples of pilot projects and ideas developed in the Portuguese ceramic industry.

The examples presented in Table 2 are not exhaustive; its purpose is to illustrate examples of small scale projects that are accessible to traditional industries. Lessons learned include (1) the need to joint efforts between multiple entities, (2) think about digital ecosystems and not merely internal applications of technology or isolated B2B – Business-to-Business channels, (3) explore pilot projects for horizontal/ vertical integration, increase trust amongst supply chain elements, and introduce improvements that redesign business processes. It seems very appropriate to quote Michael Hammer in this case: "don't automate, obliterate" (Hammer, 1990).

Table 2. Industry 4.0 in the ceramic industry: pilot projects for traditional products

Industry 4.0 Pilot Project	Description
Mobile Manufacturing Execution Systems (mMES)	This project started by CTCV – Technological Center for Ceramics and Glass, aims at the creation of a cloud-based MES for mobile devices. The main purpose of this system is to assist small ceramic industries (that use paper records in almost all the production stages) in digital production records allowing real-time information to the company partners (including suppliers and customers). The use of mobile technologies reduces the financial investment and simplifies the system adoption
Cloud Laboratory Information Management System (cLIMS)	Ceramic products require (internal/external) testing to ensure that product complies with regulations and customer requirements. Although large companies usually have laboratorial data digitalized, small companies continue to use disconnected spreadsheets and paper. Cloud-based LIMS allow to integrate data from external laboratories with the company data, also having specific interfaces for customers and to follow nonconformance actions
Mould digitalization	Moulds are one of the main tools for ceramic production but highly demanding of storage space. Each mould is specific to a product reference and companies consider them a valuable asset to keep. For example, if a customer asks for additional quantities, the ceramic company must use the same mould or create exact replicas. We can find cases of moulds kept for decades that are never needed again. What if we digitalize the moulds and then recreate them when needed using 3D printing devices? The idea is already under development but present difficulties, for example, there are complex moulds with multiple parts that make the "digitalization" – "printing" process costly, requiring algorithms to decide which moulds are economically viable to digitalize – recreate or keep in its materialized form
Digital energy management using simulation, cloud and IoT	Energy management systems are not new. However, most of the existing systems mainly acquire data and raises alerts of energy consumption. Simulation and integration with the production lines will provide additional functionalities to current energy management systems that now include affordable sensors and actuators
Big Data for Marketing and Internationalization	Analysis of Big Data can provide valuable information for designers. The design requirements change across the globe and over time. The potential of Big Data is now being used to assist design trends and market trends (e.g. potential increase of construction in specific countries that justify new local factories or new internalization actions)
Sensing Ceramics	Traditional products such as ceramic tiles can include sensors (e.g. temperature) to assist intelligent houses. It is possible to include new materials in ceramic, for example, sensors and solar systems, as we detail later in this chapter. The incorporation of new elements in ceramic will generate added volumes of digital information
Traceability systems	The use of QR codes in ceramic can assist production control, quality control, and provide digital information to the consumer of traditional products. Recently, a Portuguese ceramic company inserted QR codes in the final product to present specific details that may increase product value (e.g. lot information and quality characteristics, and a video about the production of that specific product reference)
Ceramic Industry 4.0 Maturity model	The creation of a sectorial maturity model for ceramic industry is under development in Portugal, involving universities, companies and the ceramic industry association. The project aims at the development of a tailored model that includes a portfolio of solutions for each ceramic subsector

The social and environmental aspects are equally important in ceramic industry 4.0, for example, to ensure the protection of human workers in an increasingly automated production setting and protect the environment. This is the challenge that we address in the next section.

SAFETY AND SECURITY

Industry 4.0 will promote changes on the way occupational safety and health (OSH) is considered nowadays. If some aspects are predictable, like safety aspects related with the increasing of robots in workplaces, others questions arising are more difficult to predict or could be speculative. In the later, it could be mentioned the psychological impact of disruptive changes in work or the decrease of human error as cause of accidents. Additionally, the dissemination of new technologies in manufacture such as additive manufacturing and nanotechnologies, occurring side-by-side with digitalization, are also factors affecting OSH.

In Table 3 some aspects related to Industry 4.0 and their possible impacts on OSH are presented.

Digitalization can contribute to increase safety and health in workplaces, reducing heavy tasks, eliminating hazardous operations and creating new prevention opportunities, like sensors or training tools. On the other hand, new risks will appear or existing ones will increase, particularly, those related with the robotization and increasing de-humanization of the work. A third category of impacts are the new challenges to OSH, in this particular, the security-safety relation. In the following items, the possible negative impacts and challenges are discussed.

Security of IT Systems and Relation With Safety

The Smart Factory concept (Lucke, Constantinescu, & Westkämper, 2008) poses challenges related to the safety and security. These factors are considered critical to the success of the manufacturing digitalization and, consequently, of business success, since neither processes nor products should represent a risk to persons (including workers and consumers) or environment (Acatech, 2013). In ceramic industry, process safety is a key element and should not be jeopardized by IT security breaches. It is necessary that IT systems are secure against misuse and unauthorized access to prevent modifications or destruction.

Table 3. Industry 4.0 technologies/enablers and their relation to occupational safety

Industry 4.0 Technology/Enabler	Impact on OSH
Robotization	Collaborative robot safety Human training relating to robots Decrease in manual work Hazardous tasks performed by robots
Internet of things	Cyber-security related to safety of cyber-physical systems
Complex embedded software	Hackers or malware as cause of malfunction leading to accidents
Big-data	Increase of capacity to deal with OSH knowledge
Sensors (widespread use)	Workplace environment continuously monitored
Augmented reality	Reduce hazards in maintenance tasks OSH information and warnings in real-time

Besides industrial piracy or information confidentiality, issues like sabotage or terrorist acts are crucial. Even unintended acts could cause accidents if equipment controls are improperly accessed (Acatech, 2013).

As a result of integration between Information and Communication Technologies (ICT) and Industrial Control Systems (ICS) vulnerability to cyber-attacks is an workplace safety issue (Steijn, Vorm, & Luiijf, 2016). Besides the known possible risk emerging from Industry 4.0 there are also other unknown risks that could emerge (Steijn et al., 2016). From the previously exposed, it is clear that the security-safety relation is more complex with Industry 4.0.

Interaction/Integration Human-Robots in the Workplace

Robots are already present in industry, operating in isolated cells, both with physical and/or virtual barriers, usually avoiding the contact with workers. Considering the increase of robots in manufacturing, and most of all the increasing interaction human-robots as one of the main characteristics of Industry 4.0 (Erol, Jäger, Hold, Ott, & Sihn, 2016; Zhou, 2013) the occupational safety aspects are crucial (Fryman & Matthias, 2012). New risks are identifiable while robots will be no longer confined to a location or operating inside a protective cage (segregation paradigm) and collisions between robots and people or other hazardous events could occur (Bicchi, Peshkin, & Colgate, 2008). In a recent study, Dutch researchers identified a set of threats and vulnerabilities related with collaborative robots: change of task, unforeseen situations trust in machines, shared responsibility, regulatory gaps, non-compliance and cyber security (Steijn & Luiijf, 2016). In the same research several control measures are suggested to face the risks arising from the foreseeable interactions.

The basis for safety rules and future regulations are the three laws of robotics[1] defined by Isaac Asimov in 1942 (Magruk, 2016), although those could be adapted or complemented (Steijn & Luiijf, 2016).

Nanotechnologies

In parallel with digitalization of industry it is also expected the increase of the use of nanomaterials. Ceramic industry is already one of the industrial sectors using several nanomaterials (DECHEMA/VCI, 2011) to achieve products with improved proprieties, in particular photo catalytic ceramic tiles and bactericide sanitaryware (Chen & Poon, 2009; van Broekhuizen, van Broekhuizen, Cornelissen, & Reijnders, 2011).

Since a nanomaterial is, in general, more hazardous than the bulk form of same chemical compound, special care should be considered during its use. There are numerous studies in nanotoxicology field pointing to possible harmful effects of nanomaterials to human health and environment (Bleeker et al., 2015). Considering the uncertainty related with nanomaterials, it is important to act with specific concerns on OSH. In short, nanotechnology should be side-by-side with nanosafety (Savolainen et al., 2013) and occupational risk management is crucial (Technical Committee ISO/TC 229, 2012).

Information about OSH aspects related to nanotechnology and, in particular, exposure to nanomaterials is available from several international organizations, among others the European Commission and National Institute for Occupational Safety and Health (NIOSH) in the United States of America (European Commission, 2013; NIOSH, 2009).

There are several methods for exposure and risk assessment methods available, both based on qualitative (Vervoort, 2012) and quantitative methodologies (Duarte, Justino, Freitas, Duarte, & Rocha-Santos, 2014), and the tiered approach for exposure assessment should be followed (Environment Directorate OECD, 2015). Recommendations for occupational risk control during nanomaterials handling are also published (Cornelissen, Jongeneelen, van Broekhuizen, & van Broekhuizen, 2011; NIOSH, 2013), being highlighted the hierarchy of controls (Technical Committee ISO/TC 229, 2012) presented in Figure 5.

Additive Manufacturing

The risks related to additive manufacturing depend most of all of the technology used. Emission of particles and/or vapours from the used materials are concerns common to other manufacturing technologies and the risk will depend, not only of the process but most of all, the hazardous nature of the materials. Considering different additive manufacturing technologies (Afshar-Mohajer, Wu, Ladun, Rajon, & Huang, 2015; Wong & Hernandez, 2012) it is possible to identify several risk factors, such as high temperatures, lasers of different types and hazardous chemicals (both in vapour or particulate form). Since the use of additive manufacturing technologies in ceramic industry is already a reality, in particular 3D printing of models and it is expected to increase, special attention should be given to these aspects. In recent research about exposure to both ultra-fine particles (UFP) and volatile organic compounds (VOC's) there were found exposures of concern (Afshar-Mohajer et al., 2015; Yi et al., 2016).

Psychosocial Aspects and Risks

It is expected that jobs will change and new jobs will emerge with the technological evolution (World Economic Forum, 2016). These changes will raise new challenges related to OSH, not only in consequence of differences in tasks and operations but also the way existing tasks will be performed. In fact, workers will interact with new interfaces and deal with increasing information. It is expected that

Figure 5. Hierarchy of controls in nanotechnology indicating the order to be followed (Technical Committee ISO/TC 229, 2012)

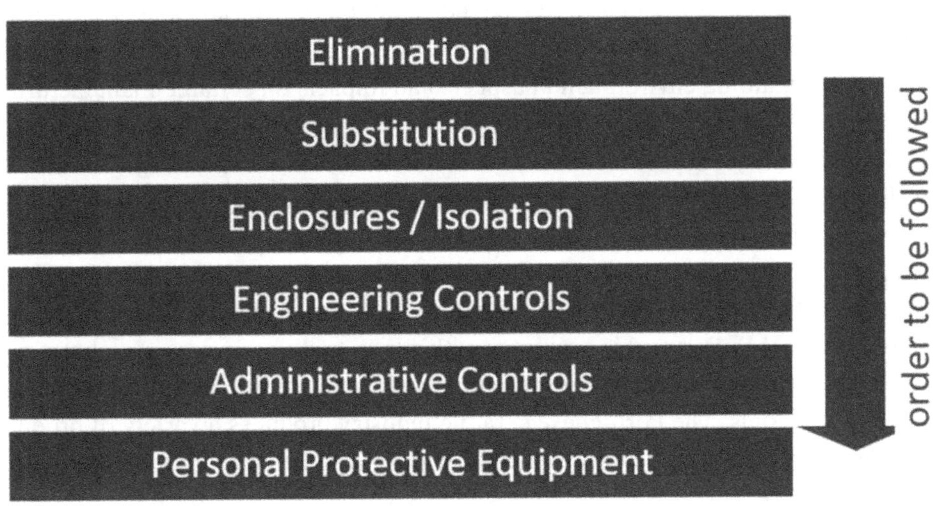

physical demand of work will drop but psychological stress will increase (Gabriel & Pessl, 2016). The emotional and mental stress may raise with increasing flexibility in work and a diminution of communication and cooperation between employees since interaction between humans and machines will increase (Degryse, 2016; Gabriel & Pessl, 2016). Also, the constant work performance evaluation could be cause for increase of stress (Degryse, 2016). On the other hand, employees' work-life balance could improve (Gabriel & Pessl, 2016).

Safety-By-Design

Considering that process automation has been an increasing safety factor (Fadier & De la Garza, 2006), it is expected, and it is important, that safety aspects will be considered during design of Industry 4.0 factories. For the last decades, several researchers in the Safety Science field are calling the attention for the need to improve safety in workplaces considering the hazards in the design phase, developing safer equipment's, processes or products (Fadier & De la Garza, 2006; Hale, Kirwan, & Kjellén, 2007). NIOSH, launched a campaign to reduce risks in workplaces called Prevention Through Design (Schulte, Rinehart, Okun, Geraci, & Heidel, 2008) and more recently, several researchers and institutions raised the importance of safety-by-design in the nanotechnologies field (Morose, 2010; Silva, Arezes, & Swuste, 2016). Furthermore, it is relevant to consider the concept Security by Design in development of IT systems (Acatech, 2013), for the reasons already mentioned.

This section highlights the importance of combining technical and social aspects in the paths of the ongoing industrial revolution. The next section reinforces this need, including the sustainability element that affects the entire society.

Digital Sustainability

The implementation of Industry 4.0 implies a digital sustainability strategy, which aims to minimize environmental impacts while improving operational measures and supporting sustainable growth. Cyber-physical systems require the analysis of ceramic production life cycle in order to optimize the economic, social and environmental risks and opportunities, with the purpose of reduce environmental impacts in all the life cycle at the previous design stage. Additionally, it is necessary to improve the operational efficiency of ceramic processes in order to reduce the depletion of natural resources, pollution and associated ecological impacts. On the other hand, it is expected to acquire more accurate and real-time information and data regarding environmental aspects and impacts on ceramic manufacturing.

According to the European Commission, the construction sector is considered the highest energy consumer in EU, accounting for almost 40% of the total energy consumption and contributing almost 36% to the EU's total greenhouse gas (GHG) emissions and produces 15% of the total industrial waste. Among the most commonly used construction materials, cement and ceramic materials are two of the most energy intensive construction materials. Ceramic products are one of the oldest building materials and generate a series of environmental impacts over their life cycle (Almeida, Dias, Demertzi, & Arroja, 2016; Quinteiro et al., 2014). Portugal is a country with a long tradition in ceramics, both in production and consumption, and is ranked as one of the top European manufacturers of ceramic products due to the high quality of raw materials. The Portuguese ceramic industry produces a variety of products adapted to building works, such as bricks, covering materials, flooring tiles, etc. This industry is responsible for a number of environmental aspects like energy consumption (energy intensive), gaseous emissions

(CO_2, NOx, HF, HCl, heavy metals, particulate matter, etc), liquid emission (suspended solid, chemical oxygen demand, heavy metals) and wastes (broken ware, packaging waste, plaster moulds, sludge, etc).

Assessing the environmental impacts of the different types of ceramic products has become crucial to improving the environmental performance of this sector. Such assessment can be achieved through Life Cycle Assessment (LCA) studies based on ISO 14040 (ISO, 2006a) and ISO 14044 (ISO, 2006b) standards, applied to the different stages of a product's life cycle. Ceramic products have significant impacts during the manufacturing process, namely the firing stage of production is one of the most relevant in terms of environmental impacts. But the type of ceramic raw materials, the kiln operation conditions, electricity consumption, raw materials used and type and distance of transport are key elements that justify the variability in impacts like global warming (climate change), ozone layer, abiotic and fossil resources depletion, eutrophication, acidification, and photochemical oxidation (Almeida et al., 2016; Quinteiro et al., 2014).

Cerame-Unie, representing the ceramic industry in Europe, stresses that resource efficiency requires a Life Cycle Assessment (LCA) approach that takes into account all stages of the product, including its durability, lifespan and reduction of resource consumption over the use phase. Although LCA is a recognized instrument to assess environmental impacts in industry including ceramic one, it requires a lot of environmental data, major collected manually, validation of the quality data and it is very time consumer (Almeida, Barata, Dias, & Arroja, 2015).

Industry 4.0 and digital systems provide a new approach in which physical production processes and ICT grow more closely together (Gabriel & Pessl, 2016). Embedded sensors, systems, mobile devices and production facilities are integrated and able to communicate with each other via the internet, in order to monitoring and controlling the ceramic process in a transparent way. The cyber-physical systems represent a further evolution from the existing embedded systems. The same priority can be identified in direct digital manufacturing (Chen et al., 2015).

On the other hand, the digital systems and the construction and operation of robots will have new environmental aspects and risks, namely the new materials that may be hazardous and potential emissions to environment (e.g. particles and volatile organic compounds (VOC's)). These aspects must be monitored and controlled.

Table 4 presents examples of possible Digital Sustainability Strategies that can be applied in the ceramic industry to achieve a sustainable development.

The integration of digital sustainability strategies is a key dimension for sustainable ceramic industry and the development of "smart ceramic products", changing to a dynamic process that entails continued improvement, diversification and industrial upgrading, and technological eco-innovation throughout the value chain.

DEVELOPING CERAMIC INDUSTRY 4.0: STRATEGIC RECOMENDATIONS

A workshop involving 120 participants from ceramic industry was scheduled to February 2nd 2017, three days after the announcement of Industry 4.0 strategy by the Portuguese government. The morning session had presentations from governmental entities, Industry 4.0 experts, and industry associations. The afternoon aimed to evaluate the perception of maturity in specific dimensions of Industry 4.0 (e.g. vertical integration, competencies) and understand priorities for key processes of the industry (e.g. energy

Table 4. Examples of digital sustainability strategies applied in ceramic process and products

Digital Sustainability Strategy	Key Aspects	Description
Ecodesign or design for sustainability	Sustainability Life cycle thinking	Life cycle thinking – aims at the prevention of environmental impacts throughout their life cycle while eco-innovation and new business opportunities are encouraged and potential cost savings arise
Selection of low-impact materials	Digitalization of quarries Incorporation of waste on a circular economy perspective	Digitalization of quarries (clay, feldspar, Kaolin, etc) in order to have precise information on the quantity and quality and the best way to explore the quarry and at same time the environmental recovery
		The use of by-products by the industry or as raw material for other industries allows cost reduction facilitating the supply of raw materials and the elimination of waste deposit. Main constraints pointed out are related to quality control due to the not homogeneous composition – use of sensors and embedded systems Incorporation of industrial waste for the production of ceramic tiles, in a way that the by-product quality and quantity is continuously monitored and interact with production line in order to avoid defects. For example, a ceramic tile can reach up to 80% of recycled material by weight while retaining the strength and versatility (example: see InEDIC, 2011)
Reduction of the material use	Reduction in products thickness	There are several examples of tiles with reduced thickness and mechanical performance in buildings. IoT can contribute in this purpose, reducing tile thickness from 12 mm to 4 mm (Light ceramic floor tile by Revigres is one example, with resource efficiency through the all life cycle)
Reduction of the environmental impact in the production phase	IT network systems High efficiency process and resource control Innovations and ceramic materials	The optimization of the industrial production process throughout Industry 4.0 and digital sustainability can lead to a reduction of energy, CO_2 and other combustion pollutant emissions, as detailed information on each point of the ceramic production process, resource (e.g. clay, sand, feldspar) and energy use can be monitored and optimized over the entire value chain
		Controlling consumptions (e.g. electricity, natural gas consumption) as well as emissions (gaseous emission, liquid emission) in main unit process like spray-driers, driers and kilns in a continuously optimized IT network system
Reduction of the environmental impact in the use phase	Eco-innovations for energy performance or multifunctionality	Innovation regarding new bricks with high thermal, mechanical and acoustic performances that improves the energy performance of the building or multifunction products. Example: SolarTiles – Integrated photovoltaic into ceramic products for high efficiency for building coatings (roof and facade claddings) incorporating thin-film photovoltaic cells (COMPETE2020, project 3380)
		New materials like ceramic tiles with phase change materials (PCM) to improve thermal characteristic and energy efficiency in buildings. Project examples include ThermoCer developed by Cinca and CTCV in Portugal (QREN, project number 23143) or Selfclean (QREN 21533) - Ceramic coating with self-cleaning properties, purifying functions, high efficiency and durability, by modifying its surface with nanostructured photocatalytic materials
Transport and logistics	Information systems for the distribution of ceramic products Smart mobility, smart logistics	New information systems for the distribution phase of ceramic products will make "smart mobility" and "smart logistics"
Optimizing the installations of end-of-life systems of ceramic materials	Information systems for installation and dismantling	The installation of ceramic products into buildings (e.g. brick or ceramic tile) or the final disposal of products can be simplified and optimized

management involving IoT and predictive algorithms). A mobile app was created to allow interaction between researchers and the workshop participants, as presented in Figure 6.

The mobile app included the possibility to evaluate company maturity on the topics presented and to define the strategic priority of the organization, for Industry 4.0. key processes were selected (e.g. energy management, digitalization) and a discussion about possible solutions occurred (e.g. IoT for energy management, Big data for marketing). The results of voting were used for the debate and to define strategies to develop Industry 4.0 in ceramic industry, presented in Figure 7.

Figure 6. Industry 4.0 Workshop: Mobile app support (extract)

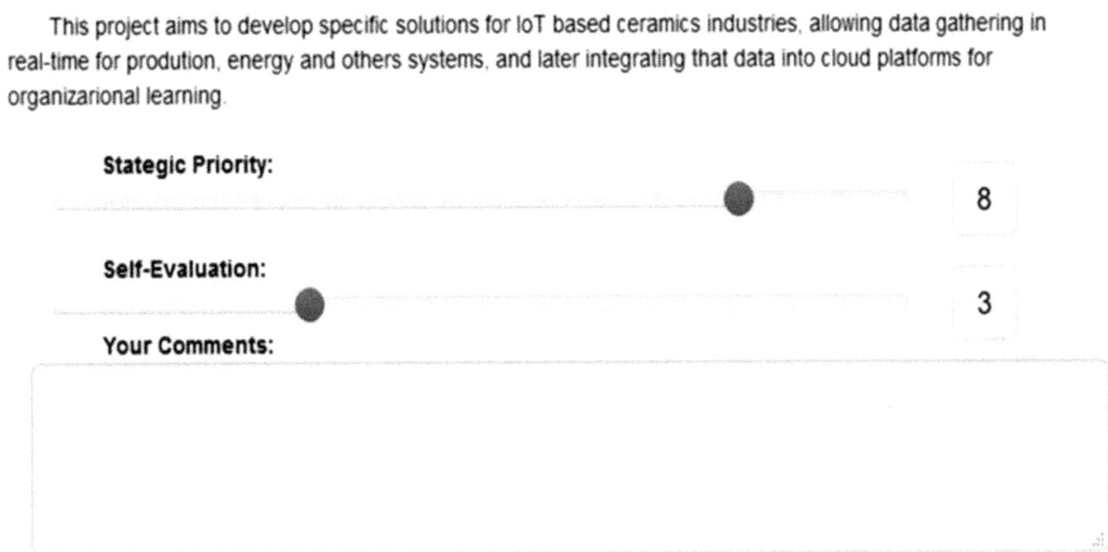

Figure 7. Industry 4.0 Workshop: Strategic priorities for the ceramic industry (extract)

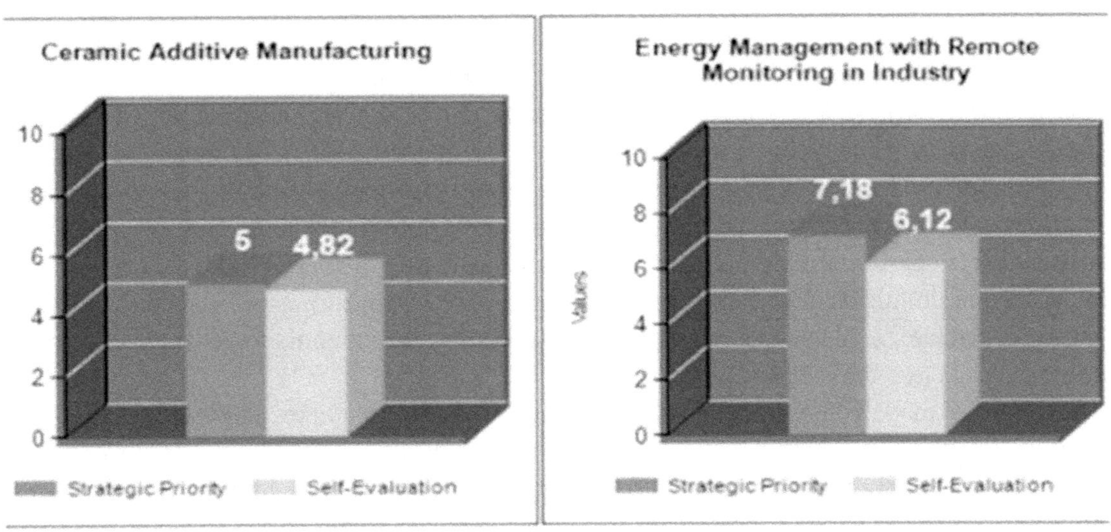

The most disrupting topics of robotization and additive manufacturing were not considered as top strategic priorities for the ceramic industry. The workshop participants were more enthusiastic about the creation of a digital services cloud, energy management solutions, new marketing strategies using Big Data (e.g. design trends for ceramic products). The workshop discussion allowed us to confirm that:

- Not all technological components associated with Industry 4.0 are relevant for their digital strategy;
- A higher maturity stage does not mean that the component has low priority (e.g. the participants considered to be well prepared in terms of energy management, but this topic is so important for ceramic and glass industries that they still consider it a top priority for all Industry 4.0 actions);
- The majority of respondents considered themselves in a positive stage (above 5 in a scale ranging from o to 10) but our field studies show a less developed scenario;
- To be successfully, Industry 4.0 in traditional sectors of the economy should not have a mere technological focus. It is necessary to consider social and organizational aspects, for example, safety and sustainability that we also addressed in this chapter.

As a result, we defined a strategic model towards Ceramic Industry 4.0. Our purpose with this high-level representation is to (1) highlight the importance of cooperation between different elements of the supply chain in traditional products and (2) include social, organizational, and technological elements in the company roadmap for industry 4.0. We used the model to communicate the results of our workshop with the industry. The model suggests that managers must define a multidisciplinary team and create pilot projects that address five interrelated dimensions (1) the industry context (e.g. relation with stakeholders, horizontal integration), (2) people (e.g. safety, competencies), (3) industrial process (e.g. energy reduction via IoT, digitalization via 3D printing), (4) Industry 4.0 Technologies (e.g. cloud platforms and augmented reality systems), and (5) information/data (e.g. data quality and protection. The model is outlined in Figure 8.

The model presented in Figure 8 suggests that digitalization involves several (internal/external) stakeholders of the organization. Therefore, managers should involve different experts in their pilot projects, develop internal competencies for Industry 4.0, and cooperate with their customers, suppliers, and research institutions. Industry 4.0 complexity does not recommend "Industry 4.0 ready solutions" offered by single suppliers.

The model also suggests developing pilot projects within the ceramic process supply chain. For example, (1) using IoT to monitor raw material inventory directly by the suppliers, (2) adopting new robots for manufacturing, (3) use Big Data potential to identify trends in global markets (e.g. design trends that are so crucial to create products that adhere to architecture movements), (4) implement augmented reality to support product selection by the end users or cloud platforms to simplify data integration.

Industry 4.0 technologies will generate large amounts of information/data and use a plethora of technologies (e.g. sensors, algorithms, standards) that evolve faster than ever. More data also represent additional responsibilities such as data protection measures and responsible use of resources (e.g. energy reduction). Therefore, academic partnerships with the industry are essential to get access to emerging developments and lead the industrial revolution.

The simplified model offers guidelines for industrial managers. First, it recommends creating knowledge networks with multiple stakeholders. Horizontal and vertical integration is not possible to achieve in isolation from the supply chain partners and regulatory bodies. Second, it highlights the need to address multiple points of the value chain, once again, involving internal and external stakeholders, for example,

Figure 8. Strategic model for the development of Ceramic Industry 4.0

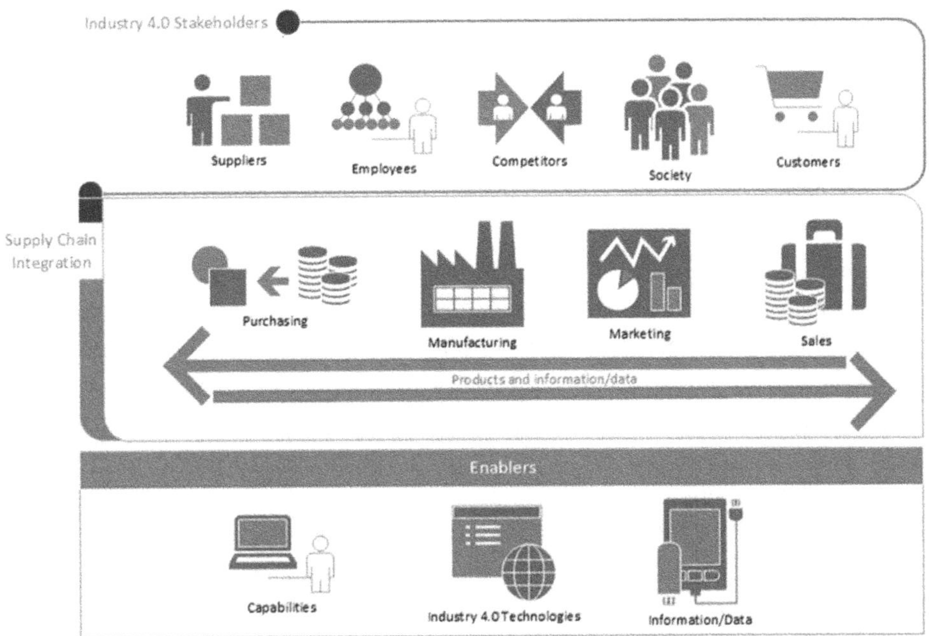

when drastic changes in the process creates the possibility to decentralize production in multiple small units around the globe. Third, Industry 4.0 enablers are complex and require specialized staff in the organization. Examples include new skills in Industry 4.0 technologies and a professional structure to deal with information/data (e.g. protects data vulnerabilities, comply with regulations, generate value with the data, for example, with data scientists). The non-strict vision of industry 4.0 as a mere technological investment can help managers to avoid what Arvidsson, Holmström, and Lyytinen (2014) named as "… strategic blindness: organizational incapability to realize the strategic intent of implemented, available system capabilities…" Next, we explain the avenues for future research that captured the attention from the managers in the Portuguese ceramic industry.

FUTURE RESEARCH DIRECTIONS

Digital ecosystems can determine the survival of the factories of the future. Current research in this area is changing the landscape of production processes, using decentralization (Brettel & Friederichsen, 2014) and new platforms connecting multiple elements of the supply chain. To compete in global markets, companies must prepare their digital infrastructure and create digital services and processes to comply with the requirements of their customers. There are also challenges that include data quality (essential in integrated systems), data protection and privacy, and regulations that improve interoperability between different systems.

Current maturity models to assess Industry 4.0 are too generic to be useful and most of these tools are still under development. It is necessary to create tailored maturity models for different sectors of the economy (Barata & Cunha, 2017). This evidence emerges when we address specificities of each economic sector; one approach to Industry 4.0 in automotive or aeronautical does not necessarily represent an optimum solution in traditional sectors of the economy. Future research is needed to test the effectiveness of the maturity models in practice, the managers' adherence and the utility of the tools to assist roadmaps implementations - not merely as a diagnosis tool.

There are many challenges and opportunities in the security and safety fields in the ceramic industry. Cooperative robot safety and cyber-security are already identified as relevant issues. Also, the additive manufacturing occupational risks and nanotechnologies need special attention from the OSH researchers. The development of sensors to monitor chemical, physical and biological contaminants in workplace environment and to monitor workers biological parameters is another relevant area of future research. The uncertainty related to the role human work in manufacturing industry in the future poses questions in research fields like ergonomics and human factors or psychosocial risks. The new professions arising from the industrial changes will need new OSH paradigms.

The emergence of the Internet of Everything (connected objects and people) will generate vast amounts of data that will help us understand more about the way we interact with each other and improve sustainability in ceramic processes. Moreover, the increasing number of sensors and connected devices puts energy management on the top of researchers' priorities. The challenge to digital sustainability is to identify transformation potential including ways of interacting in a manner that promotes the reduction of the energy consumption, gaseous and liquid emission and waste prevention. On the other hand, relevant research, education, and political constraints (e.g. limiting energy production up to the precise needs of consumers, forecasts of natural events or disasters, preventive maintenance) regarding the sustainability impacts of Industry 4.0 will have to be developed. If sustainable practices are seen as value added to the ceramic process that will act as drivers to sustainable development and eco-innovation, On the contrary, if they are merely seen as a cost burden or a constraint to business and innovation, they will not be successful. New economic and social models will be developed based on a key principle of sustainability.

CONCLUSION

Industry 4.0 affects all sectors of the economy, yet, traditional manufacturing industries face additional challenges and require specific solutions. This chapter addresses three key dimensions of Industry 4.0 in the context of ceramic production, namely, digital ecosystems, safety and security, and digital sustainability. In each case we include examples that are under development in industry, progressively changing the business strategy, the social, and the technological landscape of ceramic production. Additionally, strategic recommendations for Ceramic Industry 4.0 are provided. The guidelines emerged from a recent workshop with 120 participants and preliminary results of research projects in ceramic industry.

There are also limitations to our study that must be stated. Industry 4.0 is a vibrant area of research and many other examples could be important for traditional products. The challenges and recommendations are well supported by theory and practice in the Portuguese ceramic industry, but each sector of the economy can have specificities to consider. For this reason, it is advised to tailor generic models to the need of each economic sector and also to the need of each company. The strategic model presented in this chapter aims to assist industry managers to evolve in Industry 4.0, ensuring that a comprehensive approach is selected, yet, it is a simplification of the complex reality of Industry 4.0.

This chapter can assist researchers in the identification of new opportunities for the industry of traditional products. Small changes can have a major impact in less developed industries. For managers, this chapter presents a multidimensional perspective of Industry 4.0 challenges in ceramics, real examples that can inspire pilot projects, and strategic recommendations for the vision of Ceramic Industry 4.0.

REFERENCES

Acatech. (2013). *Recommendations for implementing the strategic initiative INDUSTRIE 4.0*. Retrieved from http://www.acatech.de/fileadmin/user_upload/Baumstruktur_nach_Website/Acatech/root/de/Material_fuer_Sonderseiten/Industrie_4.0/Final_report__Industrie_4.0_accessible.pdf

Afshar-Mohajer, N., Wu, C. Y., Ladun, T., Rajon, D. A., & Huang, Y. (2015). Characterization of particulate matters and total VOC emissions from a binder jetting 3D printer. *Building and Environment*, *93*(P2), 293–301. doi:10.1016/j.buildenv.2015.07.013

Almeida, M., Barata, J., Dias, A., & Arroja, L. (2015). Environmental product declaration of ceramic roof tiles: a case study of information requirements within the life cycle. In *EfS 2015 – 2nd Energy for Sustainability Multidisciplinary Conference*.

Almeida, M., Dias, A. C., Demertzi, M., & Arroja, L. (2016). Environmental profile of ceramic tiles and their potential for improvement. *Journal of Cleaner Production*, *131*, 583–593. doi:10.1016/j.jclepro.2016.04.131

Andersen, P., & Ross, J. W. (2016). Transforming the LEGO Group for the Digital Economy. In ICIS 2016 Proceedings, Dublin, Ireland.

Arvidsson, V., Holmström, J., & Lyytinen, K. (2014). Information systems use as strategy practice: A multi-dimensional view of strategic information system implementation and use. *The Journal of Strategic Information Systems*, *23*(1), 45–61. doi:10.1016/j.jsis.2014.01.004

Barata, J., & Cunha, P. R. (2017). Climbing the Maturity Ladder in Industry 4.0: A Framework for Diagnosis and Action that Combines National and Sectorial Strategies. In *Twenty-third Americas Conference on Information Systems,* Boston, MA (pp. 1–10).

Barata, J., Cunha, P. R., Gonnagar, A., & Mendes, M. (2017). A Systematic Approach to Design Product Traceability in Industry 4.0: Insights from the Ceramic Industry. In *Twenty-sixth International Conference on Information Systems Development*, Larnaca, Cyprus (pp. 1–12).

Bauer, H., Baur, C., Mohr, D., Tschiesner, A., Weskamp, T., & Mathis, R. (2016). *Industry 4.0 after the initial hype - Where manufacturers are finding value and how they can best capture it. McKinsey Digital* (Whitepaper). *McKinsey*. Retrieved from https://www.mckinsey.de/files/mckinsey_industry_40_2016.pdf

Becker, J., Knackstedt, R., & Pöppelbuß, J. (2009). Developing Maturity Models for IT Management. *Business & Information Systems Engineering*, *1*(3), 213–222. doi:10.100712599-009-0044-5

Bharadwaj, A., El Sawy, O., Pavlou, P. A., & Venkatraman, N. (2013). Digital Business Strategy: Toward a Next Generation of Insights. *Management Information Systems Quarterly*, *37*(2), 471–482. doi:10.25300/MISQ/2013/37:2.3

Bicchi, A., & Peshkin, M. a, & Colgate, J. E. (2008). Safety for Physical Human–Robot Interaction. In Springer Handbook of Robotics (pp. 1335–1348). Springer.

Bleeker, E. A. J., Evertz, S., Geertsma, R. E., Peijnenburg, W. J. G. M., Westra, J., & Wijnhoven, S. W. P. (2015). *Assessing health & environmental risks of nanoparticles*. Bilthoven.

Brettel, M., & Friederichsen, N. (2014). How virtualization, decentralization and network building change the manufacturing landscape: An Industry 4.0 Perspective. *International Journal of Mechanical, Aerospace, Industrial, Mechatronic and Manufacturing Engineering*, *8*(1), 37–44.

Chen, D., Heyer, S., Ibbotson, S., Salonitis, K., Steingrímsson, J. G., & Thiede, S. (2015). Direct digital manufacturing: Definition, evolution, and sustainability implications. *Journal of Cleaner Production*, *107*, 615–625. doi:10.1016/j.jclepro.2015.05.009

Chen, J., & Poon, C. (2009). Photocatalytic construction and building materials: From fundamentals to applications. *Building and Environment*, *44*(9), 1899–1906. doi:10.1016/j.buildenv.2009.01.002

Cornelissen, I. R., Jongeneelen, F., van Broekhuizen, P., & van Broekhuizen, F. (2011). *Guidance working safely with nanomaterials and nanoproducts, the guide for employers and employees*. Amesterdam, The Netherlands: IVAM.

National Institute for Occupational Safety and Health (NIOSH). (2013). Current Intelligence Bulletin 65 Occupational Exposure to Carbon Nanotubes and Nanofibers. Cincinnati: NIOSH.

DECHEMA/VCI. (2011). 10 Years of research: risk assessment, human and environmental toxicology of nanomaterials. Frankfurt.

Degryse, C. (2016). Digitalisation of the Economy and its Impact on Labour Markets. European Trade Union Institute.

Duarte, K., Justino, C. I. L., Freitas, A. C., Duarte, A. C., & Rocha-Santos, T. A. P. (2014). Direct-reading methods for analysis of volatile organic compounds and nanoparticles in workplace air. *TrAC Trends in Analytical Chemistry*, *53*, 21–32. doi:10.1016/j.trac.2013.08.008

Environment Directorate OECD. (2015). *Harmonized tiered approach to measure and assess the potential exposure to airborne emissions of engineered nano-objects and their agglomerates and aggregates at workplaces*.

Erol, S., Jäger, A., Hold, P., Ott, K., & Sihn, W. (2016). Tangible Industry 4.0: A Scenario-Based Approach to Learning for the Future of Production. *Procedia CIRP*, *54*, 13–18. doi:10.1016/j.procir.2016.03.162

European Commission. (2013). Guidance on the protection of the health and safety of workers from the potential risks related to nanomaterials at work - Guidance for employers and health and safety practitioners. Brussels.

European Commission. (2016a). *Digital Single Market: Two years on*. Brussels. Retrieved from https://ec.europa.eu/priorities/sites/beta-political/files/2-years-on-dsm_en_0.pdf

European Commission. (2016b). *IP/16/1407 - Commission sets out path to digitise European industry*. Brussels. Retrieved from http://europa.eu/rapid/press-release_IP-16-1407_pt.htm

European Commission. (2016c). *The Fourth Industrial Revolution*. Retrieved from https://ec.europa.eu/digital-single-market/en/fourth-industrial-revolution

Fadier, E., & De la Garza, C. (2006). Safety design: Towards a new philosophy. *Safety Science*, *44*(1), 55–73. doi:10.1016/j.ssci.2005.09.008

FIR. (2017). *Industrie 4.0 maturity index. RWTH Aachen University*. Aachen. Retrieved from http://www.fir.rwth-aachen.de/en/research/research-projects/i40-maturity-index

Fryman, J., & Matthias, B. (2012, March). Safety of Industrial Robots: From Conventional to Collaborative Applications. *Robotics; Proceedings of the 7th German Conference on ROBOTIK 2012* (pp. 1–5).

Gabriel, M., & Pessl, E. (2016). Industry 4.0 and Sustainability Impacts: Critical discussion of sustainability aspects with a special focus on future of work and ecological consequences. *Annals of the Faculty of Engineering Hunedoara*, *14*(2), 131.

Ganzarain, J., & Errasti, N. (2016). Three Stage Maturity Model in SME's towards Industry 4.0. *Journal of Industrial Engineering and Management*, *9*(5), 1119–1128. doi:10.3926/jiem.2073

Grzybowska, K., Kovács, G., & Lénárt, B. (2014). The supply chain in cloud computing. *Research in Logistics & Production*, *4*(1), 33–44.

Hale, A., Kirwan, B., & Kjellén, U. (2007). Safe by design: Where are we now? *Safety Science*, *45*(1–2), 305–327. doi:10.1016/j.ssci.2006.08.007

Hammer, M. (1990). Reengineering Work: Don't Automate, Obliterate. *Harvard Business Review*, *68*(4), 104–112.

Huson, D., & Hoskins, S. (2014). 3D Printed Ceramics for Tableware, Artists/Designers and Specialist Applications. *Key Engineering Materials*, *608*, 351–357. doi:10.4028/www.scientific.net/KEM.608.351

Immonen, A., Ovaska, E., Kalaoja, J., & Pakkala, D. (2016). A service requirements engineering method for a digital service ecosystem. *Service Oriented Computing and Applications*, *10*(2), 151–172. doi:10.100711761-015-0175-0

IMPULS. (2017). *Industry 4.0 Readiness Online Self-Check for Businesses. VDMA*. Retrieved from https://www.industrie40-readiness.de/?lang=en

INEDIC. (2011). INEDIC-Innovation and Ecodesign in the Ceramic Industry. Retrieved from http://www.prepare-net.com/project/inedic-innovation-and-ecodesign-ceramic-industry

IPH. (2017). *4.0 Ready. Institut für Integrierte Produktion Hannover (IPH)*. Hannover. Retrieved from https://www.iph-hannover.de/en/research/research-projects/?we_objectID=2319

Isaka, H., Nagayoshi, H., Yoshikawa, H., Yamada, T., & Kakeno, N. (2016). Next Generation of Global Production Management Using Sensing and Analysis Technology. *Hitachi Review*, *65*(5), 47–52.

ISO. (2006a). *ISO 14040:2006, Environmental Management - Life Cycle Assessment - Principles and frameworks*. Geneva: International Organization for Standardization.

ISO. (2006b). *ISO 14044:2006, Environmental Management - Life Cycle Assessment – Requirements and guidelines*. Geneva: International Organization for Standardization.

KIT. (2016). *Introduction strategies of industry 4.0 methodology and technology for smes. Karlsruhe Institute Of Technology*. Karlsruhe. Retrieved from http://www.eurekanetwork.org/project/id/10389

Knoke, B., Missikoff, M., & Thoben, K.-D. (2017). Collaborative open innovation management in virtual manufacturing enterprises. *International Journal of Computer Integrated Manufacturing*, *30*(1), 158–166. doi:10.1080/0951192X.2015.1107913

Lasi, H., Fettke, P., Kemper, H. G., Feld, T., & Hoffmann, M. (2014). Industry 4.0. *Business & Information Systems Engineering*, *6*(4), 239–242. doi:10.100712599-014-0334-4

Leyh, C., Schäffer, T., Bley, K., & Forstenhäusler, S. (2017). Assessing the IT and Software Landscapes of Industry 4.0-Enterprises: The Maturity Model SIMMI 4.0. In Information Technology for Management: New Ideas and Real Solutions, LNBIP (Vol. 277, pp. 103–119). Springer.

Lucke, D., Constantinescu, C., & Westkämper, E. (2008). Smart Factory - A Step towards the Next Generation of Manufacturing. In M. Mitsuishi, K. Ueda, & F. Kimura (Eds.), *Manufacturing Systems and Technologies for the New Frontier* (pp. 115–118). London: Springer London. doi:10.1007/978-1-84800-267-8_23

Magruk, A. (2016). Uncertainty in the Sphere of the Industry 4.0 – Potential Areas to Research. *Business. Management in Education*, *14*(2), 275–291.

Menon, K., Kärkkäinen, H., & Lasrado, L. A. (2016). Towards a Maturity Modeling Approach for the Implementation of Industrial Internet. In *Proceedings of the Pacific Asia Conference on Information Systems (PACIS)*, Chiayi, Taiwan (pp. 1–11).

Moore, J. F. (1996). *The death of competition: Leadership and strategy in the age of business ecosystems*. New York: Harper Collins.

Morose, G. (2010). The 5 principles of "Design for Safer Nanotechnology." *Journal of Cleaner Production*, *18*(3), 285–289. doi:10.1016/j.jclepro.2009.10.001

National Institute for Occupational Safety and Health (NIOSH). (2009). *Approaches to Safe Nanotechnology: An Information Exchange with NIOSH*.

Oesterreich, T. D., & Teuteberg, F. (2016). Understanding the implications of digitisation and automation in the context of Industry 4.0: A triangulation approach and elements of a research agenda for the construction industry. *Computers in Industry*, *83*, 121–139. doi:10.1016/j.compind.2016.09.006

Paulk, M. C., Curtis, B., Chrissis, M. B., & Weber, C. V. (1993). Capability Maturity Model for Software, Version 1. 1. *Carnegie Melon University*, *10*(4), 18–27.

Prifti, L., Knigge, M., Kienegger, H., & Krcmar, H. (2017). A Competency Model for "Industrie 4.0" Employees. In *13th International Conference on Wirtschaftsinformatik*, St. Gallen, Switzerland (pp. 46–60).

PwC. (2016). *Industry 4.0: Building the digital enterprise - Engineering and construction key findings*. Retrieved from https://www.pwc.com/gx/en/industries/industries-4.0/landing-page/industry-4.0-building-your-digital-enterprise-april-2016.pdf

PwC. (2017). *Industry 4.0 - Enabling Digital Operations Self Assessment*. Retrieved from https://i40-self-assessment.pwc.de/i40/

Quinteiro, P., Almeida, M., Dias, A. C., Araújo, A., & Arroja, L. (2014). The Carbon Footprint of Ceramic Products. In Assessment of Carbon Footprint in Different Industrial Sectors (Vol. 1, pp. 113-150). doi:10.1007/978-981-4560-41-2_5

Rockwell Automation. (2014). *The Connected Enterprise Maturity Model* (CIE-WP002-EN-P). Retrieved from http://www.google.pt/url?sa=t&rct=j&q=&esrc=s&source=web&cd=1&cad=rja&uact=8&ved=0ahUKEwjvs5Dm75vSAhVIbBoKHS5VDTUQFgglMAA&url=http%3A%2F%2Fliterature.rockwellautomation.com%2Fidc%2Fgroups%2Fliterature%2Fdocuments%2Fwp%2Fcie-wp002_-en-p.pdf&usg=AFQjCNFIehm

Savolainen, K., Brouwer, D., Fadeel, B., Fernandes, T., Kuhlbusch, T., & Landsiedel, R. … Pylkkänen, L. (2013). Nanosafety in Europe 2015-2025: Towards Safe and Sustainable Nanomaterials and Nanotechnology Innovations. Helsinki: Finnish Institute of Occupational Health.

Schulte, P. A., Rinehart, R., Okun, A., Geraci, C. L., & Heidel, D. S. (2008). National Prevention through Design (PtD) Initiative. *Journal of Safety Research*, *39*(2), 115–121. doi:10.1016/j.jsr.2008.02.021 PMID:18454950

Schumacher, A., Erol, S., & Sihn, W. (2016). A Maturity Model for Assessing Industry 4.0 Readiness and Maturity of Manufacturing Enterprises. *Procedia CIRP*, *52*, 161–166. doi:10.1016/j.procir.2016.07.040

Schwab, K. (2015). Will the Fourth Industrial Revolution have a human heart? Retrieved December 9, 2016, from https://www.weforum.org/agenda/2015/10/will-the-fourth-industrial-revolution-have-a-human-heart-and-soul/

Silva, F., Arezes, P., & Swuste, P. (2016). Systematic design analysis and risk management on nanoparticles occupational exposure. *Journal of Cleaner Production*, *112*, 3331–3341. doi:10.1016/j.jclepro.2015.11.001

Smit, J., Kreutzer, S., Moeller, C., & Carlberg, M. (2016). *Industry 4.0 - Study for the ITRE Committee*. Retrieved from http://www.europarl.europa.eu/RegData/etudes/STUD/2016/570007/IPOL_STU(2016)570007_EN.pdf

Soldatos, J., Gusmeroli, S., Malo, P., & Di Orio, G. (2016). Internet of Things Applications in Future Manufacturing. In *Digitising the Industry. Internet of Things Connecting the Physical and Virtual World* (pp. 153–183). River Publishers.

Steijn, W., & Luiijf, E. (2016). *Emergent risk to workplace safety as a result of the use of robots in the work place. No. TNO 2016 R11488*. TNO.

Steijn, W., Van Der Vorm, J., & Luiijf, E. (2016). *Emergent risks to workplace safety as a result of IT connections of and between work equipment. No. TNO 2016 R11143*. TNO.

Technical Committee ISO/TC 229. (2012). ISO/TS 12901-1: Nanotechnologies — Occupational risk management applied to engineered nanomaterials — Part 1: Principles and approaches. Geneva: ISO.

van Broekhuizen, P., van Broekhuizen, F., Cornelissen, R., & Reijnders, L. (2011). Use of nanomaterials in the European construction industry and some occupational health aspects thereof. *Journal of Nanoparticle Research, 13*(2), 447–462. doi:10.100711051-010-0195-9

Vervoort, M. B. H. J. (2012). *A comparison of risk assessment methods in order to determine the risk of occupational used nanomaterials in a research environment*. NSPOH.

Vidgen, R., & Wang, X. (2006). From business process management to business process ecosystem. *Journal of Information Technology, 21*(4), 262–271. doi:10.1057/palgrave.jit.2000076

Wang, J., Huang, J., Chen, W., Liu, J., & Xu, D. (2016). Design of IoT-based energy efficiency management system for building ceramics production line. In *Proceedings of the 2016 IEEE 11th Conference on Industrial Electronics and Applications, ICIEA 2016* (pp. 912-917). 10.1109/ICIEA.2016.7603712

Wang, L., & Alexander, C. (2016). Additive Manufacturing and Big Data. *International Journal of Mathematical, Engineering and Management Sciences, 1*(3), 107–121. Retrieved from http://ijmems.in/assets/2.-ijmems-16-015-vol.-1,-no.-3,-107–121,-2016.pdf

Weill, P., & Woerner, S. L. (2015). Thriving in an Increasingly Digital Ecosystem. *MIT Sloan Management Review, 56*(4), 27–34. doi:10.1287/isre.1100.0318

Westermann, T., Anacker, H., Dumitrescu, R., & Czaja, A. (2016). Reference Architecture and Maturity Levels for Cyber-Physical Systems in the Mechanical Engineering Industry. In *International Symposium on Systems Engineering (ISSE)* (pp. 1–6). Edinburgh, Scotland: IEEE. 10.1109/SysEng.2016.7753153

Wong, K. V., & Hernandez, A. (2012). A Review of Additive Manufacturing. *ISRN Mechanical Engineering, 2012*, 1–10. doi:10.5402/2012/208760

World Economic Forum. (2016). *The Future of Jobs*.

Yi, J., LeBouf, R. F., Duling, M. G., Nurkiewicz, T., Chen, B. T., Schwegler-Berry, D., ... Stefaniak, A. B. (2016). Emission of particulate matter from a desktop three-dimensional (3D) printer. *Journal of Toxicology and Environmental Health. Part A., 79*(11), 453–465. doi:10.1080/15287394.2016.1166467 PMID:27196745

Zhou, J. (2013). Digitalization and intelligentization of manufacturing industry. *Advances in Manufacturing, 1*(1), 1–7. doi:10.100740436-013-0006-5

KEY TERMS AND DEFINITIONS

Ceramic Industry 4.0: The ongoing initiatives for digital transformation in the Portuguese ceramic sectors of the economy. The roadmap includes social, technical, and organizational changes that are necessary to compete in global supply chains.

Digital Ecosystem: Socio-technical system inspired in natural ecosystems that connects a group of companies/people/things via digital platforms. It requires a digital infrastructure and digital services to interact with external parties of the organization. Similarly to natural ecosystems, sustainability and safety are critical aspects.

Digital Sustainability: The opportunities raised by digital transformation to meet the sustainability goals and reduce the carbon footprint.

Life Cycle Assessment: Assessment of the environmental impacts applied to the different stages of a product's life cycle.

Maturity Model: A tool used to assess the current state of an organization in a specific context of analysis. This type of models is also used to communicate best practices and guide organizational improvements.

Nanosafety: The different techniques, tools, and approaches related to the safety of nanotechnology. It involves the policies, standards, and research needed to ensure the proper development and use of nanomaterials in the factory of the future.

Safety-By-Design: The use of methods in early stages of the product life cycle to minimize hazards and comprehensively improve health and safety.

ENDNOTES

1 First Law: a robot may not injure a human being or, through inaction, allow a human being to come to harm. Second Law: a robot must obey the orders given it by human beings, except where such orders would conflict with the First Law. Third Law: a robot must protect its own existence as long as such protection does not conflict with the First or Second Laws.

Chapter 13
Digital Transformation in the Utilities Industry:
Industry 4.0 and the Smart Network Water

Marcelo Teixeira de Azevedo
Universidade de São Paulo, Brazil

Alaide Barbosa Martins
Universidade de Sao Paulo, Brazil

Sergio Takeo Kofuji
Universidade de Sao Paulo, Brazil

ABSTRACT

Nowadays, there is a digital transformation in industry, which is being referred to as a new revolution, known as the fourth industrial revolution. Today, we are in the fourth industrial revolution, which strongly supports itself at enabling technologies, such as: Internet of Things (IoT), big data, cyber-physical systems (CPS) and cloud computing. These technologies are working cooperatively to promote a digital transformation in the forms of: Industry 4.0, Industrial Internet Consortium and Advanced Manufacturing. To meet these needs and as a proof of concept, a platform for digital transformation for a water issue is proposed with the objective of achieving an efficient management of resources linked to rational use of water. For the platform definition, a survey was performed of the process of a water treatment plant from the third industrial revolution and improved the process by applying the concepts of digital transformation to improve the new platform.

1. INTRODUCTION

Digital transformation is part of a major ongoing technological process and it is associated with the application of digital technology in all aspects of human society. According to Khan (2016) and Westerman (2014), digital, digitization and digital transformation are concepts that are commonly confused, and there is no clear or broadly accepted definition of digital transformation.

DOI: 10.4018/978-1-5225-4936-9.ch013

Also, according to Khan (2016), it is first necessary to establish the concepts of digital and digitization in order to be able to define digital transformation. Digital is associated with the process of converting analogical information into digital format; in Collin et al. (2015), it is defined as a condition for converting, for example, a print photograph into a digital format. Meanwhile, according to Bounfour (2016), digitization is related to the issue of transforming physical processes into virtual processes using specifically tailored technologies. In accordance with Khan (2016) and Collin et al. (2015), digital transformation is the effect of this digitization, and thus not only pertains to the adoption of new technologies, but also to new ways of making business operations more efficient and competitive. However, it is still not known how digital transformation will be implemented, its degree of complexity, the nature of its interconnections among different sectors and, above all, its impact on society.

The new technologies involved in this transformation will bring far-reaching changes to all sectors of society, affecting current business, production, consumption, transportation and delivery methods, as well as the very way people live. Therefore, it is necessary to obtain a comprehensive vision of how this digital transformation will change all aspects (social, cultural or economic) of the world in which people live.

In recent centuries, industry has undergone vast operational changes, generating innovation and far-reaching social and economic shifts. The first industrial revolution took place in England at the end of the 18th Century and beginning of the 19th Century, and it was characterized by coal as the primary energy source, leading to the development of the steam engine and, consequently, the locomotive. One of the first areas to implement this new technology was the textile industry, which, prior to this revolution, was intensively manual. This new industrial model accelerated the rural exodus, thus leading to burgeoning urban centers. The second industrial revolution triggered from the second half of the 19th Century as part of the quest for continual improvement. This basically comprised the discovery of fossil fuels and electromagnetism as energy sources, with electricity arising as an alternative to steam power. Use of these resources accelerated the industrial steamroller with the discovery of the potential for reducing both costs and product manufacturing time, with the result that goods could be consumed on an ever-greater scale. One of the most evident examples of this revolution is the introduction of the assembly line, the most traditional reference of which is the Ford Model T automobile. The third industrial revolution began in the middle of the 20th Century, characterized by the intensive use of electronics and information technology (IT) in order to achieve the objective of automation in manufacturing. In the industrial area, Supervisory Control and Data Acquisition (SCADA) systems, in conjunction with the programmable logic controller (PLC) were its most significant feature, and these systems are currently converging to open-source systems that, in some cases, are connected to corporate networks or even the internet. The use of telecommunications resources and technological advances have enabled remote access, sharing, integration and, consequently, long-distance data processing using these resources. Integration with different systems is aimed at increasing productivity and efficiency in decision-making processes.

Currently, according to Schwab (2016), we are currently on the cusp of a fourth industrial revolution, the so-called Cyber-Physical Systems (CPS) systems-based industrial revolution. In addition to Cyber-Physical Systems, new technologies are arising in the form of the current enabling technologies, whose use is leading to the breakdown of the current paradigm. Such technologies include Internet of Things (IoT), big data analytics and cloud computing, intended to address the requirements of an advanced manufacturing system through integration of the entire supply chain. This new revolution is characterized by the intensive use of digital technologies with the purpose of manufacturing products rapidly, with an agile response to demands and real-time production and supply-chain optimization. Also, according to

Schwab (2016), the fourth industrial revolution basically comprises three pillars, namely: the physical, digital and biological, which have already impacted the economy, business, society, private individuals and government.

This research is intended to contribute to the understanding of digital transformation and its impacts. This new paradigm shift affects both existing business models and the entire supply chain. Considering the current moment in time, this study identifies enabling technologies that form the technological pillars on which this new revolution is to be based, in order to do the same work as global visions that attempt to comprehend, qualify and guide companies, the government and civil society towards this transformation.

Given that digital transformation affects the entire production chain in its different branches and sectors, it has become a very wide-ranging and complex issue. Accordingly, this research attempts to contextualize the issue based on a single niche, that of the Utilities industry and, specifically, the issue of rational water use in Brazil. The Utilities industry is defined by Mcnabb (2005) as companies operating in the production, transportation, distribution and commercialization of energy (electricity and gas) and water and its main attribute is that it is essential and, in the case of failure, critical for society.

1.1. Objectives

The objective of this chapter is to understand the impacts of digital transformation on the industry and to improve the currently existing architecture for an all-new platform. The current 5-layer pyramid automation architecture is transformed into a new model that makes use of enabling technologies to digitally transform the entire supply chain of a certain market niche. This research is supported by digital transformation in the context of the Utilities industry, in order to promote efficient management with rational water use linked to existing risks and threats.

1.2. Rationale

Digital transformation is becoming a reality, although studies on the impact of the adoption of technologies that enable digital transformation to be implemented have not yet matured to the degree that they inspire the confidence required to make it a reality.

The choice of industrial segment was made based on the fact that no balance between the knowledge of the managers of these industries and visions of digital transformation has been identified in research conducted to date. At the current moment, industries are operating with SCADA systems in the 5-layer pyramid model, in order to continue executing vital functions for society. With the recent appearance of enabling technologies, change in this architecture is now possible and, consequently, it is also possible to achieve the gains expected from digital transformation.

The definition and concept of digital transformation is not unanimous, and it requires investigation in several fields, such as the creation of standards, reference architecture, training, adaptation to legislation, platform and regulatory framework, among others. The absence of digital transformation policies linked to the criticality of the industries provides the grounds for this research.

Furthermore, specifically with regard to the water issue, according to Torres (2008), there is an executive order in the United States identified under the number 13,010, which names the water supply system as one of the eight critical infrastructures most targeted by terrorists and, accordingly, it requires greater protection. Although, as the author himself states, the probability of contamination is slight, attempts and actual attacks on the water supply system in the United States have proven to be a real issue, as shown in Table 1.

Although there is very extensive literature available on mitigation, response and recovery, this work is most related to natural disasters. However, it is possible to observe a gradual increase in attacks on the water distribution system this decade. The American Water Works Association (AWWA), the American Society of Civil Engineers (ASCE) and the Water Environment Federation (WEF) recently drafted an experimental common standard on guidelines for improvement of the physical safety of facilities used for the treatment and distribution of potable water.

2. THEORETICAL REFERENCE

In this chapter, the bibliographical references that serve as the basis for the development, structure and execution of this study are analyzed and discussed. The proposed subject, in its entirety, is very wide-ranging and may cover many areas of knowledge; accordingly, the references also cover several correlated areas, including concepts, applications, protocols and theoretical and practical examples. Furthermore, IoT visions in industry and a description of emerging technologies are presented, such as IoT, machine learning, M2M, CPS, big data, analytics and risk analysis. The main purpose of this item is to contextualize the research by exploring approaches to IoT in industry, emerging technologies that contribute to these visions and the issue of risk (security and safety).

2.1. Related Works

In "*Security and Privacy Challenges in Industrial Internet of Things*" (Sadeghi, Wachsmann & Waidner, 2015), the authors base their work on the principle that, due to the strong tendency of connectivity and IoT-based initiatives, everyday equipment is becoming more and more connected, from vehicles to critical infrastructure and, accordingly, it is more exposed to safety and privacy issues. Thus, this article briefly introduces the systems that use the industrial IoT concept related to safety and privacy challenges, presenting a holistic perspective on the potential solutions to these issues. Furthermore, the authors research details pertaining to attacks on an IoT industrial system and relate the existing architecture to

Table 1. Attacks against the American water treatment system

Year	Attacker	Event	Agent
1984	Religious Cult	Contamination of municipal potable water tanks in the state of Oregon/USA	Salmonella
2001	Bin Laden sympathizers	Detection of a potential contamination event in the water distribution network in 28 American states.	N/A
2002	Moroccans	Development of an initiative to contaminate water pipes at the American embassy in Rome.	Cyanide
2002	Al Qaeda	Resident arrested with plans and procedures for contaminating the water distribution network.	N/A
2003	Iraqi agents	Poisoning of food and water distribution networks.	Botulinum toxin
N/A	Religious Cult	Acquisition of cyanide to contaminate some of the reservoirs in Minneapolis	Cyanide

Source: Torres (2008)

CPS and protection techniques. Finally, they conclude that current systems are inadequate and conclude that a holistic approach is necessary for the development of a framework for cybersecurity that takes into consideration all the abstract layers of the IoT system, as well as the frontiers that interact with the platform; however, there is no practical demonstration of the concepts and solutions presented, only a conceptual consideration.

In the study entitled "*Internet of Things Applications in Future Manufacturing*" (Soldatos et al., 2016), the authors base their work on the principle that the factory of the future will be governed by flexibility in manufacturing and rapid responses to market demands, thus providing a new production model that goes beyond mass production. Accordingly, manufacturing will migrate from the stock production model to a new demand-oriented model. This also means that a new platform to support this new model will be required. The study discusses initiatives, IoT technologies and applications related to future internet technologies, including the FITMAN project. One of the purposes of the article is to demonstrate how IoT can transform manufacturing by using emerging technologies and thus to illustrate the potential factory of the future.

The article entitled "*Towards Manufacturing 2.0: an Innovative Architecture for the Factory of the Future*" (Fiaschetti, Pietrabissa, & Priscoli, 2015) proposes an information management system for the new paradigm of the factory of the future. The authors start with the initial idea that the new production model incorporated by the factory of the future will produce a huge amount of data to be processed by operators, and with the current tools available, data management is not very effective for decision making. With this in mind, the authors conclude that new technologies are necessary in order to deal with innovative manufacturing processes. Accordingly, the article demonstrates an innovative architecture solution based on use and integration in ICT, based on the Future Internet framework, specifically, the PLATINO project.

On the other hand, the starting point for "*A Cyber-Physical Systems Architecture for Industry 4.0-based manufacturing systems*" (Lee, Bagheri, & Kao, 2015) is the evolution of the manufacturing industry; the authors propose a unified architecture based on five levels working as a guide for the implementation of CPS (Cyber-Physical Systems), with the purpose of improving the quality of products and the reliability of the system with more intelligent and resilient equipment. Accordingly, five steps are proposed, including: intelligent connection, conversion of data to information, cyber, cognition and configuration.

In the article entitled "*Implementing Smart Factory of Industry 4.0: An Outlook*" (Wang, Li, & Zhang, 2015), the authors describe the concept of Industry 4.0 and the requirement for horizontal, *end-to-end* and vertical integration in order to fulfill the Industry 4.0 proposal. The primary objective of the article is to address the issues of vertical integration; in order to fulfill this objective, the authors propose a framework for control systems engineering that incorporates industrial wireless networks, cloud computing, big data and intelligent terminals, in order to form a self-organized system assisted by cloud-implemented feedback blocks and coordination, based on big data analysis.

In the article entitled "*Smart Factories in Industry 4.0: A Review of the Concept and of Energy Management Approached in Production Based on the Internet of Things Paradigm*" (Shrouf, Ordieres, & Miragliotta, 2014), the authors present a reference architecture for intelligent factories based on the internet of things, defining the main characteristics of these factories with a focus on sustainability perspectives. Subsequently, they propose an approach for energy management, wherein decision making is carried out using energy consumption data from the factory floor (production line), providing *on-line* consumption in real time for any device.

The integration of these data improves efficiency, because they can be used to discover and reduce waste, and to allow conscientious decision making at the production management level (for example, maintenance, production programming and management).

The study entitled "*Service innovation and smart analytics for Industry 4.0 and big data environment*" (Lee; Kao, & Yang, 2014) is based on the principal that, in Industry 4.0, machines are connected like a collaborative community and its evolution requires use of advanced prediction tools. This article addresses the tendencies in the transformation of production services in a *big data* environment, as well as the availability of tools to manage *big data*, in order to achieve transparency and productivity.

In the article entitled "*Classification Model for Water Quality using Machine Learning Techniques*" (Muhammad et al., 2015), machine learning techniques were employed with the purpose of classifying water quality, revealing the best model for undertaking this classification. Accordingly, five models were applied: Naive Bayes, Conjunctive Rule, J48, Kstar and Bagging. In the tests proposed by the authors, the Kstar algorithm proved to be the most accurate.

The research outlined in the article entitled "Cloud based machine learning approaches for leakage assessment and management in smart water networks" (Mounce et al., 2015) is based on the problem that one third of public services worldwide report losses of almost 40% of clean water due to leaks, and this loss could be minimized by means of an intelligent water system. In order to prove this concept, a loss analysis in a controlled environment was proposed using machine learning techniques to predict leaks, with historical flows and pressure data as parameters.

In the study entitled "*Mathematical Model for Detection of Leakage in Domestic Water Supply Systems by Reading Consumption from an Analogue Water Meter*" (Oren & Stroh, 2013), a mathematical model is presented to detect domestic leaks in real time. The model uses data measured from a water meter and it employs two criteria simultaneously, namely: deviations in mean consumption and comparisons of water consumption over time. The simulation of a domestic consumption model was implemented in a device called Antileaks that was constructed by the authors. This device transforms analogical consumption data into digital format in real time.

In the study entitled "A Medição Setorizada Como Instrumento De Gestão Da Demanda De Água Em Sistemas Prediais – Estudo De Caso: Programa De Uso Racional Da Água Da Universidade De São Paulo" [Sectorized Measurement as an Instrument for Managing Water Demand in Buildings - Case Study: University of São Paulo Program for Rational Water Use] (Tamaki, Sanches, & Gonçalves, 2006), concluded that many benefits obtained and reported in the case of the University of São Paulo can be extrapolated to similar situations involving the implementation of sectorized measurement in order to manage water demand, taking into account considerations raised and the peculiarities of each application.

In the article entitled "*Towards an IoT Based Water Management System for a Campus*" (Verma et al., 2015), the design and results of an IoT based system for water distribution management on a university campus are discussed. The scope is focused on two main components, namely: 1) low-cost ultrasound water level measurement sensors; and 2) a sub-GHz wireless network to connect the sensors. The network is connected to a gateway that can update the data online for viewing and analytics.

In the study entitled "*Managing the water distribution network with a Smart Water Grid*" (Walter, 2018), the author base their work on the implementation of a Smart Water Grid system supports the Public Utilities Board (PUB)'s mission to supply good water 24/7 to its customers. With sensors and analytic tools deployed island-wide to provide a real-time monitoring and decision support system.

In the other study "LIFE SWSS - Smart Water Supply Systems", (SWSS, 2018) is based on the propose to decrease to 15% the energy consumption and CO2 emission associated with the water supply for the Singapore customers based on the use of enable technologies.

In line with the aforementioned studies, this study intends to contribute with projects in the area of intelligent cities and, specifically, the issue of water treatment through the use of enabling technologies and approaches to digital transformation.

3. THEORETICAL GROUNDS

The architecture of industrial automation is represented by Moraes and Castrucci (2007) by means of a pyramid called the Pyramid of Automation, whose primary objective is to divide equipment levels involved in accordance with their role in industry. This pyramid can be observed in Figure 1, where:

Level 1: The level of machines.
Level 2: The level of digital, dynamic and logic controllers.
Level 3: Enables the production process in the plant to be controlled.
Level 4: Responsible for production programming and planning.
Level 5: Responsible for management of company resources.

The next items identify new technologies that led changes in this traditional pyramid toward a new concept based on enabling technologies, which are discussed below.

Figure 1. Automation pyramid
Source: Moraes and Castrucci (2007)

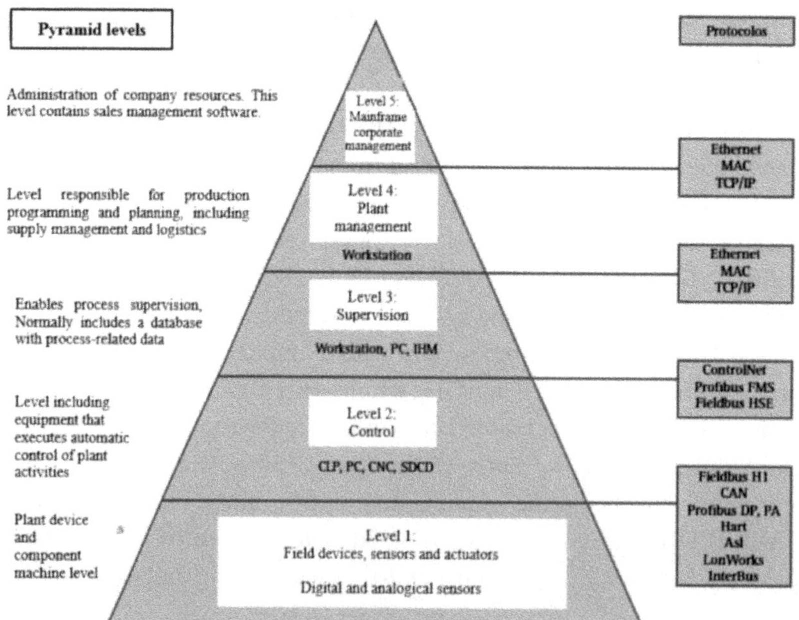

Enabling technologies are the technological pillars for digital transformation, permitting the development of new processes and products. According to the European Commission, the formalization of enabling technologies, called KETs (Key Enabling Technologies) is characterized by the huge potential of knowledge and abundance of research and development (R&D), containing rapid innovation cycles and intensity in capital and qualification (Key Enabling Technologies, 2016). According to the European Commission, key technologies are:

- Advanced Materials
- Nanotechnology
- Micro and nanoelectronics
- Industrial Biotechnology
- Photonics
- Advanced Manufacturing Systems

Advanced manufacturing is based on the use and coordination of automation, information, computation, software, sensing and networking, and it requires intensive use of enabling technologies (Loural, 2014). The North American government drafted a document entitled A National Strategic Plan for Advanced Manufacturing, which defined advanced manufacturing as follows:

... is a family of activities that depend on the use and coordination of automation, information, computation, software, sensing, and networking, and/or make use of cutting edge materials and emerging capabilities enabled by the physical and biological sciences, for example nanotechnology, chemistry, and biology. It involves both new methods for manufacturing existing products, and the manufacture of new products emerging from new advanced technologies. (NSTC, 2012)

3.1. IoT (Internet of Things)

The first reference to the concept to what is now known as IoT - Internet of Things was in 1982, when researchers from Carnegie Mellon University developed the first Coca-Cola machine with IoT characteristics (Oldest Internet of Things Objects in the World, 2017). Subsequently, at the beginning of the 1990s, Mark Weiser developed the widest ranging-concept of these characteristics, describing specialist hardware and software elements connected by cables, radio and infrared waves whose presence could not be perceived, thus introducing the term "Ubiquitous Computing" (Weiser, 1992). The term Internet of Things was defined around 1999 by Kevin Ashton, during a presentation at Procter & Gamble (P&G), in which it was connected to the RFID concept within the supply chain, with the subject in focus at the time, which was the Internet (Ashton, 2009). Subsequently, IoT was formally introduced by the International Telecommunication Union (ITU) through the 2005 ITU Internet Report. Then, in 2012, IoT was defined by the ITU-T Y.2060 recommendation as a global infrastructure for the information society, enabling advanced services to be supplied through the interconnection of things (physical and virtual) based on the already existing information and communication technology, as well as its evolution.

The term Internet of Things (IoT) is a new paradigm that has been gaining space in the modern wireless telecommunications scenario, wherein the main idea is the pervasive presence around individuals by means of a variety of things or objects, such as Radio-Frequency Identification (RFID) tags, sensors, actuators, mobile telephones, etc.

3.2. Big Data and Analytics

Nist (2015) defines the term big data as a huge data set, in which the great mass of data is not structured and requires real-time analysis. The term big data addresses the inability of traditional data architecture to efficiently handle large quantities of data, requiring a scalable platform to effectively store, process and analyze them. The characteristics of big data imply a new architecture known as the "V's" of Big Data, which are:

- Volume (size of the data set);
- Variety (data from multiple repositories, domains or types);
- Velocity (data flow rates);
- Variability (coherence in the data set).

3.3. Cloud Computing and Fog Computing

The definition of cloud computing given by Nist (2011) is a model that permits ubiquitous, convenient and on-demand access to shared and configurable resources, which can be rapidly provided and released with minimal management effort. It comprises five essential characteristics, three service models and four implementation models, and its essential characteristics are as follows:

1. **On-demand and self-service:** Consumers may provision computational capacity without requiring human interaction with the service provider;
2. **Wide-ranging access to the network:** The resources are available through the network and can be accessed by means of heterogenic platforms, such as: mobile phones, tablet computers, laptop computers and work stations;
3. **Set of resources:** The access provider's computer resources are grouped to serve several consumers with different physical and virtual resources, which are dynamically assigned according to consumer demand;
4. **Rapid Elasticity:** The resources may be elastically provisioned (in any quantity and at any moment) and released according to demand;
5. **Measurement Service:** The cloud service automatically manages and optimizes use of the resources, and these are monitored, controlled and reported in a manner that is transparent both to the provider and to the consumer of the acquired service.

The three service models are:

1. **Software as a Service (SaaS):** Use of applications in a manner that is accessible by different platforms. Consumers have no managerial control whatsoever over computer resources;
2. **Platform as a Service (PaaS):** Consumers have control over the implementation of the application and the possibility of configurations in the application environment; however, they do not manage the infrastructure of the cloud computing resources;
3. **Infrastructure as a Service (IaaS):** Provides consumers with the capacity to manage cloud-computing resources and to implement and execute applications of their choosing.

The four implementation models are:

1. **Private Cloud:** Cloud infrastructure is provisioned for exclusive use of a single organization, comprising several consumers;
2. **Community Cloud:** Cloud infrastructure is provisioned for community use and specific use by certain users with shared interests;
3. **Public Cloud:** Cloud infrastructure is provisioned for open use and for the general public;
4. **Hybrid Cloud:** The cloud infrastructure comprises two or more cloud infrastructures (for example, private, public or community), which remain single entities, but which are linked according to standardization.

3.4. Machine Learning

The first definition of machine learning was made by Samuel (1959), who defined it as a field of study that provides computers with the ability to learn without being specifically programmed to do so. Later, it was redefined by Mitchell (1997) as techniques capable of improving performance in a given task using prior experience. Further according to Mitchell (1997), the machine learning algorithms can be divided in 04 ways, namely: supervised, unsupervised, semi-supervised and reinforcement learning.

In supervised learning, data (inputs and outputs) are presented in a model, providing the desired output for each input, where the primary objective is to learn a general rule in order to understand the inputs and outputs. In unsupervised learning, no type of information is given to the algorithm, and the algorithm itself is left to understand the data structure. In the semi-supervised algorithm, a training set with several missing outputs is provided. Reinforcement learning involves interaction with a dynamic environment in which a certain objective is proposed; autonomous vehicles are the classic example.

Learning algorithms can be divided into predictive (classification and regression) and descriptive (grouping, summarization and association) algorithms. Algorithms are used according to the objective and the issue to be resolved. For example, it is possible to highlight analyses of temporal series that use algorithms with the purpose of learning a function that maps a real value example. Examples of algorithms for this class of problems are regression trees, artificial neural networks, linear regression and support vector machines. On the other hand, the purpose of learning a function that associates the description of an example to a class can be exemplified by Bayesian, logistic regression and K-NN networks, among others.

3.5. Information Security and Process Safety

The importance of defining the two areas of activity in this study (security and safety) is obvious, because both are included in the risk analysis and verification of failures for the improvement of process safety.

In order to obtain a better definition of the terms security and safety, we should first understand the term 'risk'. According to Almeida Jr. (2003):

...risk can be viewed as being directly related to the frequency of potentially hazardous events and with the severity of their consequences. The greater the frequency of a hazardous event, the greater the risk and severity of its outcomes, and, thus, the greater the risk assumed. ...any system has inherent associated

risks, and the risk presented by one system may or may not be accepted by society, which, ultimately, will define the concept of acceptable risk. (p. 15)

With regard to the nomenclatures of the terms security and safety, Almeida Jr. (2003) presents the following definitions:

Although there is a difference, in Portuguese, these two factors have the same denomination; i.e. 'segurança', which in English can mean either 'safety' or 'security'. The term safety is used to refer to the safety of Critical Systems, while the term security is used to refer to the security of Information Systems. (p. 19)

In this study, the term security is used for the definitions of procedures, structures and processes for the improvement of information security systems, data and architecture of the communication network; while the term safety refers to the state of protection of a system, person, object or process, and it covers several types of error, failure and problem. The term is differentiated in a way similar to that adopted by Almeida Jr. (2003) in his research: safety for process safety (the safety of critical systems) and security for information security (information systems).

Process safety correlates quality, reliability, maintenance and safety of people and the environment. The primary objective is to know and manage risks.

Risk analysis means the process of understanding the nature of the risk and ascertaining risk levels. According to Braendeland, Refsdal and Stolen (2010), risk is defined as a combination of the consequences and probability of an undesired event.

According to the CETESB P4.261(2003) standard:

Recommendations and measures resulting from the risk analysis and assessment study for the reduction of the frequency and consequences of potential accidents, must be considered as integral parts of the risk management process; however, regardless of the adoption of these measures, premises containing hazardous substances or processes must be operated and maintained throughout their useful lives within standards considered tolerable, which is why the Risk Management Program [Programa de Gerenciamento de Riscos, PGR] must be implemented and used in all the activities, both routine and otherwise, of an industrial plant. (p. 34)

The requirement for more and more reliable systems is included in a context that demands a point of equilibrium to be reached, since poor reliability may result in loss of human life and material losses; however, on the other hand, extreme reliability may not be financially feasible.

In the process of this study, several standards and measurement and risk analysis techniques are addressed, such as IEC 61508, Reliability, FMEA, CORAS, QSP 31000 (ISO/IEC 31000), CETESB P4.261 and others.

3.6. Consultative Objective Risk Analysis Systems (CORAS)

CORAS is a methodology that includes several techniques to appraise risk based on Unified Modeling Language (UML). The project was initiated in 2001 by four European Union countries (Greece, Germany, Norway and the United Kingdom), with support from IT companies (Intracom, Solinet and Telenor) and seven institutes (CTI, FORTH, IFE, NCT, NR, RAL and Sintef), as well as the QMUL

University in the United Kingdom. The project was conducted in two stages: the main task in the first phase (2001-2003) was to develop an information evaluation platform and the primary objective of the second phase (2004-2007) was to improve CORAS functionalities.

CORAS complies with international standards, such as Austria Standard/New Zealand Standard (AS/NZS) 4360:1999 and ISO/IEC 17799-1:2000. However, its main basis lies in the risk management norm that establishes the international standard, ISO/IEC 31000:2009 – General Guidelines for Principles and Implementation of Risk Management, which, in turn, was based on the AS/NZS 4360 standard. The risk management process according to the ISO/IEC 31000 standard has the scope presented in Figure 2, which includes the CORAS stages.

Phase 1 - Identify the context: Aims to characterize the target of the analysis, and the focus and scope of the risk analysis. To what level is risk tolerable?

Phase 2 - Identify the risks: Identify threats and vulnerabilities to assets.

Phase 3 - Estimate the level of risk, evaluate risks: Not all risks can be eliminated, and thus we need to decide which risks need to be processed.

Phase 4 - Process the risks: Identify the processing of undesirable risks. Evaluate and prioritize different methods of processing.

CORAS uses identification and risk analysis tools originating from process accident prevention (safety), such as:

Figure 2. ISO/IEC 31000 diagram
Source: IEC (2027)

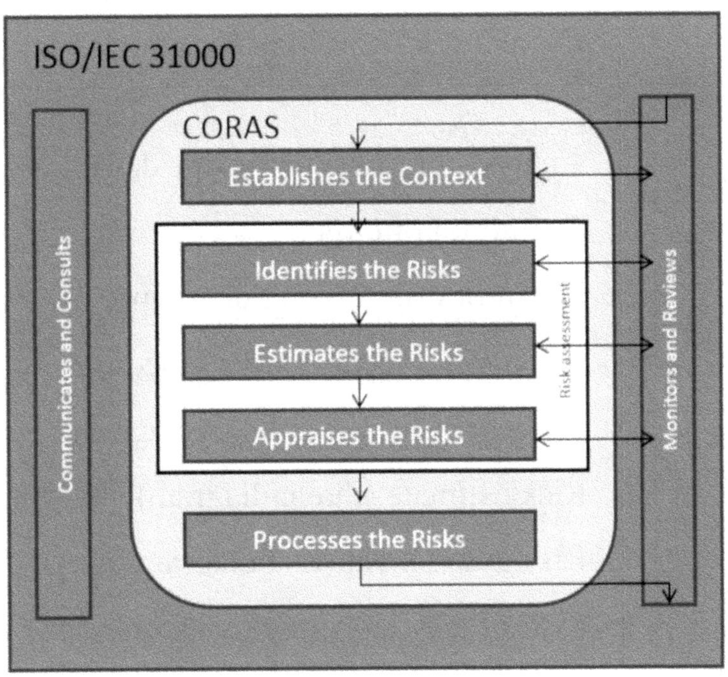

1. Hazard and Operability Study (HAZOP);
2. Failure Tree Analysis (FTA);
3. Failure Mode and Effect Criticality Analysis (FMECA);
4. Markov quantitative analysis methods (Markov);
5. CCTA risk analysis and management (CCTA *Risk Analysis and Management Method* – CRAMM).

The primary objective of CORAS is to improve traditional risk assessment methodologies in order to achieve better outcomes. Analysis is performed over eight stages.

CORAS is the risk analysis method that identifies, analyzes and describes risks in standardized language. The CORAS structure can be split into three different instruments:

1. **Risk modeling language:** Graphs and textual syntax;
2. **Method:** Description of the safety analysis process and guidelines for creating CORAS diagrams;
3. **Tool:** For documentation, maintenance and communication of the risk analysis outcomes.

CORAS method resources:

1. Open-source solution;
2. Free updates and software;
3. Supplied with five risk assessment methods, including HAZOP, FMECA, Markov analysis and CRAMM;
4. Portable with XML standard;
5. In modules, which facilitate the inclusion of new functionalities;
6. Integrated with three main eXtensible Markup Language (XML) data models, including:
    a. XML Metadata Exchange (XMI) standardized per Object Management Group (OMG);
    b. Compatible with XML, formatted risk assessment indicator data;

Table 2. 8-stage analysis

Context	1. Analysis preparation (scope definition);
	2. Definition of targets;
	3. Detail of origin (asset diagram);
	4. Approval of detail of the objective;
Risk assessment	5. Risk identification (threat diagram);
	6. Risk estimate (threat diagram);
	7. Risk assessment (risk diagram);
	8. Risk processing (processing diagram).

c. Intrusion Detection Messaging Exchange Format (IDMEF) by the Internet Engineering Task Force (IETF).

Furthermore, CORAS UML and Unified Process (UP) are applied in risk assessments for information security and process safety.

The CORAS language is a graphic modeling language for documentation, communication, security threat and risk scenario analysis. In risk analysis, it is common to use the structured brainstorming technique in meetings to obtain information. In these brainstorming sessions, the CORAS language is applied to create threat and risk scenario models. One of the advantages of using the CORAS language in brainstorming session is to promote standardization in a heterogenic environment, in which many people with specific knowledge are involved. The CORAS language improves the efficiency of the analysis process, as well as the quality of the outcomes obtained. According to Lund, Solhaug and Stolen (2011), standardization is facilitated mainly by the limited use of icons to represent the different environments, such as described in the Table 3.

3.7. MTBF

Mean time between failures (MTBF) is the reliability parameter of a component, measured between the initial functioning timepoint and the occurrence of an incident that causes a stoppage in the regular functioning of this system (unavailability). The value is normally given by the manufacturer of the component in its technical specifications and it indicates, in accordance with the testing procedure employed, the mean time between failures of that product in the manufacturer's laboratories.

In this study, an assessment of the failure probability and unavailability time was made for the system under analysis, taking into account the data provided by the suppliers of the components and the parameters referenced by the IEEE 493-2007 standard.

4. PROPOSED SOLUTION

Considering the entire process pertaining to the Utilities industry for water treatment and based on the FINESCE Smart Energy platform, we propose a new architecture involving enabling technologies and approaches to digital transformation. As stated in the introduction, the pyramid architecture for automation is based on the 5-layer concept. For the purposes of illustration, Figure 3 demonstrates the water treatment plant inside this current model.

Using enabling technologies and approaches to digital transformation, we propose a new collaborative architecture based on 3 layers, rather than the pyramid-based scheme. The new high-level architecture can be seen in Figure 4, which works as described below:

Layer 1: The edge layer, called edge computing. This layer contains the cyberphysical and IoT systems, as well as their communication with higher layers.

Layer 2: This layer contains plant operation and control, which can be executed manually – by operator decision – or automatically, by means of a machine learning decision algorithm. This layer also contains the data for managerial decisions.

Table 3. Icons representation

Icon	Description
	Assets: something of value for the interested party that requires protection.
	Internal threats: unconscious – a potential cause of an undesirable incident.
	External threats: conscious.
 Threat (non-human)	Non-human threats.
	*Stakeholders:* people or organizations that may be affected, or that may influence an assessment or activity under analysis.
	Vulnerability: weakness that may be exploited by threats.
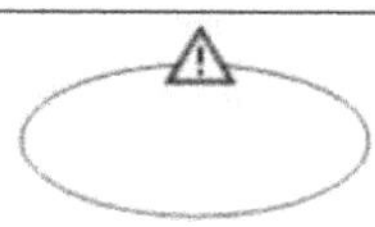	Threat scenario: something that may occur because of a threat that exploits a particular vulnerability.
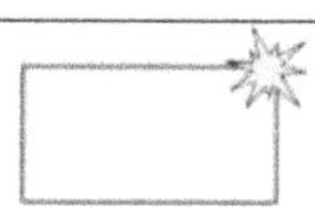	Undesirable incident: something that we want to avoid, because it may cause damage to assets.
	Risk: the possibility of something happening that will impact assets.
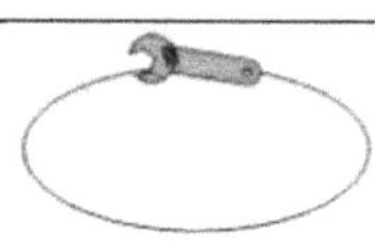	Treatment scenario: failure correction tools and techniques.

Source: Author.

Layer 3: Applications with real-time interaction on the factory floor and reorganization according to demand or climate conditions.

In order to enable a discussion regarding the new model, migration of the water treatment system to this new architecture was undertaken whilst taking into account the following levels of automation described in Table 2.

Figure 3. Pyramid architecture of an ETA (water treatment station)
Source: Author

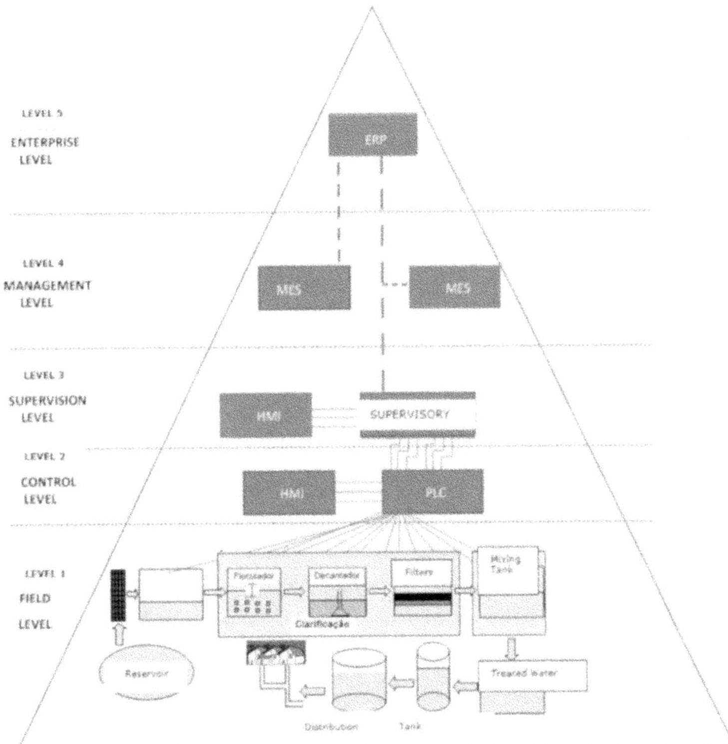

Figure 4. New proposed high-level architecture
Source: Author

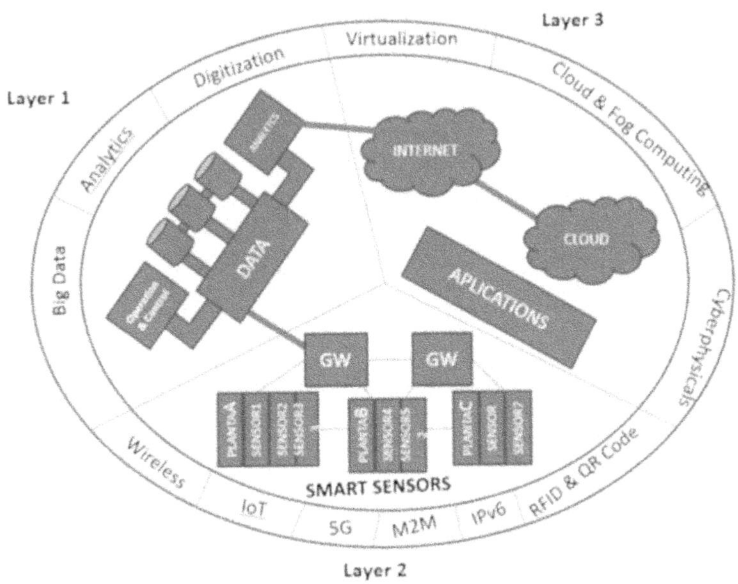

Table 2. Activation of the ETA

Id	Stages	Physical Action	Analysis	Non-Conformity	Action
01	Catchment	Activation of the pump, valve, level sensor and dosing	Turbidity, pH and DO	Provide a corrective solution	Manage pre-dosing of calcium hydroxide Manage dosing of Aluminum Polychloride Manage pre-chlorination
02	Flocculation	Pump, Level sensor, dosing and frequency inverter	Loss of flocculation	Provide a corrective solution	Apply normalization solution by identifying which dose is required (calcium hydroxide, aluminum polychloride or pre-chlorination)
03	Decanting	Pump, level sensor and dosing	Loss in decanting	Provide a corrective solution	Apply normalization solution by identifying which dose is required (calcium hydroxide, aluminum polychloride or pre-chlorination)
04	Filtering	Pump and level sensor	Filtering within the norm	Wash filter	Stop pump, empty filter and start manual washing
05	Disinfection	Pump, level sensor and dosing	Water quality	Control addition of fluorine and pH	Dose recommended value
06	Distribution	Pump and level sensor	Water loss	Identify leak	Activate maintenance team
07	Consumption	Consumer's meter	Consumption	Identify losses and/ or leaks	Activate maintenance team

Source: Author.

The choice of FIWARE among CRYSTAL and SOFIA2 was due to the fact that it is an open-source innovation ecosystem that provides programmers with the functionalities and properties necessary for the development of cloud solutions and applications, and it enables the provision of open data and communication of its specific requirements. Furthermore FIWARE is partnering with the Global City Teams Challenge 2016 programmer, GCTC2016, which is being organized under the lead of the U.S. Department of Commerce's National Institute of Standards and Technology (NIST). Based on these definitions and in the FINESCE energy platform, the following architecture was devised (Figure 5).

Compared to traditional architecture, the gains from use of this architecture are:

1. Use of historical data to predict necessities, collaborating with the entire supply chain in order to fulfill demands;
2. Management of water sources and exert control and use defined by automated processes from a remote location;
3. Integrate non-water related data sources, such as: government, research and industry, in a general manner, in order to accompany the conservation and continued improvement efforts of the existing operation;
4. Increase the visibility of installations of assets and improve performance and data analysis using dashboards;
5. Use of real-time data regarding climate events in order to anticipate changes in resource availability or requirements;
6. Provide consumers with greater control and visibility of consumption and residential and municipal resources in order to manage resource conservation efforts;

Figure 5. Architecture using FIWARE for water treatment based on Industry 4.0
Source: Author

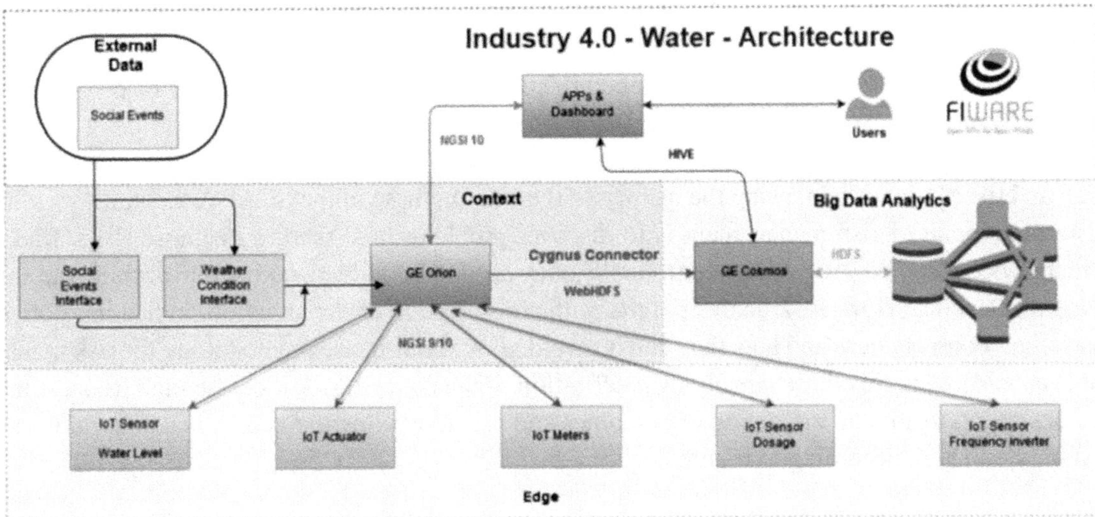

7. Use of data to discover improvements in procedures and thus serve citizens more effectively;
8. Improvement in the accuracy of water meter readings with greater frequency.

5. METHODOLOGY

Methodology is based on the following stages, and research planning includes theoretical and documentary study; consequently, an empirical study with the development of a case study for intelligent water use employing enabling technologies:

- Identification of the critical systems and components, assessment, quantification and modeling of a water treatment and distribution plant following the classic 5-layer pyramid model;
- In the theoretical and documentary study, an analysis of the enabling technologies and digital transformation was undertaken in light of Industry 4.0 (Schwab, 2016);
- Definition of the platform of use, with enabling technologies as a reference, visions from around the world for digital transformation and the FINESCE energy platform (Finesce, 2016);
- Validation of the proposal through use of the platform in experiments;
- Risk analysis, proposing a methodology for risk management, with the CORAS method as the main basis. During this research, it was noted that CORAS does not include several important stages for full risk management, especially with regard to the risk estimate, quantitative method and inspection with auditing of the process risk management system.

This study assumes that the concessionaire of the case study has an operational automated sanitation plant; therefore, the methodology for risk and failure management of the process in this environment will be validated, with emphasis on automated technology.

The conventional analysis of reparable system failures establishes the study of two random variables of interest: Time Between Failures (TBF) and repair time (RT). It is assumed in this work that any failure will be immediately repaired, and that this maintenance is capable of returning the item that failed (component or system) back to its original condition, with the same characteristics as a new component. As a consequence, the distribution of the random Time to Failure (TTF) variable will be considered equal to the distribution of the random TBF variable.

The proposed methodology performs the risk/failure analysis based on mathematical modeling for MTBF and the Markov chain, using the history of the existing base in the SCADA system.

The main point of risk management is to discover problems, and then to diagnose risks, find out where they occur and how to resolve them; finally, they need to be audited, taking into account the entire process as a cyclical flow. Risk analysis starts with a description of the situation, indicating potential threats, their consequences and how they can be avoided. A seven-phase methodology for risk management is considered in this thesis, involving a definition of the context, risk identification, risk estimate, risk assessment, treatment, inspection of outcomes and periodic follow-up audits, in a continuous flow based on the PDCA, according to Figure 6.

Figure 6. Risk management methodology cycle
Source: Author

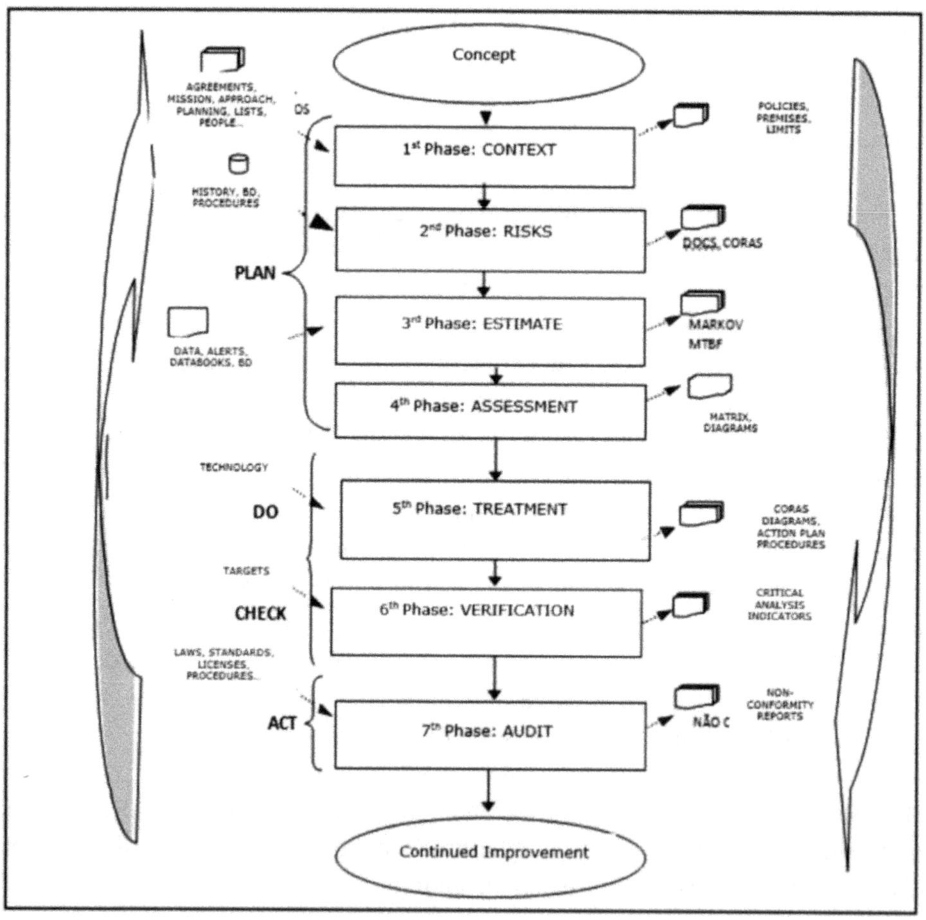

The methodology begins with a characterization of the process by describing the environment it is linked to, its mission, cultural issues and its requirements. Phase one (1st Phase) is the context definition stage, the time point at which the basic parameters are defined, through which risks that need to be managed and the scope of the rest of the risk management process will be identified. The criteria to be used in the identification, assessment, impact and acceptance of the risks will also be defined. The definition of the context takes into account all the relevant information about the organization that is relevant for defining the context of safety risk management.

The main step towards obtaining the context is the correct application of techniques such as HAZOP, which enable a description of the objectives, premises and criteria of the environments that will be used to determine the risks pertaining to the process. These criteria involve determining the safety consequences and the methods used for risk analysis and assessment.

The next step (2nd Phase), pertains to identification of the risks, according to Brandão and Fraga (2008) and this is one of the most critical stages, because any risks that are not identified will not subsequently be analyzed or treated. The purpose of this stage is to determine the events that may cause potential losses and it should be clear how, where and why a loss may occur. Identification must include the risks that are and are not under control of the study process. In practice, risk identification is much more complex, because the information is almost always based on subjective experience and criteria of those responsible for conducting the project. (Brandão & Fraga, 2008). Risk identification involves identifying threats, existing controls, vulnerabilities and consequences. It was in this phase of the project that the CORAS method made the most significant contribution to the development of the risk analysis, as outlined in Section 4.4 of this thesis.

The third phase (3rd Phase) of the methodology pertains to estimates; in this stage, data are produced that will support the decision about which risks will be treated and the treatment methods to be adopted, taking cost efficiency into consideration. Accordingly, it involves considerations regarding the origin of the risks, their consequences and their probability of occurrence, as well as a reliability study and process availability. The estimate may be either qualitative or quantitative, or even a combination of the two.

The qualitative estimate uses a scale with qualifying attributes that describe the magnitude of the potential consequences and the probability that these consequences will occur, and they are duly represented in the CORAS tool, while the quantitative estimate adopts mathematical modeling in order to determine risks, with failure and process reliability estimates and an analysis of the availability of process components. A combination of two types of estimates is adopted for this study. Outputs from this phase of the risk management methodology are recorded in FMEA forms, MTBF spreadsheets, Markov models and consolidation in the CORAS, as detailed in the following sections.

The fourth phase (4th phase) of the methodology pertains to assessments; the purpose of the risk assessment is to facilitate decision making, based on outcomes of the risk analysis in phases 2 and 3 of the methodology. It is necessary to define the priorities and the real necessity for treating the risks analyzed. This assessment involves a comparison of the risk levels encountered, with the criteria established when the context was considered. In this phase, it is relevant to evaluate corporate security, the impacts of the decision made on the selection of the accepted risks. However, nothing will prevent a new assessment from being made at any time and new risks being treated, because the risk management process is a dynamic one.

The fifth phase (5th phase) pertains to the risk treatment, including identification of the treatment options, assessment of the options and the costs of preparing for implementation of the treatments in the study scenario. This stage is equivalent to the state of risk mitigation of the SP800-30 specification or even the ISO/IEC 31000. Treatment is identified in the CORAS treatment diagram, associated with the possible scenarios of incidents that cause them. Treatment options are related to each of the risks. There are four potential options for the risk treatment: reduction, retention, prevention and transfer. The residual risks are identified at the end of the treatment process. If these risks are not acceptable, treatments may be reworked or, alternatively, the entire risk management process may be revised. (Brandão & Fraga, 2008). The implementation of treatment may or may not involve the acquisition of technologies, or it may simply involve the creation of standards, training of personnel and development of internal rules to be followed.

The verification phase (6th phase) involves monitoring and verification of the outcome; therefore, it is essential for the risk management process, because risks are not static and must be monitored in order to ensure the efficacy of implementation strategies and the outcome of the management mechanisms used in risk treatment. Thus, the monitoring process must be ongoing and dynamic. In addition to continuous monitoring, organizational or external changes may alter the context of the analysis, leading to the necessity for a full revision of risk management.

The last phase of the methodology (7th phase) pertains to the audit, which we consider to be very important for ensuring the efficacy of the risk management and that the process is kept up-to-date. The audits may be periodically initiated by an internal team or undertaken by third parties. The purpose of internal audits is to verify, based on objective evidence, whether the following conditions occur satisfactorily:

- The operational procedures and instructions are adequate and effective;
- The areas have been acting in accordance with the instructional documents and checklists have been completed;
- The alert reports and logs are being assessed.

Any non-conformities (real and potential) detected must be recorded in accordance with a specific procedure, including initiatives for registration and actions to conclude them. A critical analysis of the non-conformities is indicated and, if necessary, their causes are investigated, corrective actions are defined and implemented and amendments to procedures are recorded. After the implementation of corrective actions, their efficacy must be evaluated before they are concluded.

Therefore, the first four phases are actually sufficient to determine the impact of risks and they are included in the CORAS method, or even by making an analogy with the PDCA quality cycle, pertaining to the planning phase, while phases five and six (execution) correspond to the process of implementing treatment, risk mitigation and verification of the outcome, and in the last phase, periodic auditing is indicated, which must be performed in order to ensure that the risk analysis system is always up-to-date and that it reflects the actual status of the operational system.

Each stage of the risk management process must be properly documented. Suppositions, methods, origins of data, analyses, outcomes and rationale for decisions made must be recorded. The reports produced must be as succinct and objective as possible. (Brandão & Fraga, 2008).

The suggested cycle in this work, if implemented, will lead to true risk management, and not just to a one-off assessment of the risk in a system. The difference is that, in a one-off risk assessment, normally a risk is only identified and estimated, whereupon its existence is considered, whereas, in risk management, in addition to being identified, the risk is treated and efforts are made to prevent it from happening.

6. OUTCOMES AND CONCLUSION

This research project was based on the necessity for assessing digital transformation in order to implement improvements in the traditional automation and control architecture of Utilities industries and the associated risk analysis. Up-to-date bibliographical research was undertaken in order to implement this proposal, covering a wide range of subjects related to digital transformation; accordingly, enabling technologies were studied that allow for this paradigm shift. It has been concluded that the new technologies that enable this change must work in a collaborative and integrated manner, because the complete solution for digital transformation depends on correct functioning of enabling technologies, including: Internet of Things, Machine Learning, Cloud and Fog Computing, cyberphysical systems, M2M, 5G and big data analytics. The approaches that guide this digital transformation around the world have also been studied, as well as research into work related to the subject in question. Special attention was paid to the Industry 4.0 approach, due to its pioneering and unique nature.

Based on this initial analysis, it was decided to apply the concepts of digital transformation to the Utilities industry. The choice of this segment was due to the fact that water treatment and distribution is more widespread among citizens, rather than being restricted to a segment of traditional industry. Additionally, and during this study, a platform was encountered for energy, where the concept of intelligent cities was the starting point; however, no platforms were found for the processing and/or management of water consumption. Accordingly, this study intends to contribute to the issue of water treatment and distribution in intelligent cities, combining with other equally important priorities.

Based on this lack and the need for contribution, a visit was made to a water treatment and distribution unit. The purpose of this visit was to conduct interviews regarding the process of generating the documentation, flow charts and steps for a correct understanding of the entire water treatment and distribution cycle. The study is intended to contribute to knowledge regarding the operation of a water distribution station and the characterization process enables knowledge to be gained regarding industrial equipment and the system for managing and acquiring data, as well as the protocols used.

The classic 05-layer architecture has been improved based on this research regarding water treatment and distribution linked to studies of enabling technologies and approaches to digital transformation, and it now comprises 03 layers employing enabling technologies by means of an open-source innovation ecosystem called FIWARE, which provides the functionalities and properties necessary for the development of cloud solutions and applications, permitting the availability of open-source data and communication of its specific necessities.

The improvement in the architecture was put to the test within the concept of intelligent cities and the improvements outlined below were attained:

- All decision making was automated, based on decisions made by machines with known values as parameters;
- Integration of the different components and manufacturers, because this is an open-source architecture with an equally open communications protocol;
- Use of historical data to predict necessities, collaborating with the entire supply chain in order to fulfill demands;
- From a remote location, plant management, control and use was defined by automated processes;

- General integration of data sources not related to water, such as: government, research and industry, in order to keep up with conservation efforts and continued improvement of the existing operation;
- Using dashboards, it is possible to increase visibility of the asset installations and improve data performance and analysis;
- Use of real-time data regarding climate events in order to anticipate changes in resource availability or requirements;
- Provide consumers with greater control and visibility of consumption and residential and municipal resources in order to manage resource conservation efforts;
- Use of data to discover improvements in procedures and, thus, provide better services to citizens;
- Accuracy improvements by more frequent water meter readings.

Risk management in a sanitation plant focused on process safety, information security and continued improvement has been applied to the new model. However, it was ascertained that use of the CORAS method alone was not sufficient for the development of risk management, because CORAS is limited to a diagnosis of risk analysis and documentation of the treatment; i.e., it is very occasional and it does not permit monitoring, verification and auditing of risk management in the environment.

However, for the purpose of risk diagnosis, CORAS provided excellent outcomes, enabling a reduction in analysis time due to ease of use of the tool and standardization of process implementation. In addition, CORAS permitted documentation with the use of several techniques to be consolidated in a single tool.

The development of a seven-phase risk management methodology enabled implementation of a targeted, objective and ongoing risk management program without any complications, in addition to facilitating the consolidation of several characteristics of already existing tools.

The operational safety and efficiency of sanitation systems are significantly affected by their automation bases and consequently the availability and vulnerabilities of their equipment. Understanding the interface between information technologies (IT) and automation technologies (AT) was of particular relevance in achieving the conclusions obtained in this study, because the risk management developed included the different stages of the plant process, with security and safety analysis.

It was ascertained that a basic vulnerability could cause major losses to an organization or even to the community surrounding a plant; however, these risks are predictable and quantifiable in a risk management project, as presented in the Figure 7, during one of the stages of implementation of the methodology.

The simulation of reliability of several plant components was relevant for decision making, especially for maintenance management regarding the list of replacement parts or even the technology replacement process, such as the change in the CLP model, because the studies of wear failures and failures that occur before wear present a mathematical model that quantifies the unavailability time of equipment in hours.

Modeling with the Markov chain for verification of the state of components and the behavior of the process, developed during validation of the methodology, presented favorable outcomes, because the flexibility in executing the simulation including new states, and consideration of the interdependence between plant equipment favored comprehension and quantification of the connection between field instruments and the pumping system, including in terms of stoppage time and the frequency quantification process, through the execution of probability analyses.

Figure 7. Risk analysis
Source: Author

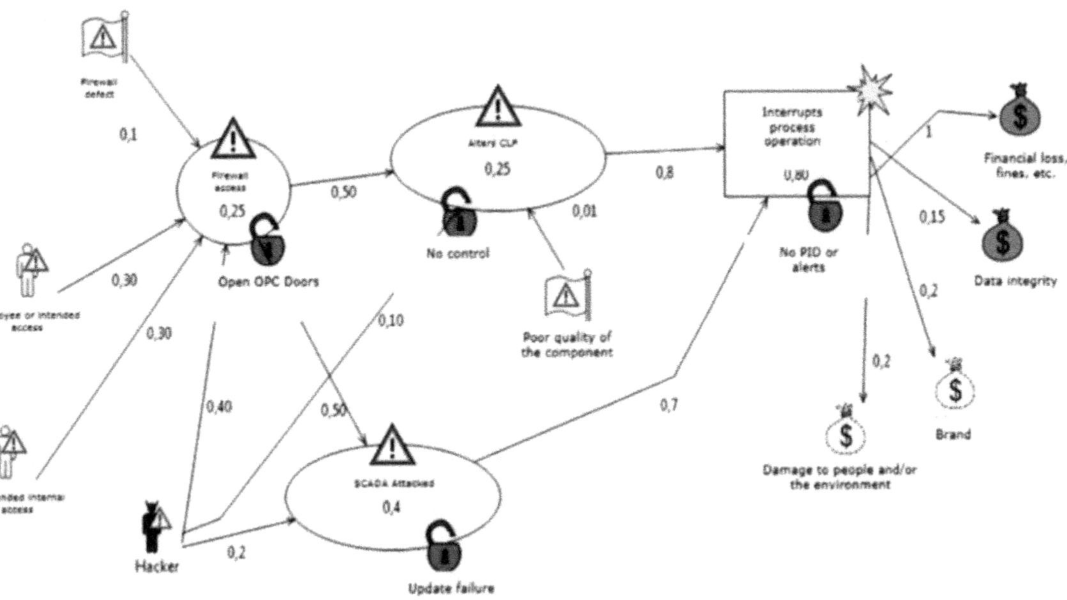

A positive outcome was also ascertained in the use of tools such as the CARMS software program for simulation of the Markov chain, enabling use of this mathematical modeling in systems of greater complexity, and in many states, as well as providing the necessary information regarding the idle period for the environment under analysis in the risk management methodology.

The FMEA stages executed consolidated in CORAS were relevant in the verification of the subjective items and in the qualitative analysis of the operational process of the sewage treatment station.

With the application of the risk management methodology, it was easy to ascertain in the simulations executed by the mathematical models that small investments can produce significant returns, in large part due to a reduction in costs with operational unavailability or even reduced billings (which, in this case, due to the evaluable QID grade, enables higher financial penalties to be levied under the agreement). One of the outcomes of the implementation of the risk management methodology was precisely the improvement of performance indicators over the last 12 months, because a considerable margin of progression has been recorded in this indicator with a mean increase of 22% in plant availability, resulting solely from one improvement program defined by risk management.

Among the outcomes ascertained during validation of the methodology, the supply of quantitative information that aids in decision making was observed; for example, in the case study, there was a necessity to assess the time information between failure, criticality and repair time in order to aid in decision making, with regard to the option for treatment architecture adopted in the risk mitigation of harmonics in the study environment. Another favorable attitude observed in risk management was the flexibility that the methodology offers in emphasizing the most relevant phase according to the profile of the study scenario, and including a social risk analysis. It should also be emphasized that the methodology flow is continuous, thus favoring revisions of maintenance, safety and quality, as well as supervision and control efforts, without forgetting the motivational and cultural aspects that, as is well known, induce a potential for continued improvement that is difficult to quantify.

Therefore, the risk management flow developed throughout this study demonstrates that it is also able to assess systems dominated by uncertainty and scarcity of reliable data. It is also apparent that there is a window of possibilities to conduct risk management in similar plants using analysis codes, reusing models defined for equipment and diagrams developed in the CARMS and CORAS, thus facilitating the reliability analysis process of the different processes in sanitation plants.

Finally, it is possible to use the new platform obtained in this study to generate benefits with regard to conscientious consumption and management of water resources. It may be extended, with adaptations, to other equally critical environments, such as energy supply, nuclear power plants and the petrochemical industry, among others.

REFERENCES

Almeida, J. R., Jr. (2003). Segurança em sistemas críticos e em sistemas de informação: um estudo comparativo [Doctoral dissertation]. Escola Politécnica, Universidade de São Paulo, São Paulo.

Ashton, K. (2009). That 'internet of things' thing. *RFiD Journal, 22*(7), 97-114. Retrieved from http://www.rfidjournal.com/articles/view?4986

bism@rt. (2017). Oldest Internet Of Things Objects In The World. Retrieved from https://bismart.com/en/oldest-internet-of-things-objects-in-the-world/

Bounfour, A. (2016). *Digital Futures, Digital Transformation, Progress in IS: From Lean Production to Acceluction.* Cham, *Switzerland*: Springer International Publishing. doi:10.1007/978-3-319-23279-9

Brandão, J. E. M. S. (2008). *FRAGA.J.S. Gestão de risco. In SBSEG2008 VIII Simpósio Brasileiro em Segurança da Informação de Sistemas Computacionais.* Santa Catarina: UFSC.

Collin, J., Hiekkanen, K., Korhonen, J. J., Halén, M., Itälä, T., & Helenius, M. (2015). *It leadership in transition-the impact of digitalization on finnish organizations.* Helsink-Finland: Unigrafia Oy.

Fiaschetti, A., Pietrabissa, A., & Delli Priscoli, F. (2015). Towards manufacturing 2.0: An innovative architecture for the Factory of the Future. In *2015 European Conference on Networks and Communications, EuCNC 2015* (pp. 450-454). 10.1109/EuCNC.2015.7194116

FINESCE. (2016). The FINESCE Smart Energy Platform Seventh Framework (n. 604677).

Wang, S., Wan, J., Li, D., & Zhang, C. (2016). Implementing Smart Factory of Industrie 4.0: An Outlook. *International Journal of Distributed Sensor Networks, 12*(1).

International Electrotechnical Commission (IEC). (2009). *31000: general guidelines for principles and implementation of risk management.* Switzerland: ISO/IEC.

Key Enabling Technologies. (2016). European Commission. Growth. Internal Market, Industry, Entrepreneurship and SMEs. Retrieved from https://ec.europa.eu/growth/industry/key-enabling-technologies_en

Khan, S. (2016). *Leadership in the Digital Age - a study on the effects of digitalization on top management leadership.* Shahyan Khan.

Lee, J., Bagheri, B., & Kao, H. A. (2015). A Cyber-Physical Systems architecture for Industry 4.0-based manufacturing systems. Manufacturing Letters, *3*, 18–23. doi:10.1016/j.mfglet.2014.12.001

Lee, J., Kao, H. A., & Yang, S. (2014). Service innovation and smart analytics for Industry 4.0 and big data environment. *Procedia CIRP*, *16*, 3–8. doi:10.1016/j.procir.2014.02.001

Loural, C. de A. (2014, June 26). Um panorama de novas tecnologias e seus impactos na indústria. Retrieved from https://www3.eco.unicamp.br/neit/images/stories/arquivos/Novas_tecnologias_e_seu_impacto_na_industria_-_v140626.pdf

Companhia Ambiental do Estado de São Paulo. (2003). Manual de orientação para a elaboração de estudos de análise de riscos. São Paulo, Spain: Companhia Ambiental do Estado de São Paulo.

Mcnabb, D. E. (2005). *Public utilities: management challenges for the 21st century*. Edward Elgar Publishing. doi:10.4337/9781845426811

Mitchell, T. M. (1997). *Machine Learning*. New York, NY: McGraw Hill.

Moraes, C. C., & Castrucci, P. D. L. (2007). *Engenharia de automação industrial*. Rio de Janeiro: LTC.

Mounce, S. R., Pedraza, C., Jackson, T., Linford, P., & Boxall, J. B. (2015). Cloud based machine learning approaches for leakage assessment and management in smart water networks. *Procedia Engineering*, *119*, 43–52. doi:10.1016/j.proeng.2015.08.851

Muhammad, S. Y., Makhtar, M., Rozaimee, A., Aziz, A. A., & Jamal, A. A. (2015). Classification Model for Water Quality using Machine Learning Techniques. *International Journal of Software Engineering and Its Applications*, *9*(6), 45–52. doi:10.14257/ijseia.2015.9.6.05

NIST. (2011). The NIST Definition of Cloud Computing Recommendations of the National Institute of Standards and Technology.

Chang, W. L. (2015). NIST Big Data Interoperability Framework: Volume 1. *Definitions*.

Oren, G., & Stroh, N. Y. (2013). Mathematical Model for Detection of Leakage in Domestic Water Supply Systems by Reading Consumption from an Analogue Water Meter. *International Journal of Environmental Sciences and Development*, *4*(4), 386–389. doi:10.7763/IJESD.2013.V4.377

Public Utilities Board Singapore. (2016). Managing the water distribution network with a Smart Water Grid. *Smart Water*, *1*, 1-13. Retrieved from http://smartwaterjournal.springeropen.com/articles/10.1186/s40713-016-0004-4

Sadeghi, A. R., Wachsmann, C., & Waidner, M. (2015). Security and privacy challenges in industrial internet of things. In *2015 52nd ACM/EDAC/IEEE Design Automation Conference (DAC)* (pp. 1-6). IEEE. 10.1145/2744769.2747942

Samuel, A. L. (1959). Some Studies in Machine Learning Using the Game of Checkers. *IBM Journal*, *3*(3), 535–554. doi:10.1147/rd.33.0210

Schwab, K. (2016). The Fourth Industrial Revolution (VINT research report).

Shrouf, F., Ordieres, J., & Miragliotta, G. (2014). Smart Factories in Industry 4.0: A Review of the Concept and of Energy Management Approached in Production Based on the Internet of Things Paradigm. In *2014 IEEE International Conference on Industrial Engineering and Engineering Management (IEEM)* (pp. 697-701). IEEE.

SMART CITY. (2016). Plataforma para transformação digital. Retrieved from http://www.pad.lsi.usp.br/ptd

Soldatos, J., Gusmeroli, S., Malo, P., & Di Orio, G. (2016). Internet of Things Applications in Future Manufacturing. In *Digitising Industry-Internet of Things Connecting the Physical* (pp. 153–183). Digital and Virtual Worlds.

SWSS. (2018). Smart Water Supply Systems. Retrieved from http://www.adp.pt/pt/sustentabilidade/inovacao-e-desenvolvimento/life-swss/?id=51

Tamaki, H., Sanches, G., & Gonçalves, O. (2006). A medição setorizada como instrumento de gestão da demanda de água em sistemas prediais – Estudo de caso: programa de uso racional da água da Universidade de São Paulo. Associação Nacional de Tecnologia do Ambiente Construído.

Torres, J. M. (2008). Analyzing risk and uncertainty for improving water distribution system security from malevolent water supply contamination events [Master's Thesis]. Office of Graduate Studies of Texas A&M University, TX.

Verma, P., Kumar, A., Rathod, N., Jain, P., Mallikarjun, S., Subramanian, R., ... & Sundaresan, R. (2015, October). Towards an IoT based water management system for a campus. In *2015 IEEE First International Smart Cities Conference (ISC2)* (pp. 1-6). IEEE.10.1109/ISC2.2015.7366152

Weiser. (1992). Ubiquitous Computing. Retrieved from http://www.ubiq.com/hypertext/weiser/UbiHome.html

Westerman, G., Bonnet, D., & McAfee, A. (2014). *Leading Digital: Turning technology into business transformation*. Boston, MA: Harvard Business Press.

Chapter 14
Industry 4.0 From the Supply Chain Perspective:
Case Study in the Food Sector

Andrés Boza
Universitat Politècnica de València, Spain

Faustino Alarcón
Universitat Politècnica de València, Spain

David Perez
Universitat Politècnica de València, Spain

Pedro Gómez-Gasquet
Universitat Politècnica de València, Spain

ABSTRACT

Industry 4.0 proposals have an important impact not only in companies but also in the complete supply chain. The effect in the supply chain is different depending on each Industry 4.0 proposal and each supply chain configuration. This chapter deals with Industry 4.0 from the supply chain perspective. Four levels of scope on the supply chain and a classification scheme based on three blocks (technologies, features and extensions) have been defined. Finally, the case study in the food sector gives a real vision of some previously introduced concepts about Industry 4.0 from the supply chain perspective.

INTRODUCTION

Industry 4.0 promotes, among other things, autonomous decision-making, interoperability, agility, flexibility, efficiency and cost reductions. This is the reason by which many companies are aiming to implement the technologies and concepts related to "Industry 4.0" (Perez et al., 2017). Industry 4.0 holds huge potential. Smart factories allow individual customer requirements to be met and mean that even one-off items can be manufactured profitably. In Industry 4.0, dynamic business and engineering

DOI: 10.4018/978-1-5225-4936-9.ch014

processes enable last-minute changes to production and deliver the ability to respond flexibly to disruptions and failures on behalf of suppliers, for example. End-to-end transparency is provided over the manufacturing process, facilitating optimised decision-making. Industry 4.0 will also result in new ways of creating value and novel business models (Kagermann, 2013). The importance of these advances is recognized by the European Commision promoting proposals in this line such as Factories of Future (European Commission, 2013; 2016).

These technical and organizational advances in each company due to Industry 4.0 may have an impact on the Supply Chain (SC) performance. Industry 4.0 concepts have been usually treated from an intra-enterprise viewpoint. However, industries are involved in (more or less) organized supply chains, so that the success of the company depends on the success of the supply chain. The inclusion of industry 4.0 proposals in a company in the supply chain may have different repercussion on the SC depending on different factors. This chapter deals with Industry 4.0 proposals from the SC perspective analyzing the Industry 4.0 concept, studying the different supply chain configurations, defining the scope of Industry 4.0 on the SC and making a classification scheme based on technologies, features and extensions. Furthermore, a case study in the food sector is presented as an example of inter-enterprise collaboration.

INDUSTRY 4.0

Industry 4.0 and the Industrial Internet of Things (IIoT) has become one of the most talked about industrial business concepts in recent years. The IIoT has brought about a new strategy, which has arisen in industry, especially within manufacturing, and it is based on the producer focusing on what the customer actually wants rather than the product they buy. According with Gilchrist (2016) the four main characteristics of Industry 4.0 include:

- Vertical integration of smart production systems
- Horizontal integration through global value chain networks
- Through-engineering across the entire value chain
- Acceleration of manufacturing

Reviewing these main characteristics, it is easy to identify the important contribution of Industry 4.0 for the value chain or SC.

On the other hand, according to Gilchrist (2016) the nine technological trends more important in an Industrial 4.0 context are: Big Data and Analytics, Autonomous Robots, Simulation, Horizontal and Vertical System Integration, The Industrial Internet of Things, Cyber-Security, The Cloud, Additive Manufacturing and Augmented Reality.

Within these trends, IoT is possibly the most important and far-reaching technological trend. In that sense, Gubbi et al. (2013) indicates that the IoT is a radical development of today's Internet towards a network of interconnected objects that not only collects information from environment and interacts in the physical world (action/command/control), but one that also uses existing standards on the Internet to provide information transfer services, analyses, applications and communications. An important component of Industry 4.0 is the fusion of the physical and the virtual world (Kagermann, 2013). This fusion is made possible by Cyber Physical Systems (CPS). CPS are "integrations of computation and physical processes. Embedded computers and networks monitor and control the physical processes, usually with feedback loops where physical processes affect computations and vice versa." (Lee, 2008).

Furthermore, IoT protocols are fundamentals and key elements from a technical point of view. They are in constant evolution thanks to different working groups. For example, The Internet Engineering Task Force (IETF) started with three working groups (WG) (Keränen & Bormann, 2016): 6LoWPAN, which defined IPv6 adaption layer and header compression suitable for constrained radio links; ROLL, which focuses on routing protocols for constrained-node networks; and CoRE, which aims to extend the Web architecture to most constrained networks and embedded devices. The activity around IoT has increased and today seven WGs actively look into various aspects of IoT such as Home networking WG, Authentication and Authorization for Constrained Environments or IP Wireless Access in Vehicular Environments WG (Olshansky, 2018). Also, oneM2M is leading a global standardization body for M2M (Machine to Machine) and the IoT. It was established to develop a single horizontal platform for exchange and sharing of data among all applications (oneM2M, 2015).

Different protocols participate in the complex architecture of these systems to go from a physical level (p.ej. IEEE 802.15), network level (p.ej. 6LoWPAN, TCP/UDP) and application level (p.ej. CoAP) to achieve new functionalities. Evidently, these functionalities that IoT provides will have an important impact in the manner in which people, data and things are connected in the company and hence, in its processes. By this, the processes play an important role in how each of these entities (people, data, and things) works with the others to deliver value in the connected world of IoT and Industry 4.0. With the correct process, connections become relevant and add value because the right information is delivered to the right person at the right time in the appropriate way.

For the purpose of this chapter, we put the focus on the new functionalities that improve processes in the SC. Thus, the planning and control process across the SC is a key aspect of the supply chain management (SCM) (Dudek & Stadder, 2005). The work of Meyer et al. (2011) studies the IoT from the production control perspective to overcome drawbacks or alterations in the production plant, such as product errors, production errors, machine failure, quality problems or delivery errors.

Alarcón et al (2016) analyze the IoT applications in a production planning context and the potential benefits obtained.

Some IoT applications cited in the literature are:

- Monitoring product quality, which will allow a better control of damaged, lost and stolen products;
- Improving and ensuring product safety, especially the safety of perishable products, such as food or pharmaceutical products, by means of a powerful real-time traceability system;
- Making real-time product information available and for all the supply chain nodes, including customers. This will permit better product follow-up all along the transport route, the exact location and recovery of products. Also, the participants in the supply chain will be synchronized with changes in demand;
- Identifying products automatically and accurately classify them;
- Knowing the equipment status and its performance in real time, identifying and managing emergences more quickly, and conferring autonomy to "things" so they can react.

On the other hand, the benefits that using the IoT can provide companies can be grouped as follows:

- **More efficiency:** The IoT can improve system efficiency in general, and production efficiency in particular, by allowing production to be done according to customer requirements. Furthermore, this may improve the ratio of equipment usage and obtain more realistic plans;
- **Increased quality and safety:** The traceability offered by the IoT allows the exact conditions that products must pass through to be known and will, therefore, prevent quality and safety problems;
- **Facilitate integration:** The facilities that the IoT offers for supply chain members include information exchange, which favors the chain's integration, coordination and interoperability;
- **Costs reduction:** The IoT helps reduce inventory levels, improves the control of spoilt products and increases system efficiency, including production, storage, distribution and transport. It improves system management as it makes it easier. It also helps cut down the number of errors and allows less dependence on operators. Allocating routine activities at the lowest operational levels will allow humans to be capable of concentrating on strategic matters;
- **Improved customer response:** The IoT can help achieve more reliable deliveries and a better service level. It is also possible to detect changes in demand more quickly and to make the adjustments required to better cover demand.

The IoT can be characterized by the following four basic aspects (Alarcon et al., 2016):

1. **Reliable and accurate data collection:** The sensors and elements (things), designed according to the environment in which they must work, are capable of collecting very reliable and accurate data;
2. **Capacity to collect a huge quantity of data:** The development of information technologies enables huge quantities of data to be stored and processed;
3. **Rapid data transmission:** Today's communication technologies enable huge quantities of data to be transported in a very short time (e.g. wireless technologies), and confer "things" with a real-time response capacity;
4. **Automated processes:** "Things" can become completely autonomous, and entire activities (data collection, calculations and data processing, sending signals with messages, etc.) and whole processes can be performed, including decision making.

The IoT's capacity to collect and work with huge quantities of data will be fundamental to process information on the company's capacities (depending on the number of sections, human resources, machines or the complexity of the production process). The possibility that the IoT offers for automating these tasks must be suitably assessed and used to improve efficiency and to reduce errors and costs. As far as rapid data transmission is concerned, the IoT characteristic is considered of little relevance to develop plans if we bear in mind the horizons and dates that are usually dealt with, and it is not strictly necessary to acquire input data for real-time plans. However, for executing and controlling the plan, the IoT characteristic will be extremely valuable for rapid data transmission. To execute and control the plan, it will be necessary to know the plan itself in adequate detail and the performance of all our resources, preferably in real time, so that we can react as soon as possible to incidences or deviations from the plan. Likewise, the reliable and accurate data collection that the IoT offers can be used to send plans and to execute them, and to also follow-up and exactly control their execution. Automation is also considered most important for this case, and the IoT performs tedious, hazardous, complex or repetitive tasks. Perhaps the least relevant IoT characteristic to execute and control the plan is the capacity to collect a large quantity of data, but what is really necessary in this case is to follow-up what occurs in the plant in real time.

Some studies about IoT applications in the food or agricultural sectors than can help to understand better the case study described at the end of this paper. Yue (2011), where it is demonstrated that the IoT can improve the chain's efficiency and the degree of food safety because its use allows real-time communication and it shares accurate information to carry out smart recognition of products, their location, follow-up and monitoring. Also, Gu and Jing (2011) presents three specific IoT applications and their benefits in a fresh agricultural products SC. The work of these authors includes a table that reflects the improvements obtained from using the IoT in the links of this particular SC. More recent proposals describe the IoT's key enabling technologies from 'Farm to Fork'. These technologies impact in the i) production phase (smart farming and Livestock management), ii) transport and distribution phase, and iii) retail phase (Nukala et al., 2016). Companies as Accenture combine digital technologies such as the Internet of Things (IoT) with its big data analytics, visualization capabilities, and industry knowledge to create a Digital Agriculture Service and a Connected Crop Solution. This service helps large-scale farms gather and cross-correlate a wide range of critical data to make meaningful and timely operating decisions that can enhance yield and profitability. Also, the Connected Crop Solution can enable agro-input providers to boost field agent productivity and help smallholder farmers improve yields by providing fertilizer, pesticide, and seed recommendations personalized for each farmer's land and needs (Accenture, 2017). About food traceability, Chen (2017) proposes a food traceability cyber physical system by fog computing. The IoT- enabled cyber physical system concept can enhance the efficiency of food traceability system by means of the highly value-added traceable process.

SUPPLY CHAIN CONFIGURATIONS

A supply chain is referred to as an integrated system which synchronizes a series of inter-related business process: 1) acquire raw materials and parts; 2) transform these raw materials and parts into finished products; 3) add value to these products; 4) distribute and promote these products to either retailers or customers; and 5) facilitate information exchange among various business entities. The ultimate success of a firm will depend on its managerial ability to integrate and coordinate the intricate network of business relationships among supply chain members. Successful supply chain integration depends on the supply chain partners' ability to synchronize and share `real-time' information. Such information encompasses data, technology, know-how, designs, specifications, samples, client lists, prices, customer profiles, sales forecasts, and order history (Min & Zhou, 2002). Supply chains and networks are now looking to enforce collaborative agreements, which would produce more efficient workflow, flexibility, effectiveness, agility and coordination between chain links. In order to achieve proper synchronization and organization, every chain link must be provided with sufficient information and communication technology to further facilitate their operational integration (Vargas, 2013).

As stated before, a key aspect of SCM is the planning and control process across the SC (Dudek & Stadder, 2005). Planning in the SC is an important process for which all those involved in covering demand are organized with time to optimize their work and to obtain the biggest benefit possible for the company. Its main objective is to respond to customer requirements in quality, quantity and due date terms (Meyer et al., 2011) at the lowest possible cost so that available resources are used in the best possible way. The monitoring and control task covers the activities performed to react to alterations, which enables deviations from the original plan.

Industry 4.0 is an opportunity to improve these supply chain processes. These new information technologies facilitate new information flows and greater integration of business processes across the supply chain. Information sharing and collaboration in the IIoT can occur between people, between people and things, and between things. Sensing a predefined event is usually the first step for information sharing and collaboration. In the supply chain area, information sharing and collaboration enhance situational awareness and avoid information delay and distortion (Lee & Lee, 2015). The scope of this shared knowledge in the CS must include according with Simatupang, Wright and Sridharan (2002):

1. **Logistics synchronization:** Facilitates the value creation process that is designed to improve both individual and overall supply chain performance through implementing improvement initiatives (Value creation loop);
2. **Information Sharing:** Provides visibility enabling consideration of a global scope to make better decisions that optimise supply chain performance (Facilitation loop);
3. **Incentive Alignment:** Designed to motivate the chain members to align decisions and actions with supply chain profitability (Motivation loop);
4. **Collective Learning:** Combines fragmented skills and to enable chain members to acquire new skills from one another (capability loop).

The Enterprise Collaboration Maturity Model (ECMM) has the main objective of assessing organizations that desire to know their collaboration and interoperability maturity level with respect to a set of best practices. Four maturity levels were defined: 1. Performed: Collaboration with external entities is done, but in an ad-hoc and chaotic manner. Collaborative tasks and processes usually exceed budget and schedule, their past success cannot be repeated, and the potential of the technology is not used properly. 2. Managed: The objective is to create a management foundation for collaboration. Network technologies are used to collaborate and interoperate. 3. Standardized: The objective is to establish a common business strategy and business process infrastructure for collaboration. Business collaboration is facilitated through interoperability technologies and use of standards. 4.Innovating: The objective is to manage and exploit the capability of the networked organization process infrastructure to achieve predictable results with controlled variation. Additionally, another objective is to continuously improve the networked organization processes and the resulting products and services through continuous capability, and planned innovative improvements (Alonso et al. 2010). This approach is also followed by Vargas (2013) where, in addition to business collaboration, their proposal includes strategic alignment and enterprise architecture as tools for change.

This evolution in the collaboration allows the creation of different configurations in the SC. Lejeune and Yakova (2005) distinguished among four supply chain configurations: communicative, coordinated, collaborative and co-opetitive. They described these configurations as the 4 C's in supply chain management: The communicative configuration is characterized by shallow dependence. The only form of integration takes place within each entity that develops cross-functional collaboration. The coordinated configuration is characterized by deep dependence. This configuration can be viewed as a hierarchy of entities in the supply chain, which is captained or dominated by a supply chain leader. The collaborative configuration is characterized by shallow interdependence. Entities agree on a set of commonly defined objectives, and use their complementary assets to gain long-term competitive advantage. Finally, the co-opetitive supply chain configuration is characterized by deep interdependence. This concept is presented as a combination of cooperation and competition and rest on the assertion that competitors can benefits when they work together.

Industry 4.0 proposals can improve the current supply chain (4 C's), it can be used as a tool to "re-configurate" the SC, or it can be used to move along the maturity level. In this sense, the identification of "coordination mechanisms" can help in this purpose. Gaudreault et al. (2009) consider that the coordination mechanisms state how the SC members' relationship is operationalized (e.g. rules and protocols to exchange information, which decisions must be taken, by whom and when). However, the identification of these coordination mechanisms to be improved can be difficult. Cuenca et al. (2013) identified nine structural elements that characterize coordination mechanisms. 1. Number of decision-makers. 2. Collaboration level. 3. Interdependence relationships nature (Hierarchical/Non-hierarchical). 4. Interdependence relationships type (Temporal integration/Spatial integration). 5. Number of coordination mechanisms (Unique/Various). 6. Information exchanged. 7. Information processing. 8. Decision sequence characteristics (Beginning of the coordination mechanism/Sequence of decisions). 9. Stopping criteria. The identification of these mechanisms in a SC allows recognizing areas where Industry 4.0 can be a tool for improvement of the SC.

INDUSTRY 4.0 PROPOSALS ON THE SUPPLY CHAIN

Two perspectives have been defined in this point in order to locate industry 4.0 proposals on supply chain: i) the scope on the supply chain and ii) a classification scheme.

Scope on the SC

Depending on the scope on the SC, the activities carried out to implementing Industry 4.0 concept, can be classified into four levels:

Level 0: Focus on intra-enterprise without impact in the SC.
Level 1: Focus on intra-enterprise with direct impact in the SC.
Level 2: Focus on inter-enterprise with impact in a limited number of members of the SC.
Level 3: Focus on inter-enterprise with impact in the complete SC (from Industry 4.0 to SC 4.0).

Level 0: Intra-Enterprise Focus Without Impact on the SC

Applications as early Enterprise Resource Planning (ERP) systems or, more simply, enterprise systems (ES), were not primarily focused on the SC. Their initial focus was to execute and integrate such internally-oriented applications that support finance, accounting, manufacturing, order entry, and human resources (Davenport & Brooks, 2004). In this first level of Industry 4.0 concept application, new developments and applications are developed in an intra-enterprise context. The distinct actions carried out would have the purpose of reduce cost, time or improve the internal information flow, but focusing inside the company and, hence, without a SC vision.

This case may correspond, for example, with a company that implements a mobile phone application that allows consulting, online, the amount of liquid in a deposit located in their facilities. The use of this new application would mean a reduction of the movements of the operators to the tank to check the existing amount of liquid and, therefore, would result in a cost savings. Obviously, although this application is only used by the company, cost savings and improved "local" efficiency could indirectly suppose a "global" improvement for SC, but this is not really the objective and, in any case, the impact would be minimal.

Level 1: Intra-Enterprise Focus With Direct Impact on the SC

At this level of Industry 4.0 concept application, also is considered that the new developments and applications could be implemented in an intra-enterprise context but, in this case, with a clear impact or influence in the SC. In this level, the internal developed actions seek to improve inside activities that are directly related with suppliers or customers and, therefore, we can say that it exist a positive impact in the SC.

Adapting the previous example, the amount of liquid of a deposit with product for a customer can be motorized with a mobile phone application. This implementation can be considered at this level since this application can reduce stockouts and improve the service to the customer, that is, the next company on the SC. This "local" action, in short, would imply a clear "global" improvement for the SC.

Level 2: Focus on Inter-Enterprise With Impact in a Limited Number of Members of the SC

The main instrument of change of the revolution in SC Management is the Internet (Davenport & Brooks, 2004). The information exchange among companies is allowing improving the operational transactions as well as strategic analyses and tactical decisions. Industry 4.0 proposals help in this line with operational and analytical tools. Therefore, at this level, two companies in the SC (or a limited number of them) define joint actions within the Industry 4.0 concept. The outcome of the adopted agreements could mean that only one company will incorporate Industry 4.0 elements, but the decision will have been taken collaboratively. However, in most cases, this level of collaboration has as consequence the incorporation of technology and the development of applications that will facilitate the exchange of information between the collaborative companies. The main idea is inter-operate with suppliers or customers, focus on external integration to complete inter-operation in a SC framework. In any case, this level of Industry 4.0 concept application is based on inter-enterprise relations and has a positive impact in the SC. Following with the deposit case, this level of application of the Industry 4.0 concept could be considered when, for example, two or more SC companies agree to implement a mobile phone application to share information about the deposit. In this case, both the complete SC, and especially the involved links, would have benefits.

Level 3: Focus on Inter-Enterprise With Impact in the Complete SC (From Industry 4.0 to Supply Chain 4.0)

In this level, all companies in the SC define joint actions to apply the Industry 4.0 concept on the whole SC. This is the highest level of application of Industry 4.0 principles to the whole SC and, therefore, introduces the idea of SC 4.0. At this level, a high degree of collaboration is required, which involves: coordination, trust, exchange of information, etc.

In the case of the deposit, this level of application of the Industry 4.0 concept would require that the whole SC uses new applications to know the level of liquid in the deposit and to use this shared information in a collaborative way with jointed actions. Knowing the level of liquid in the deposits of other links in the SC (downstream or upstream) would achieve better activity plans and save costs or/and improve the service level to their customers. Obviously, at this level, the whole SC would have benefits.

Impact in the SC Configuration

The scope level has a different impact depending on the SC Configuration (Table 1). The Scope Level 0 has an internal focus, as a consequence, there is no important effects in the SC. Level 1 has an internal point of view but with clear consequences for the SC. This scope level can improve the communicative SC Configurations because their shallow dependence. Coordinate, Collaborative and Co-opetitive SC Configurations has not prominent effects because their higher participation in the SC. Level 2 and 3 are focus on inter-enterprise relationships and as a consequence Industry 4.0 projects in this level impact in the SC configuration with higher participation: Coordinate, Collaborative and Co-operative.

Classification Scheme

It is said that while Industry 3.0 automated the factory, Industry 4.0 automates the Supply Chain. But in which extent has the broad term of Industry 4.0 led to this new form of managing the Supply Chain, also known as "Supply Chain Management 4.0" (SCM 4.0).

A classification scheme, based on Perez et al. (2017) is proposed, leading to a better understanding of the concept of "SCM 4.0".

This classification scheme is based on three blocks:

1. Which technologies have emerged in the last years as a consequence of the "Industry 4.0" revolution;
2. Which features characterize what is known as "SCM 4.0";
3. In which extent these "Industry 4.0 technologies" are supporting this new paradigm of "SCM 4.0".

Table 1. Main impact of the scope level for the different SC configurations

SCOPE LEVEL / SC CONFIGURATION	Communicative	Coordinated	Collaborative	Co-opetitive
0. Intra-Enterprise focus without impact on the SC				
1. Intra-enterprise focus with direct impact on the SC	✓			
2. Focus on inter-enterprise with impact in a limited SC members		✓	✓	✓
3. Focus on inter-enterprise with impact in the complete SC		✓	✓	✓

Industry 4.0: Related Technologies

First, a list of different technologies that support the concept of "Industry 4.0" is considered.

They have been grouped under five large clusters: internet of things (IoT), cyber-physical Systems (CPS), smart data, advanced processing analytics and humans-machines interaction.

These five selected clusters representing all the technologies under the scope of the "Industry 4.0" term are reflected and described in detail in the following:

1. **Internet of Things (IoT):** IoT may be defined as everyday objects which can be equipped with identifying, sensing, networking and processing capabilities that will allow them to communicate with one another and with other devices and services over the Internet to achieve some useful objective (Atzori et al. 2010).

Another definition is given by Shrouf et al. (2014) "IoT is a system a system where the physical items are enriched with embedded electronics and connected to the Internet. Thanks to IoT physical objects are seamlessly integrated into the information network where they can become active participants in business processes, communicate information about their status, surrounding environment, production processes, maintenance schedule and even more".

Relevant IoT-based technologies are: RFID, GPS chips, sensors, smart cards, actuators, etc.

2. **Cyber-Physical Systems (CPS):** CPS are defined as transformative technologies for managing interconnected systems between its physical assets and computational capabilities. By integrating CPS with production, logistics and services in the current industrial practices, today's factories will be transformed into Industry 4.0 factories with significant economic potential (Lee et al., 2015).

The combination of these technologies not only enables the translation of the physical into the virtual world but also facilitates the link back from the virtual to the physical world.

With the growth of IoT the number of CPS systems continues to increase as they allow IoT applications to be put into place (Evans, 2014). This is mainly due to three factors:

- **Interoperability:** Made possible by new communication protocols designed especially for seamless machine-to-machine (M2M) interaction;
- **Connectivity:** Enabled by LPWA technologies that provide the wireless infrastructure to connect thousands of IoT nodes;
- **Affordability:** Hardware prices are decreasing.

Relevant CPS-based technologies are: M2M, machine vision, 3D-printing (additive manufacturing), etc.

3. **Smart Data:** Future plants will produce a large amount of data that will need to be saved, processed and analyzed.

This data management will be possible due to some factors (RolandBerger, 2014):

- Increasing storage capacities, mainly for the use of cloud computing, a general term for the delivery of hosted services over the Internet. It enables companies to consume compute resources as a utility rather than having to build and maintain computing infrastructures in-house, promising several attractive benefits for businesses and end users;
- Lower computation and data-storing costs;
- Advanced methods and tools to handle big data processing and analyzing.

Relevant Smart Data-based technologies are: cloud computing, big data, wireless networks, etc.

4. **Advanced Processing Analytics:** Although closely related with smart data, it has the enough relevance to handle it separately as another cluster of "Industry 4.0 - related technologies". Significant knowledge advances have taken place in this area over the last few years. A typical enterprise generates very large and diverse data sets coming from its distributed business locations. These massive amounts of detailed data can be combined and analyzed by predictive analytics, data mining, simulation or statistics. Doing this process in real-time creates a business advantage for the company by giving insight into the real-world dynamics of their business (Thoma et al., 2013). This cluster also encompasses the advances in artificial intelligence and machine learning which have permitted the machines not only execute repetitive tasks (as in the past) but also have a certain degree of autonomization.

Relevant Advance Processing Analytics-based technologies are: artificial intelligence, predictive analytics, simulation models, machine learning, automated robots, etc.

5. **Humans/Machines Interaction:** Human-machine interaction is described as the interaction and communication between human users and a machine, a dynamic technical system, via a human-machine interface. It studies the ways in which humans make, or do not make, use of computational artifacts, systems and infrastructures.

Human-machine interaction with industrial plants and other dynamic technical systems has nowadays been recognized as essential for process safety, quality, and efficiency. It comprises all aspects of interaction and communication between human users and their machines via human-machine interfaces. The term "machine" indicates any kind of dynamic technical system (or real-time application), including its automation and decision support equipment and software and it relates to many diverse application domains (Johannsen, 2007).

Relevant Humans/Machines interaction-based technologies are: smart mobile devices (phones, tablets, etc.), embedded computation, augmented reality, smart glasses, touch interfaces…

Features Characterizing a Supply Chain 4.0.

As pointed out Industry 4.0 is changing the technology landscape and increasing connectivity between digital and physical worlds. This fact is leading the companies to new forms of managing themselves and, in turn, the Supply Chains they take part.

This revolution is therefore changing the way products are designed, created, and delivered to customers and it has tremendous implications for the supply chain. New access to information, computational abilities, and innovative technologies have collapsed and connected the formerly linear supply chains. Now real-time information and insights can be shared across the entire supply network to drive actionable decisions (Mussomeli et al., 2016).

So...What it is expected from the management of future SC's?

Based on some studies (Cortés et al., 2015; Herman et al., 2015; Pfohl et al., 2015; Pérez et al., 2017), eight characterizing features were selected as the most representative ones. The description is as follows:

1. **Virtualization:** It means that companies are able to monitor physical processes. Sensors data are linked to virtual plant models and simulation models. Thus, a virtual copy of the physical world is created. For that, captured data by sensors must be previously digitalized;
2. **Transparency:** The systems create a virtual copy of the physical world through sensor data in order to contextualize information, providing instant visibility across multiple aspects of the supply chain all at once, so that companies can see how all components interact and relate to each other. It must be enabled to companies to track material flow, synchronize schedules and balance supply and demand;
3. **Socializing:** The collaboration in networks is enabling machines (not only smartphones) to start communicating and interacting with other machines and/ or humans in a socialized manner. Herewith, the collaboration with machines is socialized, since humans are able to get into a conversation with the machines;
4. **Autonomization:** It is the ability of machines and algorithms to make decisions and perform learning-activities autonomously. This autonomous decision-making and learning is based on man-made algorithms and enables whole factories and manufacturing facilities to work with minimum human-machine interaction with the machines;
5. **Mobility:** It refers to human–computer interaction by which a computer is expected to be transported during normal usage, which allows for transmission of data, voice and video. Mobile computing involves mobile communication, mobile hardware, and mobile software. Communication issues include ad hoc networks and infrastructure networks as well as communication properties, protocols, data formats and concrete technologies. Hardware includes mobile devices or device components. Mobile software deals with the characteristics and requirements of mobile applications;
6. **Energy Efficiency:** Climate change and scarcity of resources are megatrends that will affect to future industry players. These megatrends leverage energy decentralization for plants, triggering the need for the use of carbon-neutral technologies in manufacturing. Using renewable energies will be more financially attractive for companies. Energy efficiency and renewable energy are said to be the twin pillars of sustainable energy policy and are high priorities in the sustainable energy hierarchy;
7. **Flexibility:** Due to new and more complex demands requirements, processes such as products development, products production processes or decision-making procedures need to be performed faster. This aspect is directly related with mass-customization, so that large volumes are produced but with a tremendous increase in variety and customization without a corresponding increase in costs.

Impact of Industry 4.0-Related Technologies on SCM 4.0

The technical and organizational advances due to Industry 4.0 are leading the companies to new forms of managing themselves and, in turn, the Supply Chain/s they are involved.

It must be remarked that implementing an Industry 4.0-related technology doesn't necessarily mean that it will go beyond an intra-enterprise scope. A certain technology could be implemented intra-enterprise without any impact in the SC. Actually, when the term "Industry 4.0" was first introduced it just focused on advanced manufacturing.

At this point, only "Industry 4.0 technologies" which in some extent have led to draw this new Supply Chain Management paradigm, coined as SCM 4.0, are described. SCM is referred in the new era of Industry 4.0 as "SCM 4.0" in which digital relationships within and between companies become the core point of supply chain management. However, just a few works have addressed in which extent Industry 4.0 – related technologies have influenced SCM 4.0. In this work it is aimed to explore how some of the aforementioned "Industry 4.0 technologies" have brought important and relevant impacts to SCM 4.0. in the practical arena.

Table 2 shows each one of the eight SCM 4.0 characterizing features. For each of them one example of Industry 4.0 – related technology is assessed but distinguishing among the different SC activities. These activities are those concerning to "source", "make" and "deliver" according with the SCOR Model (Supply Chain Council, 2010). As it can be observed, "source" and "deliver" activities are the ones which better reflect SC interrelationships, and therefore representing those companies placed in the second level (inter-enterprise scope). However, first level companies would be represented by "make" activity (intra-enterprise scope) as far as improvements had some influence on immediate upstream or downstream SC actors.

CASE STUDY IN THE FOOD SECTOR

This section aims to approach some of the previously exposed concepts through its implementation in a case. It should be emphasized that the application of the Industry 4.0 framework not only implies a change in the way some operations are carried out within the production plants, but also expand and improve the management of the operations. The case study allows observing how to coordinate several actors of the same supply chain in an example of inter-enterprise collaboration.

The case study is framed in the food sector, in particular the processing of vegetables. It is a sector in which actors are combined with very traditional processes such as farmers, industrial players with extensive capacity for automation and logistics platforms with extensive experience in the implementation of communications technologies.

The case study presented is generated from the relationship of the authors with a company belonging to the multinational company ARDO. This international group is dedicated to commercialize different types of products, each with its own added value and its own function in the kitchen, from the range of pre-cooked dishes «Express» to organic products «Bio». The productive plant (in which this case study is focus on) has a department of TPM (Total Productivity Management) with a long experience in projects of continuous improvement, recognized with international awards.

Table 2. Contribution of the different relevant Industry 4.0 – based technologies to the development of the features characterizing "SCM 4.0"

SCM 4.0 Characterizing Features	SCM 4.0 ACTIVITIES		
	Source	**Make**	**Deliver**
Virtualization	Cloud Computing platform to share data with suppliers and perform holistic and optimized decisions	Sensors to have an on-line monitoring so that problems can be reported and rectified during the manufacturing process	RFID technology for products digitization in a warehouse to have an on-line stock control
Transparency	Sensors providing real-time information to monitor product temperature to ensure its stability during the shipment	Cloud Computing solution to host and make efficient use of big data generated by decentrally networked smart production systems	Embedded sensors providing visibility into customer behavior and product usage, gaining user insight and fostering anticipatory shipping.
Socializing	GPS chips to check when a truck rolls past a transportation point so the arrival can be predicted.	M2M technology for monitoring machinery for not only how much it's producing, but also for certain predictors of mechanical failure like vibration, temperature and pressure	Augmented reality to help a driver rapidly identify exactly where the shipment is located within the truck, cutting the amount of time spent not driving drastically.
Autonomization	Artificial Intelligence for driverless vehicle routes to enhance stock management of components in a warehouse	Machine Learning for machines which can predict failures and trigger maintenance processes autonomously	Intelligent Algorithms to minimize the shipments distance to save cost and time. Traffic jam or an accident may be anticipated with CPS technologies so these algorithms are updated to re-calculate the routes.
Mobility	Sensors to capture data on the identity, status or condition of physical assets and make that data available to smart-phone app.	Smart-Tablet App to check the pressure variation in a process and take it under control (some actuators can be activated if necessary acting over the physical system)	Smartphone Apps, that may allow a quick response with targeted advertising to ecological and social trends discussed in social networks
Energy Efficiency	Autonomous Vehicles (Av´s) centrally controlled by some mining companies (on private roads) to save fuel emissions.	Automated Systems that allow equipment to run faster or slower depending on the demand saving energy.	Smart-Data captured by sensors to have an on-line monitoring of the fleet energy consumption
Flexibility	3D printing to increase complexity in collaborative product design with suppliers with no additional cost	Advanced Robotics may reduce the costs of production that arise from a reduced dwell-time and automated processes	Artificial Intelligence in drones to make inventories of warehouse stock levels and deliver spare parts anytime.

The purpose of the supply chain is to obtain high quality products in different farms, transport them to the factory (where they are processed and stored) and serve the processed product, usually in the form of a bag with deep-frozen vegetables, to distribution centres or supermarkets. The industrial process can be divided basically into two stages:

1. Treatment of vegetables to be stored in frozen rooms;
2. Cooking of prepared dishes.

In both cases, the process ends with a packaging for distribution. These operations are apparently simple but require efficient coordination with the other actors in their supply chain. On the one hand, upstream suppliers are farmers whose products are sent to the factory on the basis of maturity criteria of the harvest and not plant capacity, which generates moments of saturation in the reception of vegetables and instants of no activity. On the other hand, downstream, are the large distribution chains that demand products that may have seasonal demand and commodity products. Seasonal products are the ones that generate greater stress in the productive system.

Management techniques are applied to get maximum efficiency in the use of resources at all times. An example of this approach is the "lean" philosophy in which it is tried that the operations are put at the service of the demand to attend them with the minimum waste. This implies generating the least amount of product in stock of any material or final product, the least amount of waste, the least amount of reprocessing, etc. In the ideal of this philosophy the product would circulate by the factory just in time to be processed and sent to the customer so that it reaches its destination just at the time of being consumed. Although obviously this cannot be applied to all industries is a concept to introduce efficiency. A challenging sector to introduce these improvements is the agrifood sector because its dependence of natural factors.

In the case of the company under study, efficiency is pursued through production planning in the medium term and in the short term. The objective is to meet the demand considering the periods in which the different vegetables are harvested in order to adjust the resources and coordinate their use. In this way, when a bean or pea crop is produced, the plant must be prepared to execute the set of operations necessary for its processing depending on the set of final products that each item allows. In this scenario planners must know how to distribute resources between seasonal products and no seasonal. It seems logical to ask what the limits of a collaboration between farmers (producers of the raw material) and industry are, since everything suggests that raw material depends only on nature. It would be great news that somehow the technology, which we commented in previous sections, would allow us to implement some kind of improvement.

This scenario is quite challenging and there are several improvement areas that must be addressed collaboratively between different stakeholders. The case is going to deal specifically with the supply chain of corn products: corn cobs and grain.

The corn harvest season normally starts in the middle of July and ends up in October. In a typical season, the enterprise collects approximately 13.000 tons of corn from several farms, approximately 120 km from the factory plant. The farming is managed by enterprise, dealing with aspects like the selection of seeds and other agricultural products, the distribution of seeds in the field or the quality of the corn. On the other hand, the farmer pays for the seeds and deals with all operative aspects of farming, like watering.

In general, corn is a product that requires some synchronization among the actors in the supply chain if it is to obtain the highest quality. The time that passes between harvesting and the freezing of corn is one of the factors on which the final quality of the product depends. If the treatment of maize is late, the frozen product is suitable for consumption but does not offer the consumer the taste and texture that it requires. However, it must be taken into account that the harvesting time is marked by environmental conditions (soil conditions, humidity, etc ...), whose evolution is difficult to predict accurately. In other words, it is complicated to set the picking time and therefore industrial processing of the product months or weeks in advance, which makes it difficult to fit this process into the usual production planning means based on previously known static data.

In this environment of uncertainty in which the expected harvest dates can vary over several days or weeks, and more in a situation of evident climate change, the company proposes a first step in deploying the technology to help stakeholders to have:

1. The information on the desired date to carry out the harvest flows from the factory to the plantation;
2. The information on the state of the crop, and its possible date of harvest, can flow from the plantation to the factory.

The benefits that would be obtained with this improvement in the flow of the information are evident. The industrial company could plan the production and establish which are the days in which it would be desirable to harvest the corn to have more resources and, therefore, a faster processing and thus obtain the best quality of each corn grain. With this information the farmer could control certain actions that facilitate delay or advance the time of harvest. But since not all the parameters are under the control of the farmer through the second flow of information would allow the industrial company to have the real state of maturity of the corn plantation to be able to recalculate the forecast on the real time of the harvest.

To carry out this idea has been proposed the implementation of an information system in the plantations that allows:

1. Capture the data of temperature, humidity, and other environmental parameters that affect the development of the crop;
2. Control the irrigation of the plantation;
3. Calculate a forecast based on historical and actual information of the current crop of the most probable date for harvesting.

This information system is intended for the primary user to be the farmer. The system consists of a set of electronic devices, designed for use in extreme climatic conditions, which are distributed in plantations. They collect data and control irrigation systems. These devices communicate using 4G technology via telephone with a data centre that accumulates and processes the data to become useful information for the farmer. The person in charge of the plantation can receive the information and make the corresponding decisions through a simple application for mobile, and therefore at any time or location. This example of IoT allows the farmer to define the value of a set of parameters that he considers suitable for a crop and to review its evolution or be warned by a message just in case of abnormal event. It also allows the farmer to act on the irrigation system to modify some of these parameters (humidity, for example). In the future the control of the farmer could be extended to other actions like fertilizer control or the control of enclosures in the case of greenhouses.

The project extends a second step to expand the number of participants in the supply chain. In this second phase it should be possible to have a platform called GREEN-CLOUD based on data in the cloud that facilitates sharing information not only between farmer and factory but also with the logistics platform. In future projects it is planned to include large areas in the system, as shown in figure 1.

In this context the project requires the inclusion of some changes in the management in the plant. These are specific aspects of the Industry 4.0 proposal, in order to obtain control over the consumption of different types of resources: machinery, labour, energy consumption and raw materials. For this purpose, a Manufacturing Execution System (M.E.S.) application known as "mesview" has been developed. This allows control over the consumption of resources is based on the following points:

Figure 1. GREEN-CLOUD platform for the integration of information in the supply chain

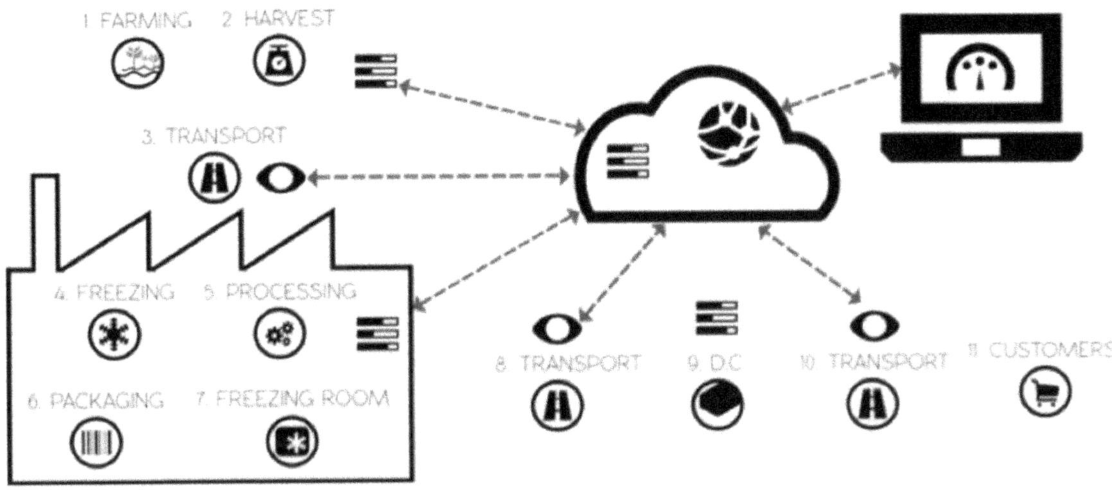

- Monitor the consumption of resources in real time through a data capture system;
- Calculate key indicators (KPIs) that represent the consumption of resources;
- Compare the indicators obtained with a previously set objective;
- Show information (indicators and deviations with objectives) in real time and historical to all personnel involved in the process.

In addition, it is expected to have a digital support to automate the flow of information with the operational level at the plant level and eliminate formalities in paper (quality controls, incidents, roadmaps for the process) (Figure 2). By computerizing these processes, the exchange of information is much faster (in real time), safe and reliable. Not only for corn processing but also for other crops.

The ultimate goal is to align plant operations with the rest of the supply chain and integrate information into a global system. The idea is that if the GREEN-CLOUD platform is deployed that allows better decision making in the field of operations planning the management of plant resources should be excellent thanks to mesview. Otherwise, you will not be able to take advantage of the start-up of the GREEN-CLOUD platform.

Some of the benefits that can be quantified as a result of the implementation of the project are:

- The expected impact in the farm is a reduction of at least 15% of the costs of water and products, i.e. and a proportional reduction of the environmental performance of corn farming;
- Improve the energy efficiency and reduce the amount of water used in the freezing process and the freezing rooms. The expected impact is a reduction of approximately 25% of the costs of energy;
- Make packaging more compact to optimize transport. The project is expected to yield a reduction of 15% of the cost of packaging and lowering the shortage of packaging to less than 1%.

In conclusion, it is possible to highlight how the project has linked the basic ideas of the Industry 4.0 proposal, such as the intelligent automation of plant resources, human-machine interaction, connected devices, etc. with the improvement in the decision making in the field of the supply chain. In

Figure 2. Data analysis panel (I) of mesview software (www.mesview.com)

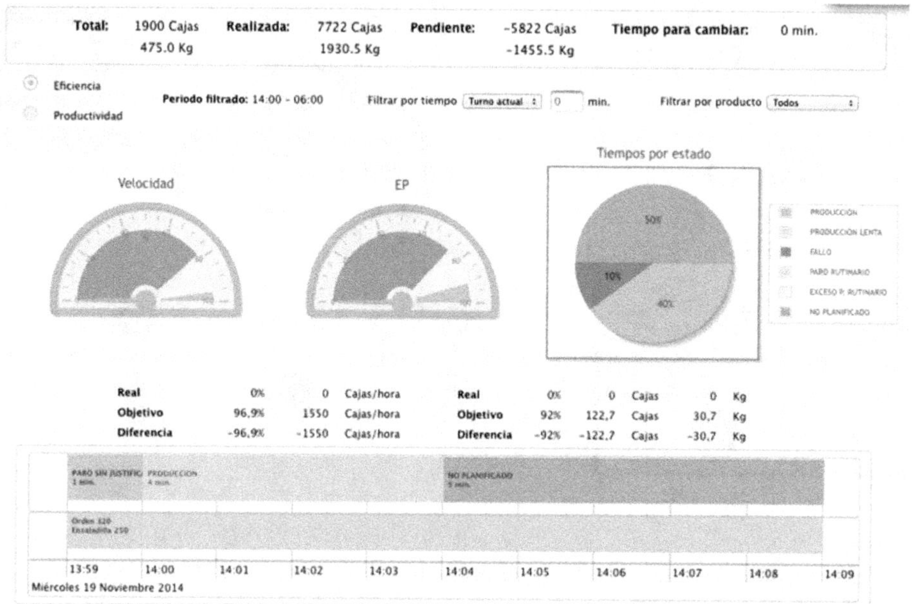

this framework of action there is an economic and work effort for the different actors in the chain, but as you can see there are also interesting returns on investment.

For all this, this case study is currently an example of level one and its evolution towards level two is clearly perceived.

CONCLUSION

Supply Chains have undergone profound changes in the last decade. However, new technical and organizational advances, due to Industry 4.0 proposals, are contributing in the continuity of this path of improvement. Industry 4.0 promotes integration through global value chain networks.

The impact of these advances is not only conditioned by the technologies but also by the supply chain configuration. The 4 C's in supply chain (communicative, coordinated collaborative and co-opetitive configuration) establishes different scenarios for the development of Industry 4.0 proposals in a greater or lesser degree.

The scope on the supply chain of the Industry 4.0 proposals has been classified into four levels according with their focus (intra-enterprise or inter-enterprise) and their impact in the SC:

Level 0: Focus on intra-enterprise without impact in the SC.
Level 1: Focus on intra-enterprise with direct impact in the SC.
Level 2: Focus on inter-enterprise with impact in a limited number of members of the SC.
Level 3: Focus on inter-enterprise with impact in the complete SC (from Industry 4.0 to SC 4.0).

Furthermore, this classification scheme has been proposed for a better understanding of the treated concepts based on related technologies for Industry 4.0, features characterizing a Supply Chain 4.0 and contribution of relevant technologies to the development of features characterizing "SCM 4.0".

Industry 4.0 includes a new milestone in these significant changes in SC's. The scope of these changes is not only intra-enterprise, but also inter-enterprise. Thus, Industry 4.0 contributes in the conception of new Smart Supply Chains or Supply Chain 4.0.

REFERENCES

Accenture (2017). *Digital Agriculture: Improving Profitability*. Retrieved from http://accenture.com

Alarcón, F., Perez, D., & Boza, A. (2016). Using the internet of things in a production planning context. *Brazilian Journal of Operations & Production Management, 13*(1), 72–76. doi:10.14488/BJOPM.2016.v13.n1.a8

Alonso, J., Martínez de Soria, I., Orue-Echevarria, L., & Vergara, M. (2010). Enterprise Collaboration Maturity Model (ECMM): Preliminary Definition and Future Challenges. In K. Popplewell, J. Harding, R. Poler, & R. Chalmeta (Eds.), *Enterprise Interoperability IV*. London: Springer. doi:10.1007/978-1-84996-257-5_40

Atzori, L., Iera, A., & Morabito, G. (2010). The internet of things: A survey. *Computer Networks, 54*(15), 2787–2805. doi:10.1016/j.comnet.2010.05.010

Chen, R. Y. (2017). An intelligent value stream-based approach to collaboration of food traceability cyber physical system by fog computing. *Food Control, 71*, 124–136. doi:10.1016/j.foodcont.2016.06.042

Cortés, B., Boza, A., Pérez, D., & Cuenca, L. L. (2015). Internet of Things Applications on Supply Chain Management. In *17th International Conference on Industrial, Information and Engineering Systems (ICIIES)* (pp. 2304-3309).

Cuenca, L., Boza, A., Alemany, M. M. E., & Trienekens, J. J. (2013). Structural elements of coordination mechanisms in collaborative planning processes and their assessment through maturity models: Application to a ceramic tile company. *Computers in Industry, 64*(8), 898–911. doi:10.1016/j.compind.2013.06.019

Davenport, T. H., & Brooks, J. D. (2004). Enterprise systems and the supply chain. *Journal of Enterprise Information Management, 17*(1), 8–19. doi:10.1108/09576050410510917

Dudek, G., & Stadder, H. (2005). Negotiation-based collaborative planning between supply chains partners. *European Journal of Operational Research, 163*(3), 668–687. doi:10.1016/j.ejor.2004.01.014

European Commission. (2013). Factories of the future. Multi-annual roadmap for the contractual PPP under. *Horizon*.

European Commission. (2016). *Factories 4.0 and Beyond*. Luxembourg: Publications Office of the European Union.

Evans, D. (2014). *The Internet of Things: How the Next Evolution of the Internet is Changing Everything* (White paper). CISCO. Retrieved from https://www.cisco.com/web/about/ac79/docs/innov/io-t_ibsg_0411final.pdf

Gaudreault, J., Frayret, J. M., & Pesant, G. (2009). Distributed search for supply chain coordi-nation. *Computers in Industry, 60*(6), 441–451. doi:10.1016/j.compind.2009.02.006

Gilchrist, A. (2016). Introducing Industry 4.0. In Industry 4.0 (pp. 195-215). Apress.

Gu, Y., & Jing, T. (2011). The IOT research in supply chain management of fresh agricultural products. In *2011 2nd International Conference on Artificial Intelligence, Management Science and Electronic Commerce (AIMSEC)* (pp. 7382-7385). IEEE.

Gubbi, J., Buyya, R., Marusic, S., & Palaniswami, M. (2013). Internet of things (IoT): A vision, architectural elements, and future directions. *Future Generation Computer Systems, 29*(7), 1645–1660. doi:10.1016/j.future.2013.01.010

Herman, M., Pentek, T., & Otto, B. (2015). *Design Principles for Industrie 4.0 Scenarios: A Literature Review*. Technische Universität Dortmund.

Johannsen, G. (2007). Human-Machine interaction. In H. Unbehauen (Ed.), *Control Systems, Robotics, and Automation. In Encyclopedia of Life Support Systems (EOLSS).* Oxford, UK: EOLSS Publishers.

Kagermann, H., Helbig, J., Hellinger, A., & Wahlster, W. (2013). *Recommendations for implementing the strategic initiative INDUSTRIE 4.0: Securing the future of German manufacturing industry; final report of the Industrie 4.0 Working Group*. Forschungsunion.

Keränen, A., & Bormann, C. (2016). *Internet of Things: Standards and Guidance from the IETF. IETF Journal.*

Lee, E. A. (2008). Cyber physical systems: Design challenges. In *2008 11th IEEE international symposium on Object oriented real-time distributed computing (ISORC)* (pp. 363-369). IEEE.

Lee, I., & Lee, K. (2015). The Internet of Things (IoT): Applications, investments, and challenges for enterprises. *Business Horizons, 58*(4), 431–440. doi:10.1016/j.bushor.2015.03.008

Lee, J., Bagheri, B., & Kao, H.-A. (2015). A cyber-physical systems architecture for industry 4.0-based manufacturing systems. *Manufacturing Letters, 3*, 18–23. doi:10.1016/j.mfglet.2014.12.001

Lejeune, M. A., & Yakova, N. (2005). On characterizing the 4 C's in supply chain management. [Elsevier]. *Journal of Operations Management, 23*(1), 81–100. doi:10.1016/j.jom.2004.09.004

Meyer, G.G., Wortmann, J.C. & Szirbik, N.B. (2011). Production monitoring and control with intelligent products. *International Journal of Production Research, 49*(5), 1303–1317. doi:10.1080/00207543.2010.518742

Min, H., & Zhou, G. (2002). Supply chain modeling: Past, present and future. *Computers & Industrial Engineering, 43*(1-2), 231–249. doi:10.1016/S0360-8352(02)00066-9

Mussomeli, A., Gish, D., & Laaper, S. (2016). *The rise of the digital supply network: Industry 4.0 enables the digital transformation of supply chains. Deloitte*. University Press.

Nukala, R., Panduru, K., Shields, A., Riordan, D., Doody, P., & Walsh, J. (2016). Internet of Things: A review from 'Farm to Fork.' In *2016 27th Irish Signals and Systems Conference (ISSC)* (pp. 1-6). IEEE. 10.1109/ISSC.2016.7528456

Olshansky, S. (2018) Rough Guide to IETF 101: Internet of Things. *IETF Journal.* Internet Society. oneM2M (2015). *The interoperability enabler for the entire M2M and IoT ecosystem* (White Paper). Retrieved from http://www.onem2m.org

Perez, D., Alarcón, F., & Boza, A. (2017). *Industry 4.0: A classification scheme.* Springer.

Pfohl, H.-C., Yahsi, B., & Kuznaz, T. (2015). The impact of Industry 4.0 on the Supply Chain. In C. M. R. Wolfgang Kersten, & T. Blecker (Eds.), *Proceedings of the Hamburg International Conference of Logistic (HICL)-20* (pp. 32-58).

Roland Berger. (2014). Industry 4.0 The new industrial revolution. *Roland Berger Strategy Consultants.* Retrieved from https://www.rolandberger.com

Shrouf, F., Ordieres, J., & Miragliotta, G. (2014). Smart factories in Industry 4.0: A review of the concept and of energy management approached in production based on the Internet of Things paradigm. In *2014 IEEE International Conference on Industrial Engineering and* Engineering Management (IEEM) (pp. 697-701).

Simatupang, T. M., Wright, A. C., & Sridharan, R. (2002). The knowledge of coordination for supply chain integration. *Business Process Management Journal, 8*(3), 289–308. doi:10.1108/14637150210428989

Supply Chain Council (2010). Supply Chain Operations Reference Model Version 10.0

Thoma, M., Antonescu, A. F., Mintsi, T., & Braun, T. (2013). Linked services for enabling interoperability in the sensing enterprise. In *International IFIP Working Conference on Enterprise Interoperability* (pp. 131-144). Springer. 10.1007/978-3-642-36796-0_12

Vargas, A., Boza, A., Cuenca, L., & Sacala, I. (2013). Inter-Enterprise Architecture and Internet of the Future. In L. M. Camarinha-Matos, S. Tomic, & P. Graça (Eds.), *Technological Innovation for the Internet of Things. DoCEIS 2013. IFIP Advances in Information and Communication Technology* (Vol. 394). Springer. doi:10.1007/978-3-642-37291-9_3

Yue, T. (2011). Internet of things technology application in the food supply chain management. In *2011 International Conference on E-Business and E-Government (ICEE)* (pp. 1-4). IEEE. 10.1109/ICEBEG.2011.5882613

Chapter 15
Enterprise Integration With the Structural Services Architectural Style

José Carlos Martins Delgado
Instituto Superior Técnico, Universidade de Lisboa, Portugal

ABSTRACT

The Service-Oriented Architecture (SOA) and Representational State Transfer (REST) architectural styles are the most used for the integration of enterprise applications. Each is more adequate to a different class of applications and exhibits advantages and disadvantages. This chapter performs a comparative study between them. It is shown that SOA and REST are dual architectural styles, one oriented towards behavior and the other towards state. This raises the question of whether it is possible to combine them to maximize the advantages and to minimize the disadvantages. A new architectural style, Structural Services, is proposed to obtain the best characteristics from SOA and REST. As in SOA, services are able to offer a variable set of operations and, as in REST, resources are allowed to have structure. This style uses structural interoperability, based on structural compliance and conformance. A service-oriented programming language is also introduced to instantiate this architectural style.

INTRODUCTION

After three industrial revolutions, focusing on mechanization, mass production and digitization, respectively, the industry has set its goals on a fourth revolution, commonly known as Industry 4.0 (Liao, Deschamps, Loures, & Ramos, 2017), entailing a vision of an intelligent factory where people, machines, processes, customers and suppliers are streamlined to produce and maintain smart products and services, thereby contributing to an improved society.

One of the main challenges that need to be overcome to turn this vision into a reality is *integration* (Panetto & Molina, 2008), the ability to meaningfully and efficiently cooperate with other subsystems in order to pursue the goals of the system as a whole. Integration can be seen at all levels of abstraction and complexity, from low-level cyber-physical systems (Zanero, 2017) to high-level enterprise value chains

DOI: 10.4018/978-1-5225-4936-9.ch015

targeting the capabilities required by Industry 4.0 (Schumacher, Erol, & Sihn, 2016). This chapter focus on the integration of enterprise information systems (or simply enterprise integration), a major goal to support the requirements of Industry 4.0. The EN/ISO 19439 standard (ISO, 2006) refers to enterprise integration as the process of ensuring the interaction between enterprise entities necessary to achieve domain objectives.

Enterprise integration is a very complex issue, spanning from lower levels, such as data and service interoperability, to the highest levels, including strategy and governance alignment. The former typically resort to technologies based on architectural styles such as SOA (Erl, 2016) and REST (Pautasso, 2014). The latter are mostly dealt with in a tacit way or at the documentation level, in coordination with the enterprise architectures.

The ISO/IEC/IEEE 24765 standard (ISO, 2010) provides the seemingly most cited definition of *interoperability*, as "the ability of two or more systems or components to exchange information and to use the information that has been exchanged".

Interoperability asserts the ability of two systems to understand each other's messages, whereas integration requires collaboration to achieve common goals. Interoperability is thus necessary but not sufficient to achieve enterprise integration (Chen, Doumeingts, & Vernadat, 2008), which usually entails cooperation and coordination at higher abstraction levels.

Just ensuring interoperability is already a daunting task, given the complex, heterogeneous and highly variable enterprise collaboration networks that characterize today's fast-paced enterprise landscape, namely Industry 4.0 scenarios. This is particularly true when we consider the most recent, game-changing developments, such as cloud computing (Mezgár, & Rauschecker, 2014), big data (Marz, & Warren, 2015; Reed, & Dongarra, 2015) and applications in the context of the Internet of Things (Want, Schilit, & Jenson, 2015).

However, this is not the full picture. In general, the integration problem revolves around two conflicting goals:

- **Interoperability:** Enterprises need to interact to accomplish collaboration, either designed or emergent. This necessarily entails some form of mutual knowledge and understanding, but this creates dependencies that may hamper the evolution (changes) of enterprise information systems;
- **Coupling:** Enterprises should not have dependencies on others, in order to evolve freely and dynamically. Unfortunately, independent enterprises do not understand each other and are not able to interact, which means that some form of coupling is unavoidable.

Therefore, the *fundamental problem of integration* is to provide (at most) the *minimum coupling* possible while ensuring (at least) the *minimum interoperability* requirements. In other words, the main goal is to ensure that each interacting party knows just enough about the others to be able to interoperate with them but no more than that, to avoid unnecessary dependencies and constraints. This is an instance of the *principle of least knowledge* (Hendricksen, 2014).

Enterprises are complex systems and suffer exactly from the same problem, with the additional concern that agility (which translates to fast evolution) is a critical requirement for the survival of even the largest enterprises. These are in fact quite vulnerable, because their complex architecture has typically woven a large web of dependencies. Current integration technologies, based on SOA and REST, do not necessarily comply with the principle of least knowledge.

SOA is guided by behavior and REST by state. SOA defines which resource types are used and establishes the set of operations provided by each, whereas REST starts with a state diagram, without relevant concern about distinguishing which state belongs to which resource.

REST is more flexible, in the sense that, if the server changes the links it sends in the responses, the client will follow this change automatically by using the new links. However, this is not as general as it may seem, since the client must be able to understand the structure of the responses. It is not merely a question of following all the links in a response. This is why REST favors standardized media types, to reduce the coupling between the client and the server. However, instead of reducing coupling, it merely transfers it from runtime to design time, when media types are defined.

SOA is more static, since the service contracts are defined beforehand and there is no provision for resource structure, but benefits from the support provided by tools in detecting errors. This is the classical tradeoff that a programmer has to face when choosing between a dynamic or static typed language.

In REST, there is an interesting side effect from the fact that the client uses a response to perform the next state transition. Only the links corresponding to allowed transitions are there, in line with the *affordances* approach (Amundsen, 2014). This is equivalent to dynamically adjusting the interface available to the client, according to the current state of the application. This is good protection mechanism, already used in human computer interfaces, with menus that become enabled and disabled, according to context. In SOA, this is not easy to achieve. Once a link to a service is available, all its operations can be invoked.

The flexibility of REST comes at a price, however, since it lies at a much lower level than SOA and the architecture of the problem does not map as easily onto state diagrams as onto resource class diagrams and processes. Nevertheless, it is simpler than SOA, requiring only basic Web technologies, and scales much better, but only as long as the application complies with the constraints imposed by REST. Complex functionality and bidirectional interactions (requests from the server to the client) are not a good match for REST.

This means that the ranges of types of applications that are more adequate for SOA and REST are not the same. REST is a better match for Web style applications (many clients, stateless interactions), which provide a simpler interface. This is why all the major Internet application providers, including cloud computing, now have REST interfaces. SOA is a better match for functionally-complex, enterprise-level distributed applications, in which REST exhibits a higher semantic gap (Ehrig, 2007) between the problem and solution spaces.

Ideally, we would like:

- To combine the structural properties of REST's resources and links with the richness of generic service interfaces;
- To reduce the semantic gap of REST, without losing its dynamic structural characteristics;
- To reduce the coupling between the provider and consumer of a service (increasing its range of applicability and adaptability), without losing its contract-based support.

Unfortunately, the technologies currently used to implement SOA and REST provide no hint or support on how to do this. These include data description languages such as Extensible Markup Language (XML) and JavaScript Object Notation (JSON), protocols such as Hypertext Transfer Protocol (HTTP) and SOAP (formerly an acronym for Simple Object Access Protocol), and service description languages, such as Web Services Description Language (WSDL).

This is a direct consequence of the fact that these technologies evolved from the original Web of Documents, made for stateless browsing and with text as the dominant media type. That web was made for people, not for enterprise integration. Today, the world is rather different. The web is now the Web of Services, in which the goal is to provide active functionality, not just documents. There are now more computers connected than people, with binary data formats (computer data, images, video, and so on) as the norm rather than the exception.

Yet, the evolution has been to map the abstraction of Web of Services onto the Web of Documents, with the major revolution of XML and its schemas. The document abstraction has been retained, with everything built on top of it. The lack of support of XML for behavior (service paradigm) is one of the main sources of complexity.

The main goals of this chapter are:

- To contribute to the field of enterprise integration, particularly in Industry 4.0 contexts, by showing that revisiting the interoperability problem can yield an alternative solution that is both simpler and more effective than existing ones;
- To describe a framework that explores the various abstraction levels of enterprise interoperability, to better dissect and understand the problem;
- To assess the potential of reducing coupling to increase adaptability and changeability, while maintaining interoperability requirements;
- To propose a new architectural style that combines the best characteristics from SOA and REST, endowing resources with behavioral, structural and hypermedia capabilities;
- To introduce a new interoperability language, which considers resources and their services natively, contemplating both data and behavior instead of building everything on top of a data description language.

The chapter is organized as follows. The Background section describes some of the existing technologies relevant to the context of this chapter. Next, the main modeling paradigms are reviewed, followed by a presentation of a hierarchy of architectural styles, including a comparison between the main characteristics of SOA and REST. The chapter continues by proposing two frameworks, one focusing on interoperability and the other on coupling. A fundamental issue is to reduce coupling, for which this chapter proposes compliance and conformance as the main underlying concepts. A new architectural style, Structural Services, is proposed and compared with SOA and REST. Finally, a programming language, SIL, is briefly illustrated to show how this style can be used in practice and a brief discussion on the usefulness of the approach is presented.

BACKGROUND

Although the Industry 4.0 concept (Lu, 2017) is very recent and no generally accepted definition exists yet, integration is recognized as one of its main requirements (Mrugalska, & Wyrwicka, 2017).

Interoperability, a necessary condition to achieve integration, has been studied in domains such as enterprise cooperation (Popplewell, 2014; Rezaei, Chiew, & Lee, 2014), e-government services (Sharma & Panigrahi, 2015), military operations (Hussain, Mehmood, Haq, Alnafjan, & Alghamdi, 2014), cloud computing (Zhang, Wu, & Cheung, 2013), healthcare applications (Robkin, Weininger, Preciado, & Goldman, 2015), digital libraries (Agosti, Ferro, & Silvello, 2016) and metadata (Chen, 2015).

Historically, interoperability has been the main goal in Web-based distributed systems, whereas coupling has been one of the top concerns in software engineering, when developing an application, along with other metrics such as cohesion (Saxena & Kumar, 2012).

This dichotomy can be easily understood. Enterprise software development methods emphasize changeability and agility, which means structuring components of an application so that a change somewhere affects as little as possible the remaining components and can be implemented in a very short time. Interoperability within an application is the easy part. Type names and inheritance are shared and there is usually a single programming language.

Distributed enterprise systems are completely different. Interoperability is hard, since interacting applications are developed, compiled, and linked independently. Most likely, this means different type names, inheritance hierarchies, programming languages, execution platforms and data formats. In this context, coupling has been treated as a side issue, a best-effort endeavor after achieving interoperability.

The two main technological solutions for distributed interoperability, Web Services (Zimmermann, Tomlinson, & Peuser, 2012) and RESTful APIs (Pautasso, 2014), are based on data description languages, such as XML and JSON. Although they have been able to connect independent and heterogeneous systems, supporting enterprise interoperability, they are not brilliant solutions from the point of view of coupling.

The scientific community has engaged in a continuing debate over which architectural style, SOA or REST, is more adequate for specific classes of applications. The literature comparing these styles is vast (Kumari & Rath, 2015; Bora & Bezboruah, 2015), usually with arguments based more on technological issues than on conceptual and modeling arguments. Peng, Ma, and Lee (2009) and Sungkur and Daiboo (2015) have made proposals to integrate SOA and RESTful services.

XML-based interoperability, essential for Web Services, assumes that both interacting resources share the same schema, namely in WSDL documents and other data schemas. XML Schema and other schema languages are complex enough to award additional coupling concerns. This means that changing something in a schema affects all resources that use it, both service providers and consumers, entailing a significant coupling and reducing agility.

Services based on RESTful APIs follow the REST architectural style, developed by Fielding (2000), who claims that this reduces coupling because the consumer does not depend on the provider's specification since all it needs to do is to traverse the structured data representation of the server's resources, by following its links (URIs). If a resource changes at the server, it changes the data structured returned but, as long as the data types are the same, the consumer will still be able to traverse it without modification. However, this decoupling is limited to protocol and data syntax. The interacting resources still depend on the shared knowledge about the data types, not to mention higher-level semantics and behavior.

To understand coupling, interoperability must be understood first, which requires a framework to systematize the various aspects of interoperability. Several interoperability frameworks have been proposed, some following a layered approach, with just one dimension, and others organizing the interoperability aspects in several dimensions.

The layered frameworks essentially follow the approach of the venerable OSI Reference Model (ISO, 1994), in which higher-level layers build on concepts of lower-level layers. For example, the C4IF framework, proposed by Peristeras and Tarabanis (2006), is based on four layers: Connection (basic use of a channel), Communication (data formats), Consolidation (meaning through semantics) and Collaboration (through compatible processes). Other frameworks, with differences in the layer organization but with a similar approach, include LCIM (Wang, Tolk & Wang, 2009), the European Interoperability Framework (EIF, 2010), and the frameworks proposed by Lewis, Morris, Simanta, and Wrage (2008) and Monfelt, Pilemalm, Hallberg, and Yngström (2011).

The multidimensional frameworks, usually applied at the enterprise level, recognize that a single set of layers is not enough and add other dimensions, to cater for other aspects of interoperability, including lifecycle and non-functional concerns. Examples include SOSI (Morris, Levine, Meyers, Place, & Plakosh, 2004), the ATHENA Interoperability Framework (Berre et al., 2007), the Framework for Enterprise Interoperability (FEI) defined by the CEN EN/ISO 11354-1 standard (ISO, 2011), and an interoperability framework conceived for ultra large scale systems, proposed by Ostadzadeh, Shams, and Badie (2015).

The European project ENSEMBLE (Charalabidis, Lampathaki, & Jardim-Goncalves, 2014) embodies an effort to formulate a science base for enterprise interoperability (Popplewell, 2014). This involves discovering which interoperability problems exist in domains related to interoperability, such as social, applied and formal sciences, and identifying which are the relevant scientific areas underlying interoperability.

These frameworks and technologies emphasize interoperability and treat coupling as a second-class problem. Other approaches, such as the Enterprise Service Bus (ESB) (González, Laborde, Galnares, Fenoglio, & Ruggia, 2014) and Event-based Programming (Faison, 2006), try to provide an additional emphasis on coupling, but not with as good results as it may seem, since interacting resources are essentially decoupled temporally, which does not improve data or behavior dependencies. ESBs also provide adapters and transformers, but these are no more than a centralization of interoperability solutions.

Many metrics have been proposed to assess maintainability for service-based computing, based essentially on structural features. Babu and Darsi (2013) present an extensive set of metrics for service coupling, cohesion and complexity. Other authors focus on dynamic, rather than static, coupling. Geetika and Singh (2014) present a survey of metrics for assessing coupling during program execution. Although centered on object-oriented programming, many of these metrics can also apply to distributed services. There are also approaches trying to combine structural coupling with other levels of coupling, such as semantics. The work presented by Alenezi and Magel (2014) is an example. Hock-Koon and Oussalah (2010) had already presented a study on coupling metrics at the semantic, syntactic, and physical levels.

CHOOSING A MODELING PARADIGM

To study interoperability, we need models, architectures, frameworks, and so on. This implies settling on a modeling paradigm, on which enterprise interaction can be based. Two of the most important and common paradigms in distributed interaction are processes and services. Choosing one or the other does not imply that processes and services are mutually exclusive. In fact, all active systems have both. Processes cannot exist without services (the units of behavior that processes invoke) and services originate processes when the chain of service-to-service invocations is unraveled at any service's entry point. The relevant issue is which concept is at the top, process (activity perspective) or service (architectural perspective).

For example, the Zachman framework (O'Rourke, Fishman & Selkow, 2003) separates data (what) from function (how). TOGAF (Desfray & Raymond, 2014) separates the data architecture from the application architecture. Other enterprise architecture frameworks (Greefhorst & Proper, 2011) do essentially the same. They are recursive, in the sense that they can be applied to a large enterprise as well as to a small module in an application. In practice, however, granularity tends to be very high. The complexity of the entire set of considerations and rationale is enough to grant its applicability only to the entire enterprise information systems or to reasonably sized units within them.

In this context, and by separating data from applications, these frameworks implicitly choose the process paradigm and assume that:

- There is a shared data repository, with the existence of a canonical data model as a typical architectural principle (Greefhorst & Proper, 2011);
- Applications read, process and update the data stored in this shared repository.

This separation is a good match to the N-tier architectural style, illustrated by Figure 1, in which the application logic (typically, based on processes) is separated from the database. This is not merely a question of paradigm, but also a result of technological issues (such as database implementation and optimization). The presentation tier is also separated, but for different reasons, such as interface independence and security. In logical terms, the architecture exposes processes, which span the various tiers.

This centralization of data and its separation from applications is a classical, but constitutes an unnecessary assumption and restriction that can raise many problems in enterprise interoperability, when business processes cross the boundaries of the architecture of an enterprise. Each tier is, in practice, a functional silo.

The same can be said about the maintainability and adaptability of an enterprise's architecture, in which unintended effects of changes easily propagate throughout the shared data. It is hard to keep track of which data is used by or affects which processes and CRUD matrices do not scale well with complexity.

Another disadvantage of this paradigm is that the architecture development method has to deal with the semantic gap (Ehrig, 2007) between the business, seen as a set of interacting entities, both inter- and intra-enterprise, and the data/application architecture, which removes the boundaries of these entities instead of modeling them as they appear at the business level.

Figure 1. Example of a typical N-tier system

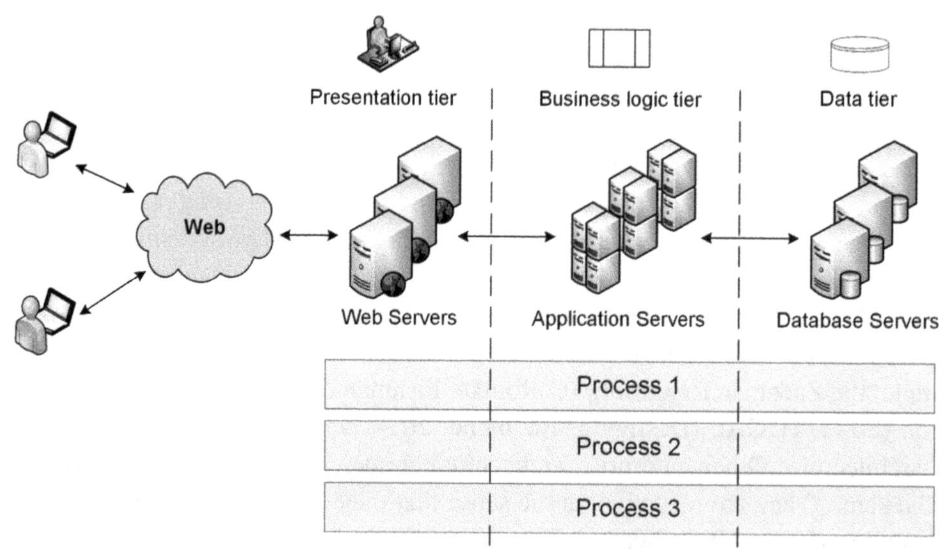

Object-oriented programming has long abandoned the shared data model of structured programming, by structuring programs according to some principles that tend to limit the impact of a change (affecting fewer objects) and to increase the adaptability (allowing objects to be adapted, even without the source code), namely:

- **Least Semantic Gap Principle:** The entities in the problem and solution spaces should have a one to one mapping, by modeling each problem entity with a solution object, with includes both state and related behavior. Any model will be easier to change if its abstractions match the entities it models. A sale, for example, will be the result of the cooperation of several entities (users, employees, information system, logistics, etc.). These entities are much closer to the service paradigm, since processes do not exist as real world entities. A small semantic gap reduces the maintainability and adaptability efforts. A small change in the system requirements translates to a small change in the system's architecture, due to that close match;
- **Information Hiding Principle:** There must be a clear separation between interface (syntax and semantics) and its implementation (both code and data). This reduces the change propagation effect. A change in one object affects those that use it only if its interface is affected. Processes are usually modeled after use cases (global uses of a system) and therefore tend to span across a large part of that system. Several processes will use the same service. A change in one service has potential impact in many processes. In contrast, services use an outsourcing mechanism (when invoking other services), but hide it from the services that have invoked them.

Services are essentially defined by their abstract interface, but need to be implemented by concrete entities, usually designated resources. In terms of modeling, resources correspond to distributed objects and these basic principles should also apply. The goal is to benefit from the advantages heralded by object-oriented programming in the enterprise architecture context, without being restricted to a centralized application.

The relationship between resources, services and processes can be described in the following way:

- A service is a set of logically related operations of a resource. In other words, it is a facet of that resource that makes sense in terms of modeling of the envisaged resource. A service is pure behavior, albeit the implementation of concrete operations may depend on state, which needs a resource as an implementation platform;
- Resources interact by sending each other messages. A resource *A* that sends a message to a resource *B* is in fact invoking the service of *B*. This constitutes a service transaction between these two resources, in which the resources *A* and *B* perform the role of service consumer and service provider, respectively. A service transaction can entail other service transactions, as part of the chain reaction to the message on part of the service provider. A service should be defined in terms of reactions to messages (external view) and not in terms of state transitions or activity flows (internal view);
- A process is a graph of all service transactions that are allowed to occur, starting with a service transaction initiated at some resource *A* and ending with a final service transaction, which neither reacts back nor initiates new service transactions. The process corresponds to a use case of resource *A* and usually involves other resources as service transactions flow.

We need an abstract model to express the relationships between the entities described above, so that we can deal with enterprise artifacts and their interoperability in a uniform way. Figure 2 depicts such a model, using UML (Unified Modeling Language) notation. A resource provides implementation for a service, namely its operations. Services invoke each other through service transactions, forming a process. Interaction is message based, assuming that an adequate channel and message protocol are available. Resources can be composed of other resources (composition) and have references to other resources (aggregation). References themselves are resources and included as components (but not the resources they reference) in their containers.

Association classes describe relationships (such as Transaction and the message Protocol) or roles, such as Message, which describes the roles (specialized by Request and Response) performed by resources when they are sent as messages.

The interaction between resources has been represented as an interaction between services, since the service represents the active part (behavior) of a resource and exposes its operations. When we mention resource interaction, we are actually referring to the interaction between the services that they implement.

Resources entail structure, state and behavior. Services refer only to behavior, without imposing a specific implementation. Processes are a view on the behavior sequencing and flow along services, which are implemented by resources. Services and processes are dual of each other, having resources as the structural entities. Where there is a service, there is a process and vice-versa. The question is which paradigm should be used first and foremost in modeling an enterprise's architecture, with the other derived from it.

Figure 2. A simple model of system interaction

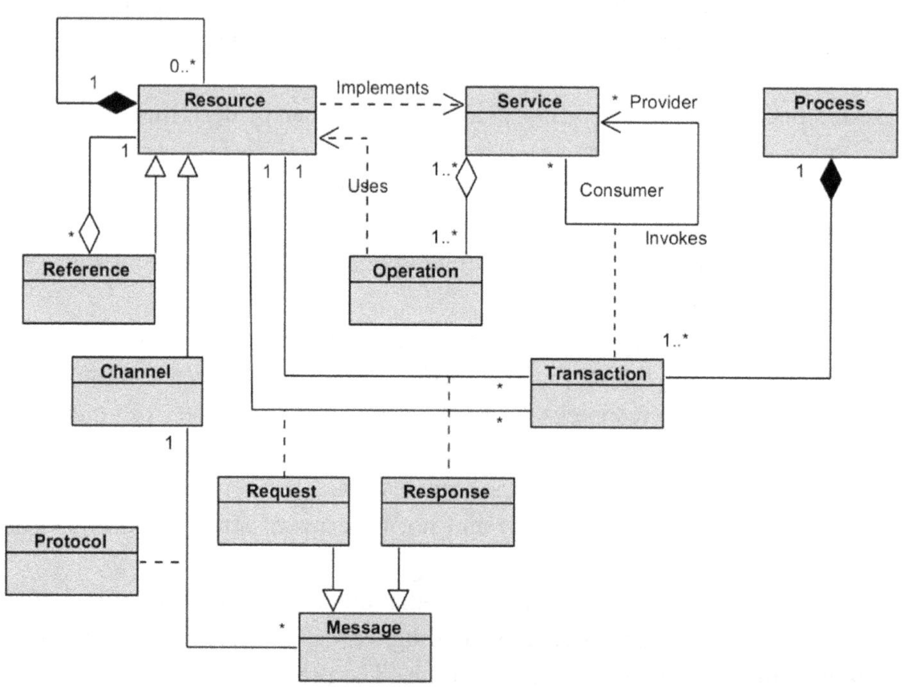

We contend that, in terms of modeling, adaptability and changeability, the service paradigm is much better than the process paradigm, and a decisive contribution to the agility of enterprises (Baghdadi, 2014).

Figure 3 illustrates this. In a classical N-tier architecture, each process is spread throughout the various tiers, as depicted in Figure 1. By changing to a service-oriented perspective, we group all the pieces of a given process in each tier into one resource, which becomes a first class citizen in a tierless graph of interacting services, the interfaces implemented by those resources.

Each resource includes (conceptually) its own independent database, holding the permanent data. In terms of implementation, the resources that reside in the same server can map their conceptual databases onto the same DBMS (Data Base Management System), but with logically separate records (Ahmad & Bowman, 2011), as depicted in the left part of Figure 3, which can be tables or objects, depending on the underlying paradigm of the DBMS.

This is a flexible arrangement, since data is logically partitioned and bound with the behavior that manages that data, which exploits the essential advantages of services referred to above in a distributed environment, either between enterprises or within the enterprise itself. The object-oriented paradigm has evolved into the service-oriented paradigm.

ARCHITECTURAL STYLES FOR INTEROPERABILITY

A Hierarchy of Architectural Styles

An *architectural style* is a collection of design patterns, guidelines and best practices (Dillon, Wu, & Chang, 2007; Sharma, Kumar, & Agarwal, 2015), in a bottom-up approach, or as a set of constraints on the concepts and on their relationships (Erl, Carlyle, Pautasso, & Balasubramanian, 2012), in a top-down approach. Each architectural style has an underlying modeling paradigm, which serves as a guiding tenet.

Figure 3. A distributed tierless system, based on the service paradigm

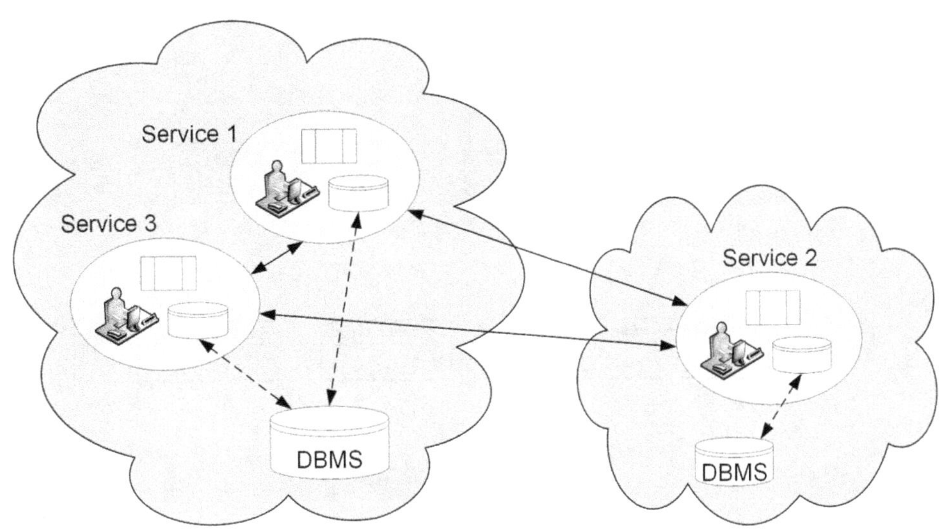

We follow the top-down approach because we believe that a rationale is needed before recognizing good design patterns and elaborating on best practices. Therefore, instead of enumerating the characteristics that define an architectural style, we describe it by additional constraints with respect to another one, with fewer restrictions. This allows us to establish a hierarchy of architectural styles, as illustrated by Figure 4.

In this diagram, using UML notation, each class represents an architectural style, which has a derivation dependency on the architectural style above by imposing some additional constraints. An architectural style can also use elements of another, as illustrated by the two "«use»" dependencies.

At the top, we have what Fielding and Taylor (2002) designated the Null style, which corresponds to asserting that there are no constraints at all, not even the existence of the modeling artifacts depicted in Figure 2.

The other architectural styles progressively refine and specialize the Null style by including more and more constraints, while reducing the spectrum of entities to which they can apply. For example, the System style identifies the existence of resources of any kind, including software applications, people, organizations and other non-computer based entities, but the Software style assumes that all resources are software components.

Figure 4. A hierarchy of architectural styles

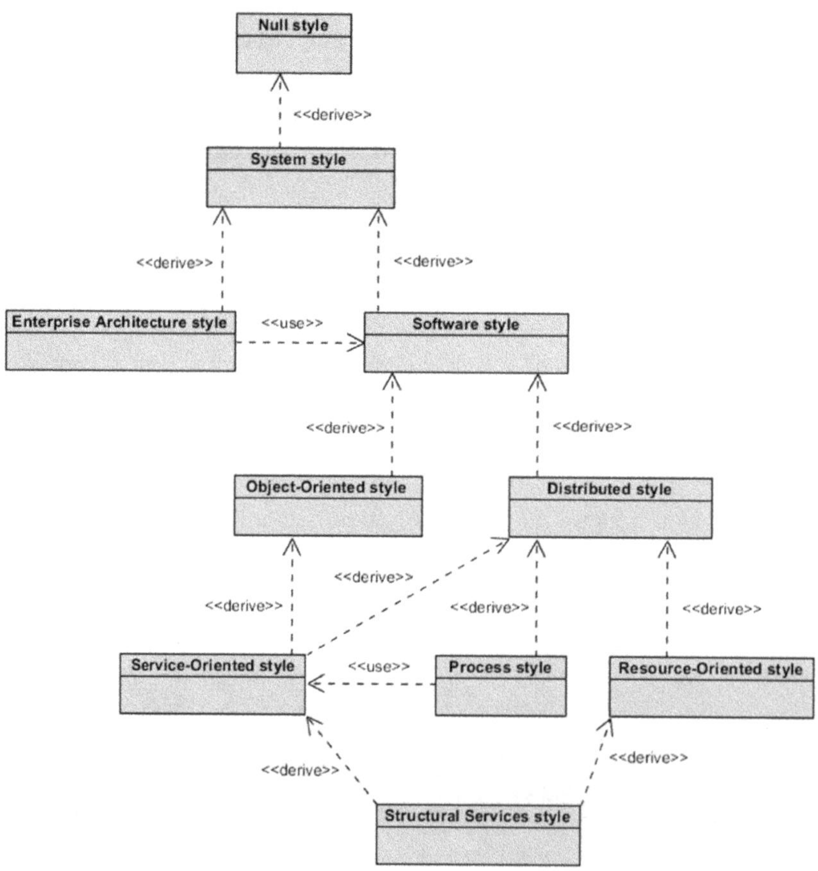

The Software style is further specialized by the Object-Oriented and Distributed styles, which in turn are the basis for the most used architectural styles in the Web and cloud contexts, the Service-Oriented and Resource-Oriented architectural styles, usually expressed by their most common instantiations, SOA (Erl, 2016) and REST (Pautasso, 2014), respectively. The Process style, with workflows that invoke services, is also rather popular, in particular in business applications (Jeston & Nelis, 2014).

The Enterprise Architecture style illustrates composition of architecture styles and shows that an architecture under consideration does not have to be completely modeled according to a single, pure style. The same happens with the Process style, which is not complete and needs to resort to the Service-Oriented style, in the sense that processes require services as the basic units of behavior.

On the other hand, one style can be a specialization of more than one style, as illustrated by the Service-Oriented style, which combines object oriented and distribution constraints, and the Structural Services style, proposed in this chapter, which specializes both the Service-Oriented and Resource styles.

Multiple derivation in Figure 4 applies to constraints, with the less stringent one passed on to the derived style if a constraint is required by both base architectural styles. The goal is to obtain the best of both worlds. For example, with respect to the Structural Services style (further described below):

- The SOA style allows any number of operations for each resource, whereas REST specifies that all resources must have the same operations. In this case, the flexibility of SOA in services is passed on to the Structural Services style;
- The REST style allows structured resources and references (links), whereas SOA does not. In this case, the flexibility of REST regarding resources is passed on to the Structural Services style.

In addition, the Structural Services style relaxes a constraint common to both SOA and REST: the types (schemas) of the messages exchanged between two interacting systems must be shared between them. Typically, XML-based systems share a link to a schema, whereas JSON-based systems need to use a common media type (standardized or previously agreed upon). This introduces unnecessary coupling, since it spans all the possible documents satisfying those schemas or types, even if not all are actually used by the interacting systems. The Structural Services style does not require schema or type sharing, resorting instead to partial interoperability based on structural compliance and conformance, thereby minimizing coupling, as described below.

Table 1 summarizes the main characteristics of all these architectural styles. Delgado (2012) presents a more detailed description of each one.

Comparing SOA and REST

SOA and REST are the most popular today in distributed interoperability and the ones we want to compare and to analyze in further detail. The first aspect to realize is that comparing architectural styles in abstract terms is not at all the same thing as comparing their practical instantiations.

SOA is easy to grasp, since it constitutes a natural evolution of the object-oriented style, with which analysts and programmers are already familiar. However, Web Services, with WSDL, XML Schema and SOAP, are complex to use, especially without adequate tools that usually automate or hide most of the development process.

Table 1. Main characteristics of the various architectural styles

Style	Brief Description	Examples of Characteristics or Constraints
Null	• No style.	• No constraints, model not applicable, no model elements identifiable.
System	• Identifies resources, services and processes (model of Figure 2).	• Resources are discrete (individualized) and identified by references. • Services described in terms of reactions. • Service interact by stimuli (can be analog and continuous).
Enterprise Architecture	• Resources can be people, software and physical.	• Governance and strategy most relevant. • Socio-economic factors. • Processes adequate for human workflow.
Software	• All resources are software related.	• Resources are digital. • Stimuli are finite-sized and time-limited messages. • Services obey a software development lifecycle.
Object-Oriented	• Resources and services follow the Object-Oriented paradigm.	• Resources are classes. • Services are interfaces. • References are pointers. • Reactions discretized into operations. • Polymorphism based on inheritance.
Distributed	• Interacting resources have independent lifecycles.	• References cannot be pointers. • Type checking cannot be name-based. • Services and resources may use heterogeneous technologies.
SOA	• Style similar to Object-Oriented, but with distributed constraints.	• Resources have no structure. • The interface of services is customizable. • No polymorphism. • Integration based on common schemas and ontologies.
Process	• Behavior modeled as orchestrations and choreographies.	• Services are the units of behavior. • A process is an orchestration of services. • Processes must obey the rules of choreographies.
REST	• Resources have structure but a fixed service.	• Client and server roles distinction. • Resources have a common set of operations. • Stateless interactions.
Structural Services	• Structured resources, customizable services, minimum coupling.	• Resources are structured and can reference others (like REST). • Services have a customizable interface (like SOA). • Interoperability based on structural compliance and conformance (new feature).

REST can be comparatively harder to master at first, in particular for people with an object-oriented mindset, who tend to slide to the probably more familiar RPC (Remote Procedure Call) style (Kukreja & Garg, 2014), with operation parameters instead of dynamic resource references (Fielding, 2008). On the other hand, its instantiation with HTTP is relatively straightforward to use, even without elaborate tools, since what is involved is essentially plain HTTP messaging.

We are more interested in the architectural styles themselves than on their instantiations, since these are heavily dependent on syntactic and protocol issues of existing interoperability tools and technologies. We want to compare styles, not their instantiations, since a style can be implemented in several ways.

Both SOA and REST have advantages and limitations. Our goal is to derive an architectural style that emphasizes the advantages and reduces the limitations. The best way to achieve this is to understand the motivations behind the appearance of each style:

- SOA appeared in the context of enterprise integration, with the main goal of achieving interoperability between existing enterprise applications with the then emerging XML-based technologies, more universal than those used in previous attempts, such as CORBA (Common Object Request Broker Architecture) (Henning, 2008) and RPC (Kukreja & Garg, 2014). It was a natural evolution of the object-oriented style, now in a distributed environment and with large-grained resources to integrate. Therefore, the emphasis was put in modeling applications as black boxes, exposing their external functionality with an interface composed of a set of operations, application dependent. Internal structure was avoided, as a measure to reduce coupling by applying the information hiding principle;
- REST appeared in the context of Web applications, as a systematization of the underlying model of the Web and with scalability as the most relevant goal, since interoperability had already been solved by HTML (and, later, XML) and HTTP. As originally proposed by Fielding (2000), REST is based on six constraints. The most relevant are stateless interactions (only the client maintains the interaction state) and uniform interface (all resources have a service with the same set of operations, with variability on structure and with links to resources, rather than on service interface). These constraints have decoupling as their main motivation, to support scalability. When potentially thousands of clients connect to a server, we want server applications to scale and to evolve as needed and as independently as possible from the clients.

There has been a lively debate between SOA and REST proponents. In particular, REST has been heralded as much simpler and providing a better decoupling between interacting systems (Pautasso, Zimmermann, & Leymann, 2008). A simple example, however, indicates that these two architectural styles are not that different, in essence.

Figure 5 describes a typical online product purchase scenario, using the process paradigm and Business Process Modeling Notation (BPMN) (Chinosi & Trombetta, 2012) to illustrate the interactions between a customer, a seller, and a logistics company (to deliver the product).

We used processes and BPMN because they constitute a popular approach to orchestrate distributed interactions, but each of the three roles in Figure 5 needs to be conceived, designed and implemented as much independently as possible, since each role can be performed by systems with different goals and

Figure 5. An example of business-level distributed interactions

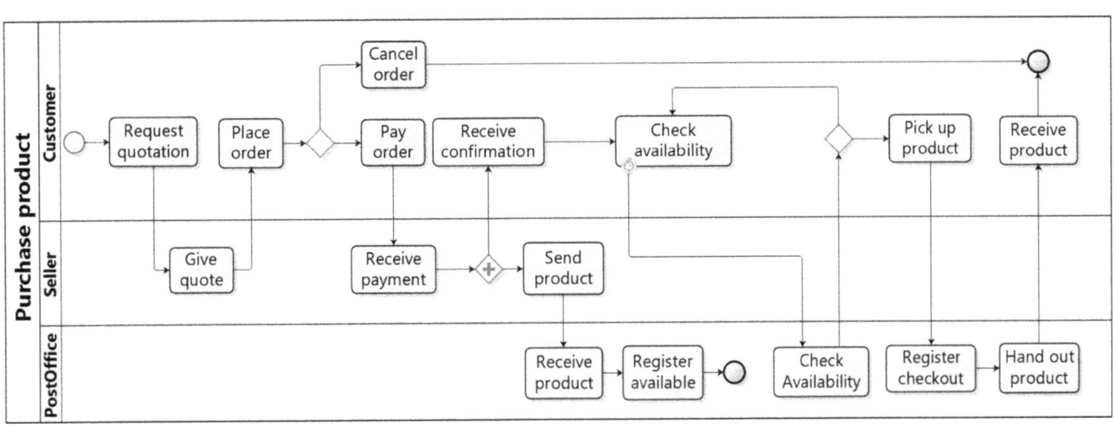

characteristics. In other words, we need to apply the fundamental problem of integration, as stated in the Introduction, to minimize coupling.

This example can be used to compare the approaches based on SOA and REST. Figure 6 depicts just the Customer process, for simplicity, using UML. Figure 6a represents the SOA approach, with a resource structure mimicking the three actors and the corresponding sequence diagram, reflecting their interactions. The REST approach goes directly to modeling interaction states, which is reflected in Figure 6b by a state diagram.

Apart from details, it is noticeable that the flow of activities of the customer in SOA is dual of the flow of states in REST. Figure 6a (SOA) uses activity-based flow graph, with activities in the nodes and state changed in the edges, whereas Figure 6b (REST) makes transitions from state to state, implemented by the activities in the edges.

SOA is guided by behavior and REST by (representation of) state. In UML terminology, SOA uses a static class diagram and a sequence diagram (Figure 6a) as a first approach, to specify which resource types are used, to establish the set of operations provided by each, and their interactions, whereas REST starts with a state diagram (Figure 6b), without relevant concern about distinguishing which state belongs to which resource. In the end, they perform the same activities and go through the same states, which should not be unexpected since, after all, the original problem, described by Figure 5, is the same. The difference is in the approaches used to solve it.

It is also clear that both SOA and REST:

- Tackle the fundamental problem of integration: to achieve interoperability while trying to minimize coupling;
- Try to comply with the information hiding principle by hiding the implementation of individual resources.

However:

Figure 6. Modeling the customer process in (a) SOA and (b) REST styles

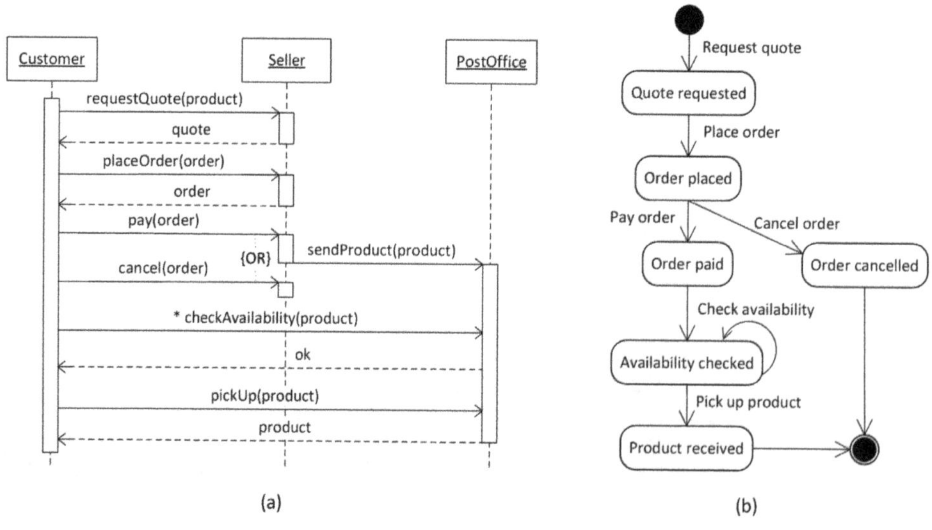

- SOA's approach is to hide resource structure (considering it part of the implementation), whereas REST's approach is to hide resource behavior (its service) by converting it into resource structure (each operation is modeled as a resource);
- SOA tries to minimize the semantic gap (Ehrig, 2007) by modeling resources as close as possible to real world entities. Since these are different from each other, offering a specific set of functionalities, the result is a set of resources with different interfaces (different set of operations). The consumer must know the specific interface of the provider. This entails coupling and may only be acceptable if the number of consumers for a given provider is usually reasonably small and the system evolves slowly (to limit the maintenance effort);
- The interface coupling is precisely what REST tries to avoid, with a simple idea: to decompose complex resources into smaller ones (with links to other resources), until they are so primitive and/or atomic that they can all be treated in the same way, with a common interface. The initial complexity is transferred to the richness of the structure of the links between resources. The most distinguishing feature of REST is this uniform interface for all resources, with a common set of operations. This corresponds to separating the mechanism of traversing a graph (the links between resources) from the treatment of each node (resource). The overall idea is that a universal link follow-up mechanism coupled with a universal resource interface leads to decoupling between server and clients, allowing servers to change what they send to the clients because they will adapt automatically, by just navigating the structure and following the links to progress in the interaction process;
- SOA has no dynamic visibility control mechanism for operations. In contrast, REST limits what the client can do to the links contained in the resource representations retrieved from the server, which helps to increase the robustness of applications. The idea of providing only the currently allowable commands has been borrowed from the field of user interfaces and is known as affordances (Amundsen, 2014). This is a strong point in favor of REST.

Unfortunately, both SOA and REST have fallacies and limitations:

- SOA assumes that all resources are at the same level and that exposing their services is just providing some interface and the respective endpoint. The fallacy is to assert that resource structure is unimportant. This may be true when integrating very large-grained resources, but not in many applications that include lists of resources that need to expose their services individually. The limitation is in changeability. Changing the interface of the provider will most likely require changes in the consumers as well;
- REST imposes a uniform interface, which allows treating all the resources in the same way. The fallacy is to consider only a syntactic interface, or even semantic (Verborgh, Harth, Maleshkova, Stadtmüller, Steiner, Taheriyan, & Van de Walle, 2014), but forgetting about the pragmatic (behavior) and symbiotic (collaboration) levels (described in Table 2). The reaction of resources need to be considered, which means that following a link cannot be done blindly. The client needs to know which kind of reaction the server is going to have. In the same way, traversing a structured resource implies knowing the type of that resource, otherwise we may not know which link should be followed next. This only means one thing: since there is no declaration of resource types, all types of resources to be used must be known in advance, either standardized or previously agreed upon (with custom media types). This is a relevant limitation of REST. What happens in practice

is that developers are more than happy to agree on simple resource structures and on an aligned API to match. REST advocates contend that an API should be represented simply by its initial link (an URI – Universal Resource Identifier –, for example) and all the functionality should be dynamically discovered by the client by exploring links. This can only be done with a set of fixed resource types (predefined or previously agreed upon). Resource structure can vary, a good advantage of REST, but only within the boundaries of resource types known in advance.

In summary, SOA is a good option for large-grained, slowly-evolving complex resources. REST is a good option for small-grained, structured resources which are relatively simple. In search for simplicity and maintainability, many providers (including cloud computing providers, such as Rackspace) have stopped using SOA APIs in favor of REST, by modeling operations as resources, using naturally occurring lists of resources as structured resources with links, and dynamically building URIs instead of resorting to operation parameters. Decoupling in REST, however, is not as good as its structure dynamicity seems to indicate. All the resource types (media types, in REST and Internet parlance) must be known in advance and the client may break if a new media type unknown to the client is used by the server. In addition, even when the type is known, the client cannot invent what to do and which link to choose when it receives a resource representation of a given type. Collaboration requires that both server and client implement behavior and deal with its effects, which means that they cannot be completely agnostic of each other.

AN ENTERPRISE INTEROPERABILITY FRAMEWORK

Interoperability can be used in the context of enterprises to adapt existing applications or to help designing enterprise architectures, which should be conceived for integration right from the start. We are more interested in the latter case, since adapting existing applications, although a real and fundamental need, is heavily conditioned by existing specifications and does not allow us to use the full range of integration aspects and possibilities.

In addition, SOA and REST are constrained by existing technologies and consider just a reduced set of interoperability aspects. The following sections define an interoperability framework that considers a wider perspective on enterprise interoperability. A framework is a set of principles, assumptions, rules and guidelines to analyze, to structure and to classify the concepts and concerns of a system or its domain. In this case, a service-oriented paradigm is assumed.

A Basic Interoperability Model

Figure 7 illustrates a basic interaction scenario. Resource *A*, in the role of consumer, sends a request to another resource *B*, in the role of provider, which executes the request and eventually returns a response (see Figure 2).

Messages, requests or responses, are resources that migrate from the context of the sender to that of the receiver, becoming part of its structure. The goal of achieving such a simple interaction can be decomposed into the following objectives:

1. The request message reaches *B*, through an adequate channel, such as a network;
2. *B* validates the request, according to its requirements for requests;

Figure 7. Basic interaction between two resources

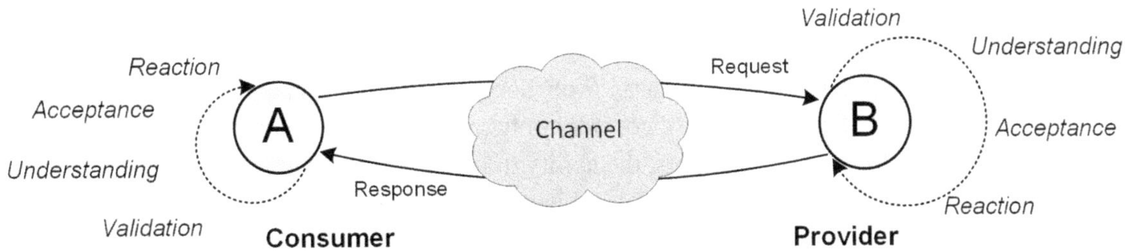

3. *B* understands what *A* is requesting;
4. *B* is willing to accept and honor the request;
5. The reaction of *B* and the corresponding effects, as a consequence of honoring the request message are consistent with the expectations of *A* regarding that reaction;
6. The response message reaches *A* (probably through the same channel as the request message);
7. *A* validates the response, according to its requirements for the response;
8. *A* understands what *B* is responding;
9. *A* is willing to accept the response;
10. *A* accepts the response and reacts appropriately, as long as it is consistent with its expectations for that response.

This means that is not enough to send a request to a provider and hope that everything goes well (and the same regarding the response). We need to ensure that the request and response are validated and understood by the service that receives it (ISO, 2010). In general, meaningfully sending a message (request and response reverse the sender and receiver roles) entails the following aspects:

- **Willingness** (objectives 4 and 9). Both sender and receiver are resources that implement services and, by definition, are willing to accept requests and responses. However, non-functional aspects such as security can impose constraints;
- **Intent** (objectives 3 and 8). Sending the message must have a given intent, inherent to the interaction to which it belongs and related to the motivation to interact and the goals to achieve with that interaction;
- **Content** (objectives 2 and 7). This concerns the generation and interpretation of the content of a message by the sender, expressed by some representation, in such a way that the receiver is also able to interpret it;
- **Transfer** (objectives 1 and 6). The message content needs to be successfully transferred from the context of the sender to the context of the receiver;
- **Reaction** (objectives 5 and 10). This concerns the reaction of the receiver upon reception of a message, which should produce effects according to the expectations of the sender.

Levels of Interoperability

In the example of Figure 5, interactions can occur horizontally, in each lane, or vertically, between lanes. In the former case, the lifecycles of resources are either synchronized or at least under the control of the same entity (e.g., different tiers, as in Figure 1). However, in the latter case, the lifecycles of the resources are necessarily independent and should exhibit the minimum possible coupling. These distributed interactions are the most relevant to this chapter.

Most interoperability technologies, such as Web Services (Zimmermann, Tomlinson, & Peuser, 2012) and RESTful APIs (Pautasso, 2014), consider only the syntactic format of messages, or at most the semantics of its terms. As Figure 7 hints, the ability to meaningfully interact involves higher levels, including naturally behavior. Table 2 establishes a linearized systematization of interoperability in several levels of abstraction. The higher levels are particularly relevant when complex resources, such as enterprise information systems (EIS), are involved.

The first column of Table 2, as well as its subdivisions in the second column, should be interpreted as follows:

- **Symbiotic:** Expresses the purpose and intent of two interacting resources to engage in a mutually beneficial agreement. This can entail a tight coordination under a common governance (if the resources are controlled by the same entity), a joint-venture agreement (if the two resources are substantially aligned), a collaboration involving a partnership agreement (if some goals are shared), or a mere value chain cooperation (an outsourcing contract). Enterprise engineering is usually the topmost level in resource interaction complexity, since it goes up to the human level, with governance and strategy heavily involved. Therefore, it maps mainly onto the symbiotic layer, although the same principles apply (in a more rudimentary fashion) to simpler resources;

Table 2. Levels of interoperability in an interaction

Category	Level	Main Artifact	Description
Symbiotic (purpose and intent)	Coordination	Governance	• Motivations to have the interaction, with varying levels of mutual knowledge of governance, strategy and goals.
	Alignment	Joint-venture	
	Collaboration	Partnership	
	Cooperation	Outsourcing	
Pragmatic (reaction and effects)	Contract	Choreography	• Management of the effects of the interaction at the levels of choreography, process and service.
	Workflow	Process	
	Interface	Service	
Semantic (meaning of content)	Inference	Rule base	• Interpretation of a message in context, at the levels of rule, known EIS components and relations, and definition of concepts.
	Knowledge	Knowledge base	
	Ontology	Concept	
Syntactic (notation of representation)	Structure	Schema	• Representation of EIS components, in terms of composition, primitive components and their serialization format in messages.
	Predefined type	Primitive resource	
	Serialization	Message format	
Connective (transfer protocol)	Messaging	Message protocol	• Lower level formats and network protocols involved in transferring a message from the context of the sender to that of the receiver.
	Routing	Gateway	
	Communication	Network protocol	
	Physics	Media protocol	

- **Pragmatic:** The effect of an interaction between a consumer and a provider is the outcome of a contract, which is implemented by a choreography that coordinates processes, which in turn implement workflow behavior by orchestrating service invocations. Languages such as Business Process Execution Language (BPEL) (Juric & Weerasiri, 2014) support the implementation of processes and Web Services Choreography Description Language (WS-CDL) is an example of a language that allows choreographies to be specified;
- **Semantic:** Interacting resources must be able to understand the meaning of the content of the messages exchanged, both requests and responses. This implies interoperability in rules, knowledge and ontologies, so that meaning is not lost when transferring a message from the context of the sender to that of the receiver. Semantic languages and specifications, such as OWL and RDF, map onto this category;
- **Syntactic:** Deals mainly with form, rather than content. Each message has a structure, composed of data (primitive resources) according to some structural definition (its schema). Data need to be serialized to be sent over the channel as messages, using data description languages such as XML or JSON;
- **Connective:** The main objective is to transfer a message from the context of one resource to the other's, regardless of its content. This usually involves enclosing that content in another message with control information and implementing a message protocol (such as SOAP or HTTP) over a communications network, according to its own protocol (such as the Transmission Control Protocol/Internet Protocol – TCP/IP) and possibly involving routing gateways.

It is important to recognize that all these levels are always present in all resource interactions, even the simplest ones. There is always a motivation and purpose in sending a message, an effect stemming from the reaction to its reception, a meaning expressed by the message, and a format used to send it over a channel under some protocol.

However, what happens in practice is that some of these levels are catered for tacitly (implicitly assumed) or empirically (explicitly hidden behind some existing specification):

- The Connective category is usually dealt with empirically, by assuming HTTP or SOAP;
- The Syntactic category is the main workhorse of integration technologies, with data description languages such as XML and JSON;
- The Semantic category, giving the increasing relevance of the Semantic Web (Rezk & Foufou, 2014), is beginning to be explicitly addressed with semantic annotations and languages, based either on XML or JSON, but it is still largely subject to tacit assumptions. XML's namespaces provide precious help here;
- The Pragmatic category is usually considered at the documentation level only, with many tacit assumptions. Choreography specifications and tools to verify them are seldom used;
- The Symbiotic category is considered only (and again at the documentation level) with very complex resources (such as enterprises) and at the conceptual stages of the lifecycle. In most cases, it is simply assumed that, if a resource exposes its service, it can be invoked regardless of motivations, purpose or any other high-level concerns.

Another important aspect is non-functional interoperability. It is not just a question of invoking the right operation with the right parameters. Adequate service levels, context awareness, security and other non-functional issues must be considered when resources interact, otherwise interoperability will be less effective or not possible at all. The framework of Table 2 must be considered for each of these concerns.

Finally, we must realize that all these interoperability levels constitute an expression of resource coupling, through the interactions of their services. On the one hand, two uncoupled resources (with no interactions between them) can evolve freely and independently, which favors adaptability, changeability and even reliability (if one fails, there is no impact on the other). On the other hand, resources need to interact to cooperate towards common or complementary goals, which means that some degree of previously agreed mutual knowledge is indispensable. The more they share with the other, the more integrated they are and the easier interoperability becomes, but the greater coupling gets.

What the interoperability framework expressed by Table 2 provides is a classification that allows us to better understand the coupling details, namely at which levels it occurs and what is involved in each level, instead of having just a blurred dependency notion. In this respect, it constitutes a tool to analyze and to compare different interoperability models and technologies.

A SIMPLE COUPLING FRAMEWORK

Resources can be distributed and interact by sending messages, in the roles of consumer (sends requests and eventually expects responses), provider (expects requests and eventually sends responses), or even both, in complex interactions.

Coupling Metrics

Figure 8 depicts the scenario of a resource immersed in its environment. It acts as provider for a set of resources (its consumers), from which it receives requests, and as a consumer of a set of resources (its providers), to which it sends requests.

Figure 8. Coupling between a resource and its consumers and providers

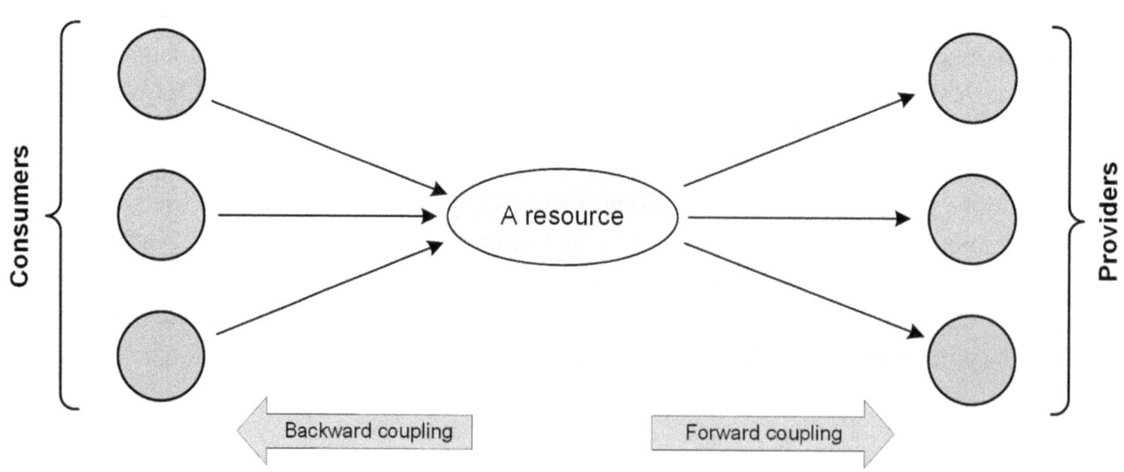

Coupling between this resource and others expresses not only how much it depends on (is affected by changes in) its providers but also how much its consumers depend on (are affected by changes in) it.

Dependency on a resource can be assessed by the fraction of its features that another resource is constrained about. A feature can be a function, a variable, a type, and so on. Two coupling metrics can thus be defined, from the point of view of a given resource:

- C_F (*forward coupling*), which expresses how much a resource is dependent on its providers, defined as:

$$C_F = \frac{\sum_{i \in P} \frac{Up_i}{Tp_i \cdot N_i}}{|P|} \quad (1)$$

where:

P: Set of providers that this resource uses, with $|P|$ as its cardinality

Up_i: Number of features that this resource uses in provider i

Tp_i: Total number of features that provider i has

N_i: Number of providers with which this resource is compatible as a consumer, in all uses of features of provider i by this resource

- C_B (*backward coupling*), which expresses how much impact a resource has on its consumers, defined as:

$$C_B = \frac{\sum_{i \in C} \frac{Uc_i}{Tc \cdot M}}{|C|} \quad (2)$$

where:

C: Set of consumers that use this resource as provider, with $|C|$ as its cardinality

Uc_i: Number of features of this resource that consumer i uses

Tc: Total number of features that this resource has

M: Number of known resources that are compatible with this resource and can replace it, as a provider

These coupling metrics yield values between 0 (completely unrelated and independent) and 1 (completely dependent, constrained by all features). The important conclusions are:

- From (1), the existence of alternative providers to a given resource reduces its forward coupling C_F, since more resources (with which this resource is compatible, as a consumer) dilute the dependencies;
- From (2), having alternatives to a resource, in its role as a provider, reduces the system dependency on it, thereby reducing the impact that resource may have on its potential consumers. Backward coupling C_B decreases;
- From both (1) and (2), the smaller the fraction of features used, the smaller coupling is.

A very interesting thought is that alternative provider resources can be found not only by designing and building them on purpose but also by reducing the fraction of features needed for compatibility to the minimum required by the interaction sought between resources, according to the overall choreography or orchestration of interacting resources. This is a consequence of the fact that the lower the number of constraints, the higher the probability of finding resources that satisfy them.

This means that both factors on which coupling depends (the fraction of features used and the number of provider alternatives available) work in the same direction, and the first reinforces the second. Therefore, reducing coupling means reducing the fraction of features used, or the knowledge of one resource about another.

Technologies such as Web Services (Zimmermann, Tomlinson, & Peuser, 2012) and RESTful APIs (Pautasso, 2014) constitute poor solutions in terms of coupling, since Web Services rely on sharing a schema (a WSDL document) and RESTful APIs require data types to have been previously agreed upon. These technologies solve the distributed interoperability problem, but not the coupling problem.

But how can coupling be reduced? Several authors researching the coupling problem try to separate several coupling cases in syntactical terms, with separate metrics for functions, arguments, types, and so on (Babu & Darsi, 2013; Geetika & Singh, 2014). Others try to add semantics (Hock-Koon & Oussalah, 2010; Alenezi & Magel, 2014), but mainly in terms of data types and their composition. We have a different approach.

Reducing Coupling With Partial Interoperability

Coupling is thus a necessary evil, without which no interaction is possible. Our goal is to minimize it as much as possible, down to the minimum level that ensures the level of interaction required by the resources. We will use Figure 9 to understand what can be done with respect to current interoperability technologies.

Figure 9a illustrates the classical document-style interaction. This is a symmetric arrangement in which a writer produces a document according to some schema and the reader uses the same schema to validate and to read the contents of the document. There is no notion of service, only the passive resource that the document constitutes. We are at the level of data description languages, such as XML or JSON.

Figure 9. Interaction styles: (a) Document; (b) Document-based service; (c) Compliance-based service

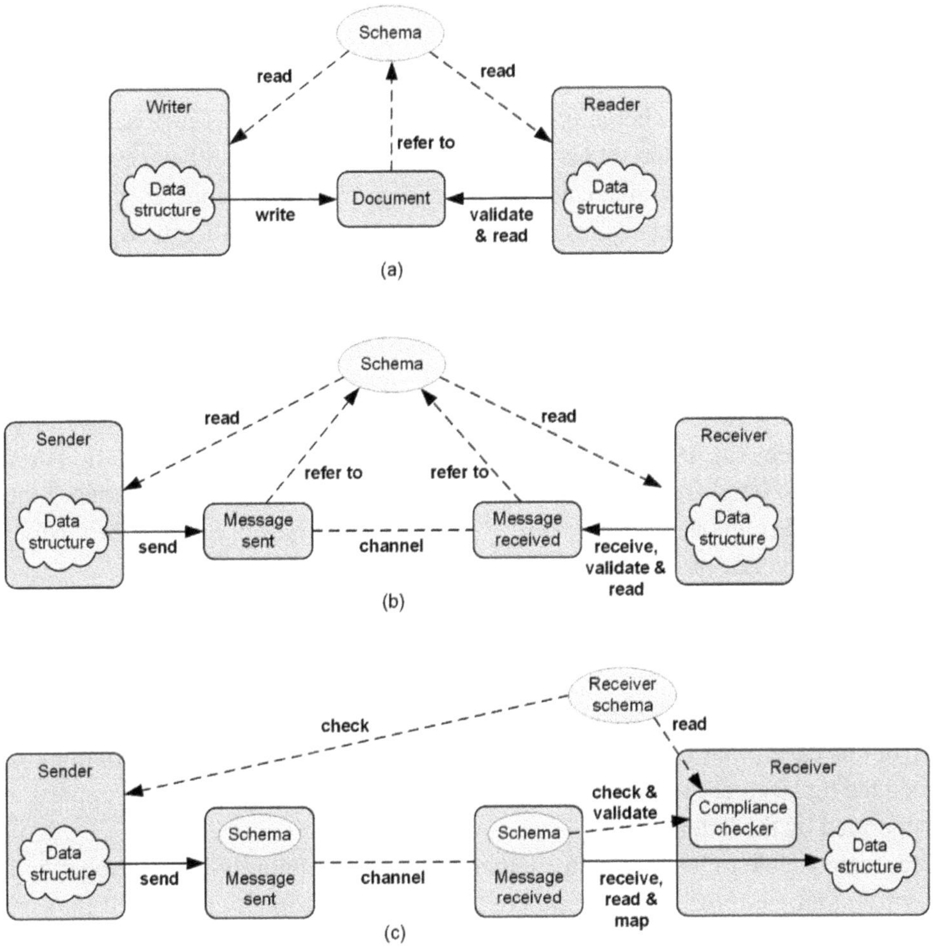

Figure 9b introduces services, in which a message is sent over a channel and received by the receiver. It is now treated as a parameter to be fed to some operation that the receiver implements, instead of just data to be read. However, the message is still a document, validated and read essentially in the same way as in Figure 9a. We are at the level of Web Services or RESTful resources. The behavior invoked can thus be exposed and implemented in various ways, but in the end the goal is similar. It is important to recognize that the schemas in Figure 9 refer to type specifications and need not be separate documents, as it is usual in XML Schema and WSDL. In the REST world, schemas are known as media types but perform the same role. The difference is that, instead of being declared in a separate document referenced by messages, they need to be previously known by the interaction resources, either by being standard or by having been previously agreed upon. In any case, the schema or media type must be the same at both sender and receiver and this imposes coupling between the resources for all its possible values, even if only a few are actually used.

Figure 9c depicts a different approach, based on *compliance* (Tran, Zdun, Oberortner, Mulo, & Dustdar, 2012). Messages do not obey some external schema. Each message has one specific value (most likely structured, but it is not a type, only one specific value), with its own exclusive schema that is nothing more than a self-description, without the value variability that a type exhibits. This value and its description can be validated against an infinite number of schemas, those that have this particular value included in the set of their instances.

The receiver in Figure 9c exposes a schema that defines the values it is willing to accept. When a message is received, its internal schema is checked against the receiver's own schema. If it *complies* (satisfies all the requirements of the receiver's schema), the message can be accepted and processed. The advantage of this is that a resource can send a message to all the resources with schemas that the message complies with and, conversely, a resource can receive messages from any resource that sends messages compliant with its receiving schema.

In other words, coupling occurs only in the characteristics actually used by messages and not in all the characteristics of the schemas used to generate the message or to describe the service of the receiving resource. Since the schemas of the message and of the receiver are not agreed beforehand, they need to be checked structurally. Resources of primitive types have predefined compliance rules. Structured resources are compared by the names of components (regardless of order of declaration or appearance) and (recursively) by the compliance between structured resources with matching names. Since the order of appearance of named component resources may different in the message and in the receiving schema, there is the need to map one onto the other (bottom of Figure 9c).

This is a form of polymorphism that increases the range of applicability of both sender and receiver, constituting a means to reduce coupling to only what is actually used. Sender and receiver no longer need to be designed for each other but, as long as compliance holds, interaction is possible.

It is also possible for one resource to replace another, as a result of a new iteration of the lifecycle, for example. In this case, what is involved is *conformance* (Khalfallah, Figay, Barhamgi, & Ghodous, 2014) between the replacement and the resource replaced, stating that the former supports all the characteristics supported by the latter.

When a resource is able to interact with another, although not entirely interoperable with it, we say that have *partial interoperability*. In full interoperability, the consumer can use all the characteristics that the provider exposes. This is what happens when schemas are shared. In partial interoperability, the consumer uses only a subset of those characteristics, which means that compliance and conformance need only be ensured for that subset.

To illustrate these concepts, consider the interoperability needed between the Customer and the Seller, in Figure 5. For a customer, this involves not only finding a suitable seller but also avoiding lock-in with it, by allowing that seller to be replaced by another with a compatible interface. Partial interoperability (compliance and conformance) allows to reduce the coupling to the bare minimum necessary.

Current technologies, such as Web Services and the tools that deal with them, were not conceived with compliance and conformance in mind, but through an example we can give an idea how that can be done. Consider the Seller as a Web Service offering several operations, including one that we will call getQuote) that allows the Customer to implement its RequestQuotation activity in Figure 5.

Listing 1 contains a fragment of a seller's WSDL file, showing only the relevant parts. A WSDL file describes the interface of a Web Service, including the operations it supports and the respective parameters and results. Only the operation getQuote is depicted, to keep it simple. It receives the specification of a product (the model and the seller's department it belongs to) and the information returned includes a quote for each product of that model found, with an optional boolean that indicates whether that product is in stock. A client program is needed at the Customer to deal with this Web Service.

Listing 1. A Fragment of the WSDL of one Seller's Web Service

```
<types>
    <xs:schema xmlns:xs="http://www.w3.org/2001/XMLSchema"
            targetNamespace="http://example.com/schema/seller1"
            xmlns="http://example.com/schema/seller1"
            elementFormDefault="qualified">
        <xs:element name="product" type="Product"/>
        <xs:element name="prodInfo" type="ProdInfo"/>
        <xs:complexType name="Product">
            <xs:sequence>
                <xs:element name="model" type="xs:string"/>
                <xs:element name="department" type="xs:string"/>
            </xs:sequence>
        </xs:complexType>
        <xs:complexType name="ProdInfo">
            <xs:sequence minOccurs="0" maxOccurs="unbounded">
                <xs:element name="ID" type="xs:string"/>
                <xs:element name="cost" type="xs:decimal"/>
                <xs:element name="inStock" type="xs:boolean"
                            minOccurs="0"/>
            </xs:sequence>
        </xs:complexType>
    </xs:schema>
</types>
<interface name="Seller_1">
    <operation name="getQuote"
                pattern="http://www.w3.org/ns/wsdl/in-out">
        <input messageLabel="In" element="seller1:product"/>
        <output messageLabel="Out" element="seller1:prodInfo"/>
    </operation>
</interface>
```

Now consider Listing 2, with the equivalent WSDL fragment of another seller's Web Service, compatible with this one but not identical to it. Its getQuote operation provides the same overall functionality and has the same name, so that a syntactical interface match is possible. For simplicity, this example does not tackle semantics. It also obtains a quote on the product, specified by a description and the product category it belongs to, and returns up to three pairs of product ID and respective price. The description can be made with one or two strings, so that brand and model, for example, can be separated if required, and the category is optional but supports two strings so that the product's division and unit, for example, can be specified.

Listing 2. A Fragment of the WSDL of Another Seller's Web Service

```
<types>
    <xs:schema xmlns:xs="http://www.w3.org/2001/XMLSchema"
            targetNamespace="http://example.com/schema/seller1"
            xmlns="http://example.com/schema/seller2"
            elementFormDefault="qualified">
        <xs:element name="productSpec" type="ProductSpec"/>
        <xs:element name="productInfo" type="ProductInfo"/>
        <xs:complexType name="ProductSpec">
            <xs:sequence>
                <xs:element name="description" type="xs:string"
                            minOccurs="1" maxOccurs="2"/>
                <xs:element name="category" type="xs:string"
                            minOccurs="0" maxOccurs="2"/>
            </xs:sequence>
        </xs:complexType>
        <xs:complexType name="ProductInfo">
            <xs:sequence minOccurs="0" maxOccurs="3">
                <xs:element name="productID" type="xs:string"/>
                <xs:element name="price" type="xs:decimal"/>
            </xs:sequence>
        </xs:complexType>
    </xs:schema>
</types>
<interface name="Seller_2">
    <operation name="getQuote"
                pattern="http://www.w3.org/ns/wsdl/in-out">
        <input messageLabel="In" element="seller2:productSpec"/>
        <output messageLabel="Out" element="seller2:productInfo"/>
    </operation>
</interface>
```

These two WSDL fragments correspond to two different schemas and the usual way to invoke the respective services would be to generate two different clients, one for each service. However, by looking at both the WSDL fragments, we notice that:

- The In element of the getQuote operation in Listing 1, of type Product, complies with the In element of the operation in Listing 2, of type ProductSpec. In other words, supplying a model and department with one string each is within the requirements of the operation in Listing 2;
- The Out element of the operation in Listing 2, of type ProductInfo, conforms to the Out element of the operation in Listing 1, of type ProdInfo. This means that all the values returned by the operation in Listing 2 will be expected as if they were returned by the operation in Listing 1.

This means that, although the services are different, a client generated to invoke the service to which Listing 1 belongs can also be used, without changes, to invoke the service to which Listing 2 belongs, as long as both getQuote operations are consistent in the interoperability levels above the Syntactic category, in Table 2. This is equivalent to saying that we are relieved of the obligation of using the same schema in both interacting parties (scenario of Figure 9b) and become entitled to simply using compatible ones, through compliance and conformance (scenario of Figure 9c). This increases the range of compatible services and translates into lower coupling.

THE STRUCTURAL SERVICES ARCHITECTURAL STYLE

We now have the necessary background to discuss the Structural Services architectural style, already introduced by Figure 4.

Main Characteristics

In a local environment (application), the object-oriented paradigm is probably the best in the non-declarative world, with its low semantic gap, information reuse (inheritance), polymorphism and ability to expose both behavior and structure. Unfortunately, in distributed environment pointers to objects can no longer be used and inheritance stops working, since the lifecycles of the objects are not synchronized (they can be modified, recompiled and linked independently).

Remote Procedure Call (Kukreja & Garg, 2014) was an early attempt to transport the object-oriented paradigm to the distributed world, but failed due to a very poor interoperability model, language specific. SOA was a much better interoperability solution (thanks to XML), but could not provide the mechanisms by which classes could expose structure and benefit from polymorphism. REST went in the opposite direction, exposing only structure (actually, converting operations into resources, or behavior into structure), but also without a solution for polymorphism (media types need to be previously known).

What we would like to have is a solution that allows resources:

- To refer to other resources by distributed references (e.g., URIs);
- To return a description of itself, under request;
- To expose operations, as needed for a low semantic gap in complex resource modeling;
- To expose structure, to easily model resources that are naturally structured;
- To use structural (rather than inheritance-based) polymorphism, for increased interoperability, adaptability and changeability, without the need to have resource types necessarily shared or known beforehand.

In a sense, we would like to have the object-oriented paradigm back, but now with distributed references, structural polymorphism and self-description. This is the essence of the *Structural Services* architectural style, proposed by this chapter.

The proposal of a new architectural style assumes that we know how to answer the question of what do we gain with it or, equivalently, which limitations of current architectural styles (namely, SOA and REST) does it solve. The discussion in the previous sections, including the critique on SOA and REST, already sheds light into this matter. Table 3 summarizes it.

Table 3. Comparison between SOA, REST and Structural Services

	SOA	REST	Structural Services
Basic tenet	• Behavior	• Hypermedia (structure + links)	• Tunable between pure behavior and pure hypermedia
Distinguishing features	• Resource-specific interface • Operations are entry points to a service • Design-time service declaration	• Uniform interface • Operations are resources • Clients react to structure received (do not invoke resource interfaces)	• Variable resource interface • Resources are structured • Operations are resources • Links are resources • Resource types need not be known at design time
Best applicability	• Large-grained resources (application integration)	• Small-grained resources (CRUD-oriented APIs)	• Wide range (small to large, behavior to structured-oriented)
Interoperability based on	• Schema sharing	• Predefined media types	• Structural polymorphism (compliance and conformance)
Self-description	• Repository (eg, WSDL document)	• Content type declaration in resource representations	• Included in each resource, message or application resource
Main advantages	• Low semantic gap (resources model closely real world entities)	• Structured resources, with links to other resources (hypermedia)	• Low semantic gap • Structured resources with links • Polymorphism • Self-description
Main fallacy	• Resource structure is unimportant	• Hypermedia increases decoupling	• None identified
Main limitation	• No polymorphism (coupling higher than needed)	• Fixed interface (semantic gap higher than needed)	• None identified

A Simple Implementation Example

For now, the Structural Services architectural style is not much more than a proposal, since it has not been tried out yet in significant scale to confirm its claims. SOA and REST have many working instantiations, which is not the case of Structural Services. However, these are limitations inherent to the development of a new architectural style, not to its intrinsic capabilities.

How can we provide an instantiation of the Structural Services style? There are some possibilities, including:

1. To extend WSDL files with another section (which we could call Structure), in which component resources and links could be exposed, thereby allowing Web Services to have structure;
2. To allow REST resources to have their own service description (e.g., WSDL file), thereby allowing the coexistence of the uniform and non-uniform interface approaches in REST;
3. To define a new resource description language, conceived specifically for the Structural Services architectural style and providing native support for its characteristics.

Solutions 1 and 2 are certainly possible, but existing tools are not designed to support the new style, which means that the benefit of using readily available software would really not exist or at least would be hampered by practical limitations. Data compliance and conformance are also possible with data description languages such as XML and JSON (these are data concepts that do not depend on notation or format), but again the available middleware does not support this kind of data mapping. In addition, compliance and conformance in Structural Services entail not only data but also operations, which neither XML nor JSON support directly (these have to be implemented by programming languages, such as Java, C# or BPEL). Generic programming languages are not universally interoperable in distributed environments and BPEL, which is platform independent by being based on XML, has a cumbersome syntax that requires visual programming and adds significantly to the complexity stack. RESTful applications, in search for simplicity, typically use only data descriptions at the platform-independent level and resort to generic programming languages to implement the required functionality, both at the server and client sides.

Therefore, we had to define and implement a language (SIL – Service Implementation Language) to describe resources and their services (Delgado, 2013), in accordance with the new style. Space limitations allow us just to present a very simple example, to provide a look and feel of its possibilities. We describe a car rental application, with the server side in Listing 3 and the client side in Listing 4.

Listing 3. Provider (Server) Side of the Car Rental Application

```
myCarRental: {
   date: definition {day: [1..31]; month: [1..12]};
   carClass: definition union {"economy"; "compact"; "standard"; minivan"};
   carInfo: definition {class: carClass; rate: float; linkCar: link to car};
   car: definition {
      class: carClass;
      occupied: list of date;        // days already reserved
      rate: float;                   // daily rate
      operation (-> carInfo) { reply {class; rate; self}; };
      reserve: operation (day: date ->) { occupied.add <- day };
      cancel: operation (day: date ->) { occupied.remove <- day };
   };
   fleet: list of car;    // exposed resource structure
   getFleet: operation (class: carclass:=any -> info: list of carInfo) {
      for (i: [0..fleet.size-1]) {
         if (fleet[i].class ==> class)   // if the car's class complies
            info.add<-(fleet[i]<-); // invoke the nameless operation in "car"
      reply info;
   };
   getFleetSize: operation (-> int) {
      reply fleet.size;
   };
};
```

Listing 3 describes a resource, myCarRental, the online application of a car rental enterprise. A resource in SIL has the look and feel of a JSON data structure (components with braces, {...}), but component names are not strings and there are operations with code.

The first declarations (date, carClass, carInfo and car) are definitions, not actual resources, and their purpose is to avoid repeating the definitions whenever a resource with these structures need to be created. Each time they are used, a resource with that structure is created. The keyword union means that carClass can have one of the string values indicated.

The first real resource component, fleet, is a list of resources describing the various cars that compose the fleet. Note that these, described the car definition, have three data resource components and three operations, one without name and two with names. The component fleet is accessible from outside myCarRental, by using the resource path myCarRental.fleet.

Operations are first-class resources and can be accessed by a resource path and be sent messages (with the operator "<-", just like any other resource. Adding a new car to the fleet can be done by using the expression myCarRental.fleet.add <- ..., sending the description of the new car as message argument. This is also illustrated in Listing 4, with access from the client side.

When a structured resource is sent a message, an operation with an argument which the message complies with is searched among the operations of the resource (which is then invoked, if it exists). If the receiving resource is an operation, then only that operation can be invoked and the message must comply with the operation's input argument (otherwise, an exception occurs). Operations can only have one input argument and one reply value, but these can be arbitrarily structured. In the operation's declaration heading, they are separated by "->". Any of them may be non-existent.

Each car offers the functionality of providing information on it (by sending it a message without an argument, as illustrated in operation getFleet in myCarRental) and this information includes the car's class and rate, as well as a link that can later be used to perform operations on the car description in the fleet, in the hypermedia style heralded by REST. The keyword self indicates a reference to the resource that contains the operation in which it appears.

The operation getFleet returns a list with information (class, rate and a link to each car descriptor in the fleet) on each car, allowing the client to navigate through this list and use the link of the chosen car to make the reservation, as illustrated by Listing 4, in which the first car with daily rate below 40 is reserved.

Note the heading of getFleet in Listing 3, which declares an input parameter (class) that is optional and has value any by default (specified by the operator ":="). This means that this argument gets the value passed, is specified, or any, if omitted. Then, the compliance check operator ("==>") indicates whether each car's class complies with the class required. If omitted, and since all resources comply with any, by definition, the entire fleet is returned.

Listing 4. Consumer (Client) Side of the Car Rental Application

```
mcrSPID: definition {
   date: definition {day: [1..31]; month: [1..12]};
   carClass: definition union {"economy"; "compact"; "standard"; "minivan"};
   carInfo: definition {class: carClass; rate: float; linkCar: link to car};
   car: definition {
      class: carClass;
      occupied: list of date;      // days already reserved
      rate: float;                  // daily rate
      operation (-> carInfo);
      reserve: operation (date ->);
```

```
      cancel: operation (date ->);
   };
   fleet: list of car;    // exposed resource structure
   getFleet: operation (carclass:=any -> list of carInfo);
   getFleetSize: operation (-> int);
};
mcr: link to mcrSPID;       // initialized with a link to myCarRental
mcr.fleet.add <- {class: "economy"; rate: 30};   // add cars
mcr.fleet.add <- {class: "standard"; rate: 50};
numberOfCars: mcr.getFleetSize <-;  // invokes operation getFleetSize
carFleet: mcr <-;                   // invokes operation getFleet and returns
                                    // the entire fleet (class in getFleet
                                    // is "any" by default)
minivanCars: mcr <- "minivan";      // returns only the fleet of minivans
smallCars: mcr <- union {"economy"; "compact"};  // returns only the fleet of
                                    // economy and compact cars
for (j: [0.. smallCars.size-1]) {
   c: smallCars [j];        // car information. Brings a link to the resource
   if (c.rate < 40){
      c.linkCar <- {23; 10};// Use the link to reserve the car for October 23
      break;
};
```

Listing 4 illustrates the usage of the myCarRental application, from the client side. We have included the SPID (SIL Public Interface Descriptor), which performs the same role as WSDL files (to provide a description of the interface). Unlike WSDL documents, SPIDs are generated automatically from the program, by eliminating private parts such as the body of operations. This is akin to SOA style, since the client can obtain this description from some resource/service registry. But this is also equivalent to declaring a media type in REST, which must be previously agreed by both client and server.

We use a link to the server application (mcr), which must be obtained by the client somehow (specified in the program or found in a registry). In SIL, a link is not necessarily a URI. It is a resource with structure previously agreed between client and server. A URI is just one of the possibilities.

We illustrate how car descriptors can be added, to grow the fleet, and how to retrieve a list of information structure on each car, as described above, both for the entire fleet and just for some of car classes. The operation invoked is the same and all is based on compliance. Finally, a small loop runs through the list and makes a reservation for October 23, for the first car with a daily rate under 40.

Naturally, these are very simple examples and many things are not tested, such as whether the car is free on this day. In a more complete and robust program, these tests would have to be performed.

SIL is a compiled language. The text source is compiled into binary form, which is used for message transmission and program execution, with an interpreter. Connection to other languages (Java, C#, Python, and so on) can be done by the usual mechanisms. There is a mapping between primitive resource types in SIL and each programming language, an adapting layer and a dynamic class load mechanism to support changes to resources.

The existence of the myCarRental's SPID in Listing 4 is not essential. At the client, mcr could have been declared as any, which switches off compiler verifications and defers compliance checks entirely to runtime, which is done structurally and without previous knowledge of the resource types. This is much slower than SPID-based compile-time verifications but, after the first time of each compliance check, the corresponding resource structural mapping (Figure 9c) can be cached and reused in subsequent messages.

Regarding the status of implementation of SIL, there is a prototype with basic functionality, with a compiler based on ANTLR (Parr, 2013) and an interpreter written in Java, with Web Sockets (Coward, 2014) as the transport protocol and a Jetty-based server to support resource links and messaging.

USEFULNESS OF THE APPROACH IN THE CONTEXT OF INDUSTRY 4.0

The Industry 4.0 concept is in line with the current trend of connecting everything, which has already been dubbed as the Internet of Everything (daCosta, 2013; Dahir, Dry, & Pignataro, 2015), with the goal of encompassing people, services, processes and cyber-physical systems in a seamless network.

This also applies to future enterprise information systems (Panetto, Zdravkovic, Jardim-Goncalves, Romero, Cecil, & Mezgár, 2016) designed with the requirements of Industry 4.0 in mind. More than basic connectivity, the goals include integration, intelligence, context-awareness, proactiveness, dynamic reconfigurability, self-organization, and self-maintenance.

Clearly, the current status of the integration technologies should be at the level of the Symbiotic category of Table 2, in which the purpose and intent to cooperate are the most important tenets. Unfortunately, the integration is done essentially at the level of the Syntactic and Semantic categories, achieving the Pragmatic and Symbiotic categories by successive layers of software and inherent complexity.

SOA has been used to easily model real world entities, with a low semantic gap (Ehrig, 2007), which is fundamental at the high levels of interaction that characterize value chains and generically Industry 4.0 cooperative and decentralized production processes. However, SOA-based technologies became too complex and ill-adapted to smaller cyber-physical systems. REST made interactions simpler, in technological terms, but increased the semantic gap, making it harder to model and to implement high-level, complex interactions between intelligent and self-organizing subsystems. In addition, neither SOA nor REST solved the always-present coupling problem.

This chapter contends that it is possible to combine the best features of both SOA and REST, while reducing coupling to the minimum required by the specifications. Unlike SOA and REST, in which interacting systems need to share the schema of the interface, the Structural Services approach proposed by this chapter uses compliance and conformance to match just the minimum set of required features of the interacting schemas.

In the context of Industry 4.0, this can be useful in the following circumstances:

- Supporting integration, even between subsystems that were not designed to interoperate, as long as one complies with the requirements of the other for an interlocutor (Figure 9c). It is very important to realize that this can apply to anything, such software applications, cyber-physical systems or even people (in terms of characteristics or skills). Compliance is a universal concept;
- Discovery of a subsystem that can be integrated with another one, through compliance. This is particularly useful for service discovery;

- Creation of a list of potential customers and suppliers, by using compliance and conformance (Figure 8), enabling construction of value chains.

FUTURE RESEARCH DIRECTIONS

The approach that we have followed, proposing the Structural Services architectural style, which can bridge the characteristics of SOA and REST and increase decoupling and the range of applicability, has the potential to improve substantially the current technological scenario and Insdustry 4.0 contexts, but much remains to be done before the architectural style is realistically tested and its capabilities demonstrated. Future work will mainly entail the following activities:

- Completing the implementation. For example, the language SIL supports concurrency and asynchronous communication, based on futures, and delegation, but these features are not implemented yet. Semantics is another aspect to complete;
- The current compliance and conformance algorithms are implemented just at design time, by the compiler. These algorithms also need to be implemented at the binary level to be checked at message arrival time, with cache-based mechanisms to optimize message exchange and dynamic resource type checking. The effectiveness of these mechanisms needs to be assessed;
- Compliance and conformance are basic concepts in interoperability and can be applied to all domains and levels of abstraction and complexity. Although work exists on its formal treatment in specific areas, an encompassing and systematic study needs to be conducted on what is the formal meaning of compliance and conformance in each level of Table 2;
- Carrying out a comparative study, with qualitative and quantitative assessment, between the Structural Services architectural style, which is instantiated with the SIL language and platform, and SOA and REST, with their instantiations with Web Services and HTTP-based APIs, respectively, with emphasis on the Industry 4.0 domain.

CONCLUSION

Enterprises need to interact to be able to pursue common or complementary goals. Their information systems need to be integrated in the best possible way, which means providing the minimum interoperability requirements while minimizing resource coupling.

SOA and REST are the most common architectural styles that try to address this problem, but they focus more on the interoperability side of the issue than on coupling. Both require previous knowledge of the types of the resources involved in the interaction, which entails more coupling than actually required. This chapter contends that both aspects, interoperability and coupling, need to be dealt with in a balanced way.

In addition, neither SOA nor REST provide a complete interoperability solution. SOA emphasizes behavior and does not contemplate resource structure. REST emphasizes structure and is not a good modeling match for complex resources, since these must be decomposed into simpler resources with a uniform syntactic interface (but different at the semantic, pragmatic and symbiotic levels).

We have described and proposed a new architectural style, Structural Services, which combines the behavior flexibility of SOA with the structural hypermedia capabilities of REST and structural interoperability based on compliance and conformance. In particular, we have shown that the SOA and REST architectural styles are dual of each other, with complementary characteristics, which combined together provide an enhanced support for a broader range of applications.

The fundamental advantage of the Structural Services style is that it not only supports both SOA and REST extremes, with one set of operations for each resource or for all resources, but also any intermediate combination, with fixed set of operations for only the resources that conform to a given specification and that should be treated in the same way, and a variable interface for resources that are specific. This means that different sets of operations can be used, instead of just one as REST imposes, and that we can choose the combinations that provide a better modeling match to the problem entities and/or to the architectural solution envisaged.

On the coupling side of the issue, our solution is to use structural compliance and conformance as the basis for relaxing the constraint of having to share previously known resource types. Any two resources that share at least the characteristics required for interoperability can interoperate, independently of knowing the rest of the characteristics, at design, compile or runtime. This minimizes coupling with regard to both SOA and REST. Reducing the number of features of a resource on which another depends not only reduces the perceivable width of the interface but also increases the number of providers that are compatible with a consumer, and vice-versa. Coupling metrics have been defined, which show that these effects reinforce each other in the goal of coupling reduction.

An implementation of this style could be conceived with Web Services, for example, by including a structural section in WSDL documents, but this would add up to the already very complex Web Services technology. Instead, we opted to use a resource and service description language that natively supports both data components and operations, with structural interoperability, supported on structural compliance and conformance, which have been informally introduced. The goal is not to replace SOA or REST, but rather to evaluate what can be gained, particularly in the context of the Industry 4.0 approach, by using a technology that natively supports the architectural style that we have proposed, instead of increasing the complexity resulting from the mismatch between this style and classical integration technologies.

REFERENCES

Agosti, M., Ferro, N., & Silvello, G. (2016). Digital library interoperability at high level of abstraction. *Future Generation Computer Systems*, *55*, 129–146. doi:10.1016/j.future.2015.09.020

Ahmad, M., & Bowman, I. (2011). Predicting System Performance for Multi-tenant Database Workloads. In *Fourth International Workshop on Testing Database Systems* (pp. 6.1-6.6). New York, NY: ACM Press. 10.1145/1988842.1988848

Alenezi, M., & Magel, K. (2014). Empirical evaluation of a new coupling metric: Combining structural and semantic coupling. *International Journal of Computers and Applications*, *36*(1). doi:10.2316/Journal.202.2014.1.202-3902

Amundsen, M. (2014). APIs to Affordances: A New Paradigm for Services on the Web. In C. Pautasso, E. Wilde, & R. Alarcon (Eds.), *REST: Advanced Research Topics and Practical Applications* (pp. 91–106). New York, NY: Springer. doi:10.1007/978-1-4614-9299-3_6

Babu, D., & Darsi, M. (2013). A Survey on Service Oriented Architecture and Metrics to Measure Coupling. *International Journal on Computer Science and Engineering, 5*(8), 726–733.

Baghdadi, Y. (2014). Modelling business process with services: Towards agile enterprises. *International Journal of Business Information Systems, 15*(4), 410–433. doi:10.1504/IJBIS.2014.060377

Berre, A., Elvesæter, B., Figay, N., Guglielmina, C., Johnsen, S., Karlsen, D., & ... (2007). The ATHENA interoperability framework. In R. Gonçalves, J. Müller, K. Mertins, & M. Zelm (Eds.), *Enterprise Interoperability II* (pp. 569–580). London, UK: Springer. doi:10.1007/978-1-84628-858-6_62

Bora, A., & Bezboruah, T. (2015). A Comparative Investigation on Implementation of RESTful versus SOAP based Web Services. *International Journal of Database Theory and Application, 8*(3), 297–312. doi:10.14257/ijdta.2015.8.3.26

Charalabidis, Y., Lampathaki, F., & Jardim-Goncalves, R. (2014). On the scientific foundations of enterprise interoperability: The ENSEMBLE Project and Beyond. In Y. Charalabidis, F. Lampathaki, & R. Jardim-Goncalves (Eds.), *Revolutionizing Enterprise Interoperability through Scientific Foundations* (pp. 336–355). Hershey, PA: IGI Global. doi:10.4018/978-1-4666-5142-5.ch014

Chen, D., Doumeingts, G., & Vernadat, F. (2008). Architectures for enterprise integration and interoperability: Past, present and future. *Computers in Industry, 59*(7), 647–659. doi:10.1016/j.compind.2007.12.016

Chen, Y. (2015). A RDF-based approach to metadata crosswalk for semantic interoperability at the data element level. *Library Hi Tech, 33*(2), 175–194. doi:10.1108/LHT-08-2014-0078

Chinosi, M., & Trombetta, A. (2012). BPMN: An introduction to the standard. *Computer Standards & Interfaces, 34*(1), 124–134. doi:10.1016/j.csi.2011.06.002

Coward, D. (2014). *Java WebSocket Programming*. Columbus, OH: McGraw-Hill Education.

daCosta, F. (2013). *Rethinking the Internet of Things: A Scalable Approach to Connecting Everything*. New York, NY: Apress Media, LLC. doi:10.1007/978-1-4302-5741-7

Dahir, H., Dry, B., & Pignataro, C. (Eds.). (2015). *People, Processes, Services, and Things: Using Services Innovation to Enable the Internet of Everything*. New York, NY: Business Expert Press, LLC.

Delgado, J. (2012). Bridging the SOA and REST architectural styles. In A. Ionita, M. Litoiu, & G. Lewis (Eds.), *Migrating legacy applications: challenges in service oriented architecture and cloud computing environments* (pp. 276–302). Hershey, PA: IGI Global.

Delgado, J. (2013). Service Interoperability in the Internet of Things. In N. Bessis, F. Xhafa, D. Varvarigou, R. Hill, & M. Li (Eds.), *Internet of Things and Inter-cooperative Computational Technologies for Collective Intelligence* (pp. 51–87). Berlin, Germany: Springer. doi:10.1007/978-3-642-34952-2_3

Desfray, P., & Raymond, G. (2014). *Modeling Enterprise Architecture with TOGAF: A Practical Guide Using UML and BPMN*. Waltham, MA: Morgan Kaufmann.

Dillon, T., Wu, C., & Chang, E. (2007). Reference architectural styles for service-oriented computing. In K. Li, C. Jesshope, H. Jin, & J. Gaudiot (Eds.) *IFIP International Conference on Network and parallel computing* (pp. 543–555). Berlin, Heidelberg: Springer-Verlag. 10.1007/978-3-540-74784-0_57

Ehrig, M. (2007). Ontology alignment: bridging the semantic gap (Vol. 4). New York, NY: Springer Science+Business Media, LLC.

EIF. (2010). European Interoperability Framework (EIF) for European public services Annex 2. *Communication from the Commission to the European Parliament, the Council, the European Economic and Social Committee and the Committee of Regions 'Towards interoperability for European public services'*. Retrieved December 16, 2015, from http://ec.europa.eu/isa/documents/isa_annex_ii_eif_en.pdf

Erl, T. (2016). *Service-oriented architecture: concepts, technology, and design* (2nd ed.). Upper Saddle River, NJ: Prentice Hall.

Erl, T., Carlyle, B., Pautasso, C., & Balasubramanian, R. (2012). *SOA with REST: Principles, Patterns &Constraints for Building Enterprise Solutions with REST*. Upper Saddle River, NJ: Prentice Hall Press.

Faison, T. (2006). *Event-Based Programming*. Berkeley, CA: Apress.

Fielding, R. (2000). *Architectural Styles and the Design of Network-based Software Architectures* [Doctoral dissertation]. University of California at Irvine, CA.

Fielding, R. (2008). REST APIs must be hypertext-driven. *Roy Fielding's blog: Untangled*. Retrieved December 16, 2015 from http://roy.gbiv.com/untangled/2008/rest-apis-must-be-hypertext-driven

Fielding, R., & Taylor, R. (2002). Principled Design of the Modern Web Architecture. *ACM Transactions on Internet Technology*, *2*(2), 115–150. doi:10.1145/514183.514185

Geetika, R., & Singh, P. (2014). Dynamic coupling metrics for object oriented software systems: A survey. *Software Engineering Notes*, *39*(2), 1–8. doi:10.1145/2579281.2579296

González, L., Laborde, J., Galnares, M., Fenoglio, M., & Ruggia, R. (2014). An adaptive enterprise service bus infrastructure for service based systems. In A. Lomuscio, S. Nepal, F. Patrizi, B. Benatallah, & I. Brandić (Eds.), Service-Oriented Computing Workshops, LNCS (Vol. 8377, pp. 480–491). Cham, Switzerland: Springer International Publishing. doi:10.1007/978-3-319-06859-6_42

Greefhorst, D., & Proper, E. (2011). *Architecture principles: the cornerstones of enterprise architecture* (Vol. 4). New York, NY: Springer Science & Business Media. doi:10.1007/978-3-642-20279-7

Hendricksen, D. (2014). *12 More Essential Skills for Software Architects*. Upper Saddle River, NJ: Addison-Wesley Professional.

Henning, M. (2008). The rise and fall of CORBA. *Communications of the ACM*, *51*(8), 52–57. doi:10.1145/1378704.1378718

Hock-Koon, A., & Oussalah, M. (2010). Defining metrics for loose coupling evaluation in service composition. In *International Conference on Services Computing* (pp. 362–369). Piscataway, NJ: IEEE Computer Society Press. 10.1109/SCC.2010.17

Hussain, T., Mehmood, R., Haq, A., Alnafjan, K., & Alghamdi, A. (2014). Designing framework for the interoperability of C4I systems. In *International Conference on Computational Science and Computational Intelligence* (102–106). Piscataway, NJ: IEEE Computer Society Press. 10.1109/CSCI.2014.102

ISO. (1994). *ISO/IEC 7498-1, Information technology – Open Systems Interconnection – Basic Reference Model: The Basic Model (2nd ed.).* Geneva, Switzerland: International Standards Office. Retrieved December 16, 2015, from http://standards.iso.org/ittf/PubliclyAvailableStandards/index.html

ISO. (2006). *Enterprise integration -- Framework for enterprise modelling. ISO/CEN Standard 19439:2006.* Geneva, Switzerland: International Organization for Standardization.

ISO. (2010). *Systems and software engineering – Vocabulary. ISO/IEC/IEEE 24765:2010(E) International Standard* (p. 186). Geneva, Switzerland: International Organization for Standardization.

ISO. (2011). *Advanced Automation Technologies and their Applications, Part 1: Framework for Enterprise Interoperability (CEN EN/ISO 11354-1).* Geneva, Switzerland: International Standards Office.

Jeston, J., & Nelis, J. (2014). *Business process management.* New York, NY: Routledge.

Juric, M., & Weerasiri, D. (2014). *WS-BPEL 2.0 beginner's guide.* Birmingham, UK: Packt Publishing Ltd.

Khalfallah, M., Figay, N., Barhamgi, M., & Ghodous, P. (2014). Model driven conformance testing for standardized services. In *IEEE International Conference on Services Computing* (pp. 400–407). Piscataway, NJ: IEEE Computer Society Press. 10.1109/SCC.2014.60

Kukreja, R., & Garg, N. (2014). Remote Procedure Call: Limitations and Drawbacks. *International Journal of Research, 1*(10), 914–917.

Kumari, S., & Rath, S. (2015). Performance comparison of SOAP and REST based Web Services for Enterprise Application Integration. In *International Conference on Advances in Computing, Communications and Informatics* (pp. 1656–1660). Piscataway, NJ: IEEE Computer Society Press. 10.1109/ICACCI.2015.7275851

Lewis, G., Morris, E., Simanta, S., & Wrage, L. (2008). Why Standards Are Not Enough To Guarantee End-to-End Interoperability. In C. Ncube, & J. Carvallo (Eds.), *Seventh International Conference on Composition-Based Software Systems* (pp. 164–173). Piscataway, NJ: IEEE Computer Society Press. 10.1109/ICCBSS.2008.25

Liao, Y., Deschamps, F., Loures, E., & Ramos, L. (2017). Past, present and future of Industry 4.0 - a systematic literature review and research agenda proposal. *International Journal of Production Research, 55*(12), 3609–3629. doi:10.1080/00207543.2017.1308576

Lu, Y. (2017). Industry 4.0: A survey on technologies, applications and open research issues. *Journal of Industrial Information Integration, 6*, 1–10. doi:10.1016/j.jii.2017.04.005

Marz, N., & Warren, J. (2015). *Big Data: Principles and best practices of scalable realtime data systems.* Greenwich, CT: Manning Publications Co.

Mezgár, I., & Rauschecker, U. (2014). The challenge of networked enterprises for cloud computing interoperability. *Computers in Industry, 65*(4), 657–674. doi:10.1016/j.compind.2014.01.017

Monfelt, Y., Pilemalm, S., Hallberg, J., & Yngström, L. (2011). The 14-layered framework for including social and organizational aspects in security management. *Information Management & Computer Security*, *19*(2), 124–133. doi:10.1108/09685221111143060

Morris, E., Levine, L., Meyers, C., Place, P., & Plakosh, D. (2004). *System of Systems Interoperability (SOSI): final report (Report No. CMU/SEI-2004-TR-004)*. Carnegie Mellon Software Engineering Institute. Retrieved December 16, 2015, from http://www.sei.cmu.edu/reports/04tr004.pdf

Mrugalska, B., & Wyrwicka, M. (2017). Towards lean production in industry 4.0. *Procedia Engineering*, *182*, 466–473. doi:10.1016/j.proeng.2017.03.135

O'Rourke, C., Fishman, N., & Selkow, W. (2003). *Enterprise architecture using the Zachman framework*. Boston, MA: Course Technology.

Panetto, H., & Molina, A. (2008). Enterprise integration and interoperability in manufacturing systems: Trends and issues. *Computers in Industry*, *59*(7), 641–646. doi:10.1016/j.compind.2007.12.010

Panetto, H., Zdravkovic, M., Jardim-Goncalves, R., Romero, D., Cecil, J., & Mezgár, I. (2016). New perspectives for the future interoperable enterprise systems. *Computers in Industry*, *79*, 47–63. doi:10.1016/j.compind.2015.08.001

Parr, T. (2013). *The Definitive ANTLR 4 Reference*. Raleigh, NC: The Pragmatic Bookshelf.

Pautasso, C. (2014). RESTful web services: principles, patterns, emerging technologies. In A. Bouguettaya, Q. Sheng, & F. Daniel (Eds.), *Web Services Foundations* (pp. 31–51). New York, NY: Springer. doi:10.1007/978-1-4614-7518-7_2

Pautasso, C., Zimmermann, O., & Leymann, F. (2008). Restful web services vs. "big"' web services: making the right architectural decision. In *International conference on World Wide Web* (pp. 805–814). New York, NY: ACM Press. 10.1145/1367497.1367606

Peristeras, V., & Tarabanis, K. (2006). The Connection, Communication, Consolidation, Collaboration Interoperability Framework (C4IF) For Information Systems Interoperability. *IBIS - International Journal of Interoperability in Business Information Systems, 1*(1), 61–72.

Popplewell, K. (2014). Enterprise interoperability science base structure. In K. Mertins, F. Bénaben, R. Poler, & J. Bourrières (Eds.), *Enterprise Interoperability VI: Interoperability for Agility, Resilience and Plasticity of Collaborations* (pp. 417–427). Cham, Switzerland: Springer International Publishing. doi:10.1007/978-3-319-04948-9_35

Reed, D., & Dongarra, J. (2015). Exascale computing and big data. *Communications of the ACM*, *58*(7), 56–68. doi:10.1145/2699414

Rezaei, R., Chiew, T., & Lee, S. (2014). A review on E-business interoperability frameworks. *Journal of Systems and Software*, *93*, 199–216. doi:10.1016/j.jss.2014.02.004

Rezk, E., & Foufou, S. (2014). A survey of semantic web concepts applied in web services and big data. In *International Conference on Computer Systems and Applications* (pp. 767–772). Piscataway, NJ: IEEE Computer Society Press. 10.1109/AICCSA.2014.7073278

Robkin, M., Weininger, S., Preciado, B., & Goldman, J. (2015). Levels of conceptual interoperability model for healthcare framework for safe medical device interoperability. In *Symposium on Product Compliance Engineering* (pp. 1–8). Piscataway, NJ: IEEE Computer Society Press. 10.1109/ISPCE.2015.7138703

Saxena, V., & Kumar, S. (2012). Impact of Coupling and Cohesion in Object-Oriented Technology. *Journal of Software Engineering and Applications*, *5*(9), 671–676. doi:10.4236/jsea.2012.59079

Schumacher, A., Erol, S., & Sihn, W. (2016). A maturity model for assessing industry 4.0 readiness and maturity of manufacturing enterprises. *Procedia CIRP*, *52*, 161–166. doi:10.1016/j.procir.2016.07.040

Sharma, A., Kumar, M., & Agarwal, S. (2015). A Complete Survey on Software Architectural Styles and Patterns. *Procedia Computer Science*, *70*, 16–28. doi:10.1016/j.procs.2015.10.019

Sharma, R., & Panigrahi, P. (2015). Developing a roadmap for planning and implementation of interoperability capability in e-government. *Transforming Government: People. Process and Policy*, *9*(4), 426–447.

Sungkur, R., & Daiboo, S. (2015). SOREST, A Novel Framework Combining SOAP and REST for Implementing Web Services. In *The Second International Conference on Data Mining, Internet Computing, and Big Data* (p. 22–34). Wilmington, DE: The Society of Digital Information and Wireless Communications.

Tran, H., Zdun, U., Oberortner, E., Mulo, E., & Dustdar, S. (2012). Compliance in service-oriented architectures: A model-driven and view-based approach. *Information and Software Technology*, *54*(6), 531–552. doi:10.1016/j.infsof.2012.01.001

Verborgh, R., Harth, A., Maleshkova, M., Stadtmüller, S., Steiner, T., Taheriyan, M., & Van de Walle, R. (2014). Survey of semantic description of REST APIs. In C. Pautasso, E. Wilde, & R. Alarcon (Eds.), *REST: Advanced Research Topics and Practical Applications* (pp. 69–89). New York, NY: Springer. doi:10.1007/978-1-4614-9299-3_5

Wang, W., Tolk, A., & Wang, W. (2009). The levels of conceptual interoperability model: Applying systems engineering principles to M&S. In G. Wainer, C. Shaffer, R. McGraw & M. Chinni (Eds.), *Spring Simulation Multiconference*. San Diego, CA: Society for Computer Simulation International.

Want, R., Schilit, B., & Jenson, S. (2015). Enabling the Internet of Things. *IEEE Computer*, *48*(1), 28–35. doi:10.1109/MC.2015.12

Zanero, S. (2017). Cyber-physical systems. *IEEE Computer*, *50*(4), 14–16. doi:10.1109/MC.2017.105

Zhang, Z., Wu, C., & Cheung, D. (2013). A survey on cloud interoperability: Taxonomies, standards, and practice. *Performance Evaluation Review*, *40*(4), 13–22. doi:10.1145/2479942.2479945

Zimmermann, O., Tomlinson, M., & Peuser, S. (2012). *Perspectives on Web Services: Applying SOAP, WSDL and UDDI to Real-World Projects*. New York, NY: Springer Science & Business Media.

KEY TERMS AND DEFINITIONS

Architectural Style: A set of constraints on the concepts of an architecture and on their relationships.

Compliance: Asymmetric property between a consumer *C* and a provider *P* (*C* is compliant with *P*) that indicates that *C* satisfies all the requirements of *P* in terms of accepting requests.

Conformance: Asymmetric property between a provider *P* and a consumer *C* (*P* conforms to *C*) that indicates that *P* fulfills all the expectations of *C* in terms of the effects caused by its requests.

Consumer: A role performed by a resource *A* in an interaction with another *B*, which involves making a request to *B* and typically waiting for a response.

Integration: The act of instantiating a given method to design or to adapt two or more resources, so that they become interoperable as a requisite to be able to cooperate and to accomplish one or more common goals.

Interoperability: Asymmetric property between a consumer *C* and a provider *P* (*C* is compatible with *P*) that holds if *C* is compliant with *P* and *P* is conformant to *C*.

Provider: A role performed by a resource *B* in an interaction with another *A*, which involves waiting for a request from *A*, honoring it and typically sending a response to *A*.

Resource: An entity of any nature (material, virtual, conceptual, noun, action, and so on) that embodies a meaningful, complete and discrete concept, makes sense by itself and can be distinguished from, although able to interact with, other entities.

Service: The set of operations supported by a resource and that together define its behavior (the set of reactions to messages that the resource exhibits).

Compilation of References

Aazam, M., Hung, P. P., & Huh, E. N. (2014, April). Smart gateway based communication for cloud of things. In *2014 IEEE Ninth International Conference on Intelligent Sensors, Sensor Networks and Information Processing (ISSNIP)* (pp. 1-6). IEEE.

ABB. (2017). AC 800M Controllers. Retrieved May 19, 2017, from http://new.abb.com/control-systems/system-800xa/800xa-dcs/hardware-controllers-io/ac800m-controllers

Acatech. (2013). *Recommendations for implementing the strategic initiative INDUSTRIE 4.0.* Retrieved from http://www.acatech.de/fileadmin/user_upload/Baumstruktur_nach_Website/Acatech/root/de/Material_fuer_Sonderseiten/Industrie_4.0/Final_report__Industrie_4.0_accessible.pdf

Accenture (2017). *Digital Agriculture: Improving Profitability.* Retrieved from http://accenture.com

Acs, Z. J., & Audretsch, A. (1991). *Innovation and small firms.* Cambridge: MIT Press.

AddVolt. (2017). AddVolt company institutional web site. Retrieved from http://www.addvolt.com/en/

Adlink. (n.d.). DDS and OPC UA. Retrieved from http://www.prismtech.com/vortex/technologies/dds-and-opc-ua

Administration, U. E. (2017). *International Energy Outlook 2017.* Retrieved December 17, 17, from U.S. Energy Information Administration: https://www.eia.gov/outlooks/ieo/pdf/0484/%282017/%29.pdf

Afshar-Mohajer, N., Wu, C. Y., Ladun, T., Rajon, D. A., & Huang, Y. (2015). Characterization of particulate matters and total VOC emissions from a binder jetting 3D printer. *Building and Environment, 93*(P2), 293–301. doi:10.1016/j.buildenv.2015.07.013

Agmon, N., & Stone, P. (2011). Leading multiple ad hoc teammates in joint action settings. In *Interactive decision theory and game theory* (pp. 2–8). ACM.

Agosti, M., Ferro, N., & Silvello, G. (2016). Digital library interoperability at high level of abstraction. *Future Generation Computer Systems, 55*, 129–146. doi:10.1016/j.future.2015.09.020

Ahmad, M., & Bowman, I. (2011). Predicting System Performance for Multi-tenant Database Workloads. In *Fourth International Workshop on Testing Database Systems* (pp. 6.1-6.6). New York, NY: ACM Press. 10.1145/1988842.1988848

Akanmu, A., Anumba, C. J., & Messener, J. (2013). Scenarios for cyber-physical systems integration in construction. *ITCon, 18*, 240–260.

Akinsola, A., Dawood, N., & Hobbs, B. (2000). Construction planning process improvement using information technology tool. *Proceedings of Construction Information Technology, 2000*, 40–51.

Aksnes, D. W. (2005). Citations and their use as indicators in science policy: Studies of validity and applicability issues with a particular focus on highly cited papers. Unpublished doctoral dissertation, University of Twente, Netherlands.

Akyos, M. (2005), Firma Düzeyinde Yenilikçilik (Yenilik) ve Bilgi Yönetimi. Retrieved 10.04.2017 from http://sistems.org./know_info1.htm)

Alagoz, B., Kaygusuz, A., & Karabiber, A. (2012). A user-mode distributed energy management architecture for smart grid applications. *Energy*, *44*(1), 167–177. doi:10.1016/j.energy.2012.06.051

Alarcón, F., Perez, D., & Boza, A. (2016). Using the internet of things in a production planning context. *Brazilian Journal of Operations & Production Management*, *13*(1), 72–76. doi:10.14488/BJOPM.2016.v13.n1.a8

Alenezi, M., & Magel, K. (2014). Empirical evaluation of a new coupling metric: Combining structural and semantic coupling. *International Journal of Computers and Applications*, *36*(1). doi:10.2316/Journal.202.2014.1.202-3902

Al-Fuqaha, A., Guizani, M., Mohammadi, M., Aledhari, M., & Ayyash, M. (2015). Internet of Things: A Survey on Enabling Technologies, Protocols, and Applications. *IEEE Communications Surveys and Tutorials*, *17*(4), 2347–2376. doi:10.1109/COMST.2015.2444095

Almada-Lobo, F. (2016). The Industry 4.0 revolution and the future of manufacturing execution systems (MES). *Journal of Innovation Management*, *3*(4), 16–21.

Al-Madani, B., & Ali, H. (2017). Data Distribution Service (DDS) based implementation of Smart grid devices using ANSI C12.19 standard. *Procedia Computer Science*, *110*, 394–401. doi:10.1016/j.procs.2017.06.082

Almeida, J. R., Jr. (2003). Segurança em sistemas críticos e em sistemas de informação: um estudo comparativo [Doctoral dissertation]. Escola Politécnica, Universidade de São Paulo, São Paulo.

Almeida, M., Barata, J., Dias, A., & Arroja, L. (2015). Environmental product declaration of ceramic roof tiles: a case study of information requirements within the life cycle. In *EfS 2015 – 2nd Energy for Sustainability Multidisciplinary Conference*.

Almeida, M., Dias, A. C., Demertzi, M., & Arroja, L. (2016). Environmental profile of ceramic tiles and their potential for improvement. *Journal of Cleaner Production*, *131*, 583–593. doi:10.1016/j.jclepro.2016.04.131

Alonso, J., Martínez de Soria, I., Orue-Echevarria, L., & Vergara, M. (2010). Enterprise Collaboration Maturity Model (ECMM): Preliminary Definition and Future Challenges. In K. Popplewell, J. Harding, R. Poler, & R. Chalmeta (Eds.), *Enterprise Interoperability IV*. London: Springer. doi:10.1007/978-1-84996-257-5_40

Amato, F., Mazzeo, A., Moscato, V., & Picariello, A. (2013). A Framework for Semantic Interoperability over the Cloud. In *27th IEEE International Conference on Advanced Information Networking and Applications Workshop* (pp. 1259–1264). 10.1109/WAINA.2013.218

Ammar, F. B., Hafsa, I. H., & Hammami, F. (2013). Analytic hierarchy process selection for batteries storage technologies. In *2013 International Conference on Electrical Engineering and Software Applications* (pp. 1-6).

Amundsen, M. (2014). APIs to Affordances: A New Paradigm for Services on the Web. In C. Pautasso, E. Wilde, & R. Alarcon (Eds.), *REST: Advanced Research Topics and Practical Applications* (pp. 91–106). New York, NY: Springer. doi:10.1007/978-1-4614-9299-3_6

Andersen, E.P. (1997). Conceptual Modeling of Objects [PhD Thesis]. University of Oslo, Oslo, Norway.

Andersen, P., & Ross, J. W. (2016). Transforming the LEGO Group for the Digital Economy. In ICIS 2016 Proceedings, Dublin, Ireland.

Andrew, H. (2015). *ISA-95 Unplugged: Putting MOM Things Together for IoT*. Retrieved January 06, 2018, from http://blog.lnsresearch.com/isa-95-unplugged-putting-mom-things-together-for-iot

Angular. (n.d.). Retrieved from https://angular.io/

Aritzeta, A., Swailes, S., & Senior, B. (2007). Belbin's team role model: Development, validity and applications for team building. *Journal of Management Studies*, *44*(1), 96–118. doi:10.1111/j.1467-6486.2007.00666.x

Armbrust, M., Fox, A., Griffith, R., Joseph, A., Katz, R., Konwinski, A., ... Zaharia, M. (2010). A View of Cloud Computing. *Communications of the ACM*, *4*(53), 50–58. doi:10.1145/1721654.1721672

Arnaut, D. (2010). Towards an entrepreneurial university. *International Journal of Euro-Mediterranean Studies*, *3*(1), 135–152.

Arvidsson, V., Holmström, J., & Lyytinen, K. (2014). Information systems use as strategy practice: A multi-dimensional view of strategic information system implementation and use. *The Journal of Strategic Information Systems*, *23*(1), 45–61. doi:10.1016/j.jsis.2014.01.004

Ashton, K. (2009). That 'internet of things' thing. *RFiD Journal, 22*(7), 97-114. Retrieved from http://www.rfidjournal.com/articles/view?4986

ATHENA European project (2007). *Specification of Interoperability Framework and Profiles, Guidelines and Best Practices*. ATHENA Integrated Project (Deliverable D.A4.2).

Atzori, L., Iera, A., & Morabito, G. (2010). The internet of things: A survey. *Computer Networks*, *54*(15), 2787–2805. doi:10.1016/j.comnet.2010.05.010

Azimi, R., Lee, S., Abourizk, S., & Alvanchi, A. (2011). A Framework for Automated and Integrated Project Monitoring and Control System for Steel Fabrication Projects. *Automation in Construction*, *20*(1), 88–97. doi:10.1016/j.autcon.2010.07.001

Baas, L. (2008). Industrial symbiosis in the Rotterdam Harbour and Industry Complex: Reflections on the interconnection of the techno-sphere with the social system. *Business Strategy and the Environment*, *17*(5), 330–340. doi:10.1002/bse.624

Babu, D., & Darsi, M. (2013). A Survey on Service Oriented Architecture and Metrics to Measure Coupling. *International Journal on Computer Science and Engineering*, *5*(8), 726–733.

Badam, A., Chandra, R., Dutra, J., Ferrese, A., Hodges, S., & Hu, P. … & Skiani, E. (2016). Software-defined Batteries. *Communications of the ACM*, *59*(12), 111–119. doi:10.1145/3007179

Baghdadi, Y. (2014). Modelling business process with services: Towards agile enterprises. *International Journal of Business Information Systems*, *15*(4), 410–433. doi:10.1504/IJBIS.2014.060377

Baheti, R., & Gill, H. (2011). Cyber-physical systems. The Impact of Control Technology, 12, 161-166.

Balog, R. S., & Krein, P. T. (2011). Bus Selection in Multibus DC Microgrids. *IEEE Transactions on Power Electronics*, *26*(3), 860–867. doi:10.1109/TPEL.2010.2094208

Barange, M., Kabil, A., De Keukelaere, C., & Chevaillier, P. (2014, June). Task-oriented conversational behavior of agents for collaboration in human-agent teamwork. In *International Conference on Practical Applications of Agents and Multi-Agent Systems* (pp. 25–37). Springer; . doi:10.1007/978-3-319-07551-8_3

Barata, J., & Cunha, P. R. (2017). Climbing the Maturity Ladder in Industry 4.0: A Framework for Diagnosis and Action that Combines National and Sectorial Strategies. In *Twenty-third Americas Conference on Information Systems*, Boston, MA (pp. 1–10).

Barata, J., Cunha, P. R., Gonnagar, A., & Mendes, M. (2017). A Systematic Approach to Design Product Traceability in Industry 4.0: Insights from the Ceramic Industry. In *Twenty-sixth International Conference on Information Systems Development*, Larnaca, Cyprus (pp. 1–12).

Barreto, O. E. A., Hernández, G. G. J., García, G. M. J., & Hernández, R. J. G. (2016). Security & Safety (S&S) and the Logistics Business. In *Proceedings ICIL2016*, AGH University of Science and Technology in Krakow, Krakow, Poland (pp. 12-23).

Barrett, S., Stone, P., & Kraus, S. (2011). Empirical evaluation of ad hoc teamwork in the pursuit domain. *Autonomous Agents and Multi-Agent Systems*, 2(May), 567–574. Retrieved from http://dl.acm.org/citation.cfm?id=2031678.2031698

Başalp, A., & Yazlık, B. (2006). Türkiye'de Teknoparklar ve Sorunları. In TOBB Ekonomi ve Teknoloji Üniversitesi Türkiye'de İnternet Konferansı Bildirileri (pp. 275–278).

Bauer, H., Baur, C., Mohr, D., Tschiesner, A., Weskamp, T., & Mathis, R. (2016). *Industry 4.0 after the initial hype - Where manufacturers are finding value and how they can best capture it. McKinsey Digital* (Whitepaper). *McKinsey*. Retrieved from https://www.mckinsey.de/files/mckinsey_industry_40_2016.pdf

Bausano. (2017). Arduino EasyCAT. Retrieved June 14, 2017, from http://www.bausano.net/en/hardware/ethercat-e-arduino/easycat.html

Beck, Z., Teacy, L., Rogers, A., & Jennings, N. R. (2016, May). Online planning for collaborative search and rescue by heterogeneous robot teams. In *Proceedings of the 2016 International Conference on Autonomous Agents & Multiagent Systems* (pp. 1024-1033). International Foundation for Autonomous Agents and Multiagent Systems.

Becker, J., Knackstedt, R., & Pöppelbuß, J. (2009). Developing Maturity Models for IT Management. *Business & Information Systems Engineering*, *1*(3), 213–222. doi:10.100712599-009-0044-5

Belbin, R. M. (1993). Team roles at work: A strategy for human resource management. *Zitiert in: Teamarbeit und Teamentwicklung*, *321*.

Belbin, M. (1981). *Management Teams, Why They Succeed or Fail*. London: Heinemann.

Bellavista, P., Montanari, R., & Das, S. K. (2013). Mobile social networking middleware: A survey. *Pervasive and Mobile Computing*, *4*(9), 437–453. doi:10.1016/j.pmcj.2013.03.001

Benedetti, M., Holgado, M., & Evans, S. (2017). MAESTRI Deliverable D4.2. Prototype library of case studies linked to a waste database. Retrieved from https://maestri-spire.eu/downloads/technical-materials/

Berkman, R. (2014). The social enterprise: Where does the information professional fit? *Online Searcher*, *6*, 45–48.

Berlin, J. M., Carlstrom, E. D., & Sandberg, H. S. (2012). Models of teamwork: Ideal or not? A critical study of theoretical team models. *Team Performance Management: An International Journal*, *18*(5–6), 328–340. doi:10.1108/13527591211251096

Berre, A., Elvesæter, B., Figay, N., Guglielmina, C., Johnsen, S., Karlsen, D., & ... (2007). The ATHENA interoperability framework. In R. Gonçalves, J. Müller, K. Mertins, & M. Zelm (Eds.), *Enterprise Interoperability II* (pp. 569–580). London, UK: Springer. doi:10.1007/978-1-84628-858-6_62

Bertino, E., Bernstein, P., Agrawal, D., Davidson, S., Dayal, U., Franklin, M., ... & Jadadish, H. V. (2011). Challenges and Opportunities with Big Data.

Beyer, M., & Laney, D. (2012). *The importance of big data: A definition*. Stamford, CT: Gartner.

Bharadwaj, A., & Omar, A. (2013). Digital Business Strategy: Toward a next generation of insights. *Management Information Systems Quarterly*, *37*(2), 471–482. doi:10.25300/MISQ/2013/37:2.3

Bicchi, A., & Peshkin, M. a, & Colgate, J. E. (2008). Safety for Physical Human–Robot Interaction. In Springer Handbook of Robotics (pp. 1335–1348). Springer.

Biro, D. F. (2017). Key world energy statistics. *International Energy Agency*. Retrieved December 17, 2017, from https://www.iea.org/publications/freepublications/publication/KeyWorld2017.pdf

bism@rt. (2017). Oldest Internet Of Things Objects In The World. Retrieved from https://bismart.com/en/oldest-internet-of-things-objects-in-the-world/

Björk, B. (2002). A formalised model of the information and materials handling activities in the construction processnull. *Construction Innovation*, *2*(3), 133–149. doi:10.1108/14714170210814739

Blanzieri, E., & Bryl, A. (2008). A survey of learning-based techniques of email spam filtering. *Artificial Intelligence Review*, *29*(1), 63–92.

Blanzieri, E., & Bryl, A. (2008, March). of email spam filtering. *Artificial Intelligence Review*, *29*(1), 63–92. doi:10.100710462-009-9109-6

Bleeker, E. A. J., Evertz, S., Geertsma, R. E., Peijnenburg, W. J. G. M., Westra, J., & Wijnhoven, S. W. P. (2015). *Assessing health & environmental risks of nanoparticles*. Bilthoven.

Bommadevara, N., Kaplan, J., & Starikova, I. (2016). *Leaders and laggards in enterprise cloud infrastructure adoption*. McKinsey & Company. Retrieved January 06, 2018 from http://www.mckinsey.com/business-functions/digital-mckinsey/our-insights/leaders-and-laggards-in-enterprise-cloud-infrastructure-adoption

Bonomi, F. (2011). Connected vehicles, the internet of things, and fog computing. In *The Eighth ACM International Workshop on Vehicular Inter-Networking (VANET)*, Las Vegas, NV (pp. 13-15).

Bonomi, F., Milito, R., Zhu, J., & Addepalli, S. (2012). Fog computing and its role in the internet of things. In *Proceedings of the first edition of the MCC workshop on Mobile cloud computing* (pp. 13-16). 10.1145/2342509.2342513

Bora, A., & Bezboruah, T. (2015). A Comparative Investigation on Implementation of RESTful versus SOAP based Web Services. *International Journal of Database Theory and Application*, *8*(3), 297–312. doi:10.14257/ijdta.2015.8.3.26

Borisov, O. I., Gromov, V. S., Kolyubin, S. A., Pyrkin, A. A., Bobtsov, A. A., Salikhov, V. I., … Petranevsky, I. V. (2016). Human-free robotic automation of industrial operations. In *Industrial Electronics Society, IECON 2016, 42nd Annual Conference of the IEEE* (pp. 6867-6872). 10.1109/IECON.2016.7793922

Bosch (2017). *Bosch in Portugal*. Retrieved June 25, 2017, from http://www.bosch.pt/en/pt/our_company_10/our-company-lp.html

Boschrexroth. (2016, November 21). Industry 4.0 for new and existing machines. Retrieved June 20, 2017, from https://www.boschrexroth.com/en/xc/company/press/index2-25280

Botta, A., Donato, W. D., Persico, V., & Pescape, A. (2016). Integration of Cloud computing and Internet of Things: A survey. *Future Generation Computer Systems*, *56*, 684–700. doi:10.1016/j.future.2015.09.021

Bounfour, A. (2016). *Digital Futures, Digital Transformation, Progress in IS: From Lean Production to Acceluction*. Cham, *Switzerland*: Springer International Publishing. doi:10.1007/978-3-319-23279-9

Bowling, M., & McCracken, P. (2005, July). *Coordination and adaptation in impromptu teams* (Vol. 5, pp. 53–58). AAAI.

Boyer, S. A. (2009). *SCADA: supervisory control and data acquisition*. International Society of Automation.

Bradshaw, J. M. Acquisti, A., Allen, J., Breedy, M., Bunch, L., Chambers, N., … Field, M. (2004). Teamwork-Centered Autonomy for Extended Human-Agent Interaction in Space Applications. Retrieved from http://www.aaai.org/Papers/Symposia/Spring/2004/SS-04-03/SS04-03-026.pdf

Bradshaw, J. M., Feltovich, P. J., & Johnson, M. (2012). Human–Agent Interaction. In *The Handbook of Human-Machine Interaction: A Human-Centered Design Approach*. Ashrate Publishing, Ltd.

Bradshaw, J. M., Feltovich, P. J., Johnson, M., Bunch, L., Breedy, M., Jung, H., & Uszok, A. (2008). Coordination in Human-Agent-Robot Teamwork. In *International Symposium on Collaborative Technologies and Systems CTS 2008* (pp. 467-476). IEEE.

Brandão, J. E. M. S. (2008). *FRAGA.J.S. Gestão de risco. In SBSEG2008 VIII Simpósio Brasileiro em Segurança da Informação de Sistemas Computacionais.* Santa Catarina: UFSC.

Braun, T., Glanzel, W., & Grupp, H. (1995a). The scientometric weight of 50 nations in 27 science areas, 1989-1993, 1. All fields combined, mathematics, engineering, chemistry and physics. *Scientometrics*, *33*(3), 263–293. doi:10.1007/BF02017332

Braun, T., Glanzel, W., & Grupp, H. (1995b). The scientometric weight of 50 nations in 27 science areas, 1989-1993. *Scientometrics*, *34*(2), 207–237. doi:10.1007/BF02020421

Brenkert, G. G. (2002). Entrepreneurship, Ethics, and the Good Society. *Ethics and Entrepreneurship*, *3*, 5–43.

Brettel, M., & Friederichsen, N. (2014). How virtualization, decentralization and network building change the manufacturing landscape: An Industry 4.0 Perspective. *International Journal of Mechanical, Aerospace, Industrial, Mechatronic and Manufacturing Engineering*, *8*(1), 37–44.

Brettel, M., Friederichsen, N., Keller, M., & Rosenberg, M. (2014). How virtualization, decentralization and network building change the manufacturing landscape: An Industry 4.0 Perspective. *International Journal of Mechanical, Industrial Science and Engineering*, *8*(1), 37–44.

Bridge, S., O'Neill, K., & Cromie, S. (1998). *Understanding enterprise, entrepreneurship & small business*. London: MacMillan Press; . doi:10.1007/978-1-349-26171-0

Brizzi, P., Lotito, A., Ferrera, E., Conzon, D., Tomasi, R., & Spirito, M. (2011, December). Enhancing traceability and industrial process automation through the VIRTUS middleware. In *Proceedings of the Middleware 2011 Industry Track Workshop* (p. 2). ACM.

Brüel & Kjær. S. (1982). *Measuring Vibration*. Retrieved June 1, 2017, from https://www.bksv.com/media/doc/br0094.pdf

Bucki, J. (2016). Definition and Explanation of Mobile Computing. *The Balance*. Retrieved from https://www.thebalance.com/definition-of-mobile-computing-2533640

Buhr, D. (2015). *Social innovation policy for Industry 4.0. Friedrich-Ebert-Stiftung*. Germany: Division for Social and Economic Policies.

Business Wire. (2017). Stealth Tire Monitoring Startup Neomatix to Take 1st Place in Shell's Competition for Sustainable Mobility. Retrieved from http://www.businesswire.com/news/home/20161027006866/en/

Caird, S., Hallett, S., & Potter, S. (2013). The Open2-Innovation Tool—A software tool for rating organisational innovation performance. *Technovation*, *33*(10-11), 381–385. doi:10.1016/j.technovation.2013.06.003

Camarinha-Matos, L. M., & Afsarmanesh, H. (2005). Collaborative networks: A new scientific discipline. *Journal of Intelligent Manufacturing*, *16*(4-5), 439–452. doi:10.100710845-005-1656-3

Camarinha-Matos, L. M., Afsarmanesh, H., Galeano, N., & Molina, A. (2009). Collaborative networked organizations – Concepts and practice in manufacturing enterprises. *Computers & Industrial Engineering*, *57*(1), 46–60. doi:10.1016/j.cie.2008.11.024

Cannon-Bowers, J. A., Salas, E., & Converse, S. A. (1990). Cognitive psychology and team training: Training shared mental models and complex systems. *Bulletin - Human Factors Society*, *33*, 1–4.

Cantillon, R. (1755). *Essai sur la Nature du Commerce en General, edited with an English translation and other material by Henry Higgs, C.B. Reissued for The Royal Economic Society by Frank Cass and Co.* London: LTD.

Cardenas, A. A., Amin, S., & Sastry, S. (2008). Secure control: Towards survivable cyber-physical systems. In *Proceedings of 28th International Conference* Distributed Computing Systems Workshops ICDCS '08. doi:10.1109/ICDCS

Çelik, M. (2011). *Şirketlerin İnovasyon Yapma Eğilimlerinde Üniversite Sanayi İşbirliğinin Rolü ve ODTÜ Teknokent Örneği.* İstanbul: Marmara Üniversitesi Sosyal Bilimler Enstitüsü, Yüksek Lisans Tezi.

Chang, W. L. (2015). NIST Big Data Interoperability Framework: Volume 1. *Definitions.*

Chang, Y. C., Yang, P. Y., Martin, B. R., Chi, H. R., & Tsai-Lin, T. F. (2016). Entrepreneurial universities and research ambidexterity: A multilevel analysis. *Technovation*, *54*, 7–21. doi:10.1016/j.technovation.2016.02.006

Charalabidis, Y., Lampathaki, F., & Jardim-Goncalves, R. (2014). On the scientific foundations of enterprise interoperability: The ENSEMBLE Project and Beyond. In Y. Charalabidis, F. Lampathaki, & R. Jardim-Goncalves (Eds.), *Revolutionizing Enterprise Interoperability through Scientific Foundations* (pp. 336–355). Hershey, PA: IGI Global. doi:10.4018/978-1-4666-5142-5.ch014

Chen, D., Doumeingts, G., & Vernadat, F. (2008). Architectures for enterprise integration and interoperability: Past, present and future. *Computers in Industry*, *59*(7), 647–659. doi:10.1016/j.compind.2007.12.016

Chen, D., Heyer, S., Ibbotson, S., Salonitis, K., Steingrímsson, J. G., & Thiede, S. (2015). Direct digital manufacturing: Definition, evolution, and sustainability implications. *Journal of Cleaner Production*, *107*, 615–625. doi:10.1016/j.jclepro.2015.05.009

Cheng, Z., Guo, X., Ou, W., & Han, D. (2009). Cloud Computing Service Composition and Search Based on Semantic. In M.Gilje Jaatun, G. Zhao & C. Rong (Eds.), *Proceedings of the 1st International Conference on Cloud Computing* (*Vol. 1*, pp. 290-300). Springer-Verlag, Berlin, Heidelberg.

Cheng, F. T., Tieng, H., Yang, H. C., Hung, M. H., Lin, Y. C., Wei, C. F., & Shieh, Z. Y. (2016). Industry 4.1 for wheel machining automation. *IEEE Robotics and Automation Letters*, *1*(1), 332–339. doi:10.1109/LRA.2016.2517208

Chen, J. Y., & Barnes, M. J. (2013). *Human-agent teaming for multi-robot control: A literature review (No. ARL-TR-6328).* Army research lab, Aberdeen Proving Ground, MD. doi:10.21236/ADA583900

Chen, J., & Poon, C. (2009). Photocatalytic construction and building materials: From fundamentals to applications. *Building and Environment*, *44*(9), 1899–1906. doi:10.1016/j.buildenv.2009.01.002

Chen, M., Mao, S., & Liu, Y. (2014). Big Data: A Survey. *Mobile Network*, *19*(2), 171–209. doi:10.100711036-013-0489-0

Chen, R. Y. (2017). An intelligent value stream-based approach to collaboration of food traceability cyber physical system by fog computing. *Food Control*, *71*, 124–136. doi:10.1016/j.foodcont.2016.06.042

Chen, Y. (2015). A RDF-based approach to metadata crosswalk for semantic interoperability at the data element level. *Library Hi Tech*, *33*(2), 175–194. doi:10.1108/LHT-08-2014-0078

Chen, Y., & Kamara, J. (2008). The mechanisms of information communication on construction sites. *FORUM Ejournal, 8*(1), 1–32.

Chertow, M. R. (2000). Industrial symbiosis: Literature and taxonomy. *Annual Review of Energy and the Environment, 25*(1), 313–337.

Chertow, M. R. (2007). "Uncovering" industrial symbiosis. *Journal of Industrial Ecology, 11*(1), 11–30. doi:10.1162/jiec.2007.1110

Chinosi, M., & Trombetta, A. (2012). BPMN: An introduction to the standard. *Computer Standards & Interfaces, 34*(1), 124–134. doi:10.1016/j.csi.2011.06.002

Chrisman, J., Hynes, T., & Fraser, S. (1995). Faculty entrepreneurship and economic development: The case of the University of Calgary. *Journal of Business Venturing, 10*(4), 267–281. doi:10.1016/0883-9026(95)00015-Z

Clark, B. R. (1998). *Creating entrepreneurial universities.* Oxford, England: Pergamon Press.

Cloud-standards. (2017). *The Cloud Standards Wiki.* Cloud-standards.org. Retrieved January 06, 2018, from http://cloud-standards.org

Co-FACTOR project. (2017). Retrieved May 19, 2017, from http://www.cofactor-eu-project.org/

Cohen, S. G., & Bailey, D. E. (1997). What makes teams work? Group effectiveness research from the shop floor to the executive suite. *Journal of Management, 23*(3), 239–290. doi:10.1177/014920639702300303

Collin, J., Hiekkanen, K., Korhonen, J. J., Halén, M., Itälä, T., & Helenius, M. (2015). *It leadership in transition-the impact of digitalization on finnish organizations.* Helsink-Finland: Unigrafia Oy.

Colombo, A., Karnouskos, S., Kaynak, O., Shi, Y., & Yin, S. (2017). Industrial cyberphysical systems: A backbone of the fourth industrial revolution. *Industrial Electronics Magazine IEEE, 11*(1), 6–16. doi:10.1109/MIE.2017.2648857

Compact RIO. (n.d.). Retrieved from http://www.ni.com/compactrio/esa/

Companhia Ambiental do Estado de São Paulo. (2003). Manual de orientação para a elaboração de estudos de análise de riscos. São Paulo, Spain: Companhia Ambiental do Estado de São Paulo.

Condition Monitoring, S. K. F. (1999). *Vibration Sensors.* Retrieved May 22, 2017, from http://www.exvalos.cz/soubory/File/SKF/SNIMACE_VIBRACI.pdf

Confusion Matrix. (2017, May 14). In *Wikipedia, The Free Encyclopedia.* Retrieved 15:06, June 20, 2017, from https://en.wikipedia.org/wiki/Confusion_matrix

Connection Technology Center. (2017). *Measuring Fan and Motor Vibration.* Retrieved May 22, 2017 from https://www.ctconline.com/pdf/pubTechPapers/23Measuring%20Fan%20and%20Motor%20Vibration.pdf

Conner, K., & Prahalad, C. (1996). A Resource-Based Theory of the Firm: Know ledge Versus Opportunism. *Organization Science, 5*(7), 477–501. doi:10.1287/orsc.7.5.477

Cornelissen, I. R., Jongeneelen, F., van Broekhuizen, P., & van Broekhuizen, F. (2011). *Guidance working safely with nanomaterials and nanoproducts, the guide for employers and employees.* Amesterdam, The Netherlands: IVAM.

Corsaro, A. (2016, August 15). A Tale of Two Industrial IoT Standards: DDS and OPC-UA. *RTInsigihts.* Retrieved from https://www.rtinsights.com/dds-opc-ua-industrial-iot-standards/

Cortés, B., Boza, A., Pérez, D., & Cuenca, L. L. (2015). Internet of Things Applications on Supply Chain Management. In *17th International Conference on Industrial, Information and Engineering Systems (ICIIES)* (pp. 2304-3309).

Coward, D. (2014). *Java WebSocket Programming*. Columbus, OH: McGraw-Hill Education.

Crăciun, D., & Bunoiu, M. (2015). Training Teachers for the Knowledge Society: Social Media in Science Education. *Broad Research in Artificial Intelligence and Neuroscience.*, *6*(3-4), 82–88.

Cuenca, L., Boza, A., Alemany, M. M. E., & Trienekens, J. J. (2013). Structural elements of coordination mechanisms in collaborative planning processes and their assessment through maturity models: Application to a ceramic tile company. *Computers in Industry*, *64*(8), 898–911. doi:10.1016/j.compind.2013.06.019

Cupek, R., Ziebinski, A., Huczala, L., & Erdogan, H. (2016). Agent-based manufacturing execution systems for short-series production scheduling. *Computers in Industry*, *82*(October), 245–258. doi:10.1016/j.compind.2016.07.009

daCosta, F. (2013). *Rethinking the Internet of Things: A Scalable Approach to Connecting Everything*. New York, NY: Apress Media, LLC. doi:10.1007/978-1-4302-5741-7

Dahir, H., Dry, B., & Pignataro, C. (Eds.). (2015). *People, Processes, Services, and Things: Using Services Innovation to Enable the Internet of Everything*. New York, NY: Business Expert Press, LLC.

Dahlmann, F., Brammer, S., & Millington, A. (2008). Barriers to proactive environmental management in the United Kingdom: Implications for business and public policy. *Journal of General Management*, *33*(3), 1–20. doi:10.1177/030630700803300301

Dash, S. K., Mohapatra, S., & Pattnaik, P. K. (2010). A survey on applications of wireless sensor network using cloud computing. *International Journal of Computer Science & Emerging Technologies*, *1*(4), 50–55.

Dastjerdi, A. V., Gupta, H., Calheiros, R. N., Ghosh, S. K., & Buyya, R. (2016). Fog computing: Principles, architectures, and applications. arXiv:1601.02752

Data, B. (2011). Let's start with the definition of Industry 4. 0. What does it mean? Retrieved from http://www.open-datasrl.it/index.php/en/

Davenport, T. H., & Brooks, J. D. (2004). Enterprise systems and the supply chain. *Journal of Enterprise Information Management*, *17*(1), 8–19. doi:10.1108/09576050410510917

DDS. (n.d.). Retrieved from http://portals.omg.org/dds/

De Burgos, J. J., García, G. M. J., Hernández, G. G. J., & Hernández, R. J. G. (2016). Generation and management knowledge. A view from the manager of packing. In *Proceedings ICIL2016*, AGH University of Science and Technology in Krakow, Krakow, Poland (pp. 32-44).

De Felice, F., & Petrillo, A. (2012). Hierarchical model to optimize performance in logistics policies: Multi attribute analysis. *Procedia: Social and Behavioral Sciences*, *58*, 1555–1564. doi:10.1016/j.sbspro.2012.09.1142

DECHEMA/VCI. (2011). 10 Years of research: risk assessment, human and environmental toxicology of nanomaterials. Frankfurt.

Degryse, C. (2016). Digitalisation of the Economy and its Impact on Labour Markets. European Trade Union Institute.

Delarue, A., Van Hootegem, G., Procter, S., & Burridge, M. (2008). Teamworking and organizational performance: A review of survey-based research. *International Journal of Management Reviews*, *10*(2), 127–148. doi:10.1111/j.1468-2370.2007.00227.x

Delgado, J. (2012). Bridging the SOA and REST architectural styles. In A. Ionita, M. Litoiu, & G. Lewis (Eds.), *Migrating legacy applications: challenges in service oriented architecture and cloud computing environments* (pp. 276–302). Hershey, PA: IGI Global.

Delgado, J. (2013). Service Interoperability in the Internet of Things. In N. Bessis, F. Xhafa, D. Varvarigou, R. Hill, & M. Li (Eds.), *Internet of Things and Inter-cooperative Computational Technologies for Collective Intelligence* (pp. 51–87). Berlin, Germany: Springer. doi:10.1007/978-3-642-34952-2_3

DeNisco, A. (2017). Report: 56% of CEOs say digital transformation has increased profits. *Tech Republic*. Retrieved from http://www.techrepublic.com/article/report-56-of-ceos-say-digital-transformation-has-increased-profits/

Desfray, P., & Raymond, G. (2014). *Modeling Enterprise Architecture with TOGAF: A Practical Guide Using UML and BPMN*. Waltham, MA: Morgan Kaufmann.

Deutschland. (2014). *Industry 4.0 at Hannover Messe*. Retrieved May 18, 2017, from https://www.deutschland.de/en/topic/business/globalization-world-trade/industry-40-at-hannover-messe

Dillon, T., Wu, C., & Chang, E. (2007). Reference architectural styles for service-oriented computing. In K. Li, C. Jesshope, H. Jin, & J. Gaudiot (Eds.) *IFIP International Conference on Network and parallel computing* (pp. 543–555). Berlin, Heidelberg: Springer-Verlag. 10.1007/978-3-540-74784-0_57

Domingo, G. L. (2016). *The Challenges of Logistics 4.0 for the Supply Chain Management and the Information Technology* [Master dissertation]. Norwegian University of Science and Technology (NTNU), Trondheim, Norwegian.

Dorn, J., Grün, C., Werthner, H., & Zapletal, M. (2007). A survey of B2B methodologies and technologies: From business models towards deployment artifacts. In *40th Hawaii International Conference on Systems Science*, Waikoloa, Hawaii, January3-6. 10.1109/HICSS.2007.37

Dosi, G. (1988). Sources, Procedures and Microeconomic Effects of Innovation. *Journal of Economic Literature*, *26*, 1120–1171.

Driskell, T., Driskell, J. E., Burke, C. S., & Salas, E. (2017). Team Roles: A Review and Integration. *Small Group Research*.

Drucker, P. F. (1985). *Innovation and entrepreneurship practice and principles*. New York, NY: Harper & Row.

Drucker, P. F. (1991). Knowledge-Worker Productivity: The Biggest Challenge. *California Review Management*, *41*(2), 79–80.

Drucker, P. F. (1995). *Managing in a time of great change*. Boston, MA: Harvard Business Press.

Duarte, K., Justino, C. I. L., Freitas, A. C., Duarte, A. C., & Rocha-Santos, T. A. P. (2014). Direct-reading methods for analysis of volatile organic compounds and nanoparticles in workplace air. *TrAC Trends in Analytical Chemistry*, *53*, 21–32. doi:10.1016/j.trac.2013.08.008

Duddu, P. (2015). *The 10 worst blackouts of the last 50 years*. Retrieved December 17, 2017, from http://www.power-technology.com/features/featurethe-10-worst-blackouts-in-the-last-50-years-4486990/

Dudek, G., & Stadder, H. (2005). Negotiation-based collaborative planning between supply chains partners. *European Journal of Operational Research*, *163*(3), 668–687. doi:10.1016/j.ejor.2004.01.014

Dunin-Keplicz, B., & Verbrugge, R. (2011). *Teamwork in multi-agent systems: A formal approach* (Vol. 21). John Wiley & Sons.

Duong, N. H., Maillé, P., Ram, A. K., & Toutain, L. (2017). Decentralized demand response for temperature-constrained appliances. *IEEE Transactions on Smart Grid*, 1. doi:10.1109/TSG.2017.2778225

Earley Information Science. (2016) Building a Successful Digital Transformation Roadmap. Retrieved from https://cdn2.hubspot.net/hubfs/109038/EIS_Assets/EIS-Whitepaper-Building-a-Successful-Digital-Transformation-Roadmap.pdf

Ehrig, M. (2007). Ontology alignment: bridging the semantic gap (Vol. 4). New York, NY: Springer Science+Business Media, LLC.

EIF. (2010). European Interoperability Framework (EIF) for European public services Annex 2. *Communication from the Commission to the European Parliament, the Council, the European Economic and Social Committee and the Committee of Regions 'Towards interoperability for European public services'*. Retrieved December 16, 2015, from http://ec.europa.eu/isa/documents/isa_annex_ii_eif_en.pdf

Environment Directorate OECD. (2015). *Harmonized tiered approach to measure and assess the potential exposure to airborne emissions of engineered nano-objects and their agglomerates and aggregates at workplaces.*

Erl, T. (2016). *Service-oriented architecture: concepts, technology, and design* (2nd ed.). Upper Saddle River, NJ: Prentice Hall.

Erl, T., Carlyle, B., Pautasso, C., & Balasubramanian, R. (2012). *SOA with REST: Principles, Patterns &Constraints for Building Enterprise Solutions with REST.* Upper Saddle River, NJ: Prentice Hall Press.

Erol-Kantarci, M., & Mouftah, H. T. (2015). Energy-efficient information and communication infrastructures in the smart grid: A survey on interactions and open issues. *IEEE Communications Surveys and Tutorials, 17*(1), 179–197. doi:10.1109/COMST.2014.2341600

Erol, S., Jäger, A., Hold, P., Ott, K., & Sihn, W. (2016). Tangible Industry 4.0: A Scenario-Based Approach to Learning for the Future of Production. *Procedia CIRP, 54*, 13–18. doi:10.1016/j.procir.2016.03.162

Er, P. H. (2013). Girişimcilik ve Yenilikçilik Kavramlarının İktisadi Düşüncedeki Yeri: Joseph A. Schumpeter. *Selçuk Üniversitesi Sosyal Bilimler Enstitüsü Dergisi, 29*, 75–85.

Ertugrul, A. M., & Demirors, O. (2015). An Exploratory Study on Role-based Collaborative Business Process Modeling Approaches. In *Proceedings of the 7th International Conference on Subject-Oriented Business Process Management* (pp. 14:1-14:5). 10.1145/2723839.2723857

EtherCAT. (2017). What's EtherCAT? Retrieved June 13, 2017, from https://www.automation.com/automation-news/article/whats-ethercat

Etzkowitz, H. (1983). Entrepreneurial Scientists and Entrepreneurial Universities in American Academic Science. Minerva.

Etzkowitz, H. (2004). The evolution of the entrepreneurial university. *International Journal of Technology and Globalisation, 1*(1), 64–77. doi:10.1504/IJTG.2004.004551

Etzkowitz, H., & Leydesdorff, L. (2000). The dynamics of innovation: From national systems and 'mode 2' to a triple helix of university industry-government relations. *Research Policy, 29*(2), 109–123. doi:10.1016/S0048-7333(99)00055-4

EUREKA. (2017) Annual Report 2016, Belgium.

European Commission. (2013). Factories of the future. Multi-annual roadmap for the contractual PPP under. *Horizon.*

European Commission. (2013). Guidance on the protection of the health and safety of workers from the potential risks related to nanomaterials at work - Guidance for employers and health and safety practitioners. Brussels.

European Commission. (2016). *European Commission sets out path to digitise European industry.* Retrieved January 06, 2018 from https://ec.europa.eu/digital-single-market/en/news/commission-sets-out-path-digitise-european-industry

European Commission. (2016). *Factories 4.0 and Beyond.* Luxembourg: Publications Office of the European Union.

European Commission. (2016a). *Digital Single Market: Two years on.* Brussels. Retrieved from https://ec.europa.eu/priorities/sites/beta-political/files/2-years-on-dsm_en_0.pdf

European Commission. (2016b). *IP/16/1407 - Commission sets out path to digitise European industry.* Brussels. Retrieved from http://europa.eu/rapid/press-release_IP-16-1407_pt.htm

European Commission. (2016c). *The Fourth Industrial Revolution.* Retrieved from https://ec.europa.eu/digital-single-market/en/fourth-industrial-revolution

Eurostat. (1997). OECD Proposed Guidelines for Collecting and Interpreting Technological Innovation Data - Oslo Manual. Paris: OECD.

Evans, D. (2014). *The Internet of Things: How the Next Evolution of the Internet is Changing Everything* (White paper). CISCO. Retrieved from https://www.cisco.com/web/about/ac79/docs/innov/io-t_ibsg_0411final.pdf

eWON (2018). eWON Flexy - Industrial M2M Router and Data Gateway. Retrieved January 6, 2018, from https://ewon.biz/products/ewon-flexy

Ewon. (n.d.). Ewon official web page. Retrieved from https://ewon.biz/

Fadier, E., & De la Garza, C. (2006). Safety design: Towards a new philosophy. *Safety Science, 44*(1), 55–73. doi:10.1016/j.ssci.2005.09.008

Faison, T. (2006). *Event-Based Programming.* Berkeley, CA: Apress.

Fan, X., & Yen, J. (2004). Modeling and simulating human teamwork behaviors using intelligent agents. *Physics of Life Reviews, 1*(3), 173–201. doi:10.1016/j.plrev.2004.10.001

Federal Ministry For Economic Affairs And Energy. (2016). *Plattform Industrie 4.0.* Retrieved December 17, 2017, from http://www.plattform-i40.de/I40/Redaktion/EN/PressReleases/2016/2016-03-02-kooperation-iic.html

Federal Ministry For Economic Affairs And Energy. (n.d.). *The background to Plattform Industrie 4.0.* Retrieved December 17, 2017, from http://www.plattform-i40.de/I40/Navigation/EN/ThePlatform/PlattformIndustrie40/plattform-industrie-40.html

Fercoq, A., Lamouri, S., & Carbone, B. (2016). Lean/Green integration focused on waste reduction techniques. *Journal of Cleaner Production, 137*, 567–578. doi:10.1016/j.jclepro.2016.07.107

Ferreira, R. A., Braga, H. A., Ferreira, A. A., & Barbosa, P. G. (2012). Analysis of voltage droop control method for dc microgrids with Simulink: Modelling and simulation. In *2012 10th IEEE/IAS International Conference on Industry Applications* (pp. 1-6).

Ferreira, M. J., Moreira, F., & Seruca, I. (2014). *A traditional organization towards a new dimension of labour – social business. In Information systems and Technology for Organizational Agility, Intelligence, and Resilience* (Ch. 9, pp. 180–204). Hershey, PA: IGI Global. doi:10.4018/978-1-4666-5970-4.ch009

Ferreira, M. J., Moreira, F., & Seruca, I. (2015). Social Business: A way to promote organizational transformation. *International Journal of Information System Modeling and Design, 6*(4), 57–81. doi:10.4018/IJISMD.2015100104

Ferreira, M. J., Moreira, F., & Seruca, I. (2017). Organizational training within digital transformation: The ToOW model. In *19th International Conference on Enterprise Information Systems (ICEIS)*, Porto, Portugal, April 26-29. 10.5220/0006366105260532

Ferrera, E., Rossini, R., Baptista, A. J., Evans, S., Hovest, G. G., Holgado, M., ... Estrela, M. A. (2017). Toward Industry 4.0: *Efficient and Sustainable Manufacturing Leveraging MAESTRI Total Efficiency Framework. In International Conference on Sustainable Design and Manufacturing* (pp. 624-633). Cham: Springer.

Fiaschetti, A., Pietrabissa, A., & Delli Priscoli, F. (2015). Towards manufacturing 2.0: An innovative architecture for the Factory of the Future. In *2015 European Conference on Networks and Communications, EuCNC 2015* (pp. 450-454). 10.1109/EuCNC.2015.7194116

Fielding, R. (2000). *Architectural Styles and the Design of Network-based Software Architectures* [Doctoral dissertation]. University of California at Irvine, CA.

Fielding, R. (2008). REST APIs must be hypertext-driven. *Roy Fielding's blog: Untangled.* Retrieved December 16, 2015 from http://roy.gbiv.com/untangled/2008/rest-apis-must-be-hypertext-driven

Fielding, R., & Taylor, R. (2002). Principled Design of the Modern Web Architecture. *ACM Transactions on Internet Technology*, *2*(2), 115–150. doi:10.1145/514183.514185

FINESCE. (2016). The FINESCE Smart Energy Platform Seventh Framework (n. 604677).

FIR. (2017). *Industrie 4.0 maturity index. RWTH Aachen University.* Aachen. Retrieved from http://www.fir.rwth-aachen.de/en/research/research-projects/i40-maturity-index

Fisher, F. M. (2000). *Innovation and Monopoly Leveraging. Dynamic Competition and Public Policy: Technology, Innovation, and Antitrust Issues* (J. Ellig, Ed.). Cambridge University Press.

Fisseler, D., & Reiners, R. (2015). Prozessorientierte Überwachung in der Produktion. In D. Cunningham, P. Hofstedt, K. Meer, & I. Schmitt (Eds.), *INFORMATIK2015* (pp. 917–927). Gesellschaft für Informatik.

Fisseler, D., Reiners, R., & Kemény, Z. (2016) Monitoring and control framework for business processes in ubiquitous environments. In *Proceedings of 2016 Intl IEEE Conferences on Ubiquitous Intelligence & Computing, Advanced and Trusted Computing, Scalable Computing and Communications, Cloud and Big Data Computing, Internet of People, and Smart World Congress* (pp. 863-868). 10.1109/UIC-ATC-ScalCom-CBDCom-IoP-SmartWorld.2016.0136

Fisseler, D., Schneider, A., Lourenço, E. J., & Baptista, A. J. (2017). Combining Process Based Monitoring with Multi-layer Stream Mapping. In G. Campana, R. Howlett, R. Setchi, & B. Cimatti (Eds.), *Sustainable Design and Manufacturing SDM 2017* (Vol. 68). Cham: Springer. doi:10.1007/978-3-319-57078-5_53

Fitzgerald, M., Kruschwitz, N., Bonnet, D., & Welch, M. (2013). Embracing Digital Technology: A New Strategic Imperative. *MIT Sloan Management Review*, 55(2), 1–12.

Flores, S. A. A., Li, Y., Chen, W. N., Zhan, Z. H., Zhang, J., & Chen, L. Y. (2015). Industry 4.0 with cyber-physical integration: A design and manufacture perspective. In *2015 21st international conference on automation and computing (ICAC)* (pp. 1-6). IEEE.

Ford, H., & Crowther, S. (1922). My Life and Work. Garden City, NY: Doubleday, Page & Co.

Franz, J. (2013). The Toyota Way to Continuous Improvement. In *Conference Proceedings, 13th International Lean Management Conference*. Wroclaw: Lean Enterprise Institute Poland.

Fraunhofer I.P.A. (2016a). *Manufacturing Service Bus. The Flexible Solution for Digitizing Production Processes*. Retrieved January 06, 2018, from http://www.ipa.fraunhofer.de/fileadmin/user_upload/Kompetenzen/Digitale_Werkzeuge_in_der_Produktion/Cloud-Plattformen/English_Documents/Productsheet_Manufactoring_Service_Bus.pdf

Fraunhofer I.P.A. (2016b). *Virtual Fort Knox. The Only Open It Cloud Platform For Manufacturing Companies*. Retrieved January 06, 2018, from http://www.ipa.fraunhofer.de/fileadmin/user_upload/Kompetenzen/Digitale_Werkzeuge_in_der_Produktion/Cloud-Plattformen/English_Documents/Productsheet_Virtual_Fort_Knox.pdf

Fryman, J., & Matthias, B. (2012, March). Safety of Industrial Robots: From Conventional to Collaborative Applications. *Robotics; Proceedings of the 7th German Conference on ROBOTIK 2012* (pp. 1–5).

Fuchs, S., & Scherer, R. J. (2017). Multimodels — Instant nD-modeling using original data. *Automation in Construction, 75*, 22–32.

Furht, B., & Escalante, A. (2010). A Handbook of Cloud Computing. New York: Springer.

Gabriel, M., & Pessl, E. (2016). Industry 4.0 and Sustainability Impacts: Critical discussion of sustainability aspects with a special focus on future of work and ecological consequences. *Annals of the Faculty of Engineering Hunedoara, 14*(2), 131.

Gamma, E., Helm, R., Hohnson, R., & Vlissides, J. (2004). *Design patterns: Elements of reusable object-oriented software*. Addison-Wesley.

Gangale, F., Vasiljevska, J., Covrig, C. F., Mengolini, A., & Fulli, G. (2017). JRC Publications Repository. *Smart Grid Project Outlook 2017*. Retrieved December 17, 2017, from http://publications.jrc.ec.europa.eu/repository/bitstream/JRC106796/sgp/outlook/2017-online.pdf

Gantz, J., Reinsel, D. 2011. Extracting value from chaos. *IDC iView, 1142*(2011), 1-12.

Ganzarain, J., & Errasti, N. (2016). Three Stage Maturity Model in SME's towards Industry 4.0. *Journal of Industrial Engineering and Management, 9*(5), 1119–1128. doi:10.3926/jiem.2073

García, G. M. J., Hernández, G. G. J., & Hernández, R. J. G. (2015). An approach to the ranking of projects to undertake. In A. Cobos & A. Vanti (Eds.), *Gobernanza empresarial de Tecnología de la Información* (pp. 427–441). Santander, España: Universidad de Cantabria.

García, G. M. J., Schwarz, I. L. M., Schwarz, I. T. M., Hernández, G. G. J., & Hernández, R. J. G. (2017). Inventories control, the Inventory manager and Matrixes Of Weighing with multiplicative factors (MOWwMf). *Journal of Global Business and Technology, 13*(1), 40–56.

García, M. V., Irisarri, E., Pérez, F., & Elisabet, E. (2017). An Open CPPS Automation Architecture based on IEC-61499 over OPC-UA for flexible manufacturing in Oil & Gas Industry. *IFAC-Papers OnLine, 50*(1), 1231–1238. doi:10.1016/j.ifacol.2017.08.347

Garetti, M., & Taisch, M. (2012). Sustainable manufacturing: Trends and research challenges. *Production Planning and Control, 23*(2-3), 83–104. doi:10.1080/09537287.2011.591619

Gath, M., Herzog, O., & Edelkamp, S. (2014). Autonomous and flexible multiagent systems enhance transport logistics. In *11th International Conference & Expo Emerging Technologies for a Smarter World (CEWIT) 2014* (pp. 1-6). IEEE. 10.1109/CEWIT.2014.7021143

Gaudreault, J., Frayret, J. M., & Pesant, G. (2009). Distributed search for supply chain coordi-nation. *Computers in Industry, 60*(6), 441–451. doi:10.1016/j.compind.2009.02.006

Geetika, R., & Singh, P. (2014). Dynamic coupling metrics for object oriented software systems: A survey. *Software Engineering Notes, 39*(2), 1–8. doi:10.1145/2579281.2579296

Gehbauer, F., Zülch, G., & Börkircher, M. (2007). Simulation-based analysis of disturbances in construction operations. In *Proc. of the 15th Annual Conference, International Group for Lean Construction (IGLC)* (Vol. 571, p. 579).

Geisberger, E., Cengarle, M., Keil, P., Niehaus, J., Thiel, C., & Thönnißen-Fries, H. J. (2011). Cyber-Physical Systems-Driving force for innovation in mobility, health, energy and production. Acatech-Deutsche Akademie der Technikwissenschaften.

Geldermann, C. (2017). Henry Ford. Britannica. Retrieved May 10, 2017, from https://www.britannica.com/biography/Henry-Ford

Genter, K. L., Agmon, N., & Stone, P. (2011, August). Role-Based Ad Hoc Teamwork. In Plan, Activity, and Intent Recognition (pp. 1782–1783). Association for the Advancement of Artificial Intelligence.

Germany Trade And Invest. (n.d.). *INDUSTRIE 4.0.* Retrieved December 17, 2017, from https://www.gtai.de/GTAI/Navigation/EN/Invest/Industries/Industrie-4-0/Industrie-4-0/industrie-4-0-what-is-it.html

Ghithub. (n.d.). Json rules. Retrieved from https://github.com/cachecontrol/json-rules-engine

Gibbs, D., & Deutz, P. (2007). Reflections on implementing industrial ecology through eco-industrial park development. *Journal of Cleaner Production*, *15*(17), 1683–1695. doi:10.1016/j.jclepro.2007.02.003

Gilchrist, A. (2016). Introducing Industry 4.0. In Industry 4.0 (pp. 195-215). Apress.

Gilchrist, A. (2016). *Industry 4.0. The industrial internet of things.* New York: Apress.

Golightly, D., Sharples, S., Harshada Patel, H., & Ratchev, S. (2016). Manufacturing in the cloud: A human factors perspective. *International Journal of Industrial Ergonomics*, *55*, 12–21. doi:10.1016/j.ergon.2016.05.011

Gölzer, P., Cato, P., & Amberg, M. (2015). Data Processing Requirements of Industry 4.0-Use Cases for Big Data Applications. In *Twenty-Third European Conference on Information Systems (ECIS)*, Münster, Germany (pp. 1-13).

Gonçalves, G., Reis, J., Pinto, R., Alves, M., & Correia, J. (2014, September). A step forward on Intelligent Factories: A Smart Sensor-oriented approach. In 2014 IEEE Emerging Technology and Factory Automation (ETFA) (pp. 1-8). IEEE. doi:10.1109/ETFA.2014.7005227

González, L., Laborde, J., Galnares, M., Fenoglio, M., & Ruggia, R. (2014). An adaptive enterprise service bus infrastructure for service based systems. In A. Lomuscio, S. Nepal, F. Patrizi, B. Benatallah, & I. Brandić (Eds.), Service-Oriented Computing Workshops, LNCS (Vol. 8377, pp. 480–491). Cham, Switzerland: Springer International Publishing. doi:10.1007/978-3-319-06859-6_42

Grangel-González, I., Halilaj, L., Coskun, G., Auer, S., Collarana, D., & Hoffmeister, M. (2016). Towards a semantic administrative shell for industry 4.0 components. In *IEEE Tenth International Conference on* Semantic Computing (ICSC) 2016 (pp. 230-237).

Grant, G. B., Seager, T. P., Massard, G., & Nies, L. (2010). Information and communication technology for industrial symbiosis. *Journal of Industrial Ecology*, *14*(5), 740–753. doi:10.1111/j.1530-9290.2010.00273.x

Grant, J., Kraus, S., & Perlis, D. (2005). A logic-based model of intention formation and action for multi-agent subcontracting. *Artificial Intelligence*, *163*(2), 163–201. doi:10.1016/j.artint.2004.11.003

Greefhorst, D., & Proper, E. (2011). *Architecture principles: the cornerstones of enterprise architecture* (Vol. 4). New York, NY: Springer Science & Business Media. doi:10.1007/978-3-642-20279-7

Grosz, B. J., & Kraus, S. (1996). Collaborative plans for complex group action. *Artificial Intelligence*, *86*(2), 269–357. doi:10.1016/0004-3702(95)00103-4

Grosz, B. J., & Kraus, S. (1999). The evolution of SharedPlans. In *Foundations of rational agency* (pp. 227–262). Springer Netherlands. doi:10.1007/978-94-015-9204-8_10

Growth Within: a circular economy vision for a competitive Europe . (2015, June 25). Retrieved from Ellen Macarthur Foundation. Retrieved from https://www.ellenmacarthurfoundation.org/publications/growth-within-a-circular-economy-vision-for-a-competitive-europe

Grzybowska, K., Kovács, G., & Lénárt, B. (2014). The supply chain in cloud computing. *Research in Logistics & Production*, *4*(1), 33–44.

GTAI. (2016). *Industrie 4.0 – Smart Manufacturing for the Future.* Retrieved June 15, 2017 from https://www.gtai.de/GTAI/Content/EN/Invest/_SharedDocs/Downloads/GTAI/Brochures/Industries/industrie4.0-smart-manufacturing-for-the-future-en.pdf

Gu, Y., & Jing, T. (2011). The IOT research in supply chain management of fresh agricultural products. In *2011 2nd International Conference on Artificial Intelligence, Management Science and Electronic Commerce (AIMSEC)* (pp. 7382-7385). IEEE.

Gubbi, J., Buyya, R., Marusic, S., & Palaniswami, M. (2013). Internet of things (IoT): A vision, architectural elements, and future directions. *Future Generation Computer Systems*, *29*(7), 1645–1660. doi:10.1016/j.future.2013.01.010

Guerrero, M., Urbano, D., Fayolle, A., Klotfsten, M., & Mian, S. (2016). Entrepreneurial universities: Emerging models in the new social and economic landscape. *Small Business Economics*, *47*(3), 551–563. doi:10.100711187-016-9755-4

Guiterrez, J. G., & Baquero, J. E. G. (2017). New cross-proposal entrepreneurship and innovation in educational programs in third level (tertiary) education. *Contaduría y Administración*, *62*(1), 239–261. doi:10.1016/j.cya.2016.10.005

Gür, U., Oylumlu, İ. S., & Kunday, Ö. (2016). Critical assessment of entrepreneurial and innovative universities index of Turkey: Future directions. *Technological Forecasting and Social Change*.

Hale, A., Kirwan, B., & Kjellén, U. (2007). Safe by design: Where are we now? *Safety Science*, *45*(1–2), 305–327. doi:10.1016/j.ssci.2006.08.007

Hammer, M. (1990). Reengineering Work: Don't Automate, Obliterate. *Harvard Business Review*, *68*(4), 104–112.

Hankel, M., & Rexroth, B. (2015). *The reference architectural model industrie 4.0 (rami 4.0)*. ZVEI.

Hanly, S. (2016, June 16). Vibration Analysis: FFT, PSD, and Spectrogram Basics. *Mide*. Retrieved June 6, 2017, from https://blog.mide.com/vibration-analysis-fft-psd-and-spectrogram

Hanna, N., & Richards, D. (2013). A Collaborative Activity for Evaluating HAT-COM: Human-agent Teamwork Communication Model. In *Proceedings of the 2013 International Conference on Autonomous Agents and Multi-Agent Systems*, St. Paul (pp. 1369-1370). Retrieved from http://dl.acm.org/citation.cfm?id=2484920.2485229

Hansen, R., & Sia, S. K. (2015). Hummel's Digital Transformation Toward Omnichannel Retailing: Key Lessons Learned. *MIS Quarterly Executive*, *14*(2), 51–66.

Harbers, M., Bradshaw, J. M., Johnson, M., Feltovich, P., Van Den Bosch, K., & Meyer, J. J. (2012). Explanation in human-agent teamwork. In Coordination, Organizations, Institutions, and Norms in Agent System VII (pp. 21-37). Springer. doi:10.1007/978-3-642-35545-5_2

Harmancı, M., & Önen, M. O. (1999). Dünyada ve Türkiye'de Teknopark ve Teknokent Uygulamaları. Türkiye Kalkınma Bakanlığı, GA/99-2-4.

Harris, P. R., & Harris, D. L. (1989). High Performance Team Management. *Leadership and Organization Development Journal*, *10*(4), 28–32. doi:10.1108/EUM0000000001141

Hästbacka, D., Jantunen, E., Karaila, M., & Barna, L. (2016, October). Service-based condition monitoring for cloud-enabled maintenance operations. In *42nd Annual Conference of the IEEE Industrial Electronics Society, IECON 2016* (pp. 5289-5295). IEEE. 10.1109/IECON.2016.7793470

Hauschild, M. (2015). Better – But is it Good Enough? On the Need to Consider Both Eco-efficiency and Eco-effectiveness to Gauge Industrial Sustainability. In *27The 22nd CIRP conference on Life Cycle Engineering* (pp. 1-7). Sydney: Procedia CIRP.

Helo, P., Soursa, M., Hao, Y., & Anussornnitisarn, P. (2014, May). Toward a cloud-based manufacturing execution system for distributed manufacturing. *Computers in Industry*, *65*(4), 646–656. doi:10.1016/j.compind.2014.01.015

Hendricksen, D. (2014). *12 More Essential Skills for Software Architects*. Upper Saddle River, NJ: Addison-Wesley Professional.

Henning, M. (2008). The rise and fall of CORBA. *Communications of the ACM*, *51*(8), 52–57. doi:10.1145/1378704.1378718

Herman, M., Pentek, T., & Otto, B. (2015). *Design Principles for Industrie 4.0 Scenarios: A Literature Review*. Technische Universität Dortmund.

Hermann, C., Blume, S., Kurle, D., Schmidt, C., & Thiede, S. (2015). The Positive Impact Factory – Transition from Eco-efficiency to Eco–effectiveness Strategies in Manufacturing. In *The 22nd CIRP conference on Life Cycle Engineering* (pp. 19-27). Sydney: Procedia CIRP.

Hermann, M., Pentek, T., & Otto, B. (2016). Design principles for industrie 4.0 scenarios. A literature review. In *49th Hawaii International Conference on* System Sciences (HICSS*)* (pp. 3928-3937). IEEE.

Hernández, R. J. G., Hernández, G. G. J., & García, G. M. J. (2017). Some sins in the negotiations in times of Industry 4.0: Soccer and BREXIT. In Delener et al. (Eds.), *Nineteenth annual International Conference Global Business and Technology Association* (pp. 224-232). Reading Book.

Hernández, J. G., & García, M. J. (2010). Mathematical models generators of Decision Support Systems for help in case of catastrophes. An experience from Venezuela. In E. Asimakopoulou & N. Bessis (Eds.), *Advanced ICTs for Disaster Management and Threat Detection: Collaborative and Distributed Frameworks* (pp. 201–220). Hershey, PA: IGI Global. doi:10.4018/978-1-61520-987-3.ch013

Hernández, J. G., García, M. J., & Hernández, G. J. (2016). Social aspects of reverse logistics and Knowledge Management. In A. K. Goel & P. Singhal (Eds.), *Product Innovation through Knowledge Management and Social Media Strategies* (pp. 65–93). Hershey, PA: IGI Global. doi:10.4018/978-1-4666-9607-5.ch004

Hess, T., Matt, C., Benlian, A., & Wiesböck, F. (2016). Options for formulating a digital transformation strategy. *MIS Quarterly Executive*, *15*(2), 123–139.

Hillary, R. (2004). Environmental management systems and the smaller enterprise. *Journal of Cleaner Production*, *12*(6), 561–569. doi:10.1016/j.jclepro.2003.08.006

Hinchcliffe, D. (2016). Using online community for digital transformation. Retrieved from https://dionhinchcliffe.com/

Hirsch, J. E. (2005). An index to quantify an individual's scientific research output. *Proceedings of the National Academy of Sciences of the United States of America*, *102*(46), 16569–16572. doi:10.1073/pnas.0507655102

Hisrich, R. D. (1985). The woman entrepreneur in the United States and Puerto Rico: A comparative study. *Leadership and Organization Development Journal*, *5*(5), 3–8. doi:10.1108/eb053558

Hisrich, R. D., & Peters, M. P. (2001). *Entrepreneurship*. McGraw-Hill College.

Ho, V. (2017). Design revolution: Microsoft HoloLens and mixed reality are changing how architects see the world. *Microsoft*. Retrieved from https://news.microsoft.com/transform/design-revolution-microsoft-hololens-mixed-reality-changing-architects-world/

Hock-Koon, A., & Oussalah, M. (2010). Defining metrics for loose coupling evaluation in service composition. In *International Conference on Services Computing* (pp. 362–369). Piscataway, NJ: IEEE Computer Society Press. 10.1109/SCC.2010.17

Holgado, M., Benedetti, M., Evans, S., Baptista, A. J., & Lourenço, E. J. (in press). Industrial symbiosis implementation by leveraging on process efficiency methodologies. Procedia CIRP, Life Cycle Engineering (LCE). *Conference, Copenhagen, Denmark.*

Holgado, M., Morgan, D., & Evans, S. (2016). Exploring the scope of industrial symbiosis: Implications for practitioners. *Smart Innovation. Systems and Technologies*, *52*, 169–178.

Hong, K., Lillethun, D., Ramachandran, U., Ottenwälder, B., & Koldehofe, B. (2013). Mobile fog: A programming model for large-scale applications on the internet of things. In *Proceedings of the second ACM SIGCOMM workshop on Mobile cloud computing*, (pp. 15-20). 10.1145/2491266.2491270

Ho, W., Xu, X., & Dey, P. K. (2010). Multi-criteria decision making approaches for supplier evaluation and selection: A literature review. *European Journal of Operational Research*, *202*(1), 16–24. doi:10.1016/j.ejor.2009.05.009

Hsieh, J. C., Li, A. H., & Yang, C. C. (2013). Mobile, cloud, and big data computing: Contributions, challenges, and new directions in telecardiology. *International Journal of Environmental Research and Public Health*, *11*(10), 6131–6153. doi:10.3390/ijerph10116131 PMID:24232290

Huson, D., & Hoskins, S. (2014). 3D Printed Ceramics for Tableware, Artists/Designers and Specialist Applications. *Key Engineering Materials*, *608*, 351–357. doi:10.4028/www.scientific.net/KEM.608.351

Hussain, T., Mehmood, R., Haq, A., Alnafjan, K., & Alghamdi, A. (2014). Designing framework for the interoperability of C4I systems. In *International Conference on Computational Science and Computational Intelligence* (102–106). Piscataway, NJ: IEEE Computer Society Press. 10.1109/CSCI.2014.102

Iarovyi, S., Mohammed, W., Lobov, A., Ferrer, B., & Lastra, J. (2016). Cyber-physical systems for open-knowledge-driven manufacturing execution systems. *Proceedings of the IEEE*, *104*(5). 10.1109/JPROC.2015.2509498

IDC. (2016). Third platform. Retrieved from http://www.idc.com/promo/thirdplatform

IDC. (2017) Digital Transformation. Retrieved from https://www.idc.com/promo/thirdplatform/digitaltransformation

IEEE. (2015). *Towards definition of the Internet of Things*. Retrieved December 17, 2017, from https://iot.ieee.org/images/files/pdf/IEEE_IoT_Towards_Definition_Internet_of_Things_Revision1_27MAY15.pdf

IIC & Plattform Industrie 4.0. (2017). *Industrial Internet Consortium & Platform Industrie 4.0*. Accessed January 06, 2018, from https://www.infosys.com/engineering-services/white-papers/Documents/industrial-internet-consortium-architecture.pdf

IIC Working TWG and SWG. (2015). *Industrial Internet Reference Architecture*. Industrial Internet Consortium. Retrieved January 06, 2018, from https://www.iiconsortium.org/IIRA.htm

IIC. (2015). *Industrial Internet Reference Architecture* [technical report]. Industrial Internet Consortium.

Ikebe, H. (2003). Power systems for telecommunications in the IT age. In *The 25th International Telecommunications Energy Conference INTELEC '03* (pp. 1-8).

Imache, R., Izza, S., & Ahmed-Nacer, M. (2015). Clustering-based urbanisation to improve enterprise information systems agility. *Enterprise Information Systems, 8*(9), 861–877. doi:10.1080/17517575.2013.879210

Immonen, A., Ovaska, E., Kalaoja, J., & Pakkala, D. (2016). A service requirements engineering method for a digital service ecosystem. *Service Oriented Computing and Applications, 10*(2), 151–172. doi:10.100711761-015-0175-0

IMPULS. (2017). *Industry 4.0 Readiness Online Self-Check for Businesses. VDMA.* Retrieved from https://www.industrie40-readiness.de/?lang=en

Inc. (2014). *Material Requirements Planning (MRP).* Retrieved May 10, 2017, from https://www.inc.com/encyclopedia/material-requirements-planning-mrp.html

India Smart Grid Forum. (2015). *Smart grid project book - A global snapshot April 2015.* Retrieved December 17, 2017 from http://www.indiasmartgrid.org/document/ISGF-Smart%20Grid%20Project%20Book%20-%20A%20Global%20Snapshot.pdf

Industry Week. (2016). IDC transformation [image]. Retrieved from http://www.industryweek.com/sites/industryweek.com/files/styles/article_featured_standard/public/uploads/2016/01/idc-transformation.jpg?itok=D1dkzVxX

INEDIC. (2011). INEDIC-Innovation and Ecodesign in the Ceramic Industry. Retrieved from http://www.prepare-net.com/project/inedic-innovation-and-ecodesign-ceramic-industry

Infosys & IIC. (2015a). *IIC - Asset Efficiency Testbed.* IIC. Retrieved January 06, 2018, from http://www.iiconsortium.org/asset-efficiency.htm

Infosys & IIC. (2015b). *IIC - Industrial Digital Thread Testbed.* IIC. January 06, 2018, from http://www.iiconsortium.org/industrial-digital-thread.htm

International Association of Science Parks. (2012). A Governance Model for Science and Technology Parks. In *29th IASP World Conference.*

International Electrotechnical Commission (IEC). (2009). *31000: general guidelines for principles and implementation of risk management.* Switzerland: ISO/IEC.

International Organization for Standardization. (2013). ISO 16739:2013 Industry Foundation Classes (IFC) for data sharing in the construction and facility management industries. Geneva: International Organization for Standardization.

IPH. (2017). *4.0 Ready. Institut für Integrierte Produktion Hannover (IPH).* Hannover. Retrieved from https://www.iph-hannover.de/en/research/research-projects/?we_objectID=2319

Isaka, H., Nagayoshi, H., Yoshikawa, H., Yamada, T., & Kakeno, N. (2016). Next Generation of Global Production Management Using Sensing and Analysis Technology. *Hitachi Review, 65*(5), 47–52.

i-SCOOP. (n.d.). *Industry 4.0: the fourth industrial revolution – guide to Industrie 4.0.* Retrieved December 17, 2017 from https://www.i-scoop.eu/industry-4-0/

Ismail, A., Srewil, Y., Scherer, R. J., & Mansperger, T. (2016). Semantic Enrichment and Multimodel Data Exchange Approach for CFD Analysis of Bridges. In *23rd International Workshop of the European Group for Intelligent Computing in Engineering, EG-ICE,* June 29-July 1.

Ismail, A., Srewil, Y., & Scherer, R. J. (2017). Integrated and collaborative process-based simulation framework for construction project planning. *International Journal of Simulation and Process Modelling*, *12*(1), 42–53. doi:10.1504/IJSPM.2017.082789

ISO. (1994). *ISO/IEC 7498-1, Information technology – Open Systems Interconnection – Basic Reference Model: The Basic Model (2nd ed.)*. Geneva, Switzerland: International Standards Office. Retrieved December 16, 2015, from http://standards.iso.org/ittf/PubliclyAvailableStandards/index.html

ISO. (2006). *Enterprise integration -- Framework for enterprise modelling. ISO/CEN Standard 19439:2006*. Geneva, Switzerland: International Organization for Standardization.

ISO. (2006a). *ISO 14040:2006, Environmental Management - Life Cycle Assessment - Principles and frameworks*. Geneva: International Organization for Standardization.

ISO. (2006b). *ISO 14044:2006, Environmental Management - Life Cycle Assessment – Requirements and guidelines*. Geneva: International Organization for Standardization.

ISO. (2010). *Systems and software engineering – Vocabulary. ISO/IEC/IEEE 24765:2010(E) International Standard* (p. 186). Geneva, Switzerland: International Organization for Standardization.

ISO. (2011). *Advanced Automation Technologies and their Applications, Part 1: Framework for Enterprise Interoperability (CEN EN/ISO 11354-1)*. Geneva, Switzerland: International Standards Office.

ISO/IEC. (1994). *Standard 7498-1:1994*. Retrieved June 5, 2017, from http://standards.iso.org/ittf/PubliclyAvailableStandards/s020269_ISO_IEC_7498-1_1994(E).zip

ISO/IEC/IEEE. (2011). *ISO/IEC/IEEE 42010:2011 Architecture description*. Retrieved January 06, 2018, from http://www.iso-architecture.org/ieee-1471/

Jacobs, D., & Ingwersen, P. (2000). A bibliometric study of the publication patterns in the sciences of South African scholars 1981-96. *Scientometrics*, *47*(1), 75–93.

Jekiel, C. M. (2011). *Lean Human Resources - Redesigning HR Processes for a Culture of Continuous Improvement*. New York: Productivity Press.

Jeney, A. (2014). *Impacto del Gerente de Sistemas de Información y Redes del Modelo Logístico Basado en Cargos en la gestión del conocimiento de una organización, medido a través de una Matriz De Ponderación* [Master dissertation]. Universidad Metropolitana, Caracas, Venezuela.

Jennings, N. R., Sycara, K., & Wooldridge, M. (1998). A roadmap of agent research and development. *Autonomous Agents and Multi-Agent Systems*, *38*(1), 7–38. doi:10.1023/A:1010090405266

Jeston, J., & Nelis, J. (2014). *Business process management*. New York, NY: Routledge.

Jin, C., De-Lin, L., & Fen-Xiang, M. (2009, July). An improved ID3 decision tree algorithm. In *4th International Conference on Computer Science & Education ICCSE'09* (pp. 127-130). IEEE

Johannsen, G. (2007). Human-Machine interaction. In H. Unbehauen (Ed.), *Control Systems, Robotics, and Automation. In Encyclopedia of Life Support Systems (EOLSS)*. Oxford, UK: EOLSS Publishers.

Jonker, C., van Riemsdijk, M. B., & Vermeulen, B. (2011). Shared Mental Models. In *Proceedings of 9th Workshop on COIN* (pp. 132-151). Springer. 10.1007/978-3-642-21268-0_8

Jula, A., Sundararajan, E., & Othman, Z. (2014). Review: Cloud computing service composition: A systematic literature review. *Expert Systems with Applications*, *41*(8), 3809–3824. doi:10.1016/j.eswa.2013.12.017

Juric, M., & Weerasiri, D. (2014). *WS-BPEL 2.0 beginner's guide*. Birmingham, UK: Packt Publishing Ltd.

Kadiri, E. S., Grabot, B., Thoben, K., Hribernik, K., Emmanouilidis, C., von Cieminski, G., & Kiritsis, D. (2016). Current trends on ICT technologies for enterprise information systems. *Computers in Industry*, *79*, 14–33. doi:10.1016/j.compind.2015.06.008

Kagermann, H., Wahlster, W., & Helbig, J. (2013). Recommendations for implementing the strategic initiative INDUSTRIE 4.0. (No. Final Report). Berlin, Germany: acatech – National Academy of Science and Engineering.

Kagermann, H., Helbig, J., Hellinger, A., & Wahlster, W. (2013). *Recommendations for implementing the strategic initiative INDUSTRIE 4.0: Securing the future of German manufacturing industry; final report of the Industrie 4.0 Working Group*. Forschungsunion.

Kagermann, H., Helbig, J., Hellinger, A., & Wahlster, W. (2013). *Recommendations for Implementing the strategic initiative INDUSTRIE 4.0: securing the future of German manufacturing industry; final report of the Industrie 4.0 working group*. Forschungsunion.

Kamar, E., Gal, Y., & Grosz, B. J. (2013). Modeling information exchange opportunities for effective human-computer teamwork. *Artificial Intelligence*, *195*, 528–550. doi:10.1016/j.artint.2012.11.007

Kaminka, G. A., & Frenkel, I. (2005). Flexible teamwork in behavior-based robots. In *Proceedings Of The National Conference On Artificial Intelligence* (Vol. 20, No. 1, p. 108). London, UK: AAAI Press.

Kastner, C. A., Lau, R., & Kraft, M. (2015). Quantitative tools for cultivating symbiosis in industrial parks; a literature review. *Applied Energy*, *155*, 599–612. doi:10.1016/j.apenergy.2015.05.037

Kato, J., Sakamoto, D., Igarashi, T., & Goto, M. (2014). Sharedo: to-do list interface for human-agent task sharing. In *Proceedings of the second international conference on Human-agent interaction* (pp. 345-351). ACM. 10.1145/2658861.2658894

Katz, J. A. (2003). The Chronology and Intellectual Trajectory of American Entrepreneurship Education. *Journal of Business Venturing*, *18*(2), 283–30. doi:10.1016/S0883-9026(02)00098-8

Kazik, O. (2010). Role-based Approaches to Development of Multi-Agent Systems: A Survey. In WDS' 10 Proceedings of Contributed Papers, Part I (pp. 19–24).

Kennedy, W. G., Bugajska, M. D., Adams, W., Schultz, A. C., & Trafton, J. G. (2008). Incorporating Mental Simulation for a More Effective Robotic Teammate. In *Proceedings of the AAAI Conference on Artificial Intelligence* (pp. 1300-1305).

Keränen, A., & Bormann, C. (2016). *Internet of Things: Standards and Guidance from the IETF. IETF Journal.*

Kevin, L. J. (2016, November 27). *Smart Manufacturing Is Cloud Computing*. Retrieved June 8, 2017, from http://kevinljackson.blogspot.pt/2016/11/smart-manufacturing-is-cloud-computing.html

Key Enabling Technologies. (2016). European Commission. Growth. Internal Market, Industry, Entrepreneurship and SMEs. Retrieved from https://ec.europa.eu/growth/industry/key-enabling-technologies_en

Kezar, A., & Eckel, P. (2002). The effect of institutional culture on change strategies in higher education: Universal principles or culturally responsive concepts? *The Journal of Higher Education*, *73*, 435–460.

Khalfallah, M., Figay, N., Barhamgi, M., & Ghodous, P. (2014). Model driven conformance testing for standardized services. In *IEEE International Conference on Services Computing* (pp. 400–407). Piscataway, NJ: IEEE Computer Society Press. 10.1109/SCC.2014.60

Khan, S. (2016). *Leadership in the Digital Age - a study on the effects of digitalization on top management leadership*. Shahyan Khan.

Kirby, D. A. (2006). Creating entrepreneurial universities in the UK: Applying entrepreneurship theory to practice. *The Journal of Technology Transfer*, *31*(5), 599–603. doi:10.100710961-006-9061-4

KIT. (2016). *Introduction strategies of industry 4.0 methodology and technology for smes. Karlsruhe Institute Of Technology*. Karlsruhe. Retrieved from http://www.eurekanetwork.org/project/id/10389

Kniazev, E. (2004). University Strategy in Regional Development. In *Entrepreneurialism and The Transformation of Russian Universities: New Trends in Higher Education*. International Institute for Educational Planning.

Knight, F. (1921). *Risk, Uncertainty and Profit*. Boston, MA: Houghton, Mifflin.

Knoke, B., Missikoff, M., & Thoben, K.-D. (2017). Collaborative open innovation management in virtual manufacturing enterprises. *International Journal of Computer Integrated Manufacturing*, *30*(1), 158–166. doi:10.1080/0951192X.2015.1107913

Kodym, O., & Unucka, J. (2016). Gathering Information from Transport Systems for Processing in Supply Chains. *Ocean Engineering*, *6*, 676–680.

Koh, Y. K., & Hatten, S. T. (2002). The Tourism Entrepreneur. *International Journal of Hospitality & Tourism Administration*, *3*(1), 21–48. doi:10.1300/J149v03n01_02

Köksal, Ö., & Tekinerdogan, B. (2017). Obstacles in Data Distribution Service Middleware: A Systematic Review. *Systems. Future Generation Computer*, *68*, 191–2210. doi:10.1016/j.future.2016.09.020

Kolberg, D., & Zühlke, D. (2015). Lean automation enabled by industry 4.0 technologies. *IFAC-PapersOnLine*, *48*(3), 1870–1875. doi:10.1016/j.ifacol.2015.06.359

Korsah, G. A., Browning, B., Stentz, A., & Dias, M. B. (2012). xBots: An Approach to Generating and Executing Optimal Multi-Robot Plans with Cross-Schedule Dependencies. In *2012 IEEE International Conference on Robotics and Automation (ICRA)* (pp. 115-122). IEEE.

Kortuem, G., Kawsar, F., Sundramoorthy, V., & Fitton, D. (2010). Smart objects as building blocks for the internet of things. *IEEE Internet Computing*, *14*(1), 44–51. doi:10.1109/MIC.2009.143

Kozlowski, S. W. J., Grand, J. A., Baard, S. K., Pearce, M., & Durso, F. (2015). Teams, teamwork, and team effectiveness: Implications for human systems integration. In Handbook of Human Systems Integration (pp. 1–30).

Kreps, J. (2014). Questioning the lambda architecture.

Kukreja, R., & Garg, N. (2014). Remote Procedure Call: Limitations and Drawbacks. *International Journal of Research*, *1*(10), 914–917.

Kumari, S., & Rath, S. (2015). Performance comparison of SOAP and REST based Web Services for Enterprise Application Integration. In *International Conference on Advances in Computing, Communications and Informatics* (pp. 1656–1660). Piscataway, NJ: IEEE Computer Society Press. 10.1109/ICACCI.2015.7275851

Kyrarini, M., Leu, A., Ristić-Durrant, D., Gräser, A., Jackowski, A., Gebhard, M., & Schlick, C. M. (2016). Human-robot synergy for cooperative robots. *Automatic Control and Robotics*, *15*(3), 187–204.

Labrinidis, A., & Jagadish, H. V. (2012). Challenges and opportunities with big data. *Proc VLDB Endowment*, 5(12), 2032–2033. 10.14778/2367502.2367572

Lai, C., Ma, Y., Chang, S., Chao, H., & Huang, Y. (2011). OSGi-based services architecture for cyber-physical home control systems. *Computer Communications*, *34*(2), 184–191. doi:10.1016/j.comcom.2010.03.034

Laney, D. (2001). 3D data management: Controlling data volume, velocity and variety. *META Group Research Note*, *6*(70).

Laney, D. (2001). 3-d data management: controlling data volume, velocity and variety. *META Group Research Note*.

Lasi, H., Fettke, P., Kemper, H. G., Feld, T., & Hoffmann, M. (2014). Industry 4.0. *Business & Information Systems Engineering*, *6*(4), 239–242. doi:10.100712599-014-0334-4

Laudon, K. C., & Laudon, J. P. (2010). *Management Information Systems: Managing the Digital Firm, Global Edition*. Pearson Higher Education.

Lean Enterprise Institute. (2014). *Lean Lexicon: a graphical glossary for Lean Thinkers (5th editio)*. Cambridge, USA: Lean Enterprise Institute.

Lee, E. A. (2008). Cyber physical systems: Design challenges. In *2008 11th IEEE international symposium on Object oriented real-time distributed computing (ISORC)* (pp. 363-369). IEEE.

Lee, E. A. (2008). *Cyber Physical Systems: Design Challenges*. University of California. Retrieved June 30, 2017, from https://www2.eecs.berkeley.edu/Pubs/TechRpts/2008/EECS-2008-8.pdf

Lee, E. A. (2015). The past, present and future of cyber-physical systems: A focus on models. *Sensors (Basel)*, *15*(3), 4837–4869. doi:10.3390150304837 PMID:25730486

Lee, I., & Lee, K. (2015). The Internet of Things (IoT): Applications, investments, and challenges for enterprises. *Business Horizons*, *58*(4), 431–440. doi:10.1016/j.bushor.2015.03.008

Lee, J., Bagheri, B., & Kao, H.-A. (2015). A cyber-physical systems architecture for industry 4.0-based manufacturing systems. *Manufacturing Letters*, *3*, 18–23. doi:10.1016/j.mfglet.2014.12.001

Lee, J., Kao, H. A., & Yang, S. (2014). Service innovation and smart analytics for industry 4.0 and big data environment. *Procedia CIRP*, *16*, 3–8. doi:10.1016/j.procir.2014.02.001

Lehni, M., Schmidheiny, S., & Stigson, B. (2000). Eco-efficiency: creating more value with less impact. Geneva: World Business Council for Sustainable Development.

Lejeune, M. A., & Yakova, N. (2005). On characterizing the 4 C's in supply chain management. [Elsevier]. *Journal of Operations Management*, *23*(1), 81–100. doi:10.1016/j.jom.2004.09.004

Leng, J., Fyfe, C., & Jain, L. (2006, October). Teamwork and simulation in hybrid cognitive architecture. In *International Conference on Knowledge-Based and Intelligent Information and Engineering Systems* (pp. 472-478). Springer.

Lewis, G., Morris, E., Simanta, S., & Wrage, L. (2008). Why Standards Are Not Enough To Guarantee End-to-End Interoperability. In C. Ncube, & J. Carvallo (Eds.), *Seventh International Conference on Composition-Based Software Systems* (pp. 164–173). Piscataway, NJ: IEEE Computer Society Press. 10.1109/ICCBSS.2008.25

Leyh, C., Schäffer, T., Bley, K., & Forstenhäusler, S. (2017). Assessing the IT and Software Landscapes of Industry 4.0-Enterprises: The Maturity Model SIMMI 4.0. In Information Technology for Management: New Ideas and Real Solutions, LNBIP (Vol. 277, pp. 103–119). Springer.

Liao, Y., Deschamps, F., Loures, E., & Ramos, L. (2017). Past, present and future of Industry 4.0 - a systematic literature review and research agenda proposal. *International Journal of Production Research*, *55*(12), 3609–3629. doi:10.1080/00207543.2017.1308576

Li, B. H., Zhang, L., Wang, S. L., Tao, F., Cao, J. W., Jiang, X. D., ... Chai, X. D. (2010). Cloud manufacturing: A new service-oriented networked manufacturing model. *Jisuanji Jicheng Zhizao Xitong*, *16*(1), 1–7.

LimeNET. (2017). Network-in-a-Box. Retrieved May 18, 2017, from https://www.crowdsupply.com/lime-micro/limenet

Ling, Z., Chen, W., & Yu, J. (2004, June). Research and implementation of OPC server based on data access specification. In *Fifth World Congress on Intelligent Control and Automation WCICA 2004* (Vol. 2, pp. 1475-1478). IEEE.

LioN-Power. (2017). Active I/O modules. Retrieved May 18, 2017, from http://www.lumberg-automation.com/en/Lumberg_Automation/news/lion-power/index.phtml

Li, Q., Qin, W., Han, B., Wang, R., & Sun, L. (2011). A case study on REST-style architecture for cyber-physical systems: Restful smart gateway. *Computer Science and Information Systems*, *8*(4), 1317–1329. doi:10.2298/CSIS110310062L

Liu, Q., Jiang, P., Zhao, J., Zhang, B., Bian, H., & Qian, G. (2011). Life cycle assessment of an industrial symbiosis based on energy recovery from dried sludge and used oil. *Journal of Cleaner Production*, *19*(15), 1700–1708. doi:10.1016/j.jclepro.2011.06.013

Liu, Y., & Xu, X. (2016). Industry 4.0 and Cloud Manufacturing: A Comparative Analysis. *Journal of Manufacturing Science and Engineering*, *139*(3), 034701. doi:10.1115/1.4034667

Liu, Z., & Li, M. (2014). Research on energy efficiency of dc distribution system. In *2nd AASRI Conference on Power and Energy Systems* (pp. 68-74). 10.1016/j.aasri.2014.05.031

Löffler, M., & Tschiesner, R. (2013). The Internet of Things and the future of manufacturing. *McKinsey*. Retrieved from http://www.mckinsey.com/business-functions

Lom, M., Pribyl, O., & Svitek, M. (2016). Industry 4.0 as a part of smart cities. In *Smart Cities Symposium Prague (SCSP)* (pp. 1-6). IEEE. 10.1109/SCSP.2016.7501015

Loopback. (n.d.). Retrieved from http://loopback.io/

Loural, C. de A. (2014, June 26). Um panorama de novas tecnologias e seus impactos na indústria. Retrieved from https://www3.eco.unicamp.br/neit/images/stories/arquivos/Novas_tecnologias_e_seu_impacto_na_industria_-_v140626.pdf

Lourenco, E., Baptista, A., Pereira, J., & Dias-Ferreira, C. (2013). Multi-layer stream mapping as a combined approach for industrial processes eco-efficiency assessment. In Re-engineering Manufacturing for Sustainability (pp. 427-433).

Lucero, S. (2016). IoT platforms: enabling the Internet of Things. *IHS*. Retrieved January 06, 2018, from https://cdn.ihs.com/www/pdf/enabling-IOT.pdf

Lucke, D., Constantinescu, C., & Westkämper, E. (2008). Smart Factory - A Step towards the Next Generation of Manufacturing. In M. Mitsuishi, K. Ueda, & F. Kimura (Eds.), *Manufacturing Systems and Technologies for the New Frontier* (pp. 115–118). London: Springer London. doi:10.1007/978-1-84800-267-8_23

Lu, Y. (2017). Industry 4.0: A survey on technologies, applications and open research issues. *Journal Of Industrial Information Integration*, *6*, 1–10. doi:10.1016/j.jii.2017.04.005

Madsen, H., Burtschy, B., Albeanu, G., & Popentiu-Vladicescu, F. L. (2013). Reliability in the utility computing era: Towards reliable fog computing. In *2013 20th International Conference on Systems, Signals and Image Processing (IWSSIP)* (pp. 43-46).

MAESTRI. (2018). Resource and Energy efficiency for Process Industries. Retrieved December 28, 2017, from https://maestri-spire.eu

Magruk, A. (2016). Uncertainty in the Sphere of the Industry 4.0 – Potential Areas to Research. *Business. Management in Education*, *14*(2), 275–291.

Mahnke, W., Leitner, S., & Damm, M. (2009). *OPC unified architecture*. Springer Science & Business Media. doi:10.1007/978-3-540-68899-0

Mais, J. (2002, May). Spectrum Analysis. The key features of analysing spectra. *SKF*. Retrieved June 8, 2017, from http://www.skf.com/binary/tcm:12-113997/CM5118%20EN%20Spectrum%20Analysis.pdf

Majumder, R., Ghosh, A., Ledwich, G., & Zare, F. (2010). Power management and power flow control with back-to-back converters in a utility connected microgrid. *IEEE Transactions on Power Systems*, *25*(2), 821–834. doi:10.1109/TPWRS.2009.2034666

Malente. (2016, October 20). *Kuhnke FIO Controller WL - IoT compatible Controller.* Retrieved May 19, 2017, from http://kuhnke.kendrion.com/industrial/ics/en/press-releases/iot-compatible-controller-plc-with-integrated-wifi-data-interface.html

Mamaqi, X. (2015). The efficiency of different ways of informal learning on firm performance: A comparison between, classroom, web 2 and workplace training. *Computers in Human Behavior*, *51*, 812–820. doi:10.1016/j.chb.2014.11.080

Manca, S., & Ranieri, M. (2016). "Yes for sharing, no for teaching!": Social Media in academic practices. *Internet and Higher Education*, *29*, 63–74. doi:10.1016/j.iheduc.2015.12.004

Mann, D. (2015). *Creating a Lean Culture: Tools to Sustain Lean Conversions* (3rd ed.). Boca Raton, FL: CRC Press.

Mao, Y., You, C., Zhang, J., Huang, K., & Letaief, K. B. (2017). A Survey on Mobile Edge Computing: The Communication Perspective. *IEEE Communications Surveys and Tutorials*, *19*(4), 2322–2358. doi:10.1109/COMST.2017.2745201

Margerison, C. J., & McCann, D. (1990). *Team management: Practical new approaches*. Mercury Books.

Margerison, C., & McCann, D. (1984). Team mapping: A new approach to managerial leadership. *Journal of European Industrial Training*, *8*(1), 12–16. doi:10.1108/eb002166

Marshall, A. (1890). *Principles of Economics*. London, UK: Mcmillan and Co.

Marston, S., Zhi, L., Bandyopadhyay, S., Zhang, J., & Ghalsasi, A. (2011). Cloud computing - The business perspective. *Decision Support Systems*, *51*(1), 176–189. doi:10.1016/j.dss.2010.12.006

Martini, D. D., Benetti, G., Vedova, M. L., & Facchinetti, T. (2017). Adaptive Real-Time Scheduling of Cyber-Physical Energy Systems. *ACM Transactions on Cyber-Physical Systems*, *1*(4), 20–25.

Marxt, C., & Brunner, C. (2013). Analyzing and improving the national innovation system of highly developed countries—The case of Switzerland. *Technological Forecasting and Social Change*, *80*(6), 1035–1049. doi:10.1016/j.techfore.2012.07.008

Marz, N. (2012). Big data lambda architecture.

Marz, N., & Warren, J. (2015). *Big Data: Principles and best practices of scalable realtime data systems*. Greenwich, CT: Manning Publications Co.

Mathieu, J. E., Maynard, T., Rapp, T., & Gilson, L. L. (2008). Team effectiveness 1997-2007: A review of recent advancements and a glimpse into the future. *Journal of Management*, *34*(3), 410–476. doi:10.1177/0149206308316061

Mathieu, J. E., Tannenbaum, S. I., Kukenberger, M. R., Donsbach, J. S., & Alliger, G. M. (2015). Team Role Experience and Orientation: A Measure and Tests of Construct Validity. *Group & Organization Management*, *40*(1), 6–34. doi:10.1177/1059601114562000

McClelland, D. (1987). Characteristics of Successful Entrepreneurs. *The Journal of Creative Behavior*, *21*(3), 219–233. doi:10.1002/j.2162-6057.1987.tb00479.x

Mcnabb, D. E. (2005). *Public utilities: management challenges for the 21st century*. Edward Elgar Publishing. doi:10.4337/9781845426811

Mendling, J. (2008). *Metrics for Process Models: Empirical Foundations of Verification, Error Prediction, and Guidelines for Correctness*. Springer. doi:10.1007/978-3-540-89224-3

Menicocci, G. (2015). *Disruptive technologies and their impact on the labor market* [Master dissertation]. Libera Università Internazionale degli Studi Sociali Guido Carli, Rome, Italy.

Menon, K., Kärkkäinen, H., & Lasrado, L. A. (2016). Towards a Maturity Modeling Approach for the Implementation of Industrial Internet. In *Proceedings of the Pacific Asia Conference on Information Systems (PACIS)*, Chiayi, Taiwan (pp. 1–11).

MESA International & IBM. (2010). SOA Manufacturing Guidebook. Retrieved January 06, 2018, from ftp://public.dhe.ibm.com/software/plm/pdif/MESA_SOAinManufacturingGuidebook.pdf

Meyer, S., Ruppen, A., & Hilty, L. (2015). The Things of the Internet of Things in BPMN, In J. Barjis, A. Gupta & A. Meshkat (Eds.), Advanced Information Systems Engineering Workshops (pp. 42-56). Springer. doi:10.1007/978-3-319-19243-7_27

Meyer, A., & Weske, M. (2013). Extracting data objects and their states from process models. In *Proceeding of 17th IEEE International Enterprise Distributed Object Computing Conference* (pp. 27-36). 10.1109/EDOC.2013.13

Meyer, G.G., Wortmann, J.C. & Szirbik, N.B. (2011). Production monitoring and control with intelligent products. *International Journal of Production Research*, *49*(5), 1303–1317. doi:10.1080/00207543.2010.518742

Mezgar, I., & Rauschecker, U. (2014). The Challenge of Networked Enterprises for Cloud Computing Interoperability. *Computers in Industry*, *65*(4), 657–674. doi:10.1016/j.compind.2014.01.017

Michelsen, O., Fet, A. M., & Dahlsrud, A. (2006). Eco-efficiency in extended supply chains: A case study of furniture production. *Journal of Environmental Management*, *79*(3), 290–297. doi:10.1016/j.jenvman.2005.07.007 PMID:16253417

Miler, R. 2016. Digital transformation requires total organizational commitment. Retrieved from https://techcrunch.com/2016/01/31/digital-transformation-requires-total-organizational-commitment/

MindSphere (2017). Retrieved June 12, 2017, from https://www.plm.automation.siemens.com/store/en-ie/mindsphere/index.html

Min, H., & Zhou, G. (2002). Supply chain modeling: Past, present and future. *Computers & Industrial Engineering*, *43*(1-2), 231–249. doi:10.1016/S0360-8352(02)00066-9

Mirata, M. (2004). Experiences from early stages of a national industrial symbiosis programme in the UK: Determinants and coordination challenges. *Journal of Cleaner Production*, *12*(8), 967–983. doi:10.1016/j.jclepro.2004.02.031

Mitchell, T. M. (1997). *Machine Learning*. New York, NY: McGraw Hill.

Mobley, R. K. (2002). *An introduction to predictive maintenance*. Butterworth-Heinemann.

Moed, H. F. (2002). Measuring China's research performance using the Science Citation Index. *Scientometrics*, *55*(3), 281–296. doi:10.1023/A:1014812810602

Mohsenian-Rad, A.-H., & Leon-Garcia, A. (2010). Optimal residential load control with price prediction in real-time electricity pricing environments. *IEEE Transactions on Smart Grid, 1*(2), 120–133. doi:10.1109/TSG.2010.2055903

Moment Expo. (2017). Dijital Sanayi Çağı: Endüstri 4.0. Retrieved from http://www.moment-expo.com/dijital-sanayi-cagi-endustri-4-0

Monfelt, Y., Pilemalm, S., Hallberg, J., & Yngström, L. (2011). The 14-layered framework for including social and organizational aspects in security management. *Information Management & Computer Security, 19*(2), 124–133. doi:10.1108/09685221111143060

MongoDB. (n.d.). Retrieved from https://www.mongodb.com

Monk, E., & Wagner, B. (2012). *Concepts in enterprise resource planning* (4th ed.). Cengage Learning, US.

Monostori, L. (2014). Cyber-physical production systems: Roots, expectations and R&D challenges. *Procedia CIRP, 17*, 9–13. doi:10.1016/j.procir.2014.03.115

MONSOON. (2018). Model-based control framework for Site-wide Optimization of data-intensive processes. Retrieved December 28, 2017, from https://www.spire2030.eu/monsoon

Moore, J. F. (1996). *The death of competition: Leadership and strategy in the age of business ecosystems*. New York: Harper Collins.

Moraes, C. C., & Castrucci, P. D. L. (2007). *Engenharia de automação industrial*. Rio de Janeiro: LTC.

Morose, G. (2010). The 5 principles of "Design for Safer Nanotechnology." *Journal of Cleaner Production, 18*(3), 285–289. doi:10.1016/j.jclepro.2009.10.001

Morris, E., Levine, L., Meyers, C., Place, P., & Plakosh, D. (2004). *System of Systems Interoperability (SOSI): final report (Report No. CMU/SEI-2004-TR-004)*. Carnegie Mellon Software Engineering Institute. Retrieved December 16, 2015, from http://www.sei.cmu.edu/reports/04tr004.pdf

Mortaj, H., Ow, S. H., Moghavvemi, M., & Almurib, H. A. (2017). Load shedding and smart-direct load control using internet of things in smart grid demand response management. *IEEE Transactions on Industry Applications, 53*(6), 5155–5163. doi:10.1109/TIA.2017.2740832

Motamedi, A., Setayeshgar, S., Soltani, M. M., & Hammad, A. (2013). Extending BIM to incorporate information of RFID tags attached to building assets. In *Proceeding of the 4th construction specialty conference,* Montreal Canada, May 29-June 1.

Mounce, S. R., Pedraza, C., Jackson, T., Linford, P., & Boxall, J. B. (2015). Cloud based machine learning approaches for leakage assessment and management in smart water networks. *Procedia Engineering, 119*, 43–52. doi:10.1016/j.proeng.2015.08.851

Mourad, M., Nassehi, A., & Schaefer, D. (2016). Interoperability as a Key Enabler for Manufacturing in the Cloud. In *The Sixth International Conference on Changeable, Agile, Reconfigurable and Virtual Production (CARV2016)* (pp. 30–34). Procedia CIRP. 10.1016/j.procir.2016.07.051

Mrugalska, B., & Wyrwicka, M. (2017). Towards lean production in industry 4.0. *Procedia Engineering, 182*, 466–473. doi:10.1016/j.proeng.2017.03.135

Mucha, H., Büttner, S., & Röcker, C. (2016). Application Areas for Human-Centered Assistive Systems. In Human-Computer Interaction–Perspectives on Industry 4.0. Workshop at i-KNOW 2016, Graz, Austria (pp. 1-4).

Muhammad, S. Y., Makhtar, M., Rozaimee, A., Aziz, A. A., & Jamal, A. A. (2015). Classification Model for Water Quality using Machine Learning Techniques. *International Journal of Software Engineering and Its Applications*, *9*(6), 45–52. doi:10.14257/ijseia.2015.9.6.05

Murillo-Luna, J. L., Garcés-Ayerbe, C., & Rivera-Torres, P. (2007). What prevents firms from advancing in their environmental strategy? *International Advances in Economic Research*, *13*(1), 35–46. doi:10.100711294-006-9059-6

Mussomeli, A., Gish, D., & Laaper, S. (2016). *The rise of the digital supply network: Industry 4.0 enables the digital transformation of supply chains. Deloitte*. University Press.

Nalbandyan, N. (2008). A mutual assistance protocol for agent teamwork [MSc thesis]. Yerevan State University.

National Institute for Occupational Safety and Health (NIOSH). (2009). *Approaches to Safe Nanotechnology: An Information Exchange with NIOSH.*

National Institute for Occupational Safety and Health (NIOSH). (2013). Current Intelligence Bulletin 65 Occupational Exposure to Carbon Nanotubes and Nanofibers. Cincinnati: NIOSH.

National Institute for Standard Technology. (2013a). *Strategic vision and business drivers for 21st century cyber-physical systems*. USA: NIST.

National Institute for Standard Technology. (2013b). *Foundations for innovation in cyber-physical systems*. USA: NIST.

National Instruments (2016, March 24). The CompactRIO, Retrieved May 18, 2017, from http://www.ni.com/white-paper/52251/en/

National Science Board. (1972). Patterns and perspectives in environmental science.

National Science Foundation. (2006). *Cyber-Physical Systems.* Retrieved December 17, 2017 from https://www.nsf.gov/publications/pub/summ.jsp?ods/key=nsf12520/&org=NSF

Neomatix (2007) Neomatix company institutional web site. Retrieved from http://www.neomatix.com

Neomatix. (n.d.). [Image]. Retrieved from http://neomatix.com/wp-content/uploads/2017/05/optimus-brmp.jpg

NI. (2017). National instruments official web page. Retrieved from http://www.ni.com/en-us/shop.html

NI-DAQmx (2017). Software. Retrieved June 14, 2017, from https://www.ni.com/dataacquisition/nidaqmx.htm

Ning, L., & Xiaoping, L. (2012). A Resource Virtualization Mechanism for Cloud Manufacturing Systems. In International IFIP Working Conference on Enterprise Interoperability (pp. 46-59). Springer.

Nishio, T., Shinkuma, R., Takahashi, T., & Mandayam, N. B. (2013). Service-oriented heterogeneous resource sharing for optimizing service latency in mobile cloud. In *Proceedings of the first international workshop on Mobile cloud computing \& networking* (pp. 19-26). 10.1145/2492348.2492354

NIST. (2011). The NIST Definition of Cloud Computing Recommendations of the National Institute of Standards and Technology.

NIST. (2011, October 25). The NIST Definition of Cloud Computing (NIST Special Publication 800-145). Retrieved January 8, 2018 from https://www.nist.gov/news-events/news/2011/10/final-version-nist-cloud-computing-definition-published

Node OPC-UA. (n.d.). Retrieved from http://node-opcua.github.io/

Nookhong, J., & Wannapiroon, P. (2014). Development of collaborative learning using case-based learning via cloud technology and social media for enhancing problem-solving skills and ICT literacy within undergraduate students. *Procedia: Social and Behavioral Sciences*, *174*, 2096–2210. doi:10.1016/j.sbspro.2015.02.007

NSF Program Announcements. (2017). Cyber-Physical Systems (CPS) Document Number: nsf17529. December 8, 2016. Retrieved January 06, 2018, from https://www.nsf.gov/pubs/2017/nsf17529/nsf17529.pdf

Nukala, R., Panduru, K., Shields, A., Riordan, D., Doody, P., & Walsh, J. (2016). Internet of Things: A review from 'Farm to Fork.' In *2016 27th Irish Signals and Systems Conference (ISSC)* (pp. 1-6). IEEE. 10.1109/ISSC.2016.7528456

O'Rourke, C., Fishman, N., & Selkow, W. (2003). *Enterprise architecture using the Zachman framework*. Boston, MA: Course Technology.

Ochian, A., Suciu, G., Fratu, O., Voicu, C., & Suciu, V. 2014. An overview of cloud middleware services for interconnection of healthcare platforms, In *10th IEEE International Conference on Communications (COMM)* (pp. 1-4). 10.1109/ICComm.2014.6866753

Odabaşı, Y. (2008). *Meslektaş Girişimci Üniversite ve Yönetimi, Türk Üniversiteleri İçin Bir Model Önerisi*. Girişimcilik Kongresi.

Oesterreich, T. D., & Teuteberg, F. (2016). Understanding the implications of digitisation and automation in the context of Industry 4.0: A triangulation approach and elements of a research agenda for the construction industry. *Computers in Industry*, *83*, 121–139. doi:10.1016/j.compind.2016.09.006

Olivier, J., Schure, K., & Peters, J. (2017). Trends in Global CO2 and total greenhouse gas emissions. *PBL Netherlands Environmental Assessment Agency*. Retrieved December 17, 17, from http://www.pbl.nl/sites/default/files/cms/publicaties/pbl-2017-trends-in-global-co2-and-total-greenhouse-gas-emissons-2017-report/_2674.pdf

Olshansky, S. (2018) Rough Guide to IETF 101: Internet of Things. *IETF Journal*. Internet Society. oneM2M (2015). *The interoperability enabler for the entire M2M and IoT ecosystem* (White Paper). Retrieved from http://www.onem2m.org

OPC Foundation. (n.d.). Retrieved from https://opcfoundation.org/

OPC UAFoundation. (2017). *OPC Unified Architecture*. Retrieved January 06, 2018, from https://opcfoundation.org/developer-tools/specifications-unified-architecture

OPC-UA. (2017). *UnifiedArchitecture*. Retrieved June 13, 2017, from https://opcfoundation.org/about/opc-technologies/opc-ua/

Oren, G., & Stroh, N. Y. (2013). Mathematical Model for Detection of Leakage in Domestic Water Supply Systems by Reading Consumption from an Analogue Water Meter. *International Journal of Environmental Sciences and Development*, *4*(4), 386–389. doi:10.7763/IJESD.2013.V4.377

Organization for Economic Co-operation and Development. (2005). *Handbook on Economic Globalization Indicators*. OECD, Paris Organization for Economic Co-operation and Development.

Orhan, S., Aktürk, N., & Celik, V. (2006). Vibration monitoring for defect diagnosis of rolling element bearings as a predictive maintenance tool: Comprehensive case studies. *NDT & E International*, *39*(4), 293–298. doi:10.1016/j.ndteint.2005.08.008

Ortiz-Rivera, L. A., Sanz Casado, E., & Suárez-Balseiro, C. A. (2000). Scientific production in Puerto Rico in science and technology during the period 1990 to 1998. *Scientometrics*, *49*(3), 403–418. doi:10.1023/A:1010585522899

Özdem, G., & Sarı, E. (2008). *Yükseköğretimde Yeni Bakış Açılarıyla Birlikte Yeni Kurulan Üniversitelerden Beklenen İşlevler (Giresun Üniversitesi Örneği).* Üniversite ve Toplum.

Panetto, H., & Molina, A. (2008). Enterprise integration and interoperability in manufacturing systems: Trends and issues. *Computers in Industry, 59*(7), 641–646. doi:10.1016/j.compind.2007.12.010

Panetto, H., Zdravkovic, M., Jardim-Goncalves, R., Romero, D., Cecil, J., & Mezgár, I. (2016). New perspectives for the future interoperable enterprise systems. *Computers in Industry, 79*, 47–63. doi:10.1016/j.compind.2015.08.001

Paquin, R. L., & Howard-Grenville, J. (2012). The evolution of facilitated industrial symbiosis. *Journal of Industrial Ecology, 16*(1), 83–93. doi:10.1111/j.1530-9290.2011.00437.x

Parker, G. M. (2008). *Team players and teamwork: New strategies for the competitive enterprise.* San Francisco, CA: Jossey-Bass.

Park, L., Jang, Y., Cho, S., & Kim, J. (2017). Residential Demand Response for Renewable Energy Resources in Smart Grid Systems. *IEEE Transactions on Industrial Informatics, 13*(6), 3165–3173. doi:10.1109/TII.2017.2704282

Parr, T. (2013). *The Definitive ANTLR 4 Reference.* Raleigh, NC: The Pragmatic Bookshelf.

Parwekar, P. (2011). From Internet of Things towards cloud of things. In *2011 2nd International Conference on Computer and Communication Technology (ICCCT)* (pp. 329-333).

Paulk, M. C., Curtis, B., Chrissis, M. B., & Weber, C. V. (1993). Capability Maturity Model for Software, Version 1. 1. *Carnegie Melon University, 10*(4), 18–27.

Pautasso, C. (2014). RESTful web services: principles, patterns, emerging technologies. In A. Bouguettaya, Q. Sheng, & F. Daniel (Eds.), *Web Services Foundations* (pp. 31–51). New York, NY: Springer. doi:10.1007/978-1-4614-7518-7_2

Pautasso, C., Zimmermann, O., & Leymann, F. (2008). Restful web services vs. "big"' web services: making the right architectural decision. In *International conference on World Wide Web* (pp. 805–814). New York, NY: ACM Press. 10.1145/1367497.1367606

Perez, D., Alarcón, F., & Boza, A. (2017). *Industry 4.0: A classification scheme.* Springer.

Peristeras, V., & Tarabanis, K. (2006). The Connection, Communication, Consolidation, Collaboration Interoperability Framework (C4IF) For Information Systems Interoperability. *IBIS - International Journal of Interoperability in Business Information Systems, 1*(1), 61–72.

Petrucci, A., Kaneva, P., Lebedeva, E., Feldman, L., Fried, G., & Vassiliou, M. (2015). You Have a Message! Social Networking as a Motivator for FLS Training. *Journal of Surgical Education, 7*(32), 542–548. doi:10.1016/j.jsurg.2014.10.007 PMID:25496795

Pfohl, H.-C., Yahsi, B., & Kuznaz, T. (2015). The impact of Industry 4.0 on the Supply Chain. In C. M. R. Wolfgang Kersten, & T. Blecker (Eds.), *Proceedings of the Hamburg International Conference of Logistic (HICL)-20* (pp. 32-58).

Pfrommer, J. (2017). *Semantic Interoperability at Big-Data Scale with the open62541 OPC UA Implementation. In InterOSS-IoT 2016: Interoperability and Open-Source Solutions for the Internet of Things* (pp. 173–185). Springer.

Phillips, D. T. (2005). Was Einstein a Space Alien. *NASA.* Retrieved December 17, 2017, from https://www.nasa.gov/vision/universe/starsgalaxies/23mar_spacealien.html

Pinto, R., Gonçalves, J., Cardoso, H. L., Oliveira, E., Gonçalves, G., & Carvalho, B. (2016, December). A Facility Layout Planner Tool Based on Genetic Algorithms. In *2016 IEEE Symposium Series on Computational Intelligence (SSCI)* (pp. 1-8). IEEE. 10.1109/SSCI.2016.7850280

Plattform Induistrie 4.0. (2016). Reference Architectural Model Industrie 4.0 (RAMI4.0) - An Introduction.

Pollack, M. E. (1990). Plans as complex mental attitudes. In *Intentions in communication* (pp. 77–103). London, UK: The MIT Press.

Popplewell, K. (2014). Enterprise interoperability science base structure. In K. Mertins, F. Bénaben, R. Poler, & J. Bourrières (Eds.), *Enterprise Interoperability VI: Interoperability for Agility, Resilience and Plasticity of Collaborations* (pp. 417–427). Cham, Switzerland: Springer International Publishing. doi:10.1007/978-3-319-04948-9_35

Porter, M. E. (1980). *Competitive Strategy*. New York: Free Press.

Portugal Startups. (2016) AddVolt makes Luis Simões the first company combining solar and kinetic energy in refrigerated trucks. Retrieved from http://portugalstartups.com/2016/09/addvolt-makes-luis-simoes-first-company-combining-solar-kinetic-energy-refrigerated-trucks/

Portugal Ventures. (2017) AddVolt prepares internationalisation with funds from three investors. Retrieved from http://www.portugalventures.pt/en/article/addvolt-prepares-internationalisation-funds-three-investors

Posada, J., Toro, C., Barandiaran, I., Oyarzun, D., Stricker, D., de Amicis, R., ... Vallarino, I. (2015). Visual computing as a key enabling technology for industrie 4.0 and industrial internet. *IEEE Computer Graphics and Applications*, *35*(2), 26–40. doi:10.1109/MCG.2015.45 PMID:25807506

Poslad, S., Buckle, P., & Hadingham, R. (2000). The FIPA-OS agent platform: Open source for open standards. In *Proceedings of the 5th International Conference and Exhibition on the Practical Application of Intelligent Agents and Multi-Agents* (Vol. 355, p. 368).

Post, J. E., & Altma, B. W. (1994). Managing the environmental change process: Barriers and opportunities. *Journal of Organizational Change Management*, *7*(4), 64–81. doi:10.1108/09534819410061388

Predix. (2016, November). *The Industrial Internet Platform.* Retrieved May 19, 2017, from: https://www.predix.com/

Prifti, L., Knigge, M., Kienegger, H., & Krcmar, H. (2017). A Competency Model for "Industrie 4.0" Employees. In *13th International Conference on Wirtschaftsinformatik*, St. Gallen, Switzerland (pp. 46–60).

Pritchard, A. (1969). Statistical Bibliography or Bibliometrics. *The Journal of Documentation*, *25*, 348–349.

Public Utilities Board Singapore. (2016). Managing the water distribution network with a Smart Water Grid. *Smart Water*, *1*, 1-13. Retrieved from http://smartwaterjournal.springeropen.com/articles/10.1186/s40713-016-0004-4

PWC. (2016). *Global Industry 4.0 Survey: Building the digital enterprise.* Retrieved May 18, 2017, from www.pwc.com/gx/en/industries/industries-4.0/landing-page/industry-4.0-building-your-digital-enterprise-april-2016.pdf

PwC. (2016). *Industry 4.0: Building the digital enterprise - Engineering and construction key findings.* Retrieved from https://www.pwc.com/gx/en/industries/industries-4.0/landing-page/industry-4.0-building-your-digital-enterprise-april-2016.pdf

PwC. (2017). *Industry 4.0 - Enabling Digital Operations Self Assessment.* Retrieved from https://i40-self-assessment.pwc.de/i40/

Qanbari, S., Zadeh, S. M., Vedaei, S., & Dustdar, S. (2014). CloudMan: A platform for portable cloud manufacturing services. In *IEEE International Conference on Big Data* (pp. 1006-1014). IEEE. 10.1109/BigData.2014.7004334

Quinteiro, P., Almeida, M., Dias, A. C., Araújo, A., & Arroja, L. (2014). The Carbon Footprint of Ceramic Products. In Assessment of Carbon Footprint in Different Industrial Sectors (Vol. 1, pp. 113-150). doi:10.1007/978-981-4560-41-2_5

RabbitMQ. (n.d.). Retrieved from https://www.rabbitmq.com/

Rao, B. B., Saluia, P., Sharma, N., Mittal, A., & Sharma, S. V. (2012). Cloud computing for Internet of Things & sensing based applications. In *2012 Sixth International Conference on Sensing Technology (ICST)* (pp. 374-380).

Rao, B., Saluia, P., Sharma, N., Mittal, A., & Sharma, S. (2012). Cloud computing for Internet of Things & sensing based applications. In *2012 Sixth International Conference on Sensing Technology (ICST)* (pp. 374-380). IEEE.

Rashid, M. A., Hossain, L., & Patrick, J. D. (2002). The Evolution of ERP Systems: A Historical Perspective. In Enterprise Resource Planning: Global Opportunities and Challenges. Hershey, PA: IGI Global.

Rastegari, A., Archenti, A., & Mobin, M. (2017, January). Condition based maintenance of machine tools: Vibration monitoring of spindle units. In *2017 Annual Reliability and Maintainability Symposium (RAMS)* (pp. 1-6). IEEE. 10.1109/RAM.2017.7889683

Ray, L. (2017). Cyber-Physical Systems. In Handbook of Research on Applied Cybernetics and Systems Science (p. 335).

Reddi, V., & Zhu, Y. (2017). The Red Future of Mobile Web Computing. *Communications of the ACM*, *60*(1), 52–53.

Reed, D., & Dongarra, J. (2015). Exascale computing and big data. *Communications of the ACM*, *58*(7), 56–68. doi:10.1145/2699414

Ren, F. W., Liu, X., Talari, S., Shafie-khah, M., & Catalão, J. P. (2017). Multi-objective Optimization Model of Source-Load-Storage Synergetic Dispatch for Building Energy System Based on TOU Price Demand Response. *IEEE Transactions on Industry Applications*.

Ren, L., Zhang, L., Wang, L., Tao, F., & Chai, X. (2017). Cloud manufacturing: Key characteristics and applications. *International Journal of Computer Integrated Manufacturing*, *30*(6), 501–515. doi:10.1080/0951192X.2014.902105

Rezaei, R., Chiew, T., & Lee, S. (2014). A review on E-business interoperability frameworks. *Journal of Systems and Software*, *93*, 199–216. doi:10.1016/j.jss.2014.02.004

Rezai, A., Keshavarzi, P., & Moravej, Z. (2017). Key management issue in SCADA networks: A review. *Engineering Science and Technology, an International Journal, 20*, 354-363.

Rezk, E., & Foufou, S. (2014). A survey of semantic web concepts applied in web services and big data. In *International Conference on Computer Systems and Applications* (pp. 767–772). Piscataway, NJ: IEEE Computer Society Press. 10.1109/AICCSA.2014.7073278

RightScale. (2017). *State of the Cloud Report*. Retrieved July 04, 2017, from http://assets.rightscale.com/uploads/pdfs/RightScale-2017-State-of-the-Cloud-Report.pdf

Robertson, I. (2008). Comment: How universities and graduates can thrive. *Independent*. Retrieved from http://www.independent.co.uk/student/career-planning/getting-job/comment-how-universities-and-graduates-can-thrive-918572.html

Robkin, M., Weininger, S., Preciado, B., & Goldman, J. (2015). Levels of conceptual interoperability model for healthcare framework for safe medical device interoperability. In *Symposium on Product Compliance Engineering* (pp. 1–8). Piscataway, NJ: IEEE Computer Society Press. 10.1109/ISPCE.2015.7138703

Roblek, V., Meško, M., & Krapež, A. (2016). A Complex View of Industry 4.0. *SAGE Open*, *6*(2), 1–11. doi:10.1177/2158244016653987

Rockwell Automation. (2014). *The Connected Enterprise Maturity Model* (CIE-WP002-EN-P). Retrieved from http://www.google.pt/url?sa=t&rct=j&q=&esrc=s&source=web&cd=1&cad=rja&uact=8&ved=0ahUKEwjvs5Dm75vSAhVIbBoKHS5VDTUQFgglMAA&url=http%3A%2F%2Fliterature.rockwellautomation.com%2Fidc%2Fgroups%2Fliterature%2Fdocuments%2Fwp%2Fcie-wp002_-en-p.pdf&usg=AFQjCNFIehm

Rogers, D. (2016). *The Digital Transformation Playbook: Rethink Your Business for the Digital Age*. New York: Columbia Business School Publishing. doi:10.7312/roge17544

Roland Berger. (2014). Industry 4.0 The new industrial revolution. *Roland Berger Strategy Consultants*. Retrieved from https://www.rolandberger.com

Romero, D., & Vernadat, F. (2016). Enterprise information systems state of the art: Past, present and future trends. *Computers in Industry*, *79*, 3–13. doi:10.1016/j.compind.2016.03.001

Rossi, B. (2016). Industry 4.0: top 10 critical trends for 2017. *Information Age*. Retrieved from http://www.information-age.com/123463810-123463810/

Rother, M., & Shook, J. (1999). *Learning to See: Value Stream Mapping to Add Value and Eliminate MUDA*. Lean Enterprise Institute.

Rousseau, V., Aube, C., & Savoie, A. (2006). Teamwork Behaviors: A Review and an Integration of Frameworks. *Small Group Research*, *37*(5), 540–570. doi:10.1177/1046496406293125

RS-Data-Sheet. (1999). Solar Panels, RS Data Sheet: RS stock numbers 194-082, 194-105, 194-111, 194-127,194-133, 194-149, 194-161, 194-183.

RTI. (2017). *Context DDS Professional*. Retrieved from https://www.rti.com/products/dds/benchmarks

Rueil-Malmaison. (2016, November 30). Schneider Electric delivers IOT solutions platform, education. *Waterworld*. Retrieved May 19, 2017, from http://www.waterworld.com/articles/iww/2016/11/schneider-electric-delivers-iot-solutions-platform-education.html

Russell, S., & Norvig, P. (2005). Artificial Intelligence a modern approach. *Learning*, *2*(3), 4.

Rüßmann, M., Lorenz, M., Gerbert, P., Waldner, M., Justus, J., Engel, P., & Harnisch, M. (2015). Industry 4.0: The future of productivity and growth in manufacturing industries. Boston Consulting Group.

Sadeghi, A. R., Wachsmann, C., & Waidner, M. (2015, June). Security and privacy challenges in industrial internet of things. In *Design Automation Conference (DAC), 2015 52nd ACM/EDAC/IEEE* (pp. 1-6). IEEE. 10.1145/2744769.2747942

Saedsayad. (2017). Decision Tree - Classification. Retrieved June 20, 2017, from http://www.saedsayad.com/decision_tree.htm

Saez-de-Ibarra, A., Milo, A., Gaztañaga, H., Etxeberria-Otadui, I., Rodríguez, P., Bacha, S., ... (2013). Analysis and comparison of battery energy storage technologies for grid applications. In 2013 IEEE Grenoble Conference (pp. 1-6).

Safavian, S. R., & Landgrebe, D. (1991). A survey of decision tree classifier methodology. *IEEE Transactions on Systems, Man, and Cybernetics*, *21*(3), 660–674. doi:10.1109/21.97458

Şahin, E. (2010). Teknoloji Transferi Yöntemleri Bağlamında Türkiye'de Yabancı Sermaye Yatırımları ile Üniversite Sanayi İşbirliğinin Gelişimi.

Sahlman, W. A., & Stevenson, H. H. (1991). Introduction. In W. A. Sahlman & H. H. Stevenson (Eds.), *The Entrepreneurial Venture*. Boston, MA: McGraw Hill; . doi:10.2190/Y2NP-MLD5-UQKX-XMNP

Saidani, O., & Nurcan, S. (2006). A role-based approach for modeling flexible business processes. *BPMDS*, *6*, 111–120.

Sakr, D., Baas, L., El-Haggar, S., & Huisingh, D. (2011). Critical success and limiting factors for eco-industrial parks: Global trends and Egyptian context. *Journal of Cleaner Production*, *19*(11), 1158–1169. doi:10.1016/j.jclepro.2011.01.001

Salas, E., Cooke, N. J., & Rosen, M. A. (2008). On Teams, Teamwork, and Team Performance: Discoveries and Developments. *Human Factors*, *50*(3), 540–547. doi:10.1518/001872008X288457 PMID:18689065

Salas, E., Sims, D. E., & Burke, C. S. (2005). Is there a "big five" in teamwork? *Small Group Research*, *36*(5), 555–599. doi:10.1177/1046496405277134

Salminen, L., Gustafsson, M.-L., Vilén, L., Fuster, P., Istomina, N., & Papastavrou, E. (2016). Nurse teacher candidates learned to use social media during the international teacher training course. *Nurse Education Today*, *36*, 354–359. doi:10.1016/j.nedt.2015.08.026 PMID:26358632

Salvendy, G. (2001). *Handbook of Industrial Engineering: Technology and operations management*. New York: Wiley. doi:10.1002/9780470172339

Samuel, A. L. (1959). Some Studies in Machine Learning Using the Game of Checkers. *IBM Journal*, *3*(3), 535–554. doi:10.1147/rd.33.0210

Sanayei, A., Mousavi, S. F., & Yazdankhah, A. (2010). Group decision making process for supplier selection with VIKOR under fuzzy environment. *Expert Systems with Applications*, *37*(1), 24–30. doi:10.1016/j.eswa.2009.04.063

Sanislav, T., & Miclea, L. (2012). Cyber-physical systems-concept, challenges and research areas. *Journal of Control Engineering and Applied Informatics*, *14*(2), 28–33.

SAP. (2016). *What is ERP?* Retrieved May 15, 2017, from https://www.sap.com/products/what-is-erp.html

Saravanan, S., & Babu, N. R. (2016). Maximum power point tracking algorithms for photovoltaic system – A review. *Renewable & Sustainable Energy Reviews*, *57*, 192–204. doi:10.1016/j.rser.2015.12.105

Savolainen, K., Brouwer, D., Fadeel, B., Fernandes, T., Kuhlbusch, T., & Landsiedel, R. … Pylkkänen, L. (2013). Nanosafety in Europe 2015-2025: Towards Safe and Sustainable Nanomaterials and Nanotechnology Innovations. Helsinki: Finnish Institute of Occupational Health.

Sawy, O. A., Kræmmergaard, P., Amsinck, H., & Vinther, A. L. (2016). How LEGO Built the Foundations and Enterprise Capabilities for Digital Leadership. *MIS Quarterly Executive*, *15*(2), 141–166.

Saxena, V., & Kumar, S. (2012). Impact of Coupling and Cohesion in Object-Oriented Technology. *Journal of Software Engineering and Applications*, *5*(9), 671–676. doi:10.4236/jsea.2012.59079

Say, J. B. (1803). A Treatise on Political Economy (C.R. Prinsep, trans. and Clement C. Biddle). Library of Economics and Liberty

Sayed, M. S., Lohse, N., & Madsen, A. L. (2015, July). SelSus: Towards a reference architecture for diagnostics and predictive maintenance using smart manufacturing devices. In *2015 IEEE 13th International Conference on Industrial Informatics (INDIN)* (pp. 1700-1705). IEEE.

Scherer, R., & Sharmak, W. (2011). Process Risk Management Using Configurable Process Models. In L.M. Camarinha-Matos, A. Pereira-Klen & H. Afsarmanesh (Eds.), *Adaptation and Value Creating Collaborative Networks: 12th IFIP WG 5.5 Working Conference on Virtual Enterprises PRO-VE 2011* (pp. 341-348). Springer.

Schleipen, M., Gilani, S.-S., Bischoff, T., & Pfrommer, J. (2016). OPC UA & Industrie 4.0 - Enabling Technology with High Diversity and Variability. *Procedia CIRP*, *57*, 315–320. doi:10.1016/j.procir.2016.11.055

Schmidt, R., Möhring, M., Härting, R. C., Reichstein, C., Neumaier, P., & Jozinović, P. (2015). Industry 4.0-potentials for creating smart products: empirical research results. In *International Conference on Business Information Systems* (pp. 16-27). Springer International Publishing. 10.1007/978-3-319-19027-3_2

Scholtz, J., Young, J., Drury, J. L., & Yanco, H. A. (2004, April). Evaluation of human-robot interaction awareness in search and rescue. In *2004 IEEE International Conference on Robotics and Automation ICRA'04* (Vol. 3, pp. 2327-2332). IEEE. 10.1109/ROBOT.2004.1307409

Schulte, P. A., Rinehart, R., Okun, A., Geraci, C. L., & Heidel, D. S. (2008). National Prevention through Design (PtD) Initiative. *Journal of Safety Research*, *39*(2), 115–121. doi:10.1016/j.jsr.2008.02.021 PMID:18454950

Schumacher, A., Erol, S., & Sihn, W. (2016). A Maturity Model for Assessing Industry 4.0 Readiness and Maturity of Manufacturing Enterprises. *Procedia CIRP*, *52*, 161–166. doi:10.1016/j.procir.2016.07.040

Schumpeter, J. A. (1934). *The Theory of Economic Development: an Inquiry into Profits, Capital, Credit, Interest, and the Business Cycle*. Cambridge, MA: Harvard University Press.

Schwab, K. (2015). *The Fourth Industrial Revolution: What It Means and How to Respond*. Foreign Affairs, Science & Technology. Retrieved January 06, 2018, from https://www.foreignaffairs.com/articles/2015-12-12/fourth-industrial-revolution

Schwab, K. (2015). Will the Fourth Industrial Revolution have a human heart? Retrieved December 9, 2016, from https://www.weforum.org/agenda/2015/10/will-the-fourth-industrial-revolution-have-a-human-heart-and-soul/

Schwab, K. (2016). The Fourth Industrial Revolution (VINT research report).

Schwab, L. (2016). The Fourth Industrial Revolution: what it means, how to respond, Global Agenda. *World Economic Forum*. Retrieved from https://www.weforum.org/agenda/2016/01/the-fourth-industrial-revolution-what-it-means-and-how-to-respond/

Schwartz, T., Feld, M., Burckert, C., Dimitrov, S., Folz, J., Hutter, D., & Straube, S. (2016). Hybrid teams of humans, robots, and virtual agents in a production setting. In *Proceedings - 12th International Conference on Intelligent Environments, IE 2016* (pp. 234-237). 10.1109/IE.2016.53

Schwarz, I. L. M., Schwarz, I. T. M., García, G. M. J., Hernández, G. G. J., & Hernández, R. J. G. (2016). Social Impact of Restrictions on Inventory Management. In *Proceedings ICIL2016*, AGH University of Science and Technology, Krakow, Poland (pp. 272-282).

Seitz, K. F., & Nyhuis, P. (2015). Cyber-physical production systems combined with logistic models–a learning factory concept for an improved production planning and control. *Procedia CIRP*, *32*, 92–97. doi:10.1016/j.procir.2015.02.220

SelSus. (2017, April 26). Retrieved May 19, 2017, from http://www.selsus.eu/project/

Sethi, P., & Sarangi, S. R. (2017). Internet of Things: Architectures, Protocols, and Applications. *Journal of Electrical and Computer Engineering*.

Shafiq, S. I., Sanin, C., Szczerbicki, E., & Toro, C. (2015). Virtual engineering object/virtual engineering process: A specialized form of cyber physical system for Industrie 4.0. *Procedia Computer Science*, *60*, 1146–1155. doi:10.1016/j.procs.2015.08.166

Shane, S., & Eckhardt, J. (2003). The individual-opportunity nexus. In Z. J. Acs & D. B. Audretsch (Eds.), *Handbook of Entrepreneurship Research* (pp. 161–194). Dordrecht, NL: Kluwer.

Shapero, A. (1975). The Displaced, Uncomfortable Entrepreneur. *Psychology Today*, *9*(6), 83–88.

Sharma, A., Kumar, M., & Agarwal, S. (2015). A Complete Survey on Software Architectural Styles and Patterns. *Procedia Computer Science*, *70*, 16–28. doi:10.1016/j.procs.2015.10.019

Sharma, R., & Panigrahi, P. (2015). Developing a roadmap for planning and implementation of interoperability capability in e-government. *Transforming Government: People. Process and Policy*, *9*(4), 426–447.

Shen, J., Jiang, C., & Li, B. (2015). Controllable Load Management Approaches in Smart Grids. *Energies*, *8*(10), 11187–11202. doi:10.3390/en81011187

Shi, H., Peng, S. Z., Liu, Y., & Zhong, P. (2008). Barriers to the implementation of cleaner production in Chinese SMEs: Government, industry and expert stakeholders' perspectives. *Journal of Cleaner Production*, *16*(7), 842–852. doi:10.1016/j.jclepro.2007.05.002

Shook, J., & Rother, M. (1999). Learning to See: Value Stream Mapping to Add Value and Eliminate MUDA.

Short, T. (1990). The New Industrial Engineering: Information Technology and Business Process Redesign. *MIT Sloan Management Review*.

Shrouf, F., Ordieres, J., & Miragliotta, G. (2014). Smart factories in Industry 4.0: A review of the concept and of energy management approached in production based on the Internet of Things paradigm. In *2014 IEEE International Conference on Industrial Engineering and* Engineering Management (IEEM) (pp. 697-701).

Shrouf, F., Ordieres, J., & Miragliotta, G. (2014). Smart Factories in Industry 4.0: A Review of the Concept and of Energy Management Approached in Production Based on the Internet of Things Paradigm. In *2014 IEEE International Conference on Industrial Engineering and Engineering Management (IEEM)* (pp. 697-701). IEEE.

Siemens. (2017). The Internet of Things is speaking. Retrieved May 19, 2017, from https://www.siemens.com/global/en/home/products/software/mindsphere.html

Siemens. (n.d.). Industrial Ethernet: the basis for efficient automation. Retrieved from http://w3.siemens.com/mcms/automation/en/industrial-communications/industrial-ethernet/Pages/industrial-ethernet.aspx

Silva, F., Arezes, P., & Swuste, P. (2016). Systematic design analysis and risk management on nanoparticles occupational exposure. *Journal of Cleaner Production*, *112*, 3331–3341. doi:10.1016/j.jclepro.2015.11.001

SIMATIC. (2017) Box PC. Retrieved May 18, 2017, from http://w3.siemens.com/mcms/pc-based-automation/en/industrial-pc/box-pc/Pages/Default.aspx

Simatupang, T. M., Wright, A. C., & Sridharan, R. (2002). The knowledge of coordination for supply chain integration. *Business Process Management Journal*, *8*(3), 289–308. doi:10.1108/14637150210428989

Skubch, H., Wagner, M., Reichle, R., Triller, S., & Geihs, K. (2010). Towards a Comprehensive Teamwork Model for Highly Dynamic Domains. *ICAART*, *10*(2), 121–127.

SMART CITY. (2016). Plataforma para transformação digital. Retrieved from http://www.pad.lsi.usp.br/ptd

Smit, J., Kreutzer, S., Moeller, C., & Carlberg, M. (2016). *Industry 4.0 - Study for the ITRE Committee*. Retrieved from http://www.europarl.europa.eu/RegData/etudes/STUD/2016/570007/IPOL_STU(2016)570007_EN.pdf

Smit, J., Kreutzer, S., Moeller, C., & Carlberg, M. (2016). *Industry 4.0 by Policy Department A*. Retrieved May 22, 2017, from http://www.europarl.europa.eu/studies

Smith, L. (1981). Citation analysis. *Library Trends*, *30*, 83–106.

SOCe. (2018). Industrial Networking Modules and Boards. Retrieved May 22, 2017, from http://soc-e.com/products/

Soldatos, J., Gusmeroli, S., Malo, P., & Di Orio, G. (2016). Internet of Things Applications in Future Manufacturing. In *Digitising Industry-Internet of Things Connecting the Physical* (pp. 153–183). Digital and Virtual Worlds.

Soldatos, J., Gusmeroli, S., Malo, P., & Di Orio, G. (2016). Internet of Things Applications in Future Manufacturing. In *Digitising the Industry. Internet of Things Connecting the Physical and Virtual World* (pp. 153–183). River Publishers.

Soliman, M. H. A. (2015, March 12). *Machine Vibration Analysis*. Retrieved June 8, 2017, from https://es.slideshare.net/MohammedHamedAhmedSo/machine-vibration-analysis

Solis, D. (2017). Digital Transformation – The six stages of digital transformation. Retrieved from http://www.briansolis.com/2017/01/definition-of-digital-transformation/

Sommer, L. (2015). Industrial revolution-industry 4.0: Are German manufacturing SMEs the first victims of this revolution? *Journal of Industrial Engineering and Management*, *8*(5), 1512–1532. doi:10.3926/jiem.1470

Sørensen, K. B., Christiansson, P., & Svidt, K. (2010). Ontologies to support RFID-based link between virtual models and construction components. *Computer-Aided Civil and Infrastructure Engineering*, *25*(4), 285–302. doi:10.1111/j.1467-8667.2009.00638.x

Speicher, M., Tenhaft, K., Heinen, S., & Handorf, H. (2015). Enabling Industry 4.0 with holobuilder. In Informatik 2015 (pp. 1561-1575).

Srewil, Y., & Scherer, R. J. (2013). Effective construction process monitoring and control through a collaborative cyber-physical approach. In L. M. Camarinha-Matos, & R. J. Scherer (Eds.), *Collaborative systems for reindustrialization: 14th IFIP WG 5.5 working conference on virtual enterprises, PRO-VE 2013*, Dresden, Germany, September 30-October 2 (pp. 172-179). Springer. 10.1007/978-3-642-40543-3_19

Srewil, Y., Kadolsky, M., & Scherer, R. J. (2012). Integrating RFID technology in dynamic construction process planning. In *Proceeding of Joint 29th CIB W78 the Beirut* (pp. 384–393). Lebonon.

Srikantha, P., & Kundur, D. (2017). Resilient Distributed Real-Time Demand Response via Population Games. *IEEE Transactions on Smart Grid*, *8*(6), 2532–2543. doi:10.1109/TSG.2016.2526651

Steijn, W., & Luiijf, E. (2016). *Emergent risk to workplace safety as a result of the use of robots in the work place. No. TNO 2016 R11488*. TNO.

Steijn, W., Van Der Vorm, J., & Luiijf, E. (2016). *Emergent risks to workplace safety as a result of IT connections of and between work equipment. No. TNO 2016 R11143*. TNO.

Stock, T., Bietz, T., Alva, D., Swat, M., Seliger, G., & Bähre, D. (2015). Design and Control of Manufacturing Systems for Enabling Energy-Efficiency and-Flexibility. *KES Transactions on Sustainable Design and Manufacturing*, *II*, 438–449.

Stock, T., & Seliger, G. (2016). Opportunities of sustainable manufacturing in Industry 4.0. *Procedia CIRP*, *40*, 536–541. doi:10.1016/j.procir.2016.01.129

Stojmenovic, I. (2014). Fog computing: A cloud to the ground support for smart things and machine-to-machine networks. In *Telecommunication Networks and Applications Conference (ATNAC), 2014 Australasian*, (pp. 117-122). 10.1109/ATNAC.2014.7020884

Stone, P., Kaminka, G. A., Kraus, S., & Rosenschein, J. S. (2010). Ad Hoc Autonomous Agent Teams: Collaboration without Pre-Coordination. In *Twenty-Fourth AAAI Conference on Artificial Intelligence*, July (pp. 1504-1509).

Storey, D. (1994). *Understanding the small business sector*. London, UK: Routledge.

Sungkur, R., & Daiboo, S. (2015). SOREST, A Novel Framework Combining SOAP and REST for Implementing Web Services. In *The Second International Conference on Data Mining, Internet Computing, and Big Data* (p. 22–34). Wilmington, DE: The Society of Digital Information and Wireless Communications.

Sungur, C. T., Spiess, P., Oertel, N., & Kopp, O. (2013): Extending bpmn for wireless sensor networks. In *2013 IEEE 15th Conference on Business Informatics (CBI)* (pp. 109–116). IEEE. 10.1109/CBI.2013.24

Supply Chain Council (2010). Supply Chain Operations Reference Model Version 10.0

SWSS. (2018). Smart Water Supply Systems. Retrieved from http://www.adp.pt/pt/sustentabilidade/inovacao-e-desenvolvimento/life-swss/?id=51

Sycara, K., & Sukthankar, G. (2006). Literature review of teamwork models (Technical Report CMU-RI-TR-06-50). Carnegie Mellon University.

Talcott, C. (2008). Cyber-physical systems and events. In M. Wirsing, J. Banâtre, M. Hölzl, & A. Rauschmayer (Eds.), *Software-intensive systems and new computing paradigms* (pp. 101–115). Springer. doi:10.1007/978-3-540-89437-7_6

Tamaki, H., Sanches, G., & Gonçalves, O. (2006). A medição setorizada como instrumento de gestão da demanda de água em sistemas prediais – Estudo de caso: programa de uso racional da água da Universidade de São Paulo. Associação Nacional de Tecnologia do Ambiente Construído.

Tambe, M. (1997). Towards flexible teamwork. *Journal of Artificial Intelligence Research*, *7*, 83–124.

Tannenbaum, S. I., Beard, R. L., & Salas, E. (1992). Team building and its influence on team effectiveness: An examination of conceptual and empirical developments. In K. Kelley (Ed.), *Issues, theory, and research in industrial/organizational psychology* (pp. 117–153). Amsterdam: Elsevier. doi:10.1016/S0166-4115(08)62601-1

Tao, F., Cheng, Y., Zhang, L., & Nee, A. (2017, June). Advanced manufacturing systems: Socialization characteristics and trends. *Journal of Intelligent Manufacturing*, *28*(5), 1079–1094. doi:10.100710845-015-1042-8

Tarazón, C. J. C., Hernández, G. G. J., Lee, B. C., García, G. M. J., & Hernández, R. J. G. (2016). *Innovation Manager, Management Knowledge Generation, Logistics and some Multicriteria models. In Eureka 2016* (pp. 1–13). México: Torreón.

Team, O. R. (2011). *Big data now: current perspectives*. OReilly Media.

Technical Committee ISO/TC 229. (2012). ISO/TS 12901-1: Nanotechnologies — Occupational risk management applied to engineered nanomaterials — Part 1: Principles and approaches. Geneva: ISO.

Teizer, J., Cheng, T., & Fang, Y. (2013). Location tracking and data visualization technology to advance construction ironworkers' education and training in safety and productivity. *Automation in Construction*, *35*(0), 53–68. doi:10.1016/j.autcon.2013.03.004

Thames, L., & Schaefe, D. (2016). Software-defined cloud manufacturing for Industry 4.0. *Procedia CIRP*, *52*, 12–17. doi:10.1016/j.procir.2016.07.041

Thingspeak. (2017). Retrieved May 19, 2017, from https://thingspeak.com/

Thingworx. (2017). Retrieved May 22, 2017, from https://www.thingworx.com

Thoma, M., Antonescu, A. F., Mintsi, T., & Braun, T. (2013). Linked services for enabling interoperability in the sensing enterprise. In *International IFIP Working Conference on Enterprise Interoperability* (pp. 131-144). Springer. 10.1007/978-3-642-36796-0_12

Tidd, J., Bessant, J., & Pavitt, K. (2005). *Managing innovation: integrating technological, market and organizational change*. UK: John Wiley & Sons Ltd.

Timmons, J. A. (1999). *New venture creation. Entrepreneurship for the 21st century*. Singapore: Irwin McGraw-Hill.

Torres, J. M. (2008). Analyzing risk and uncertainty for improving water distribution system security from malevolent water supply contamination events [Master's Thesis]. Office of Graduate Studies of Texas A&M University, TX.

Tran, H., Zdun, U., Oberortner, E., Mulo, E., & Dustdar, S. (2012). Compliance in service-oriented architectures: A model-driven and view-based approach. *Information and Software Technology*, *54*(6), 531–552. doi:10.1016/j.infsof.2012.01.001

Tsoutsa, P., Fitsilis, P., & Ragos, O. (2017). Role Modeling of IoT Services in Industry Domains. In *Proceedings of the 2017 International Conference on Management Engineering, Software Engineering and Service Sciences* (pp. 290–295). 10.1145/3034950.3035002

Tuegel, E., Ingraffea, A., Eason, T., & Spottswood, S. (2011). Reengineering Aircraft Structural Life Prediction Using a Digital Twin. International Journal of Aerospace Engineering.

U.S. Department of Commerce. (2014). *How does Commerce define Sustainable Manufacturing?* International Trade Administration.

Ubuntu Blog. (2017, February 16). Lime Microsystems and Canonical announce LimeNET crowdfunding. Retrieved June 20, 2017, from https://insights.ubuntu.com/2017/02/16/lime-microsystems-and-canonical-announce-limenet-crowdfunding/

Uhl, A., & Gollenia, L. (2016). *Digital Enterprise Transformation: A Business-Driven Approach to Leveraging Innovative IT*. USA: Routledge.

United Nations. (1987). Report of the World Commission on Environment and Development.

United Nations. (2015). *Paris Agreement*. Retrieved December 17, 2017, from https://unfccc.int/files/essential/background/convention/application/pdf/english/paris/agreement.pdf

Valdez, A. C., Brauner, P., Schaar, A. K., Holzinger, A., & Ziefle, M. (2015). Reducing complexity with simplicity-usability methods for industry 4.0. In *Proceedings 19th Triennial Congress of the IEA* (pp. 14-21).

van Broekhuizen, P., van Broekhuizen, F., Cornelissen, R., & Reijnders, L. (2011). Use of nanomaterials in the European construction industry and some occupational health aspects thereof. *Journal of Nanoparticle Research*, *13*(2), 447–462. doi:10.100711051-010-0195-9

Van Wissen, A., Gal, Y., Kamphorst, B. A., & Dignum, M. V. (2012). Human-agent teamwork in dynamic environments. *Computers in Human Behavior*, *28*(1), 23–33. doi:10.1016/j.chb.2011.08.006

Vandoorn, T. L., Meersman, B., Kooning, J. D., & Vandevelde, L. (2012). Analogy Between Conventional Grid Control and Islanded Microgrid Control Based on a Global DC-Link Voltage Droop. *IEEE Transactions on Power Delivery*, *27*(3), 1405–1414. doi:10.1109/TPWRD.2012.2193904

Vardakas, J. S., Zorba, N., & Verikoukis, C. V. (2015). A Survey on Demand Response Programs in Smart Grids: Pricing Methods and Optimization Algorithms. *IEEE Communications Surveys and Tutorials*, *17*(1), 152–178. doi:10.1109/COMST.2014.2341586

Vargas, A., Boza, A., Cuenca, L., & Sacala, I. (2013). Inter-Enterprise Architecture and Internet of the Future. In L. M. Camarinha-Matos, S. Tomic, & P. Graça (Eds.), *Technological Innovation for the Internet of Things. DoCEIS 2013. IFIP Advances in Information and Communication Technology* (Vol. 394). Springer. doi:10.1007/978-3-642-37291-9_3

VDI. VDE, & ZVEI. (2015). *GMA Status Report: Reference Architecture Model Industrie 4.0 (RAMI 4.0)*. Retrieved July 04, 2017, from https://www.zvei.org/fileadmin/user_upload/Presse_und_Medien/Publikationen/2016/januar/GMA_Status_Report__Reference_Archtitecture_Model_Industrie_4.0__RAMI_4.0_/GMA-Status-Report-RAMI-40-July-2015.pdf

Verborgh, R., Harth, A., Maleshkova, M., Stadtmüller, S., Steiner, T., Taheriyan, M., & Van de Walle, R. (2014). Survey of semantic description of REST APIs. In C. Pautasso, E. Wilde, & R. Alarcon (Eds.), *REST: Advanced Research Topics and Practical Applications* (pp. 69–89). New York, NY: Springer. doi:10.1007/978-1-4614-9299-3_5

Verma, P., Kumar, A., Rathod, N., Jain, P., Mallikarjun, S., Subramanian, R., ... & Sundaresan, R. (2015, October). Towards an IoT based water management system for a campus. In *2015 IEEE First International Smart Cities Conference (ISC2)* (pp. 1-6). IEEE.10.1109/ISC2.2015.7366152

Vervoort, M. B. H. J. (2012). *A comparison of risk assessment methods in order to determine the risk of occupational used nanomaterials in a research environment*. NSPOH.

Vidgen, R., & Wang, X. (2006). From business process management to business process ecosystem. *Journal of Information Technology*, *21*(4), 262–271. doi:10.1057/palgrave.jit.2000076

Vuksanović, D., Ugarak, J., & Korčok, D. (2016). Industry 4.0: The future concepts and new vision of factory of the future development. *Paper presented at International Scientific Conference on ICT and e-business related research*, Singidunum University. 10.15308/Sinteza-2016-293-298

W3C. (2016). *Extensible Markup Language (XML)*. Retrieved June 11, 2017, from https://www.w3.org/XML/

Waibel, M. W., Steenkamp, L. P., Moloko, N., & Oosthuizen, G. A. (2017). Investigating the Effects of Smart Production Systems on Sustainability Elements. *Procedia Manufacturing*, *8*, 731–737. doi:10.1016/j.promfg.2017.02.094

Wang, L., & Alexander, C. (2016). Additive Manufacturing and Big Data. *International Journal of Mathematical, Engineering and Management Sciences*, *1*(3), 107–121. Retrieved from http://ijmems.in/assets/2.-ijmems-16-015-vol.-1,-no.-3,-107–121,-2016.pdf

Wang, R., & Chen, N. (2016, December). A survey of condition-based maintenance modeling of multi-component systems. In *2016 IEEE International Conference on Industrial Engineering and Engineering Management (IEEM)* (pp. 1664-1668). IEEE. 10.1109/IEEM.2016.7798160

Wang, S., Wan, J., Li, D., & Zhang, C. (2016). Implementing Smart Factory of Industrie 4.0: An Outlook. *International Journal of Distributed Sensor Networks*, *12*(1).

Wang, W., Tolk, A., & Wang, W. (2009). The levels of conceptual interoperability model: Applying systems engineering principles to M&S. In G. Wainer, C. Shaffer, R. McGraw & M. Chinni (Eds.), *Spring Simulation Multiconference*. San Diego, CA: Society for Computer Simulation International.

Wang, J., Huang, J., Chen, W., Liu, J., & Xu, D. (2016). Design of IoT-based energy efficiency management system for building ceramics production line. In *Proceedings of the 2016 IEEE 11th Conference on Industrial Electronics and Applications, ICIEA 2016* (pp. 912-917). 10.1109/ICIEA.2016.7603712

Wang, K. (2016). Intelligent Predictive Maintenance (IPdM) System–Industry 4.0 Scenario. *WIT Transactions on Engineering Sciences*, *113*(1), 259–268.

Wang, P., Gao, R. X., & Fan, Z. (2015). Cloud Computing for Cloud Manufacturing: Benefits and Limitations. ASME. *Journal Manufacturing Scientific Engineering*, *137*(4).

Wang, S., Wan, J., Li, D., & Zhang, C. (2016). Implementing smart factory of industrie 4.0: An outlook. *International Journal of Distributed Sensor Networks*, *12*(1).

Wang, W., Xu, Y., & Khanna, M. (2011). A survey on the communication architectures in smart grid. *Computer Networks, 55*(15), 3604–3629. doi:10.1016/j.comnet.2011.07.010

Wang, X. V., & Xu, X. W. (2013). An interoperable solution for Cloud manufacturing. *Robotics and Computer-integrated Manufacturing, 29*(4), 232–247. doi:10.1016/j.rcim.2013.01.005

Want, R., Schilit, B., & Jenson, S. (2015). Enabling the Internet of Things. *IEEE Computer, 48*(1), 28–35. doi:10.1109/MC.2015.12

Wei, L., Zhu, H., Cao, Z., Dong, X., Jia, W., Chen, Y., & Vasilakos, A. V. (2014). Security and privacy for storage and computation in cloud computing. *Information Sciences, 258*(10), 371–386. doi:10.1016/j.ins.2013.04.028

Weill, P., & Woerner, S. L. (2015). Thriving in an Increasingly Digital Ecosystem. *MIT Sloan Management Review, 56*(4), 27–34. doi:10.1287/isre.1100.0318

Weiser. (1992). Ubiquitous Computing. Retrieved from http://www.ubiq.com/hypertext/weiser/UbiHome.html

Weissberger, A. (2011). *IEEE CIO Says Cloud Interoperability a Bigger Problem than Security!* Retrieved February 2, 2017, from http://community.comsoc.org/blogs/alanweissberger/ieee-cio-says-cloud-interoperability-bigger-problem-security

Weske, M. (2007). *Business Process Management: Concepts, Languages, Architectures.* Germany: Springer.

Westerman, G., Bonnet, D., & McAfee, A. (2014). *Leading Digital: Turning technology into business transformation.* Boston, MA: Harvard Business Press.

Westermann, T., Anacker, H., Dumitrescu, R., & Czaja, A. (2016). Reference Architecture and Maturity Levels for Cyber-Physical Systems in the Mechanical Engineering Industry. In *International Symposium on Systems Engineering (ISSE)* (pp. 1–6). Edinburgh, Scotland: IEEE. 10.1109/SysEng.2016.7753153

Wickham, P. A. (2001). Strategic Entrepreneurship: A Decision-Making Approach to New Venture Creation and Management. 2nd Edition, Financial Times/Prentice Hall, London

Wikipedia. (n.d.). *Industrial Internet Consortium.* Retrieved December 17, 2017, from https://en.wikipedia.org/wiki/Industrial/Internet/Consortium

Wolf-Dieter, L. (2011). Industrie 4.0: Mit dem Internet der Dinge auf dem Weg zur 4. industriellen Revolution. *VDI nachrichten, 13.* Retrieved July 7, 2017, from http://www.wolfgang-wahlster.de/wordpress/wp-content/uploads/Industrie_4_0_Mit_dem_Internet_der_Dinge_auf_dem_Weg_zur_vierten_industriellen_Revolution_2.pdf

Womack, J. P., & Jones, D. T. (1996). *Lean thinking: banish waste and create wealth in your corporation.* New York, NY: Simon & Schuster.

Womack, J. P., Jones, D. T., & Roos, D. (1990). *The Machine that Changed the World: The Story of Lean Production* (1st ed.). New York: Harper Perennial.

Wong, J. C., & Solon, O. (2017, May 12). Massive ransomware cyber-attack hits nearly 100 countries around the world. *The Guardian.*

Wong, K. V., & Hernandez, A. (2012). A Review of Additive Manufacturing. *ISRN Mechanical Engineering, 2012,* 1–10. doi:10.5402/2012/208760

World Economic Forum. (2016). *The Future of Jobs.*

Wu, D., David, W. R., Wang, L., & Schaefer, D. (2014). Cloud-Based Manufacturing: Old Wine in New Bottles? In *Proceedings of the 47th CIRP Conference on Manufacturing Systems* (Vol. 17, pp. 94-99). 10.1016/j.procir.2014.01.035

Wu, D., Greer, M., Rosen, D., & Schaefer, D. (2013). Cloud manufacturing: Strategic vision and state-of-the-art. *Journal of Manufacturing Systems, 32*(4), 564–579. doi:10.1016/j.jmsy.2013.04.008

Wu, F., Zilberstein, S., & Chen, X. (2011). Online planning for ad hoc autonomous agent teams. In *IJCAI International Joint Conference on Artificial Intelligence* (pp. 439–445). doi:10.5591/978-1-57735-516-8/IJCAI11-081

Xu, L. D. (2011). Enterprise Systems: State-of-the-Art and Future Trends. *IEEE Transactions on Industrial Informatics, 4*(7), 630–640. doi:10.1109/TII.2011.2167156

Xu, X. (2012). From cloud computing to cloud manufacturing. *Robotics and Computer-integrated Manufacturing, 28*(1), 75–86. doi:10.1016/j.rcim.2011.07.002

Yadgarova, Y., & Taratukhin, V. (2016). An Interoperable Cloud Environment of Manufacturing Control System. In Mertins K., Jardim-Gonçalves R., Popplewell K., Mendonça J. (Ed.) *Enterprise Interoperability VII. Proceedings of the I-ESA Conferences (Vol. 8)*. Cham: Springer. 10.1007/978-3-319-30957-6_1

Yang, G., & Zhou, X. (2013). Cyber-physical systems.

Yan, Y., Qian, Y., Sharif, H., & Tipper, D. (2013). A Survey on Smart Grid Communication Infrastructures: Motivations, Requirements and Challenges. *IEEE Communications Surveys and Tutorials, 15*(1), 5–20. doi:10.1109/SURV.2012.021312.00034

Yen, J., Fan, X., Sun, S., Hanratty, T., & Dumer, J. (2006). Agents with shared mental models for enhancing team decision makings. *Decision Support Systems, 41*(3), 634–653. doi:10.1016/j.dss.2004.06.008

Yi, J., LeBouf, R. F., Duling, M. G., Nurkiewicz, T., Chen, B. T., Schwegler-Berry, D., ... Stefaniak, A. B. (2016). Emission of particulate matter from a desktop three-dimensional (3D) printer. *Journal of Toxicology and Environmental Health. Part A., 79*(11), 453–465. doi:10.1080/15287394.2016.1166467 PMID:27196745

Yoshikawa, K., Kawasak, H., Yoshida, W., Irie, T., Konishi, K., & Nakano, K., … & Yamamoto, K. (2017). Silicon heterojunction solar cell with interdigitated back contacts for a photoconversion efficiency over 26%. *Nature Energy*, 2.

Yue, T. (2011). Internet of things technology application in the food supply chain management. In *2011 International Conference on E-Business and E-Government (ICEE)* (pp. 1-4). IEEE. 10.1109/ICEBEG.2011.5882613

Zanero, S. (2017). Cyber-physical systems. *IEEE Computer, 50*(4), 14–16. doi:10.1109/MC.2017.105

Zaslavsky, A., Perera, C., & Georgakopoulos, D. (n.d.). Sensing as a service and big data. arXiv:1301.0159

Zawadzki, P., & Żywicki, K. (2016). Smart product design and production control for effective mass customization in the Industry 4.0 concept. *Management and Production Engineering Review, 7*(3), 105–112. doi:10.1515/mper-2016-0030

Zerenler, M., Haslioglu, S. B., & Sezgin, M. (2008). Intellectual capital and innovation performance: Empirical evidence in the Turkish automotive supplier. *Journal of Technology Management & Innovation, 3*(4), 31–41. doi:10.4067/S0718-27242008000200003

Zhang, Q., Cheng, L., & Boutaba, R. (2011). Cloud computing: state-of-the-art and research challenges. *Journal of internet services and applications, 1*(1), 7-18.

Zhang, L., Luo, Y.-L., Tao, F., Ren, L., & Guo, H. (2010). Key technologies for the construction of manufacturing cloud. *Jisuanji Jicheng Zhizao Xitong, 16*(11), 2510–2520.

Zhang, L., Luo, Y., Tao, F., Li, B., Ren, L., Zhang, X., ... Liu, Y. (2011). Cloud manufacturing: A new manufacturing paradigm. *Enterprise Information Systems, 8*(2), 167–187. doi:10.1080/17517575.2012.683812

Zhang, Y., Zhang, G., Liu, Y., & Hu, D. (2017). Research on services encapsulation and virtualization access model of machine for cloud manufacturing. *Journal of Intelligent Manufacturing*, *28*(5), 1109–1123. doi:10.100710845-015-1064-2

Zhang, Z., Wu, C., & Cheung, D. (2013). A survey on cloud interoperability: Taxonomies, standards, and practice. *Performance Evaluation Review*, *40*(4), 13–22. doi:10.1145/2479942.2479945

Zhao, F. (2006). *Entrepreneurship and Innovations in E-business: An Integrative Perspective*. Hershey, PA: IGI Global; . doi:10.4018/978-1-59140-920-5

Zhou, J. (2013). Digitalization and intelligentization of manufacturing industry. *Advances in Manufacturing*, *1*(1), 1–7. doi:10.100740436-013-0006-5

Zhou, J., Leppanen, T., Harjula, E., Ylianttila, M., Ojala, T., Yu, C., ... Yang, L.-T. (2013). CloudThings: A common architecture for integrating the Internet of Things with Cloud Computing. In *Proceedings of the 2013 IEEE 17th International Conference on Computer Supported Cooperative Work in Design* (pp. 651-657). 10.1109/CSCWD.2013.6581037

Zhou, K., Liu, T., & Zhou, L. (2015). Industry 4.0: Towards future industrial opportunities and challenges. In *12th International Conference on* Fuzzy Systems and Knowledge Discovery (FSKD) (pp. 2147-2152). IEEE.

Zhou, L., Rodrigues, J., & Oliveira, L. (2012). QoE-driven power scheduling in smart grid: Architecture, strategy, and methodology. *IEEE Communications Magazine*, *50*(5), 136–141. doi:10.1109/MCOM.2012.6194394

Zimmermann, O., Tomlinson, M., & Peuser, S. (2012). *Perspectives on Web Services: Applying SOAP, WSDL and UDDI to Real-World Projects*. New York, NY: Springer Science & Business Media.

Zissis, D., & Lekkas, D. (2012). Addressing cloud computing security issues. *Future Generation Computer Systems*, *28*(3), 583–592. doi:10.1016/j.future.2010.12.006

About the Contributors

Luís Ferreira is currently a lecturer in the Hight School of Technology at the Polytechnic Institute of Cavado and Ave, Portugal. He holds a Dipl. in the field of Mathematics and Computer Science, an M.Sci. in the field of Distributed Systems, Computer Architecture, Communication and Networks. Integration of Information Systems and Architectures and Cloud services are his main research areas. Professionally he developed and coordinated several projects on enterprises reconfiguration and integration as CTO of Sidereus Company. He coordinated the IPCA's eU project, Portuguese wireless platform for Academic Campus, and was pivot for IPCA's VOIP project. Academically, was member of INESC Porto research team, actually is a research member of IPCA 2Ai – Artificial Intelligence Laboratory, heads the Master's Degree in Computer Engineering and teaches subjects of Computer Sciences, Software Engineering, Information Systems and Architectures, to undergraduate and pos-graduate studies.

Nuno Alberto Ferreira Lopes holds a PhD in Computer Engineering from the University of Minho and a Bachelor in Systems and Computer Engineering from the same university. He is currently an Assistant Professor at the Polytechnic Institute of Cávado and Ave, coordinating disciplines on Hardware, Communications and Operating Systems. Previously, he was the Director of the Bsc Degree in Computer Systems Engineering and member of the directing board for the Master Degree in Computer Engineering and also for the Master Degree in Integrated Management Systems - Quality, Environment and Safety. He lectured disciplines in Computer Architecture, Operating Systems, Computer Networks, Data Communications, Programming Languages, Concurrent and Distributed Programming, Information Security and Auditing, and Information Systems for Quality, Environment and Security. His areas of interest include Information Security, Business Process Management, and Parallel and Distributed Computing Systems.

Joaquim G. Pereira Silva is currently an Assistant Professor in the School of Technology at the Polytechnic Institute of Cávado and Ave, Portugal. He holds a Dipl. Eng. in the field of Systems and Informatics Engineering and an MSc degree in the field of Industrial Engineering. Currently, He is conducting research studies for completing a PhD degree in the field of Computer Science. He teaches subjects related with Information Systems, Software Engineering and Decision Support Systems to undergraduate and post-graduate studies. He worked for about 10 years as an information systems manager in industry before initiating the academic career. His research work covers the areas of information systems, business analytics, and business process management.

Goran D. Putnik, Dr. Habil., DrSc., Full Professor, Department of Production and Systems Engineering, University of Minho, Portugal. His scientific and engineering interests are distributed, agile and virtual production systems and enterprises design and management theory, ICT in manufacturing, concurrent and collaborative engineering, formal theory of production systems and enterprises, and complexity management in organizations. His publishing record comprises more than 300 publications in international and national journals and conferences, including 13 books, of which the "Encyclopedia of Networked and Virtual Organizations" is distinguished. He serves as a member of Editorial Board for several International Journals. He is an associate member of the International Academy for Production Engineering CIRP.

Maria Manuela Cruz-Cunha is currently an Associate Professor with the School of Technology at the Polytechnic Institute of Cavado and Ave, Portugal. She holds a Dipl. Eng. (5 years) in the field of systems and informatics engineering, an M.Sci. in the field of Computer Integrated Manufacturing and a Dr.Sci in of Production Systems Engineering. She teaches subjects related with Information Systems, Information Technologies and Organizational Models to undergraduate and post-graduate studies. She has authored and edited 25 books and her work appears in more than 150 papers published in journals, book chapters and conference proceedings. She is founder and conference chair of the "CENTERIS – Conference on ENTERprise Information Systems", "ViNOrg – Conference on Virtual and Networked Organizations Emergent Technologies and Tools" and "SeGAH – Serious Games and Applications for Health". She is the editor-in-chief of the "International Journal of Web Portals" and of "Information and Communication Technologies for the Advanced Enterprise: An International Journal."

* * *

Faustino Alarcón is an associate professor in Operations Management and Operations Research at the Universitat Politècnica de València (UPV) in Spain. He is an industrial engineer and received his PhD in Industrial Engineering from the UPV. He is a member of Research Centre on Production Management and Engineering (CIGIP) and has participated in different Spanish and European Projects. His research is focused on Enterprise Modelling and Integration, BPM, Collaborative OP, Collaborative Operations Planning in Supply and Distribution Networks. He has published in international journals and conference proceedings.

Duarte Almeida is currently Process Development Engineer at Bosch Termotecnologia, S.A. Former Researcher in the Robotics and Intelligent Systems Unit of INESC TEC (Institute for Systems and Computer Engineering, Technology and Science). Graduated at 2012 by the Faculty of Engineering of the University of Porto in Electrical and Computer Engineering.

Marisa Almeida is the head of the environmental and sustainability department at the Technological Centre for Ceramics and Glass Industries. Her professional experience includes: environmental sampling and assessment, environmental training, risk management, environmental system (ISO 14001), life cycle assessment (ISO 14040 and 14044), Environmental Product Declaration (EPD) (ISO 14025, ISO21930 and EN 15804), energy systems (ISO 50001). She was project leader in European project related with waste management, eco-design, environmental systems and national projects for environmental improvement, Environmental Product Declaration, in many sectors of activity like ceramics, natural stone and concrete industries. Marisa holds an MSc in Environmental engineer and is member of technical commissions including CEN/TC 350 – Sustainability of construction works, and national CT 150 (environmental management). She is certified auditor for ISO14001, ISO 50001 and EDP verifier. Member of the Technical Committee of the program operator DAP Habitat - Environmental Product Declarations.

Ander Arbelaiz studied Computer Science at the University of the Basque Country (UPV/EHU) (2008-2013). In 2014 he received a M.Sc. on "Computer Engineering and Intelligent Systems". He joined Vicomtech (http://www.vicomtech.org) as a research assistant in 2013. His research interests are computer graphics, volume rendering and GPU computing. Currently, he is working at the "Industry and advance manufacturing" department in Vicomtech.

António José Caetano Baptista has got his Mechanical Engineering degree at the Faculty of Sciences and Technology of the University of Coimbra in 2002, and his PhD degree in Production Technology, Computational Mechanics field, at the same institution, in 2007. After that, he conducted work, from January 2007 until May 2011, at an automotive industry technical center (Sodecia Group) on project design of tools for automotive sheet metal parts and in international simultaneous engineering projects (with FIAT and FORD). He had been responsible for continuous improvement programs (kaizen activities, root cause analysis, cost reduction analysis, productivity improvement). He has joined INEGI R&D Institute in 2011, where he assumed responsibilities as project manager and senior researcher. He has been responsible for several multi-disciplinary projects with different industry sectors (transports, machine-tools, metalworking, etc.). His main research topics involve product development methods and tools, as well as, sustainability oriented assessment frameworks based on lean principles. He has coordinated the creation of several original methodologies: Lean Design-for-X assessment approach; ecoPROSYS, devoted to eco-efficiency assessment; MSM-Multi-Layer Stream Mapping, devoted to multi-domain system performance assessment, which includes Resource, Energy and Operations Efficiency Assessments; and MAESTRI Efficiency Framework cross linking Eco-Efficiency with Operational Efficiency.

João Barata is the head of Information Systems and Technologies (IS&T) at CTCV - Technological Centre for Ceramic and Glass Industries. João is a researcher at CISUC, University of Coimbra, Invited Assistant Professor at ESTGOH - Polytechnic Institute of Coimbra, and Adjunct Professor at Miguel Torga Institute, Portugal. Prior to this he was lecturer at the Catholic University of Portugal, from 2002 to 2005. He has an experience of 19 years as an IS&T consultant and has worked with over 250 public and private organizations in Europe, America, and Africa, including multinationals such as KPMG, BP, Exide, Ericson, Daimler, and Renault. João is an Industrial Engineer, holds a M.Sc. in Informatics Engineering, and a Ph.D. in Information Systems from the University of Coimbra. His main research interests are IS&T and quality management systems synergies, enterprise modelling, mobile information systems, business process management, supply chain management, and Industry 4.0. ORCID: 0000-0002-7456-594X.

Andoni Bilbao is an engineer in Industrial Electronics focused on automation systems for the automotive sector, currently in charge of ICS management for custom machines manufacturing at Gaindu Automation.

Andrés Boza is an Assistant Professor at the Universitat Politècnica de València (UPV) in Spain. He teaches at the Computer Science School on Management Information Systems. He is a computer science engineer and received his Ph.D. in Supply Chain Management and Enterprise Integration from the UPV. He is member of Research Centre on Production Management and Engineering (CIGIP) where he has participated in the doctoral program and in national and international projects about information systems and decision support systems for production management, supply chain management and performance measurement. He has published several papers in books, journals and international conferences in these fields and he is member of the editorial team of the "International Journal of Production Management and Engineering."

Gonçalo Cardeal has an MSc. in Mechanical Engineering from Técnico Lisboa, Universidade de Lisboa. He is currently a PhD candidate in engineering design and advanced manufacturing (EDAM), MIT-Portugal programme, IST, Portugal. He has been involved in European research projects in the field of resource and energy efficiency, such as MAESTRI – Total Resource and Energy Efficiency in Process Industries (H2020, SPIRE 2015). His research focuses on Industrial sustainability, more specifically in the conception and development of methodologies and tools to foster energy and eco-efficiency of production systems. Furthermore, he developed a predictive model, using neural networks, to estimate energy consumption in the injection moulding industry, and has been part of the development of the eco-efficiency assessment tool used in MAESTRI (ecoPROSYS).

Francisco J. A. Cardoso has a PhD (Science) from the University of Manchester (UK), and a MSc in Electrical Engineering from IST-University of Lisbon (Portugal). Currently, holds a position as Senior Lecturer at the Faculty of Science and Technology (Physics Department) of the University of Coimbra, lecturing courses and conducting research in the areas of Embedded Systems, Wireless Sensor Networks, Intelligent Systems and M2M Integration in Industrial Instrumentation Systems. As a Faculty member, he has extensively supervised postgraduate work, the vast majority of both PhD and MSc thesis relating to innovative projects aiming at solving specific industrial problems, thus promoting technology transfer in the due course. Further to many technology transfer initiatives, given his commitment to a vision of area development based on knowledge-intensive businesses, he served as Executive Director of Instituto Pedro Nunes —— a not-for-profit University-Industry interface organization led by the University of Coimbra ——, and as CTO of Eneida-Wireless & Sensors, S.A., an industrial firm operating in the area of smart sensor networks applied to asset management in electricity grids.

José C. Delgado is an Associate Professor at the Computer Science and Engineering Department of the Instituto Superior Técnico (University of Lisbon), in Lisbon, Portugal, where he earned the Ph.D. degree in 1988. He lectures courses in the areas of Computer Architecture, Information Technology and Service Engineering. He has performed several management roles in his faculty, namely Director of the Taguspark campus, near Lisbon, and Coordinator of the B.Sc. and M.Sc. in Computer Science and Engineering at that campus. He has been the coordinator of and researcher in several international research projects. As an author, his publications include one book, 20 book chapters and more than 50 papers in international refereed conferences and journals.

Rogério Dionisio received his B.Sc., M.Sc. and Ph.D. degrees in Electronics and Telecommunications Engineering from the University of Aveiro (Portugal) in 1997, 2004 and 2014, respectively. Since 1999 he is an Assistant Professor at the Polytechnic Institute of Castelo Branco. He participates in several National and European research projects on wireless communications (AGILE, COGEU, CREW, White Spaces Access, Fed4FIRE, WISHFUL) and European networks of excellence (BONE, EURO-FOS). He received the "New Frontiers of Engineering" award and was also granted Telecommunication Specialist in 2015, by the Portuguese Engineers Institution. In 2015, He co-founded Allbesmart, a start-up and SME focused on smart-territories solutions. He is author of several journal and conference publications, and his main research interests are advanced signal processing techniques for wideband communication systems, radio spectral coexistence analysis (PHY level) and wireless sensor networks.

Marco Estrela, MSc Chemical Engineering, MBA International Business, PhD candidate in Sustainable Energy Systems, MIT-Portugal programme, IST, Portugal. Senior researcher with over 20 years' experience on environmental R&D projects and project management. He has large experience on providing technical support on sustainability assessment and environmental management of products and industrial processes, in particular, through the application of different techniques and approaches, such as LCA, LCC, Eco-design, Eco-efficiency and Environmental Footprint, among others. He has coordinated 3 European research projects (Transcat (FP5) and, Oilprodiesel and Electrovalue (LIFE)). His current duties includes the coordination of three research projects, namely MAESTRI (H2020), SCALER (H2020) and LIFE SWSS.

Zelmar Etxegoien is an engineer and PhD in Computer Science focused on Industry 4.0 for the automotive sector, currently in charge of ICT and ICS software for custom machines manufacturing at Gaindu Automation.

Maria João Ferreira is Associate Professor and Researcher in the field of Information Systems at the Economics, Management and Informatics Department at Universidade Portucalense, Porto, Portugal. Graduated in Applied Mathematics and Computation at Portucalense University, she was awarded with a M.Sc. (1993) and a Ph.D. (2003) degrees in Computation by the University of Manchester. Currently, she teaches subjects in the areas of IT in Organizations, Information Systems, Requirements Engineering and Information Systems Development at undergraduate and master level while supervising M.Sc. and Ph.D. students. She is both a researcher at REMIT - Research on Economics, Management and Information Technologies at Universidade Portucalense - and at ISTTOS - Centro Algoritmi, a research unit in Information Systems and Technologies institutionally hosted by University of Minho. Her research focuses on Information Systems, Digital Transformation in Organizations (including Higher Education). She was coordinator during 10 years of the MSc in Information Technologies in Education. She is a co-author of several scientific publications with peer-review on national and international conferences, book chapters and journals and has been involved in several conferences as a program committee member, organizer committee member and local-chair member. She is also an Editorial Board Member of the IJADS and the IJAS.

Enrico Ferrera received MSc degree in Telecommunications Engineering in 2008. From 2008 to 2010 he was Research Fellow in the Signal Analysis and Simulation Group-Image Processing Lab (SAS-IPL) of the Politecnico di Torino. Since 2010, he is working as a Senior Researcher at Istituto Superiore Mario Boella (ISMB) within the Pervasive Technologies research Area. His research interests include design and development of IoT platforms and tools for heterogeneous systems integration and secure communication. Furthermore, his skills embrace design and development of big data management platforms for Smart City and Smart Factory applications. He is involved in several European funded project operating in Internet-of-Things domain.

Panos Fitsilis is full Professor at University of Applied Sciences of Thessaly, School of Business and Economics (TEI Thessaly), Greece and Visiting Professor at Innopolis University, Kazan, Russian Federation. Prof. P. Fitsilis has extensive project management experience with the development and deployment of large IT systems and extensive management experience in various senior management positions such as: Director of School of Business and Economics at TEI of Thessaly, Vice President of TEI Thessaly Research Committee, Head of Project Management department at TEI Thessaly, Head of Business Unit at international IT company, Intrasoft International, etc. In his work as Business Unit manager at Intrasoft International he was responsible for the development, deployment, and operation of a number of prestigious IT systems for European Commission. Further, he has worked in more than 50 research or commercial projects, in various roles. He is the author of four books, the last one with title "Contemporary Business Information Systems" and author of approximately 100 papers published at scientific journals and international conferences. His research interests include Project Management, Software Engineering, Business Information Systems, Social Media Applications, Smart Cities, etc.

Javier Franco-Arroyo studied Computer Science between 2003 and 2008 at the University of the Basque Country. In 2011 he received the Master in Advance Computer Systems for accessing the PhD at the Faculty of Computer Science of San Sebastian of the University of the Basque Country. Since July of 2008 he works at Vicomtech as a researcher. Nowadays Javier belongs to the Digital Security Group focusing his research on Industrial Control Systems Security and having participated in several projects related with the IT/OT security theme such as: (1) IS_IN_CLOUD, national project for the protection of critic infrastructures, (2) SEED4C, European project for the securization of the cloud through the development of a network of secured elements; (3) Thinking Factory, regional Basque project whose objective is the development of a solution suitable for the Industry 4.0 and in which Javier takes part in the Cybersecurity package.

Ander Garcia studied Telecommunication Engineering between 1997 and 2002 at the University of the Basque Country. He obtained his PhD from the same university in 2011. His thesis, which was focused on the application of operations research optimization algorithms, was entitled "Intelligent Personalised Tourist Route Generation". During 2002 he worked at Iberdrola developing Business Intelligence applications. Since 2003 he has worked at Vicomtech as a senior researcher and project manager inside several projects. He has been an associated teacher at the Computer Science Language and Systems department of the University of the Basque Country from 2005 to 2009, and since 2015 at the University College of Business Studies of the same university.

María J. García G. has a Bachelor in Chemistry and has a Master in Operations Research. Together others authors had increase their investigations, already above two hundred and forty, mainly in the areas of Evaluation and Management of Projects, Knowledge Management, Managerial and Social Decision making, Risk Management and Operations Research, especially in making decision under uncertainty and risk, and multi-criteria decision. They have been presented or published in different countries, having publications and offering their reports, chats or conferences in: Austria, Azerbaijan, Finland, Poland, Croatia, Switzerland, Greece, Germany, Italy, Czech Republic, Iceland, Lithuania, Spain, France, Portugal, United States, Panama, Uruguay, Brazil, Mexico, Argentina and Chile besides attending as guest speaker, in reiterated occasions, in lectures to relevant events in Colombia, Peru, Spain and Venezuela. Among other works she is coauthor of: "Inventories control, the Inventory manager and Matrixes Of Weighing with multiplicative factors (MOWwMf)"; "A Methodology of the Decision Support Systems Applied to Other Projects of Investigation"; "Matrixes Of Weighing and catastrophes"; "Multiattribute Model with Multiplicative Factors and Matrixes Of Weighing and the Problem of the Potable Water"; "Mathematical models generators of Decision Support Systems for help in case of catastrophes. An experience from Venezuela."

Pedro Gómez-Gasquet is Lecturer of Operation Management and Operation Research at the Polytechnic University of Valencia and Research Centre on Production Management and Engineering (CIGIP), Spain. His main research interests are Production Scheduling using advanced IT tools as Multiagent Systems. On these topics he has published > 30 works, some of them on international publications.

Manoj Madhava Gore has done his Bachelors in Science, Bachelors in Technology (Electronics and Telecommunication), and Masters in Engineering (Computer Sc. and Engg.) from University of Allahabad respectively from Ewing Christian College, J. K. Institute of Applied Physics and Technology, and Motilal Nehru Regional Engineering College Allahabad. He got his PhD (Computer Science and Engineering) degree from Indian Institute of Technology Kanpur, India. Dr Gore has been on the faculty rolls of Motilal Nehru National Institute of Technology Allahabad since January 1991 where he is currently working as Professor in the Department of Computer Science and Engineering since 2005. He has also served Indian Institute of Information Technology Allahabad, as Associate Professor for one academic year 2003-04. Dr Gore has his research interests and publications in the areas of Database Modeling, Transaction Recovery, Wireless Sensor Networks, Machine Learning and Cyber-Physical Systems. He is a Fellow of IETE(India) and IE(India), senior member of Computer Society of India, and member of Indian Society of Technical Education (ISTE), IEEE and ACM.

Gilberto J. Hernández G. has a Bachelor in Chemistry and has a Master in Technology of foods. Together others authors had increase their investigations, mainly in the areas of Food technologies, Playful, in particular in the fantastic sports leagues, Knowledge Management, Managerial and Social Decision making, Logistics, Risk Management and Operations research, especially in multi-criteria decision and making decision under uncertainty and risk. They have been presented or published in different countries, having publications and offering their reports, chats or conferences in: Azerbaijan, Finland, Poland, Croatia, Switzerland, Greece, Czech Republic, Spain, Portugal and United States besides attending as guest speaker, in lectures to relevant events in Costa Rica and Venezuela. Among other works he is coauthor of: "Social Aspects of Reverse Logistics and Knowledge Management"; "Enterprise Logistics, Indicators and Physical Distribution Manager"; "Multiattribute Models with Multiplicative factors in the Fantasy Sports"; "The Industrial design manager of LoMoBaP and Knowledge Management"; "Dynamic knowledge: Diagnosis and Customer Service."

José G. Hernández R. is a Chemical Engineer and has a Master in Operations Research. Together other authors had increase their investigations, already above two hundred and forty, mainly in the areas of Knowledge Management, Managerial and Social Decision making, Logistics, Risk Management and Operations research, especially in multi-criteria decision. They have been presented or published in different countries, having publications and offering their reports, chats or conferences in: Azerbaijan, Finland, Croatia, Switzerland, Greece, Germany, Italy, Czech Republic, Iceland, Lithuania, Spain, France, Portugal, United States, Panama, Paraguay, Uruguay, Brazil, Cuba, Mexico, Argentina and Chile besides attending as guest speaker, in reiterated occasions, in lectures to relevant events in Colombia, Peru, Costa Rica, Brazil, Spain and Venezuela. Among other works he is coauthor of: "Teaching Enterprise Logistics through Indicators: Dispatch Manager"; "Enterprise diagnosis and the Environmental manager of LoMoBaP"; "Logistics, Marketing and Knowledge Management in the Community of Consumer."

Maria Holgado is a Research Associate at the Institute for Manufacturing, University of Cambridge. She is currently working on the Energy and Resource Management Systems for Improved Efficiency in the Process Industries (MAESTRI) project, funded by the European Commission through the Horizon 2020 programme. She is a Manufacturing Engineer, graduated at University of Seville, Spain. She has obtained a PhD in Management, Economics and Industrial Engineering at Politecnico di Milano, Italy. Her PhD research examined how industrial maintenance can serve as an enabler for practicing more sustainable manufacturing operations through provision of industrial services. Her research interests include industrial sustainability, energy and resource efficiency, industrial symbiosis, sustainable value creation, sustainable business models, product service systems, maintenance management, maintenance services and maintenance contribution to industrial sustainability. She has been a Marie-Curie Research Fellow within the European project iMaPla 'Integrated Maintenance Planning' at University Federico Santa Maria, Chile, and also a Visiting Scholar at the Intelligent Maintenance Systems Center in University of Cincinnati, US, and at the Institute for Manufacturing in University of Cambridge, UK.

Gunnar Große Hovest graduated from the University of Muenster with a degree in Information Systems. Since 2005, he is scientific staff member at the ATB Institute for Applied Systems Technology Bremen, Germany. As researcher, software developer and project manager, he was involved in multiple European research projects as well as custom software development projects for industrial customers. Currently, he is working in the MAESTRI project, which aims to advance resource and energy efficiency in manufacturing and process industries, developing software solutions to support efficiency improvements, based on Internet-of-Things technologies. He is also working in IoF2020 (Internet of Food and Farm 2020), a European Large Scale Pilot project on Internet-of-Things applications in the agri-food domain.

Zeynep Kalender is a research assistant of Industrial Engineering Department at Marmara University. She graduated with a BS from Istanbul Commerce University in 2011, a MS from Istanbul Technical University in 2013. She is continuing PhD at Marmara University Industrial Engineering Department. Her current research interests include Lean Management and Production, Quality Engineering, Managing Service Quality, Engineering Management, Business Process Management, Innovation and Change Management, Sustainability and Theory of Constraints.

Emil Lezak got his Ph.D. degree in Chemistry, focused in Complex/Polymer Materials, from the CBMiM, PAN (Centre of Molecular and Macromolecular Studies, Polish Academy of Sciences) of Lodz, Poland. He possesses also Engineer degree in Protection of Environment from Technical University of Lodz, Poland, where he had focused on Sustainable use of Renewable Energy of Source (RES). He has a good knowledge of the R&D processes and products (over 8 years in R&D departments) in the field of novel multifunctional complex nanomaterials, related to the energy and environmental sector. His interdisciplinary experience, combining experimental and theoretical efforts allow to him to integrate potential of energy sources. His professional experience in relationship between structure and dynamics at different conditions of the various materials and knowledge of the RES allow to him to use the software for simulation and optimization of various source of energy system. During his intensive professional career he has obtained multidisciplinary expertise and skills related to project management, business modelling and market analysis, IPR management. Furthermore, his strong characteristic are able to effectively take decisions and to interact with different partner profiles, including small enterprises, large industrial players, academic institutions and public bodies.

Maria João Lopes is a former student of the Integrated Master's in Physics Engineering at the University of Coimbra (2017). Interned at Bosch Thermotechnology during her senior year where she started learning about Manufacturing Execution Systems. The subject discussed on the chapter reflects a one year journey which ended with the implementation of a pilot Manufacturing Execution System in a production line together with the Master's Thesis. Currently working in Bosch Thermotechnology Industry 4.0 team which aims to provide the division with leading Industry 4.0 solutions.

Emanuel João Menezes Lourenço holds a degree in Environmental Engineering (2010) and a Master's in Environmental Management (2012) from the Escola Superior Agrária de Coimbra. He joined INEGI in 2012, where he develops his professional activity in projects related to the area of efficiency and eco-efficiency of production systems. He has participated in several projects in the field of Efficiency of production systems where tools and approaches have been developed to evaluate the performance of production systems and to characterize environmental performance.

Zofia Masluszczak joined Lean Enterprise Institute Poland in 2009. Currently she works as a lead researcher on the Energy and Resource Management Systems for Improved Efficiency in the Process Industries (MAESTRI) project, funded by the European Commission through the Horizon 2020 programme. Previously she has been involved in Eco Lean Compass project funded by The National Centre for Research and Development (Poland), focused on the development of organizational eco-innovations that enable costs savings parallel to increase of eco-efficiency. She has taken part in FutureSME project, funded by the European Commission's Seventh Framework Programme (FP7), which aimed in developing a new sustainable business model for European SMEs. Zofia is a Manufacturing Engineer, graduated at Wroclaw University of Science and Technology, Poland. She has worked as a Lean Manufacturing trainer for 4 years. In that time she was involved in consulting projects for companies from several manufacturing branches i.a. automotive, mechanical designing, remanufacturing, plastic injection, home appliance.

István Mezgár is the member of the Research Laboratory on Engineering & Management Intelligence at the Institute of Computer Science and Control, Hungarian Academy of Sciences. He was an Associate Professor at the Department of Manufacturing Science and Technology, Budapest University of Technology and Economics till the end of 2016 (Diploma in m.e: 1977, from the Technical University of Budapest (TUB), C.Sc. in tech.sci.: 1995, from the Hungarian Academy of Sciences, Dr.Habil.: 2006 from the TUB). He is Editorial Board member of numerous international journals, member of IFAC WG 5.3, has more than 150 scientific publications at international forums. He served as visiting scientist/professor several times at different universities/research institutes of Austria, Italy, Japan, New Zealand and South Korea. His current interests focus on networked organizations, wireless communication networks and their security, trust building, cloud computing and cyber physical system applications in production.

Fernando Moreira is graduated in Computer Science (1992), M.Sc. in Electronic Engineering (1997) and Ph.D. in Electronic Engineering (2003), both at Faculty of Engineering of the University of Porto. He is member of the Department of Innovation, Science and Technology at Portucalense University since 1992, currently as Associate Professor and a visiting professor at the University of Porto Business School. He teaches subjects related to undergraduate and pos-graduated studies. He supervises several Ph.D. and M.Sc. students. He is a (co-)author more than 100 scientific publications with peer-review on national and international journals and conferences. He serves as a member of Editorial Advisory Board for several journals and books. He organized several special issues from JCR journals. He has already regularly serves as a member of Programme and Scientific committees of national and international conferences. He is the coordinator of the M.Sc. of Computation since 2007. He holds editorial experience, and he is co-editor of several the books. He is associated with NSTICC, ACM and IEEE. His principal research areas are mobile computing, ICT in Higher Education, Mobile learning, Social business and Digital transformation.

Xabier Oregui Biain obtained his PhD on telecommunication engineering in the CEIT (Technical Research and Investigation Center) by the University of Navarra (Spain), more precisely on the electronic and communication area, where he researched on scaling and managing automatically multi-provider IaaS instances. After a small break from research, where he applied his acquired knowledge in the world of educational applications and "serious-games" working for the company Ikasplay, in 2016 he comes back to the research environment to Vicomtech into the industry and advanced manufacturing area. From his incorporation to Vicomtech, he has worked on different projects oriented towards the management of information on industrial environments using different protocols, and on Big-Data analysis and manipulation using this gathered data.

Gianfranco Pedone is a Research Fellow and IT Engineer at MTA SZTAKI, The Institute for Computer Science and Control of the Hungarian Academy of Sciences, Budapest, Hungary. He studied information technology and management at the University of Studies in Lecce, Italy, where he graduated in IT engineering (2004). He received a Ph.D. degree in computer science (2012) from the Eotvos Lorand University, Budapest, Hungary. His main research interests include multi-agent systems, languages, cloud and web computing for semantic interoperability in scientific and business applications.

David Pérez is an Associate Professor in Operations Management and Operations Research of the Business Department at the Polytechnic University of Valencia (UPV) in Spain. He is Industrial Engineer and received his Ph. D. in Industrial Engineering from the UPV. He is member of the Research Centre on Production Management and Engineering (CIGIP) and has participated in several Spanish Government and European Projects. His research is focused on Supply Chain Management, addressed either by conceptual or quantitative analysis (based on Mathematical Programming) and mainly applied to automotive and ceramic sectors. In these fields, he has several papers in national and international conferences and journals.

Omiros Ragos is an Assistant Professor in the Department of Mathematics of the University of Patras, Greece.

Eva Masero Rubio is a graduate engineer in Electronic and Industrial Automation by the University of Extremadura, concluded in 2016. She was born in Badajoz (Spain) and she is doing an internship in Allbesmart, Castelo Branco (Portugal), in Industrial Internet of Things (I-IOT) area. She has special interests in automation, electronics and instrumentation. She is interested in working in the I-IOT area in different countries, and to improve languages and inter-cultural skills.

Raimar J. Scherer is a Senior Professor ship for Construction Informatics at Technische Universität Dresden. He obtained his Ph.D. in stochastic earthquake engineering from Technical University of Munich in 1984. His research interest includes product and process modeling, concurrent engineering, virtual organizations, information logistics, Internet-enabled network infrastructures, Semantic Web technology, fuzzy and stochastic methods. He is the founder of the biannual European Conference on Product and Process Modeling (ECPPM) and member of the editorial board of 4 international journals.

Alexander Schneider got his diploma in business informatics from the University of Cologne where he specialized himself in software architecture design and user centered software engineering. He worked as a consultant and software architect before joining Fraunhofer FIT. He has more than 12 years experiences in European projects (ebbits, Hydra, RAFT) as well as national and industry projects (E3 production, MICA [funded by SAP], SaN, ZfS). He is responsible for Industry 4.0 in the research group User Centered Ubiquitous Computing. Currently he works on the SPIRE projects MAESTRI and MONSOON - in which he manages the technical developments as the Technical Manager. Alex main research areas include Industry 4.0, Software Engineering, User Centered Design and Internet of Things.

Isabel Seruca is an Associate Professor and a Researcher at the Economics, Management and Informatics Department of Portucalense University, Porto, Portugal, where she teaches subjects in the areas of Information Systems and Programming and she is the Department Coordinator for the IT area of the Erasmus+ Programme. She graduated in Applied Mathematics and Computation at Portucalense University and she has an MSc and a PhD degree both in Computation from the University of Manchester Institute of Science and Technology (UMIST). She is the coordinator of the Data Analysis and Decision Support Systems Group of REMIT (Research on Economics, Management and Information Technologies) and member of ISTTOS (Centro Algoritmi), a research unit in Information Systems and Technologies institutionally hosted by University of Minho. Her research interests are in the Information Systems area, namely in topics such as Business Intelligence, Data and Web Mining, Social Business, Reuse and Patterns. She has participated in several Erasmus Intensive Programmes in the topics of Business Intelligence (Erasmus IP Miss LOGO 2008, 2009 and 2010), Web Mining (Erasmus IP WISDOM 2011, 2012 and 2013) and Data Science (Erasmus IP DasBuG 2014). She is a member of several program committees of international conferences and workshops, a co-editor of books and conference proceedings in Enterprise Information Systems and she has published several papers in refereed conferences and journals in the areas of Data Mining, Semantic Web, Social Business and Patterns.

Francisco A. C. Silva is an occupational hygiene practitioner in the Portuguese Technological Centre for Ceramic and Glass Industries. He undertakes occupational exposure assessments in chemical and physical agents in industry as well as other activities as a trainer and ISO 17025 auditor. He graduated in Ceramic and Glass Engineering (1986), has a Master's in Human Engineering and received his PhD in Industrial and Systems Engineering from University of Minho (2016). He is author and co-author of several occupational safety and hygiene handbooks and other technical publications. He is a collaborator member of Ergonomics & Human Factors group, Centre Algoritmi, University of Minho. He is also a member of national and international standardization committees, including European CEN/TC 352 Nanotechnologies. ORCID: 0000-0003-4234-5640.

Bruno Simoes, with more than 7 years of research experience and a PhD in Mathematics, has developed outstanding skills in applying immature technologies, mitigating research risks, and delivering ground-breaking results to the field of User Experience Design, GeoVisual Analytics and Augmented Reality. In the last years, he has participated in several multi-disciplinary research projects, funded by the European Commission and industry, in the context of smart digital content and creative industries. He has collaborated actively with researchers in other scientific disciplines, particularly in Industrial Design and Cultural Heritage. His research interests lie in User Interaction Design, Interactive Visualization, and Data Physicalization.

Brijendra Pratap Singh was born in the Janwaar Kshatriya family in village Rithuakhor, Gorakhpur, Uttar Pradesh, India, in 1989. He received B. Tech. degree in Computer Science and Engineering in 2011, from Dr. K N Modi Institute of Engineering and Technology, Modinagar, Ghaziabad (affiliated to UPTU). He received the M. Tech. Degree in Software Engineering, in 2013, from Motilal Nehru National Institute of Technology (MNNIT) Allahabad, Allahabad - 211004, India. He is currently a Ph.D. Research Scholar in the Department of Computer Science and Engineering at Motilal Nehru National Institute of Technology Allahabad, India. His area of research is cyber-physical systems, smart grid, demand response program and machine learning.

Yaseen Srewil is a researcher at the Institute of Construction Informatics at Technische Universität Dresden. He received his BSc in Structural Engineering from Damascus University and MSc degree from the Institute for Construction Informatics at TU-Dresden. He was involved in several national and European research projects in the field of Process modelling, Open BIM and Application of RFID in construction. In his PhD research discusses linking of the digital models and physical construction and location context in the framework of CPS. His current interests include BIM and construction planning, smart construction site, Construction 4.0, CPS and IoT applications in construction industry.

Pedro Torres received his BSc and MSc degree in Electronics and Telecommunications Engineering from the Polytechnic Institute of Castelo Branco (Portugal), in 2005 and 2007, respectively. Received his PhD degree in Mechanical Engineering from the, Instituto Superior Técnico, Universidade de Lisboa (Portugal), in 2015. He is Assistant Lecturer at Polytechnic Institute of Castelo Branco, School of Technology, since 2007. He participated in several National and European research projects in the field of Robotics and Image Processing. Highlighting the EU – FP7 – ICT project, ECHORD, experiment "HipRob - Robot-assisted and ultrasound-guided navigation for hip resurfacing arthroplasty". He is author of several publications, in the fields of Robotics, Image Processing and Control. He have special interests for Robotics, Medical Image Processing Computer Vision, Control Systems, Electronics and Instrumentation. Pedro also is member of the IEEE and the order of Engineers of Portugal.

Paraskevi Tsoutsa is a Lecturer to the department of Accounting, University of Applied Sciences of Thessaly and a PHD candidate to the department of Mathematics, University of Patras in Greece.

Özalp Vayvay has taken the degree of bachelor degree in chemical engineering field from the Yıldız Technical University. Ozalp has obtained master's and doctoral degrees in the field of industrial engineering from Marmara University. His primary research interests are "supply chain management", "technology & innovation management", "operations management", "quality management", "industry 4.0." He is the chairman of Engineering Management Department at Marmara University.

Index

F

G

H

I

K

L

M

N

O

P

Q

R

S

T

U

V

W

www.ingramcontent.com/pod-product-compliance
Lightning Source LLC
LaVergne TN
LVHW062108300926
849161LV00010B/2398